Experimental Testing, Manufacturing and Numerical Modelling of Composite and Sandwich Structures

Experimental Testing, Manufacturing and Numerical Modelling of Composite and Sandwich Structures

Editor

Raul D. S. G. Campilho

Basel • Beijing • Wuhan • Barcelona • Belgrade • Novi Sad • Cluj • Manchester

Editor
Raul D. S. G. Campilho
ISEP—School of Engineering
Porto
Portugal

Editorial Office
MDPI AG
Grosspeteranlage 5
4052 Basel, Switzerland

This is a reprint of articles from the Special Issue published online in the open access journal *Materials* (ISSN 1996-1944) (available at: https://www.mdpi.com/journal/materials/special_issues/9OY45YO828).

For citation purposes, cite each article independently as indicated on the article page online and as indicated below:

Lastname, A.A.; Lastname, B.B. Article Title. *Journal Name* **Year**, *Volume Number*, Page Range.

ISBN 978-3-7258-1953-9 (Hbk)
ISBN 978-3-7258-1954-6 (PDF)
doi.org/10.3390/books978-3-7258-1954-6

© 2024 by the authors. Articles in this book are Open Access and distributed under the Creative Commons Attribution (CC BY) license. The book as a whole is distributed by MDPI under the terms and conditions of the Creative Commons Attribution-NonCommercial-NoDerivs (CC BY-NC-ND) license.

Contents

Preface vii

Raul Duarte Salgueiral Gomes Campilho
Experimental Testing, Manufacturing and Numerical Modeling of Composite and Sandwich Structures
Reprinted from: *Materials* 2024, 17, 3468, doi:10.3390/ma17143468 1

Lei Wang, Bo Gao, Yue Sun, Ying Zhang and Liang Hu
Effect of High Current Pulsed Electron Beam (HCPEB) on the Organization and Wear Resistance of CeO_2-Modified Al-20SiC Composites
Reprinted from: *Materials* 2023, 16, 4656, doi:10.3390/ma16134656 6

Julia A. Baimova and Stepan A. Shcherbinin
Strength and Deformation Behavior of Graphene Aerogel of Different Morphologies
Reprinted from: *Materials* 2023, 16, 7388, doi:10.3390/ma16237388 19

Joeun Choi, Yohanes Oscar Andrian, Hyungtak Lee, Hyungyil Lee and Naksoo Kim
Fatigue Life Prediction for Injection-Molded Carbon Fiber-Reinforced Polyamide-6 Considering Anisotropy and Temperature Effects
Reprinted from: *Materials* 2024, 17, 315, doi:10.3390/ma17020315 32

Kosuke Sanai, Sho Nakasaki, Mikiyasu Hashimoto, Arnaud Macadre and Koichi Goda
Fracture Behavior of a Unidirectional Carbon Fiber-Reinforced Plastic under Biaxial Tensile Loads
Reprinted from: *Materials* 2024, 17, 1387, doi:10.3390/ma17061387 50

Jiahao Li, Bo Gao, Zeyuan Shi, Jiayang Chen, Haiyang Fu and Zhuang Liu
Graphene/Heterojunction Composite Prepared by Carbon Thermal Reduction as a Sulfur Host for Lithium-Sulfur Batteries
Reprinted from: *Materials* 2023, 16, 4956, doi:10.3390/ma16144956 67

Angelos Ntaflos, Georgios Foteinidis, Theodora Liangou, Elias Bilalis, Konstantinos Anyfantis, Nicholas Tsouvalis, et al.
Enhancing Epoxy Composite Performance with Carbon Nanofillers: A Solution for Moisture Resistance and Extended Durability in Wind Turbine Blade Structures
Reprinted from: *Materials* 2024, 17, 524, doi:10.3390/ma17020524 80

Krzysztof Ciecieląg
Machinability Measurements in Milling and Recurrence Analysis of Thin-Walled Elements Made of Polymer Composites
Reprinted from: *Materials* 2023, 16, 4825, doi:10.3390/ma16134825 92

Tomáš Knápek, Štěpánka Dvořáčková and Martin Váňa
The Effect of Clearance Angle on Tool Life, Cutting Forces, Surface Roughness, and Delamination during Carbon-Fiber-Reinforced Plastic Milling
Reprinted from: *Materials* 2023, 16, 5002, doi:10.3390/ma16145002 106

Grzegorz Matula and Błażej Tomiczek
Manufacturing of Corrosion-Resistant Surface Layers by Coating Non-Alloy Steels with a Polymer-Powder Slurry and Sintering
Reprinted from: *Materials* 2023, 16, 5210, doi:10.3390/ma16155210 121

Kassym Yelemessov, Layla B. Sabirova, Nikita V. Martyushev, Boris V. Malozyomov, Gulnara B. Bakhmagambetova and Olga V. Atanova
Modeling and Model Verification of the Stress-Strain State of Reinforced Polymer Concrete
Reprinted from: *Materials* **2023**, *16*, 3494, doi:10.3390/ma16093494 . . . . **140**

Jimesh D. Bhagatji, Oleksandr G. Kravchenko and Sharanabasaweshwara Asundi
Mechanics of Pure Bending and Eccentric Buckling in High-Strain Composite Structures
Reprinted from: *Materials* **2024**, *17*, 796, doi:10.3390/ma17040796 . . . . **164**

Young W. Kwon, Emma K. Markoff and Stanley DeFisher
Unified Failure Criterion Based on Stress and Stress Gradient Conditions
Reprinted from: *Materials* **2024**, *17*, 569, doi:10.3390/ma17030569 . . . . **181**

Luis M. Ferreira, Carlos A. C. P. Coelho and Paulo N. B. Reis
Numerical Simulations of the Low-Velocity Impact Response of Semicylindrical Woven Composite Shells
Reprinted from: *Materials* **2023**, *16*, 3442, doi:10.3390/ma16093442 . . . . **200**

Paripat Kraisornkachit, Masanobu Naito, Chao Kang and Chiaki Sato
Multi-Objective Optimization of Adhesive Joint Strength and Elastic Modulus of Adhesive Epoxy with Active Learning
Reprinted from: *Materials* **2024**, *17*, 2866, doi:10.3390/ma17122866 . . . . **217**

Maruri Takamura, Minori Isozaki, Shinichi Takeda, Yutaka Oya and Jun Koyanagi
Evaluation of True Bonding Strength for Adhesive Bonded Carbon Fiber-Reinforced Plastics
Reprinted from: *Materials* **2024**, *17*, 394, doi:10.3390/ma17020394 . . . . **234**

Mahsa Khademi, Daniel P. Pulipati and David A. Jack
Nondestructive Inspection and Quantification of Select Interface Defects in Honeycomb Sandwich Panels
Reprinted from: *Materials* **2024**, *17*, 2772, doi:10.3390/ma17112772 . . . . **245**

Ruiqian Wang, Dan Yao, Jie Zhang, Xinbiao Xiao and Xuesong Jin
Effect of the Laying Order of Core Layer Materials on the Sound-Insulation Performance of High-Speed Train Carbody
Reprinted from: *Materials* **2023**, *16*, 3862, doi:10.3390/ma16103862 . . . . **263**

Dariusz Grzybek
Experimental Analysis of the Influence of Carrier Layer Material on the Performance of the Control System of a Cantilever-Type Piezoelectric Actuator
Reprinted from: *Materials* **2024**, *17*, 96, doi:10.3390/ma17010096 . . . . **276**

Marcela-Elisabeta Barbinta-Patrascu, Cornelia Nichita, Bogdan Bita and Stefan Antohe
Biocomposite Materials Derived from *Andropogon halepensis*: Eco-Design and Biophysical Evaluation
Reprinted from: *Materials* **2024**, *17*, 1225, doi:10.3390/ma17051225 . . . . **300**

Aleksandra Grząbka-Zasadzińska, Magdalena Woźniak, Agata Kaszubowska-Rzepka, Marlena Baranowska, Anna Sip, Izabela Ratajczak, et al.
Enhancing Sustainability and Antifungal Properties of Biodegradable Composites: Caffeine-Treated Wood as a Filler for Polylactide
Reprinted from: *Materials* **2024**, *17*, 698, doi:10.3390/ma17030698 . . . . **319**

Preface

This comprehensive scientific volume, "Experimental Testing, Manufacturing and Numerical Modelling of Composite and Sandwich Structures," serves as a state-of-the-art resource for researchers, engineers, and practitioners in the fields of materials science and engineering. The special issue brings together cutting-edge research and developments in the characterization, manufacturing, behaviour prediction, and sustainability of composite and sandwich structures. The scope of this work ranges from experimental testing to numerical modelling, aiming to provide a detailed understanding of these advanced materials.

Paper 1 introduces the editorial paper, which provides the special issue contextualization and motivation for its publication. Papers 2 through 7 focus on the detailed characterization of various composites, highlighting their unique properties and performance under different conditions. The following sections, papers 8 to 10, discuss innovative manufacturing techniques, including machining processes and coating methods that enhance the functionality and durability of these materials. The subsequent papers, 11 to 14, present sophisticated models to predict the behaviour of composites under diverse mechanical stresses, providing invaluable insights for design and application. Papers 15 and 16 concentrate on adhesive joints, optimizing their strength and elasticity through advanced learning methodologies. In papers 17 to 19, the focus shifts to the specificity of sandwich structures, emphasizing non-destructive inspection techniques and material performance in high-speed trains and piezoelectric actuators. Finally, papers 20 and 21 address the pressing need for sustainability in materials engineering, exploring biocomposite materials derived from Andropogon halepensis and the innovative use of caffeine-treated wood as a biodegradable filler.

This special issue is motivated by the growing demand for more efficient, durable, and environmentally friendly composite materials in modern engineering applications. It seeks to bridge the gap between theoretical research and practical implementation, offering detailed experimental and numerical insights. The work is intended for an audience of advanced students, researchers, and industry professionals who are engaged in the development, testing, and application of composite materials. By integrating diverse topics and innovative research, this special issue aims to advance the knowledge and application of composite and sandwich structures, fostering innovation and sustainability in material science. As Editor of this special issue, I am grateful to the esteemed contributors who have shared their expertise, insights, and innovative research.

Raul D. S. G. Campilho
Editor

Editorial

Experimental Testing, Manufacturing and Numerical Modeling of Composite and Sandwich Structures

Raul Duarte Salgueiral Gomes Campilho [1,2]

1 CIDEM, ISEP—School of Engineering, Polytechnic of Porto, R. Dr. António Bernardino de Almeida, 431, 4200-072 Porto, Portugal; raulcampilho@gmail.com
2 INEGI—Institute of Science and Innovation in Mechanical and Industrial Engineering, Pólo FEUP, Rua Dr. Roberto Frias, 400, 4200-465 Porto, Portugal

Citation: Campilho, R.D.S.G. Experimental Testing, Manufacturing and Numerical Modeling of Composite and Sandwich Structures. *Materials* 2024, 17, 3468. https://doi.org/10.3390/ma17143468

Received: 29 June 2024
Accepted: 10 July 2024
Published: 13 July 2024

Copyright: © 2024 by the author. Licensee MDPI, Basel, Switzerland. This article is an open access article distributed under the terms and conditions of the Creative Commons Attribution (CC BY) license (https://creativecommons.org/licenses/by/4.0/).

Composite materials have become indispensable in a multitude of industries, such as aerospace, automotive, construction, sports equipment, and electronics [1]. These materials are known for their exceptional strength-to-weight ratio, fatigue and corrosion resistance, and minimal thermal expansion. The ability to tailor composite materials to specific applications by varying the type, size, and orientation of the reinforcement material, as well as the type of matrix material used, increases their versatility and effectiveness [2]. However, issues such as high production costs, repair complexity, susceptibility to delamination and other damage types, and difficulties in characterization and modeling, are significant obstacles [3]. Currently, advanced manufacturing techniques, including additive and digital manufacturing, are being explored to reduce production costs and improve the precision and reliability of composite materials [4]. The development of nanocomposites aims to enhance strength, stiffness, thermal conductivity, and electrical conductivity, thereby broadening their application potential [5]. In parallel, sustainability demands triggered the creation of sustainable composites, which incorporate renewable or recycled materials to reduce environmental impact [6]. Multi-functional composites, integrating materials like shape memory alloys, piezoelectric materials, and carbon nanotubes, enable the creation of intelligent systems capable of self-monitoring, adaptive responses, and energy harvesting [7]. Bioinspired composites draw inspiration from natural materials, such as spider silk, seashells, and bone, striving to replicate their unique properties [8]. Numerical modeling has also seen substantial advancements, e.g., in simulating the complex behaviors of composite materials under various conditions [9].

Composite sandwich structures find application in industries such as aerospace, automotive, marine, and construction [10]. Their high strength-to-weight ratio, stiffness, and durability make them ideal for these demanding environments. Advanced materials, including fiber-reinforced polymers and metal matrix composites for face sheets, and innovative core materials, such as foams, honeycombs, and lattice structures, are being developed. Automated and robotic systems are employed to improve consistency, reduce labor costs, and enable the production of more complex geometries [11]. Advanced cores, such as 3D-printed lattice structures and bioinspired geometries, are being developed to improve impact resistance, energy absorption, and structural integrity [12]. Understanding the dynamic behavior of sandwich structures under impact, blast, and vibration is crucial [13].

Within the scope of this Special Issue, Wang et al. [14] explored the combined effects of high-current pulsed electron beam (HCPEB) treatment and CeO_2 denaturant on enhancing the microstructure and properties of Al-20SiC composites produced by powder metallurgy. Grazing incidence X-ray diffraction (GIXRD) revealed a selective orientation of aluminum grains, particularly with Al(111) crystal faces, following HCPEB treatment.

Baimova and Shcherbinin [15] focused on three different morphologies of graphene aerogels with a honeycomb-like structure, examining their strength and deformation behavior through a molecular dynamics simulation. Their findings advanced the understanding of the microscopic deformation mechanisms in graphene aerogels.

Choi et al. [16] examined the influence of anisotropy and temperature on short carbon fiber-reinforced polyamide-6 (CF-PA6) produced via injection molding to determine its static and fatigue characteristics. The tensile strength and fatigue life of CF-PA6 varied with changes in temperature and anisotropy. A semi-empirical strain–stress fatigue life prediction model was positively validated.

A biaxial tensile test was conducted by Sanai et al. [17] to assess the fracture behavior of a unidirectional carbon fiber-reinforced plastic (CFRP) under load in the fiber (0°) and transverse (90°) directions. This study demonstrated that the occurrence of each fracture mode is solely characterized by a normal strain in the 90° direction (ε_y).

In the work of Li et al. [18], an interlayer nanocomposite (CC@rGO) featuring a graphene heterojunction with CoO and Co9S8 was synthesized using a hydrothermal calcination technique and evaluated as a cathode sulfur carrier for lithium–sulfur batteries. The CC@rGO sulfur cathode exhibited excellent electrochemical performance, rate capability, and cycling stability across various current densities.

Ntaflos et al. [19] produced graphene nanoplatelet (GNP)-enhanced glass fiber-reinforced plastics (GFRP) produced by filament winding. The performance of these materials was tested under harsh environmental conditions. Results showed that GNPs improved the in-plane shear strength of GFRP by 200% and reduced water uptake by up to 40%.

Ciecielag [20] investigated the milling of CFRP and GFRP saturated with epoxy resin using various tools (polycrystalline diamond inserts, physically coated carbide inserts with titanium nitride, and uncoated carbide inserts). Milling thin-walled CFRP with uncoated tools at high feeds per revolution and high cutting speeds resulted in the highest forces.

Knápek et al. [21] investigated how the clearance angle of milling tools affects wear, cutting forces, machined edge roughness, and delamination during the milling of CFRP panels with a twill weave and a 90° fiber orientation. The clearance angle significantly impacted tool wear, surface roughness, and the delamination of the CFRP panel. Higher tool wear led to higher cutting forces, which increased surface roughness and delamination.

Matula and Tomiczek [22] discussed the integration of surface engineering and powder metallurgy to develop coatings with enhanced corrosion resistance and wear properties. Tensile tests indicated that both steel types exhibited higher strength after sintering in a nitrogen-rich atmosphere.

Yelemessov et al. [23] experimentally and analytically explored the potential application of polymer concrete composites in building structures through a strength analysis. The rubber concrete beam deformation at failure exceeded that of cement concrete by 2.5 to 6.5 times and by 3.0 to 7.5 times with a fiber reinforcement.

Bhagatji et al. [24] addressed the fabrication, experimental testing, and progressive failure modeling of an ultra-thin composite beam. The continuum damage mechanics (CDM) model for the beam, calibrated through experimental coupon testing, was utilized in finite element explicit analysis. The finite element model accurately predicted localized transverse fiber damage under eccentric buckling.

Kwon et al. [25] investigated specimens made from a variety of materials with different geometric features to predict failure loads using a newly proposed criterion that incorporates both stress and stress gradient conditions. The types of notches examined included cracks and holes. There was a good agreement between experimental data and theoretical predictions across all cases.

The work of Ferreira et al. [26] introduced an efficient method to study the low-velocity impact response of woven composite shells using 3D finite element models that incorporate both intralaminar and interlaminar progressive damage mechanisms. The numerical predictions closely matched experimental data regarding load and energy histories, maximum impact load, displacement, and contact time.

Kraisornkachit et al. [27] investigated adhesive joint strength and elastic modulus, critical for adhesive performance. The experiments employed a bisphenol A-based epoxy resin with a polypropylene glycol curing agent, generating initial data from 32 conditions used to train a machine learning model. Bayesian optimization identified conditions surpassing this boundary, achieving an adhesive joint strength of 25.2 MPa and an elastic modulus of 182.5 MPa.

Takamura et al. [28] performed a numerical analysis on single-lap shear joints between carbon fiber-reinforced thermoplastics (CFRTPs) to estimate the local stress state at the point of failure initiation, revealing the true bonding strength. Results indicated that the single-lap shear test underestimates the apparent bonding strength by less than 14% of the true bonding strength.

Khademi et al. [29] studied a material system comprising unidirectional carbon fiber composite face sheets with a honeycomb core, investigating various defects at this critical interface. The work introduced high-frequency ultrasound testing (UT) to detect and quantify the geometry and type of defects. The results indicated successful defect detection.

Wang et al. [30] enhanced sound insulation in composite structures through strategic material arrangement. Initially, a predictive model for sound insulation in sandwich composite plates was developed and validated. Optimization strategies for sound insulation in high-speed train composite floors were explored. Implementing this approach in high-speed train body design improved sound insulation by 1–3 dB in the 125–315 Hz band.

The study of Grzybek [31] involved experimental analysis of the control systems for piezoelectric actuators based on composite and aluminum materials, specifically examining unimorph and bimorph structures. Two piezoelectric actuators were manufactured with a cantilever sandwich beam configuration: one with a glass-reinforced epoxy composite (FR4) carrier layer and another with 1050 aluminum. The study also proposed a modification to the linear quadratic regulator (LQR) control algorithm.

Barbinta-Patrascu et al. [32] explored a sustainable approach using weeds to produce silver nanoparticles (AgNPs) with diverse potential applications. Two different types of AgNPs were generated by varying the ratio of phytoextract to silver salt solution. The resulting green composite materials were characterized and showed significant potential for designing innovative bioactive materials suitable for biomedical applications.

Grząbka-Zasadzińska et al. [33] investigated the potential of using caffeine-treated and untreated black cherry (Prunus serotina Ehrh.) wood as a filler in polylactide composites. The study highlighted the novel application of caffeine as a natural compound to modify wood, thereby altering the supermolecular structure and nucleating abilities of polylactide/wood composites.

Taking into account all of the collected contributions, the future of research in experimental testing, manufacturing, and numerical modeling of composite and sandwich structures is expected to advance towards more integrated, multi-functional, and sustainable solutions that address diverse industry needs through innovations in material science and engineering. The identified hot topics are:

- Multi-Material and Hybrid Structures: Combining different materials (composites with metals, ceramics, and polymers) to achieve superior performance in, for example, enhanced mechanical properties, tailored thermal and electrical conductivity, and improved damage tolerance [34].
- Advanced Manufacturing Techniques: Additive manufacturing (3D printing), automated layup methods (such as automated fiber placement and tape laying), and innovative curing techniques to enhance production efficiency and reduce costs [35].
- Durability and Environmental Resistance: Developing coatings, surface treatments, and materials that resist degradation from environmental factors (ultraviolet or UV exposure, moisture, and chemicals) and promote an extended service life [36].
- Damage Detection and Structural Health Monitoring: Integrating embedded sensors, data analytics, and machine learning algorithms to enable real-time monitoring and predictive maintenance of composite structures [37].

- Simulation and Modeling: Developing multi-scale models that capture the complex behavior of composites, from microstructure to macroscopic performance, and validating these models with experimental data [38].
- Functional and Smart Materials: Integration of functional materials (such as shape memory alloys, piezoelectric materials, and sensors) into sandwich structures to exhibit adaptive, self-healing, or sensing capabilities [39].
- Sustainability and Recycling: Developing eco-friendly manufacturing processes, bio-based resins, and recyclable composite and sandwich structures. Improving recycling methods to recover and reuse materials effectively [40].

Conflicts of Interest: The author declares no conflicts of interest.

References

1. Ozturk, F.; Cobanoglu, M.; Ece, R.E. Recent advancements in thermoplastic composite materials in aerospace industry. *J. Thermoplast. Compos. Mater.* **2023**. [CrossRef]
2. Rajak, D.K.; Pagar, D.D.; Kumar, R.; Pruncu, C.I. Recent progress of reinforcement materials: A comprehensive overview of composite materials. *J. Mater. Res. Technol.* **2019**, *8*, 6354–6374. [CrossRef]
3. Pervaiz, S.; Qureshi, T.A.; Kashwani, G.; Kannan, S. 3D printing of fiber-reinforced plastic composites using fused deposition modeling: A status review. *Materials* **2021**, *14*, 4520. [CrossRef] [PubMed]
4. Bi, X.; Huang, R. 3D printing of natural fiber and composites: A state-of-the-art review. *Mater. Des.* **2022**, *222*, 111065. [CrossRef]
5. Hassan, T.; Salam, A.; Khan, A.; Khan, S.U.; Khanzada, H.; Wasim, M.; Khan, M.Q.; Kim, I.S. Functional nanocomposites and their potential applications: A review. *J. Polym. Res.* **2021**, *28*, 36. [CrossRef]
6. Maiti, S.; Islam, M.R.; Uddin, M.A.; Afroj, S.; Eichhorn, S.J.; Karim, N. Sustainable fiber-reinforced composites: A Review. *Adv. Sustain. Syst.* **2022**, *6*, 2200258. [CrossRef]
7. Guan, X.; Chen, H.; Xia, H.; Fu, Y.; Qiu, Y.; Ni, Q.-Q. Multifunctional composite nanofibers with shape memory and piezoelectric properties for energy harvesting. *J. Intel. Mat. Syst. Str.* **2020**, *31*, 956–966. [CrossRef]
8. Sonia, P.; Srinivas, R.; Kansal, L.; Abdul-Zahra, D.S.; Reddy, U.; Kumari, V. Bioinspired Composites a Review: Lessons from Nature for Materials Design and Performance. In Proceedings of the E3S Web of Conferences, Ordos, China, 22–23 June 2024; p. 01024.
9. Ghatage, P.S.; Kar, V.R.; Sudhagar, P.E. On the numerical modelling and analysis of multi-directional functionally graded composite structures: A review. *Compos. Struct.* **2020**, *236*, 111837. [CrossRef]
10. Ma, W.; Elkin, R. *Sandwich Structural Composites: Theory and Practice*; CRC Press: Boca Raton, FL, USA, 2022.
11. Al-Khazraji, M.S.; Bakhy, S.; Jweeg, M. Composite sandwich structures: Review of manufacturing techniques. *J. Eng. Des. Technol.* **2023**. [CrossRef]
12. Peng, C.; Fox, K.; Qian, M.; Nguyen-Xuan, H.; Tran, P. 3D printed sandwich beams with bioinspired cores: Mechanical performance and modelling. *Thin-Walled Struct.* **2021**, *161*, 107471. [CrossRef]
13. Essassi, K.; Rebiere, J.-L.; Mahi, A.E.; Souf, M.A.B.; Bouguecha, A.; Haddar, M. Experimental and numerical analysis of the dynamic behavior of a bio-based sandwich with an auxetic core. *J. Sandw. Struct. Mater.* **2021**, *23*, 1058–1077. [CrossRef]
14. Wang, L.; Gao, B.; Sun, Y.; Zhang, Y.; Hu, L. Effect of High Current Pulsed Electron Beam (HCPEB) on the Organization and Wear Resistance of CeO2-Modified Al-20SiC Composites. *Materials* **2023**, *16*, 4656. [CrossRef]
15. Baimova, J.A.; Shcherbinin, S.A. Strength and Deformation Behavior of Graphene Aerogel of Different Morphologies. *Materials* **2023**, *16*, 7388. [CrossRef] [PubMed]
16. Choi, J.; Andrian, Y.O.; Lee, H.; Lee, H.; Kim, N. Fatigue Life Prediction for Injection-Molded Carbon Fiber-Reinforced Polyamide-6 Considering Anisotropy and Temperature Effects. *Materials* **2024**, *17*, 315. [CrossRef]
17. Sanai, K.; Nakasaki, S.; Hashimoto, M.; Macadre, A.; Goda, K. Fracture Behavior of a Unidirectional Carbon Fiber-Reinforced Plastic under Biaxial Tensile Loads. *Materials* **2024**, *17*, 1387. [CrossRef]
18. Li, J.; Gao, B.; Shi, Z.; Chen, J.; Fu, H.; Liu, Z. Graphene/Heterojunction Composite Prepared by Carbon Thermal Reduction as a Sulfur Host for Lithium-Sulfur Batteries. *Materials* **2023**, *16*, 4956. [CrossRef]
19. Ntaflos, A.; Foteinidis, G.; Liangou, T.; Bilalis, E.; Anyfantis, K.; Tsouvalis, N.; Tyriakidi, T.; Tyriakidis, K.; Tyriakidis, N.; Paipetis, A.S. Enhancing Epoxy Composite Performance with Carbon Nanofillers: A Solution for Moisture Resistance and Extended Durability in Wind Turbine Blade Structures. *Materials* **2024**, *17*, 524. [CrossRef] [PubMed]
20. Ciecieląg, K. Machinability Measurements in Milling and Recurrence Analysis of Thin-Walled Elements Made of Polymer Composites. *Materials* **2023**, *16*, 4825. [CrossRef]
21. Knápek, T.; Dvořáčková, Š.; Váňa, M. The Effect of Clearance Angle on Tool Life, Cutting Forces, Surface Roughness, and Delamination during Carbon-Fiber-Reinforced Plastic Milling. *Materials* **2023**, *16*, 5002. [CrossRef]
22. Matula, G.; Tomiczek, B. Manufacturing of Corrosion-Resistant Surface Layers by Coating Non-Alloy Steels with a Polymer-Powder Slurry and Sintering. *Materials* **2023**, *16*, 5210. [CrossRef]

23. Yelemessov, K.; Sabirova, L.B.; Martyushev, N.V.; Malozyomov, B.V.; Bakhmagambetova, G.B.; Atanova, O.V. Modeling and Model Verification of the Stress-Strain State of Reinforced Polymer Concrete. *Materials* **2023**, *16*, 3494. [CrossRef]
24. Bhagatji, J.D.; Kravchenko, O.G.; Asundi, S. Mechanics of Pure Bending and Eccentric Buckling in High-Strain Composite Structures. *Materials* **2024**, *17*, 796. [CrossRef] [PubMed]
25. Kwon, Y.W.; Markoff, E.K.; DeFisher, S. Unified Failure Criterion Based on Stress and Stress Gradient Conditions. *Materials* **2024**, *17*, 569. [CrossRef]
26. Ferreira, L.M.; Coelho, C.A.C.P.; Reis, P.N.B. Numerical Simulations of the Low-Velocity Impact Response of Semicylindrical Woven Composite Shells. *Materials* **2023**, *16*, 3442. [CrossRef] [PubMed]
27. Kraisornkachit, P.; Naito, M.; Kang, C.; Sato, C. Multi-Objective Optimization of Adhesive Joint Strength and Elastic Modulus of Adhesive Epoxy with Active Learning. *Materials* **2024**, *17*, 2866. [CrossRef]
28. Takamura, M.; Isozaki, M.; Takeda, S.; Oya, Y.; Koyanagi, J. Evaluation of True Bonding Strength for Adhesive Bonded Carbon Fiber-Reinforced Plastics. *Materials* **2024**, *17*, 394. [CrossRef]
29. Khademi, M.; Pulipati, D.P.; Jack, D.A. Nondestructive Inspection and Quantification of Select Interface Defects in Honeycomb Sandwich Panels. *Materials* **2024**, *17*, 2772. [CrossRef] [PubMed]
30. Wang, R.; Yao, D.; Zhang, J.; Xiao, X.; Jin, X. Effect of the Laying Order of Core Layer Materials on the Sound-Insulation Performance of High-Speed Train Carbody. *Materials* **2023**, *16*, 3862. [CrossRef]
31. Grzybek, D. Experimental Analysis of the Influence of Carrier Layer Material on the Performance of the Control System of a Cantilever-Type Piezoelectric Actuator. *Materials* **2024**, *17*, 96. [CrossRef]
32. Barbinta-Patrascu, M.-E.; Nichita, C.; Bita, B.; Antohe, S. Biocomposite Materials Derived from Andropogon halepensis: Eco-Design and Biophysical Evaluation. *Materials* **2024**, *17*, 1225. [CrossRef]
33. Grząbka-Zasadzińska, A.; Woźniak, M.; Kaszubowska-Rzepka, A.; Baranowska, M.; Sip, A.; Ratajczak, I.; Borysiak, S. Enhancing Sustainability and Antifungal Properties of Biodegradable Composites: Caffeine-Treated Wood as a Filler for Polylactide. *Materials* **2024**, *17*, 698. [CrossRef] [PubMed]
34. Ding, Z.; Zou, Z.; Zhang, L.; Li, X.; Zhang, Y. Multi-scale topological design of asymmetric porous sandwich structures with unidentical face sheets and composite core. *Comput. Methods Appl. Mech. Eng.* **2024**, *422*, 116839. [CrossRef]
35. Parveez, B. Rapid prototyping of core materials in aircraft sandwich structures. In *Modern Manufacturing Processes for Aircraft Materials*; Elsevier: Amsterdam, The Netherlands, 2024; pp. 63–87.
36. Duarte, C.; de Queiroz, H.; Neto, J.; Cavalcanti, D.; Banea, M. Evaluation of durability of 3D-printed multi-material parts for potential applications in structures exposed to marine environments. *Procedia Struct. Integr.* **2024**, *53*, 299–308. [CrossRef]
37. Khanahmadi, M.; Mirzaei, B.; Amiri, G.G.; Gholhaki, M.; Rezaifar, O. Vibration-based damage localization in 3D sandwich panels using an irregularity detection index (IDI) based on signal processing. *Measurement* **2024**, *224*, 113902. [CrossRef]
38. Kheyabani, A.; Ali, H.Q.; Kefal, A.; Yildiz, M. Coupling of isogeometric higher-order RZT and parametric HFGMC frameworks for multiscale modeling of sandwich laminates: Theory and experimental validation. *Aerosp. Sci. Technol.* **2024**, *146*, 108944. [CrossRef]
39. Ghalayaniesfahani, A.; Oostenbrink, B.; van Kasteren, H.; Gibson, I.; Mehrpouya, M. 4D Printing of Biobased Shape Memory Sandwich Structures. *Polymer* **2024**, *307*, 127252. [CrossRef]
40. Lv, Q.; Zhu, X.; Zhou, T.; Tian, L.; Liu, Y.; Wang, Y.; Zhang, C. Multifunctional and recyclable aerogel/fiber building insulation composites with sandwich structure. *Constr. Build. Mater.* **2024**, *423*, 135902. [CrossRef]

Disclaimer/Publisher's Note: The statements, opinions and data contained in all publications are solely those of the individual author(s) and contributor(s) and not of MDPI and/or the editor(s). MDPI and/or the editor(s) disclaim responsibility for any injury to people or property resulting from any ideas, methods, instructions or products referred to in the content.

Article

Effect of High Current Pulsed Electron Beam (HCPEB) on the Organization and Wear Resistance of CeO$_2$-Modified Al-20SiC Composites

Lei Wang [1], Bo Gao [1,*], Yue Sun [1], Ying Zhang [1] and Liang Hu [2]

[1] Key Laboratory for Ecological Metallurgy of Multimetallic Mineral (Ministry of Education), Northeastern University, Shenyang 110819, China; surfwangl@163.com (L.W.); surfsuny@163.com (Y.S.); surfzhangy@163.com (Y.Z.)
[2] Key Laboratory for Anisotropy and Texture of Materials (Ministry of Education), School of Materials Science and Engineering, Northeastern University, Shenyang 110819, China; hss5566123@163.com
* Correspondence: gaob@smm.neu.edu.cn

Abstract: This paper investigates the joint effect of high current pulsed electron beam (HCPEB) and denaturant CeO$_2$ on improving the microstructure and properties of Al-20SiC composites prepared by powder metallurgy. Grazing Incidence X-ray Diffraction (GIXRD) results indicate the selective orientation of aluminum grains, with Al(111) crystal faces showing selective orientation after HCPEB treatment. Casting defects of powder metallurgy were eliminated by the addition of CeO$_2$. Scanning electron microscopy (SEM) results reveal a more uniform distribution of hard points on the surface of HCPEB-treated Al-20SiC-0.3CeO$_2$ composites. Microhardness and wear resistance of the Al-20SiC-0.3CeO$_2$ composites were better than those of the Al matrix without CeO$_2$ addition at the same number of pulses. Sliding friction tests indicate that the improvement of wear resistance is attributed to the uniform dispersion of hard points and the improvement of microstructure on the surface of the matrix after HCPEB irradiation. Overall, this study demonstrates the potential of HCPEB and CeO$_2$ to enhance the performance of Al-20SiC composites.

Keywords: high current pulsed electron beam; Al-20SiC; cerium dioxide; microhardness; wear resistance

Citation: Wang, L.; Gao, B.; Sun, Y.; Zhang, Y.; Hu, L. Effect of High Current Pulsed Electron Beam (HCPEB) on the Organization and Wear Resistance of CeO$_2$-Modified Al-20SiC Composites. *Materials* 2023, *16*, 4656. https://doi.org/10.3390/ma16134656

Academic Editor: Raul D. S. G. Campilho

Received: 19 May 2023
Revised: 9 June 2023
Accepted: 23 June 2023
Published: 28 June 2023

Copyright: © 2023 by the authors. Licensee MDPI, Basel, Switzerland. This article is an open access article distributed under the terms and conditions of the Creative Commons Attribution (CC BY) license (https://creativecommons.org/licenses/by/4.0/).

1. Introduction

With the development of industrialization, the performance of wear-resistant materials has put forward higher requirements. Composite materials have emerged as an indispensable and important part of the wear-resistant field, such as SiC particle-reinforced aluminum matrix composite with high hardness and good wear resistance. It is a comprehensive composite of excellent friction materials with promising applications [1–7]. It has been used to some extent in engine pistons and cylinder liners in high-speed train brake discs, automobile brake discs, and sliding bearing materials. At present, the conventional methods for preparing Al-20SiC composites include the extrusion casting method, in situ reaction method, stirring casting method, spray deposition method, powder metallurgy method, and so on [8]. Among them, the powder metallurgy method has the advantages of easy control of the interfacial reaction, low preparation cost, and significantly better performance and stability of the material than those prepared by other methods [9]. However, this preparation method produces significant casting defects, such as porosity, cracking, and agglomeration, which limit the wide application of Al-20SiC composites [10]. Therefore, it is necessary to improve the properties of Al-20SiC composites using different treatment methods [11].

High current pulsed electron beam (HCPEB) surface treatment technology is an emerging high-energy beam surface treatment technology with higher energy efficiency than laser beams and is not affected by ionic impurities in ion beam technology. With high energy density and short pulse electron beam irradiation of the material surface, the electron beam carries the energy deposited to the material surface, resulting in instantaneous heating and cooling and the formation of a special surface microstructure. Then, the expected corrosion resistance, wear resistance, and other surface performance effects are obtained [12]. Walker et al. [13] treated eutectic Al-Si alloy using a pulsed electron beam and showed that the average surface roughness increased and then decreased with increasing accelerating voltage; in addition, the average dynamic friction coefficient of the treated sample was higher than that of the untreated alloy surface and increased the wear rate by 66%. Hao et al. [14] used HCPEB for the surface treatment of a peri-eutectic Al-15Si alloy, and the results of the study showed that the alloy elements Al and Si undergo mutual diffusion in the molten state of the electron beam accompanied by a transient solidification effect, and a supersaturated solid solution of aluminum is formed in the surface layer of the alloy with enhanced overall wear resistance. At the same time, some literature studies have shown that the addition of trace amounts of metamorphic agents can improve the irradiation damage of metals during HCPEB irradiation and improve the overall properties of the matrix. Shi Weixi et al. [15] added 0.3 wt.% Nd to Al-20Si alloy material, and the results show that the grain size was significantly reduced and the wear resistance was significantly improved. Hu et al. [16] studied the elimination of microcracks on the surface of Al alloy samples by rare earth elements, and the results showed that the addition of trace rare earth elements reduced the density of microcracks to some extent, which led to grain refinement and enhanced the alloy's resistance in a corrosion solution [17].

Among rare earth elements, CeO_2 is an important denaturing agent that can form a uniform distribution in aluminum matrix composites, effectively improving the strength and hardness of the material. Additionally, CeO_2 can enhance the material's crystal structure and overall performance [18,19]. Compared to other denaturing agents, CeO_2 is relatively inexpensive, which can reduce the production cost of the material and improve its economic competitiveness. In this study, the utilization of the HCPEB surface modification technique aims to address the inherent defects commonly found in powder metallurgy. Additionally, the incorporation of rare earth oxide CeO_2 as a trace additive is expected to enhance the elimination of microcracks and pores within the material. Notably, this research represents the first investigation into the wear resistance of HCPEB-modified pre- and post-Al-20SiC composites, providing a comprehensive understanding of the underlying mechanisms governing wear resistance [20,21].

2. Materials and Methods

2.1. Material Preparation

The raw materials used in this experiment were commercially available aluminum powder, SiC powder, cerium oxide powder, and hydroxypropyl methylcellulose (HPMC). The raw material composition and particle sizes are shown in Table 1.

Table 1. Composition and granularity of raw materials.

Powder	Purity/wt.%	Particle Size/μm
Al	99.9 wt.%	20–30 μm
SiC	99.9 wt.%	1–2 μm
CeO_2	99.9 wt.%	6–10 μm

The specific preparation process is as follows: The raw materials—aluminum powder, SiC powder, cerium oxide powder, and 1 wt.% hydroxypropyl methyl cellulose (HPMC)—were put into the ball mill tank. At the same time, zirconium oxide agate balls were added with a ratio of 3:1 and mixed on a roller mixer at 300 r/min for 5 h. A total of 3 g of well-mixed material was weighed on an electronic balance and loaded into a mold with specifications of $\Phi 25 \times 100$ mm and 5 mm × 5 mm × 50 mm to press the raw material into a molding system. The billets were mechanically compacted in a cold isostatic press and dried in a vacuum drying oven under vacuum at 80 °C for 3 h. Finally, the billets were sintered in a tubular resistance furnace at a heating rate of 9 °C/min for 8 h at 590 °C to produce Al-20SiC-0.3CeO$_2$ composites. Prior to HCPEB treatment, the material was cut into 10 mm × 10 mm × 5 mm samples by a metal cutter and then polished sequentially using sandpaper (100#, 240#, 400#, 800#, 1500#, 3000#) and diamond polishing paste (~1 μm).

2.2. HCPEB Treatment

The surface modification of the material was carried out by the HOPE-I type HCPEB treatment device manufactured by the Dalian University of Technology (Liaoning, China). The relevant process parameters are shown in Table 2. The number of pulses for each experiment was 5, 15, and 25.

Table 2. Working parameters of the HCPEB system.

Acceleration Voltage (kV)	Energy Density (J/cm^2)	Pulse Time (μs)	Pulse Frequency (Hz)	Vacuum Level (Pa)
24.5	2	3	0.1	6.5×10^{-3}

2.3. Microstructure Characterization and Performance Analysis

Grazing Incidence X-ray Diffraction was performed using a multifunctional X-ray diffractometer (model X'Pert PRO, Panaco, The Netherlands). The friction coefficient of the aluminum matrix composite surface before and after the electron beam modification was measured by the friction test using the reciprocating motion mode on the surface of the sample with the wear instrument "MTF-5000" (Atech Instruments Technology Co., Ltd., Nanjing, China). The friction conditions were as follows: Si$_3$N$_4$ ball, 2 N load, 7 mm friction distance, and 10 min friction test time, and a Hitachi S-4800 field emission scanning electron microscope (Hitachi High-Tech Corporation, Tokyo, Japan) was used to observe the wear morphology on the surface of aluminum matrix composites with and without the addition of rare earth oxide (CeO$_2$), before and after the HCPEB modification. For the measurement of the surface microhardness of Al-20SiC composites before and after the HCPEB treatment, a Vickers hardness tester of type LM247AT (Luotai Precision Instruments Co., Ltd., Dongguan, China.) was selected. The test parameters were: load of 200 g and holding time of 13 s.

3. Results

Figure 1 illustrates the mixing of the raw material powders after ball milling. From the figure, it can be seen that the Al powder, SiC powder and CeO$_2$ powder are uniformly mixed. After ball milling, each powder is broken and refined to some extent.

Figure 2 shows the Grazing Incidence X-ray Diffraction (GIXRD) patterns of the Al-20SiC-0.3CeO$_2$ composite samples before and after the electron beam treatment. Figure 2 shows that the electron beam treatment of the sample with the addition of rare-earth Ce did not result in the formation of new phases. The sample mainly consisted of two phases, Al and SiC, and no rare-earth Ce-rich phases were detected. This is likely due to the fact that the content of rare-earth Ce in the alloy is very small and falls below the detection limit of the GIXRD instrument. The HCPEB treatment induced rapid melting and solidification processes that altered the original oriented casting organization in the surface layer of the alloy [22]. The temperature and stress fields induced by the electron beam induce severe plastic deformation on the surface of the alloy because of the FCC structure of aluminum and the high number of slip systems. Yan et al. [23] found a significant enhancement in the intensity of the diffraction peak of Al(220) in the surface-modified layer of 2024-type aluminum alloy treated with HCPEB, showing a selective growth of Al(110) grains in the modified layer. Hao et al. [24] treated AISI 316L stainless steel with HCPEB and found plastic deformation during the modification process, with a selective orientation of the (111) grain surface. It can be seen from Figure 2 that the crystal orientation of the surface layer of the sample was changed after five pulse treatments, and the tendency of aluminum grains to grow along the Al(111) and Al(200) crystal planes in a meritocratic manner was enhanced. Since the Al(111) crystal faces are tightly packed with minimal surface energy and good stability, it is beneficial to improve the substrate microstructure and wear resistance [25]. After 25 pulse treatments, the diffraction peak of Al(111) was shifted to a high angle, attributed to the generation of residual compressive stress. Residual compressive stresses can improve the wear resistance of materials by reducing the likelihood of surface damage and wear [26].

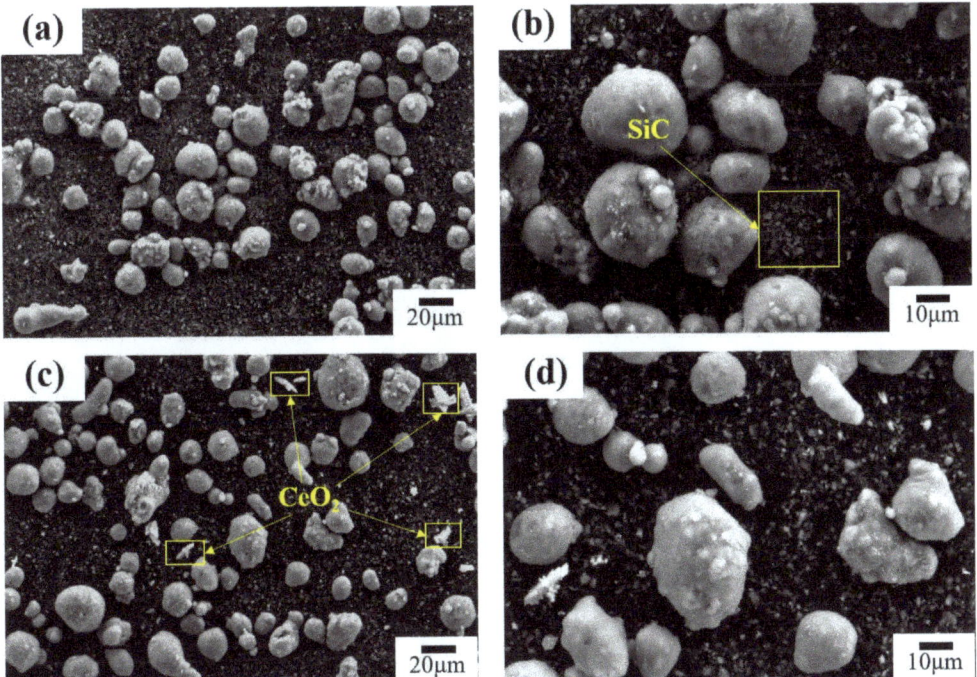

Figure 1. Mixing powder after ball milling: (**a**) Al-20SiC mixing powder, (**b**) partial enlargement, (**c**) Al-20SiC-0.3CeO$_2$ mixing powder, and (**d**) partial enlargement.

Figure 3 displays the surface microstructure morphology of the Al-20SiC composites before and after HCPEB treatment. The original histomorphology showed that the gray silicon carbide phase was uniformly distributed in the aluminum matrix, and the size of the SiC particles was around 10 μm (Figure 3a). The vicinity of the SiC particles exhibited a number of pore structures, which is believed to be due to the high viscosity of the liquid Al phase at the low sintering temperature of 590 °C [27]. This led to relatively poor mobility of the aluminum liquid and prevented the material from completing the complementary shrinkage during the subsequent solidification process, eventually resulting in the pore structure [28]. As shown in Figure 3b–d, an increase in the number of pulses results in the evaporation of SiC particles from the subsurface and their eventual eruption from the melting surface, forming a characteristic crater morphology. The EDS results (Figure 3e) indicate that the particles erupted at A are likely composed of SiC or Si. Previous studies have demonstrated that microstructural irregularities, such as grain boundaries, phase boundaries, and second-phase particles, are more prone to serve as nucleation centers for the formation of these crater-shaped features [29].

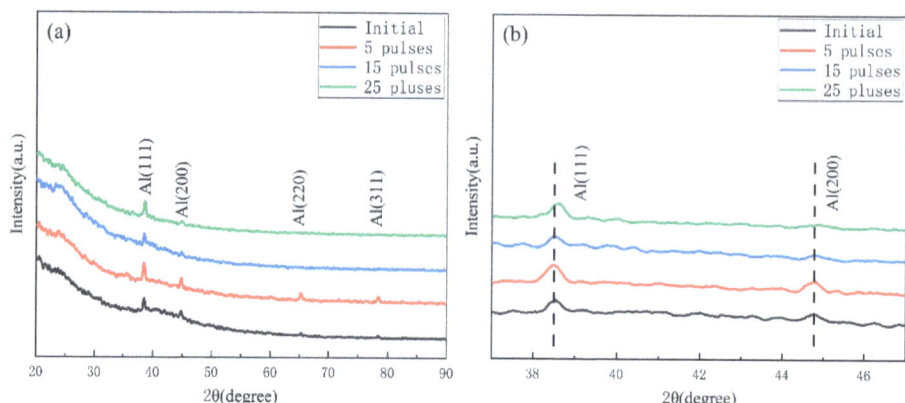

Figure 2. GIXRD patterns of Al-20SiC-0.3CeO$_2$ samples before and after HCPEB irradiation: (**a**) complete GIXRD patterns; (**b**) enlarged patterns of Al(111) crystal plane and Al(200).

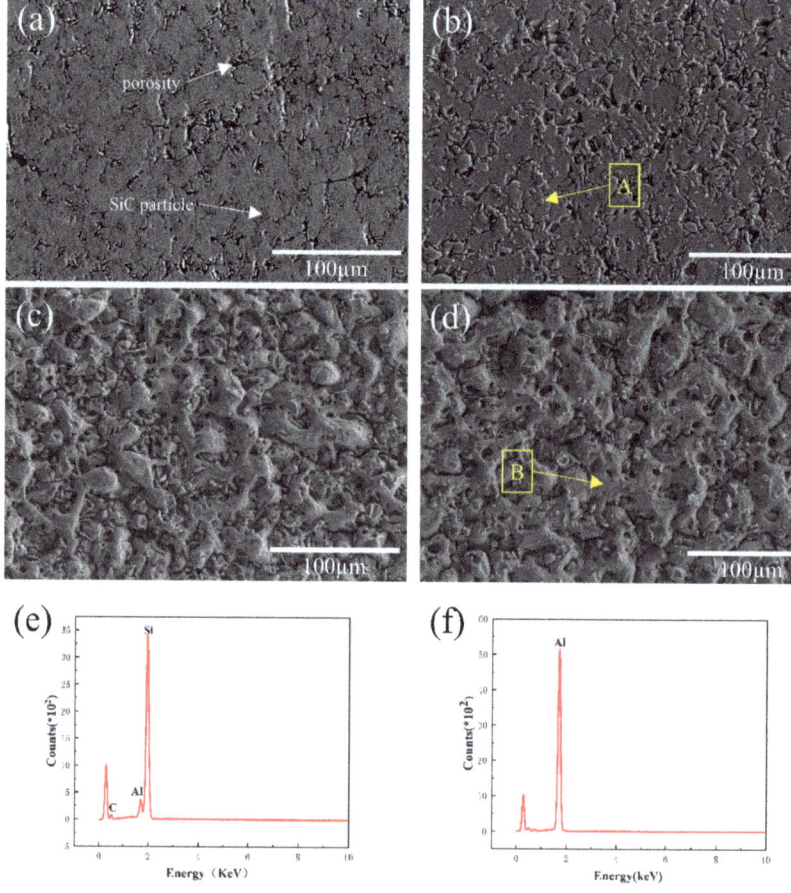

Figure 3. Surface morphology of Al-20SiC sample before and after strong current pulsed electron beam treatment. (**a**) Original sample; (**b**) 5 pulses; (**c**) 15 pulses; (**d**) 25 pulses; (**e**) EDS results of the A region in Figure 3b; (**f**) EDS results of the B region in Figure 3d.

Figure 4 depicts the surface microstructure morphology of Al-20SiC-0.3CeO$_2$ composites before and after HCPEB treatment. Compared to the sample without the addition of rare earth oxides in Figure 3a, there are fewer pores present in the original sample after the addition of CeO$_2$, as seen in Figure 4a [30]. This is mainly due to the fact that the addition of CeO$_2$ can significantly enhance the wettability between the liquid phase Al and the solid phase SiC, resulting in relatively dense samples during the sintering process. The mechanism of crater formation in Figure 4c,d is not explained as previously mentioned. Figure 4e shows that the surface of the aluminum matrix is covered with an oxide film. The oxide layer improves the wear resistance of the aluminum matrix composite surface because the oxide layer has a high hardness and resists friction and wear. In addition, the oxide layer provides lubrication to the material surface and reduces the coefficient of friction. From Figure 4f, it can be concluded that the rare earth CeO$_2$ is able to diffuse during the HCPEB treatment due to the electron beam's role in promoting the elemental diffusion effect [31]. In addition, CeO$_2$ reacts with impurities in the remelted layer in the HCPEB action zone and weakens the local stress concentration in the brittle phase melt pool. It also reduces the surface tension, leading to a smaller contact angle between Al and SiC, which is conducive to the elimination of pores and greatly improves the surface microstructure [2].

Figure 5 shows the changes in the aluminum-based Vickers hardness of Al-20SiC and Al-20SiC-0.3CeO$_2$ alloys before and after the intense current pulsed electron beam treatment. It can be observed from the figure that the Vickers microhardness of both Al-20SiC and Al-20SiC-0.3CeO$_2$ is greatly improved after the electron beam treatment, and the hardness tends to increase with the increase in the number of pulses [32]. The average value of Al matrix microhardness before HCPEB treatment was 50.2 HV, the average value of Al matrix microhardness for 5 pulses was 71.4 HV, the average value of Al matrix microhardness for 15 pulses was 109.5 HV, and the value of Al matrix microhardness for Al-20SiC-0.3CeO$_2$ composites after 25 pulses was the largest with an average value of 130.1 HV [33]. As seen in Figure 4, the HCPEB treatment of the material with the addition of the rare earth oxide CeO$_2$ increases the uniformity of the hard-phase particle distribution, which enhances the load-bearing capacity of the matrix aluminum per unit area. At the same time, the addition of CeO$_2$ improves the mobility of alloying elements and reduces tissue sparseness, resulting in better bonding of the reinforcing phase and the Al matrix, increasing the stress tolerance of the Al matrix, and increasing the hardness of the material. Ahmad et al. [34] showed that when the surface of the alloy was treated with an electron beam, the reinforcing phase SiC was dissolved and broken into fine particles uniformly distributed in the Al matrix under the action of the electron beam. The hardness of the primary phase is increased due to the uniform dispersion of the hard points, and the hardness of the coating is significantly enhanced due to the large amount of hard-phase SiC in the coating, which acts as a barrier to dislocation movement. CeO$_2$ is mostly located at grain boundaries or phase boundaries, which significantly reduces the activity of the interface and hinders the diffusive movement of aluminum grain boundaries, which can play a certain role in nailing the aluminum grain boundaries, causing the deformation of the matrix aluminum to be hindered, increasing the deformation resistance, and increasing the hardness [2]. The microhardness of the material surface is increased, thus enhancing to some extent the bearing effect of the composite material on stresses during frictional wear while causing less plastic deformation on the material surface during frictional wear and reducing the friction coefficient [35].

Figure 6 demonstrates the evolution of the friction coefficient with friction time and the corresponding friction coefficient of the Al-20SiC-0.3CeO$_2$ composite surface for different numbers of pulses [36]. From Figure 6a, it can be seen that the friction profile of the untreated and Al-20SiC-0.3CeO$_2$ composite surface under five pulses fluctuates more, and the width and depth of the profile also vary visually [37]. However, the friction curves of the Al-20SiC-0.3CeO$_2$ composite surface after 15 pulses and 25 pulses steadily floated, with small width and shallow depth of the profile. From Figure 6b, it can be seen that the friction coefficient on the surface of Al-20SiC-0.3CeO$_2$ composites shows an overall trend of decreasing all the time with the increase in the number of pulses. The friction coefficient of the specimen without HCPEB treatment is 0.520, and the friction coefficient on the surface of the specimen reaches the minimum value of 0.111 at 25 pulses, with a decrease of 78.65%. Shi Weixi [15] et al. added rare-earth Nd to an Al-Si alloy, and the results showed that the addition of rare earths resulted in a significant refinement of the primary silicon organization located in the alloy and a stronger bond with the matrix; thus, fatigue damage and chipping were greatly reduced, which in turn greatly improved the wear resistance of the material. The specific mechanism of the action of the composite surface wear resistance is further described below.

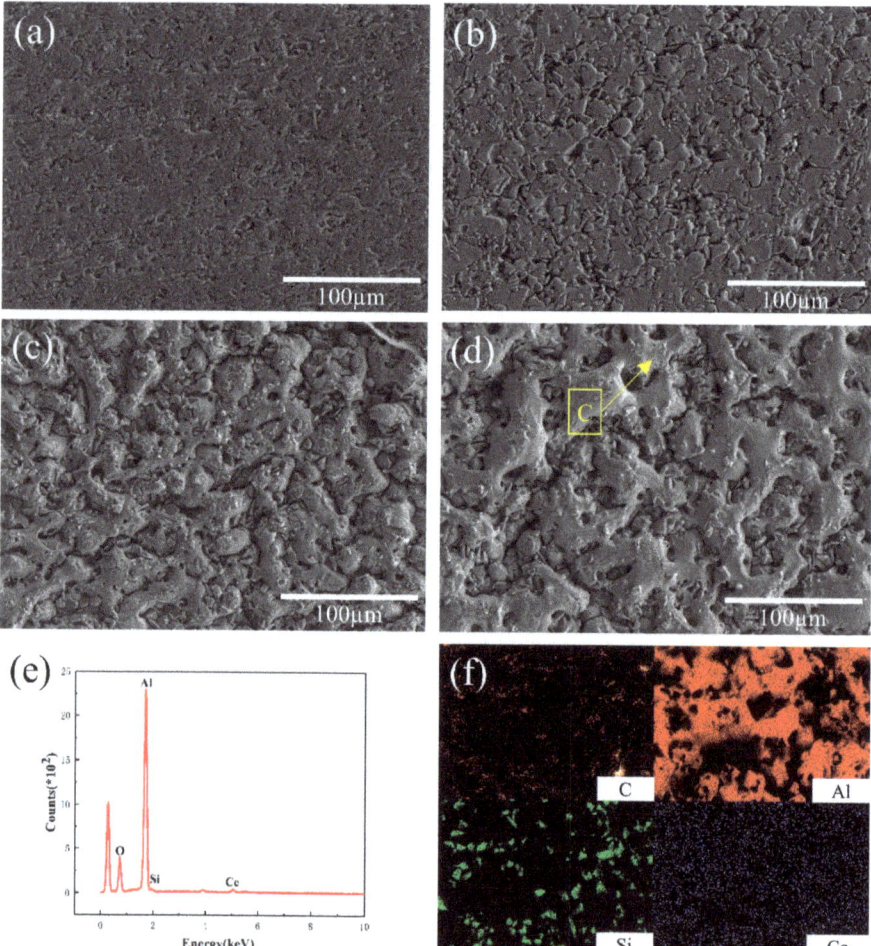

Figure 4. Surface morphology of Al-20SiC-CeO$_2$ sample before and after strong current pulsed electron beam treatment. (**a**) Original sample; (**b**) 5 pulses; (**c**) 15 pulses; (**d**) 25 pulses; (**e**) EDS results of the C region in Figure 4d; (**f**) alloy surface element distribution at 25 pulses.

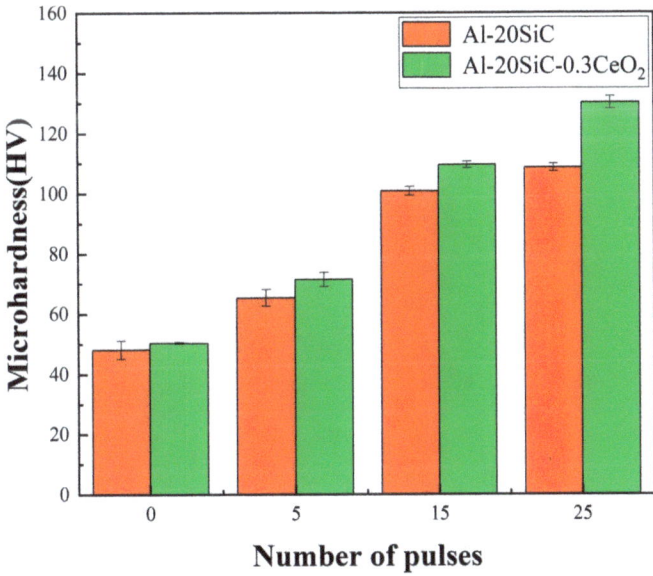

Figure 5. Microhardness measurements of Al-20SiC and Al-20SiC-0.3CeO$_2$ samples under different pulse times.

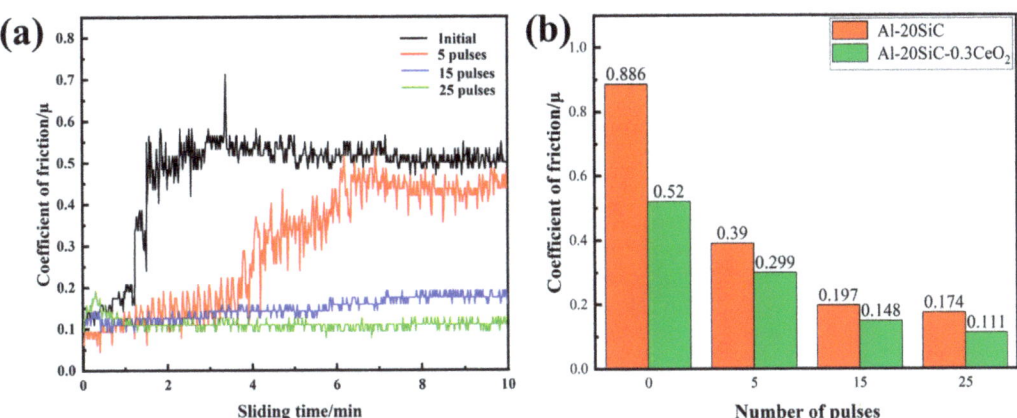

Figure 6. (**a**) Variation curves of friction coefficient of Al-20SiC-0.3CeO$_2$ with friction time for different pulse numbers; (**b**) surface friction coefficients of Al-20SiC and Al-20SiC-0.3CeO$_2$ for different pulse numbers.

Figure 7 illustrates the wear rate changes of Al-20SiC and Al-20SiC-0.3CeO$_2$ before and after HCPEB irradiation. As shown in the graph, the wear rate of the sample with CeO$_2$ addition is lower than that of the sample without CeO$_2$ at the same number of pulses; the wear rates of the samples decreased from 8.84 and 6.17 to 2.89 and 1.93, respectively, as the pulse number increased. Combined with the friction coefficient curve, the results indicate that the wear resistance of the samples is best after the 25-pulse treatment. The addition of CeO$_2$ contributes to the improvement of wear resistance.

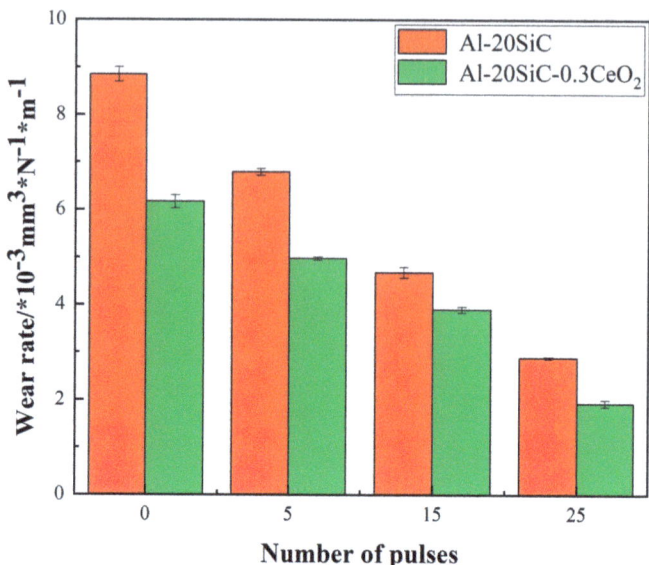

Figure 7. Variation of sample wear rate before and after HCPEB irradiation.

Figure 8 shows the wear morphology of the surface of Al-20SiC composites before and after 25-pulse treatments. During the frictional wear of the aluminum matrix composite, the SiC hard particles—as the main load-bearing phase—are subjected to both positive and tangential stresses [38]. From the wear morphology, it can be seen that on the untreated Al-20SiC composite, the wear surface showed a debris-like flaking phenomenon and a certain degree of plastic deformation occurred along the sliding direction on both sides of the plow groove. The stripped SiC constituted abrasive wear during the friction process or entered between the friction subsets, forming scratches and grooves on the wear surface [39]. At the same time, abrasive wear was further accelerated by the generation of hard particles and adherence to the surface during the sliding wear test, which produced microcuttings in the surface layer. The worn surface showed deep grooves in addition to furrow abrasions, and the wear mechanism was abrasive wear [40]. The wear surface of the samples after 25-pulse treatments was flatter, relatively smooth, and with shallow scratches, and the wear rate after modification was lower than the wear rate before modification.

Figure 9 shows the morphology of the worn surface of Al-20SiC-0.3CeO$_2$ composites before and after 25-pulse treatments. It can be seen from the figure [41] that the untreated Al-20SiC-0.3CeO$_2$ composite has obvious plow-like stripes of different widths and depths on the wear surface, and some of the wear surfaces have traces of being cut. Additionally, relatively deep grooves appear, which indicates that the matrix alloy wears relatively severely, and its wear mechanism is typical of adhesive wear [42]. This is due to the low hardness of the matrix alloy material and the tearing wear caused by the plowing action of the anti-abrasive during frictional wear. For the Al-20SiC-0.3CeO$_2$ composite treated with 25 pulses, the wear surface is relatively flat with no obvious groove-like streaks. Micron-sized particles can be seen on the matrix that were ground off and not dislodged but flatly exposed to the wear surface. This indicates that the regrind particles are directly subjected to frictional wear and play the main load-bearing role, and their wear mechanism is typical of abrasive wear [43]. It can be seen from the figure that the addition of rare earth oxides gives a lower wear rate at the same number of pulses and improves the wear resistance of the material surface.

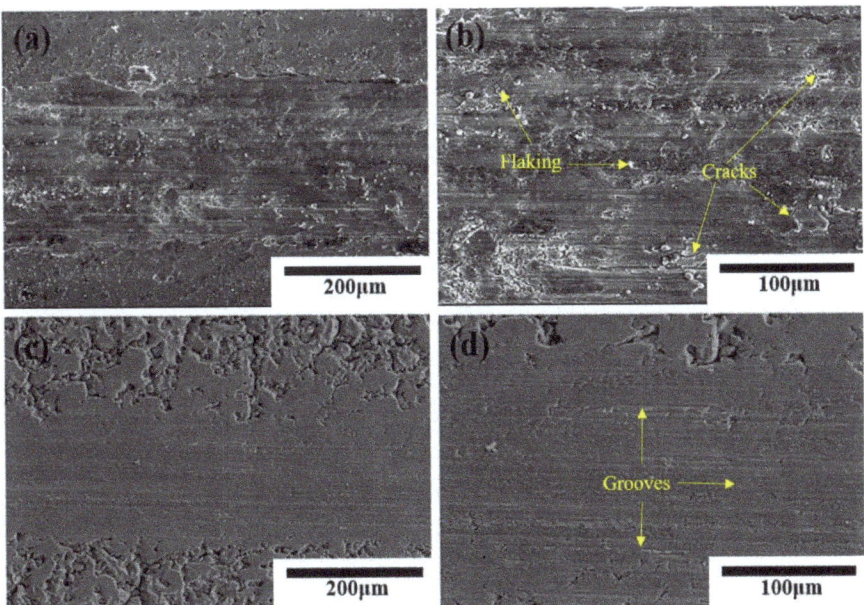

Figure 8. Surface wear morphology of Al-20SiC before and after electron beam treatment. (**a**) Original sample; (**b**) local enlarged image of original sample; (**c**) 25 pulses; (**d**) local enlarged image of 25 pulses.

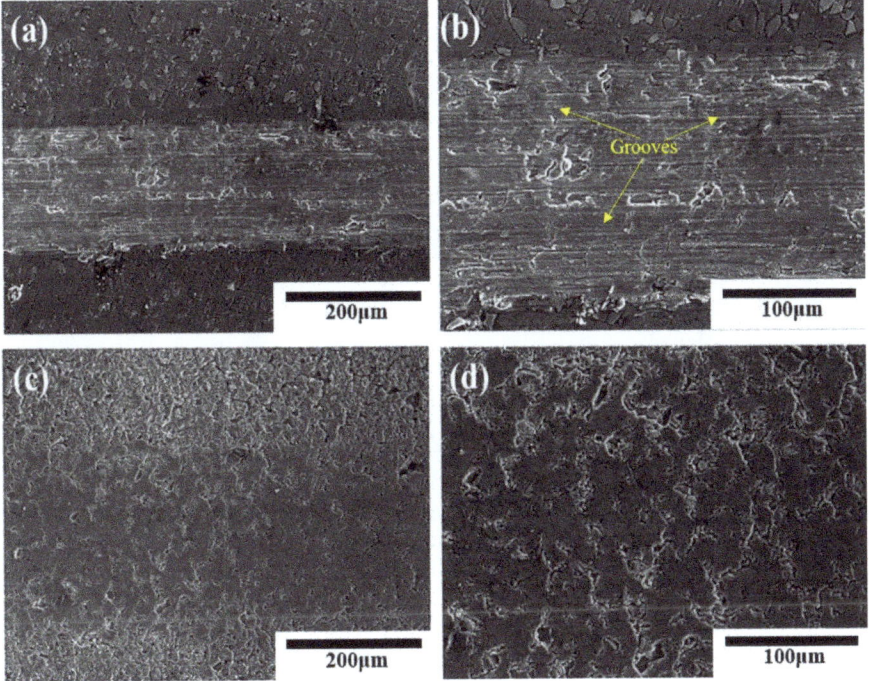

Figure 9. Surface wear morphology of Al-20SiC-0.3CeO$_2$ before and after electron beam treatment. (**a**) Original sample; (**b**) local enlarged image of original sample; (**c**) 25 pulses; (**d**) local enlarged image of 25 pulses.

The HCPEB treatment plays a pivotal role in optimizing the microstructure of Al-20SiC composites. The intense current pulses generated by HCPEB initiate rapid melting and solidification processes, resulting in improved grain orientation. This optimized microstructure significantly enhances the material's resistance to wear by minimizing crack propagation, reducing surface deformation, and mitigating fatigue. Additionally, CeO_2 strengthens the bonding between the reinforcing phase (SiC) and the aluminum matrix, effectively reducing the occurrence of defects such as pores and microcracks. This improved bonding enhances the overall integrity and strength of the composite material, resulting in enhanced wear resistance [44]. Finally, CeO_2 enhances the mobility of alloying elements within the material. This increased mobility allows for better dispersion of hard-phase particles, such as SiC, throughout the aluminum matrix. The uniform dispersion of these hard-phase particles leads to heightened hardness and improved resistance to wear and abrasion.

4. Results and Discussion

4.1. Results

In this paper, the effect of CeO_2 on the properties of Al-20SiC composites after HCPEB treatment was investigated.

(1) The results of GIXRD analysis showed that the rapid melting and solidification processes triggered by the HCPEB treatment led to selective orientation of the matrix and selective growth of Al(111) grains in the modified layer.
(2) SEM results showed that the presence of rare earth elements effectively eliminated defects such as porosity.
(3) Hardness tests showed that the addition of rare earth Ce increased the average microhardness of the matrix by 159.16%.
(4) The friction coefficient showed a reduction of 87.18% with the synergistic effect of CeO_2 and HCPEB.

4.2. Discussion

The optimal growth of the Al(111) crystal plane improves the material's crystal structure and grain orientation, benefiting from its favorable crystallization properties and high-density arrangement. This well-ordered arrangement of grains effectively withstands external stress and wear, resulting in reduced surface wear and fatigue of the material.

The increase in surface hardness can be attributed to two factors: Firstly, the addition of CeO_2 enhances the activity of alloying elements, reduces tissue sparseness, and improves the bonding between the reinforcing phase and the aluminum matrix. Secondly, the treatment of the alloy surface with an electron beam results in the dissolution and fragmentation of the reinforcing phase SiC into fine particles that are uniformly dispersed within the aluminum matrix under the beam's action.

The improvement in wear resistance is mainly attributed to the enhanced hardness of the material's surface, which enhances its resistance to scratches and wear. Additionally, the HCPEB treatment induces rapid melting and solidification, optimizing the microstructure of the material and further contributing to its enhanced wear resistance.

Author Contributions: Conceptualization, L.W., Y.Z. and B.G.; methodology, L.W. and Y.S.; software, Y.Z. and L.H.; validation, L.W., B.G. and Y.Z.; formal analysis, L.W.; investigation, L.W., L.H. and Y.S.; resources, B.G.; data curation, L.W. and Y.Z.; writing—original draft preparation, Y.Z., B.G. and Y.S.; writing—review and editing, Y.Z.; visualization, L.W.; supervision, B.G.; project administration, B.G.; funding acquisition, B.G. All authors have read and agreed to the published version of the manuscript.

Funding: The National Natural Science Foundation of China (51671052), the Fundamental Research Funds for the Central Universities (N182502042) and the Liao Ning Revitalization Talents Program (XLYC1902105).

Institutional Review Board Statement: Not applicable.

Informed Consent Statement: Not applicable.

Data Availability Statement: Data will be made available upon request from the corresponding author.

Acknowledgments: This work was supported by the National Natural Science Foundation of China (51671052), the Fundamental Research Funds for the Central Universities (N182502042), and the Liao Ning Revitalization Talents Program (XLYC1902105).

Conflicts of Interest: The authors declare no conflict of interest.

References

1. Li, L.; Ning, Z.; Huang, W.; Liao, L.; Zheng, Y.; Zhuang, K.; Lan, S.; Zhang, Y.; Yao, R. In-situ fabrication of lightweight SiC(Al, rGO) bulk ceramics derived from silicon oxycarbide for aerospace components. *J. Alloys Compd.* **2021**, *869*, 159297. [CrossRef]
2. Liu, Q.; Wang, F.; Qiu, X.; An, D.; He, Z.; Zhang, Q.; Xie, Z. Effects of La and Ce on microstructure and properties of SiC/Al composites. *Ceram. Int.* **2020**, *46*, 1232–1235. [CrossRef]
3. Ozben, T.; Kilickap, E.; Çakır, O. Investigation of mechanical and machinability properties of SiC particle reinforced Al-MMC. *J. Mater. Process. Technol.* **2008**, *198*, 220–225. [CrossRef]
4. Shaga, A.; Shen, P.; Guo, R.-F.; Jiang, Q.-C. Effects of oxide addition on the microstructure and mechanical properties of lamellar SiC scaffolds and Al–Si–Mg/SiC composites prepared by freeze casting and pressureless infiltration. *Ceram. Int.* **2016**, *42*, 9653–9659. [CrossRef]
5. Wang, L.; Fan, Q.; Li, G.; Zhang, H.; Wang, F. Experimental observation and numerical simulation of SiC_{3D}/Al interpenetrating phase composite material subjected to a three-point bending load. *Comput. Mater. Sci.* **2014**, *95*, 408–413. [CrossRef]
6. Yang, D.; Zhou, Y.; Yan, X.; Wang, H.; Zhou, X. Highly conductive wear resistant Cu/Ti_3SiC_2(TiC/SiC) co-continuous composites via vacuum infiltration process. *J. Adv. Ceram.* **2020**, *9*, 83–93. [CrossRef]
7. Zhang, Q.; Ma, X.; Wu, G. Interfacial microstructure of SiCp/Al composite produced by the pressureless infiltration technique. *Ceram. Int.* **2013**, *39*, 4893–4897. [CrossRef]
8. Ganesh, M.R.S.; Reghunath, N.; Levin, M.J.; Prasad, A.; Doondi, S.; Shankar, K.V. Strontium in Al–Si–Mg Alloy: A Review. *Met. Mater. Int.* **2022**, *28*, 1–40. [CrossRef]
9. Chak, V.; Chattopadhyay, H.; Dora, T.L. A review on fabrication methods, reinforcements and mechanical properties of aluminum matrix composites. *J. Manuf. Process.* **2020**, *56*, 1059–1074. [CrossRef]
10. Sadhu, K.K.; Mandal, N.; Sahoo, R.R. SiC/graphene reinforced aluminum metal matrix composites prepared by powder metallurgy: A review. *J. Manuf. Process.* **2023**, *91*, 10–43. [CrossRef]
11. Sun, Y.; Gao, B.; Hu, L.; Li, K.; Zhang, Y. Effect of CeO_2 on Corrosion Resistance of High-Current Pulsed Electron Beam Treated Pressureless Sintering Al-20SiC Composites. *Coatings* **2021**, *11*, 707. [CrossRef]
12. Zou, J.; Grosdidier, T.; Zhang, K.; Dong, C. Mechanisms of nanostructure and metastable phase formations in the surface melted layers of a HCPEB-treated D2 steel. *Acta Mater.* **2006**, *54*, 5409–5419. [CrossRef]
13. Walker, J.C.; Murray, J.; Narania, S.; Clare, A.T. Dry Sliding Friction and Wear Behaviour of an Electron Beam Melted Hypereutectic Al–Si Alloy. *Tribol. Lett.* **2012**, *45*, 49–58. [CrossRef]
14. Hao, Y.; Gao, B.; Tu, G.F.; Li, S.W.; Dong, C.; Zhang, Z.G. Improved wear resistance of Al–15Si alloy with a high current pulsed electron beam treatment. *Nucl. Instrum. Methods Phys. Res. Sect. B Beam Interact. Mater. At.* **2011**, *269*, 1499–1505. [CrossRef]
15. Shi, W.X.; Gao, B.; Tu, G.F.; Li, S.W. Effect of Nd on microstructure and wear resistance of hypereutectic Al–20%Si alloy. *J. Alloys Compd.* **2010**, *508*, 480–485. [CrossRef]
16. Hu, L.; Gao, B.; Zhu, G.; Hao, Y.; Sun, S.; Tu, G. The effect of neodymium on the microcracks generated on the Al–17.5Si alloy surface treated by high current pulsed electron beam. *Appl. Surf. Sci.* **2016**, *364*, 490–497. [CrossRef]
17. Ravinath, H.; Ahammed, I.; Harigovind, P.; Devan, S.A.; Senan, V.R.A.; Shankar, K.V.; Nandakishor, S. Impact of aging temperature on the metallurgical and dry sliding wear behaviour of $LM25/Al_2O_3$ metal matrix composite for potential automotive application. *Int. J. Lightweight Mater. Manuf.* **2023**, *6*, 416–433. [CrossRef]
18. Yang, J.; Bai, S.; Sun, J.; Wu, H.; Sun, S.; Wang, S.; Li, Y.; Ma, W.; Tang, X.; Xu, D. Microstructural understanding of the oxidation and inter-diffusion behavior of Cr-coated Alloy 800H in supercritical water. *Corros. Sci.* **2023**, *211*, 110910. [CrossRef]
19. Zhang, P.; Gao, Y.; Liu, Z.; Zhang, S.; Wang, S.; Lin, Z. Effect of cutting parameters on the corrosion resistance of 7A04 aluminum alloy in high speed cutting. *Vacuum* **2023**, *212*, 111968. [CrossRef]
20. Guo, K.; Gou, G.; Lv, H.; Shan, M. Jointing of CFRP/5083 Aluminum Alloy by Induction Brazing: Processing, Connecting Mechanism, and Fatigue Performance. *Coatings* **2022**, *12*, 1559. [CrossRef]
21. Wu, J.; Djavanroodi, F.; Shamsborhan, M.; Attarilar, S.; Ebrahimi, M. Improving Mechanical and Corrosion Behavior of 5052 Aluminum Alloy Processed by Cyclic Extrusion Compression. *Metals* **2022**, *12*, 1288. [CrossRef]
22. Ji, L.; Cai, J.; Liu, S.C.; Zhang, Z.Q.; Hou, X.L.; Lv, Y.P.; Guan, Q.F. Formation of Surface Nanoaustenite and Properties of 3Cr13 Steel Induced by Pulsed Electron Beam Irradiation under Melting Mode. *Adv. Mater. Res.* **2013**, *787*, 363–366. [CrossRef]
23. Yan, P.; Zou, J.; Zhang, C.; Grosdidier, T. Surface modifications of a cold rolled 2024 Al alloy by high current pulsed electron beams. *Appl. Surf. Sci.* **2020**, *504*, 144382. [CrossRef]

24. Hao, S.; Wu, P.; Zou, J.; Grosdidier, T.; Dong, C. Microstructure evolution occurring in the modified surface of 316L stainless steel under high current pulsed electron beam treatment. *Appl. Surf. Sci.* **2007**, *253*, 5349–5354. [CrossRef]
25. Bai, Q.; Wang, J.; Xing, S.; Ma, Y.; Bao, X. Crystal orientation and crystal structure of paramagnetic α-Al under a pulsed electromagnetic field. *Sci. Rep.* **2020**, *10*, 10603. [CrossRef] [PubMed]
26. Qin, Y.; Zou, J.; Dong, C.; Wang, X.; Wu, A.; Liu, Y.; Hao, S.; Guan, Q. Temperature–stress fields and related phenomena induced by a high current pulsed electron beam. *Nucl. Instrum. Methods Phys. Res. Sect. B Beam Interact. Mater. At.* **2004**, *225*, 544–554. [CrossRef]
27. Balmon, J.; Fouvry, S.; Villechaise, P.; Paturaud, J.; Tschofen, J.; Feraille, J. Influence of SiC particles orientation on fretting crack extension in an Al-SiC metal matrix composite. *Eng. Fract. Mech.* **2023**, *281*, 109091. [CrossRef]
28. Meisner, L.L.; Semin, V.O.; Mironov, Y.P.; Meisner, S.N.; D'yachenko, F.A. Cross-sectional analysis of the graded microstructure and residual stress distribution in a TiNi alloy treated with low energy high-current pulsed electron beam. *Mater. Today Commun.* **2018**, *17*, 169–179. [CrossRef]
29. Zhang, K.; Zou, J.; Grosdidier, T.; Dong, C.; Yang, D. Improved pitting corrosion resistance of AISI 316L stainless steel treated by high current pulsed electron beam. *Surf. Coat. Technol.* **2006**, *201*, 1393–1400. [CrossRef]
30. Hao, S.; Yao, S.; Guan, J.; Wu, A.; Zhong, P.; Dong, C. Surface treatment of aluminum by high current pulsed electron beam. *Curr. Appl. Phys.* **2001**, *1*, 203–208. [CrossRef]
31. Hu, L.; Gao, B.; Xu, N.; Sun, Y.; Zhang, Y.; Xing, P. Effect of Cerium and Magnesium on Surface Microcracks of Al–20Si Alloys Induced by High-Current Pulsed Electron Beam. *Coatings* **2022**, *12*, 61. [CrossRef]
32. Lu, J.; Tang, S.; Zhang, H.; Zhong, X.; Liu, Q.; Lv, Z. Preparation and mechanical properties of SiC$_w$-reinforced WC-10Ni$_3$Al cemented carbide by microwave sintering. *Ceram. Int.* **2023**, *49*, 21587–21601. [CrossRef]
33. Kaur, K.; Anant, R.; Pandey, O.P. Tribological Behaviour of SiC Particle Reinforced Al–Si Alloy. *Tribol. Lett.* **2011**, *44*, 41–58. [CrossRef]
34. Ahmad, M.; Akhter, J.I.; Iqbal, M.; Akhtar, M.; Ahmed, E.; Shaikh, M.A.; Saeed, K. Surface modification of Hastelloy C-276 by SiC addition and electron beam melting. *J. Nucl. Mater.* **2005**, *336*, 120–124. [CrossRef]
35. Camacho-Rios, M.L.; Saenz-Trevizo, A.; Lardizabal-Gutierrez, D.; Medrano-Prieto, H.M.; Estrada-Guel, I.; Garay-Reyes, C.G.; Martinez-Sanchez, R. Effect of CeO$_2$ nanoparticles on Microstructure and Hardness of A6063 Aluminum Alloy. *Microsc. Microanal.* **2018**, *24*, 2276–2277. [CrossRef]
36. Wilson, S.; Alpas, A.T. Wear mechanism maps for metal matrix composites. *Wear* **1997**, *212*, 41–49. [CrossRef]
37. Zhang, J.; Wei, Z.; Li, Z.; Hou, B.; Xue, R.; Xia, H.; Shi, Z. SiC honeycomb reinforced Al matrix composite with improved tribological performance. *Ceram. Int.* **2021**, *47*, 23376–23385. [CrossRef]
38. Birsen, D.; Tütük, İ.; Acar, S.; Karabeyoğlu, S.S.; Özer, G.; Güler, K.A. Microstructure and wear characteristics of hybrid reinforced (ex-situ SiC–in-situ Mg$_2$Si) Al matrix composites produced by vacuum infiltration method. *Mater. Chem. Phys.* **2023**, *302*, 127743. [CrossRef]
39. Yadav, P.K.; Dixit, G. Erosive-Corrosive Wear of Aluminium-Silicon Matrix (AA336) and SiC$_p$/TiB$_{2p}$ Ceramic Composites. *Silicon* **2019**, *11*, 1649–1660. [CrossRef]
40. Kumar, A.; Mahapatra, M.M.; Jha, P.K. Modeling the abrasive wear characteristics of in-situ synthesized Al–4.5%Cu/TiC composites. *Wear* **2013**, *306*, 170–178. [CrossRef]
41. Liu, S.Y.; Wang, Y.; Zhou, C.; Pan, Z.Y. Mechanical properties and tribological behavior of alumina/zirconia composites modified with SiC and plasma treatment. *Wear* **2015**, *332–333*, 885–890. [CrossRef]
42. Jiang, T.; Zhang, W.; Su, Z.; Xue, Y.; Wang, S.; Zhao, H.; Sun, Y.; Li, Y.; Xu, G. Improving the wear resistance of 50 wt% Si particle-reinforced Al matrix composites treated by over-modification with a Cu-P modifier. *Tribol. Int.* **2023**, *180*, 108247. [CrossRef]
43. Yadav, P.K.; Dixit, G.; Dixit, S.; Singh, V.P.; Patel, S.K.; Purohit, R.; Kuriachen, B. Effect of eutectic silicon and silicon carbide particles on high stress scratching wear of aluminium composite for various testing parameters. *Wear* **2021**, *482–483*, 203921. [CrossRef]
44. Manivannan, I.; Ranganathan, S.; Gopalakannan, S.; Suresh, S.; Nagakarthigan, K.; Jubendradass, R. Tribological and surface behavior of silicon carbide reinforced aluminum matrix nanocomposite. *Surf. Interfaces* **2017**, *8*, 127–136. [CrossRef]

Disclaimer/Publisher's Note: The statements, opinions and data contained in all publications are solely those of the individual author(s) and contributor(s) and not of MDPI and/or the editor(s). MDPI and/or the editor(s) disclaim responsibility for any injury to people or property resulting from any ideas, methods, instructions or products referred to in the content.

Article

Strength and Deformation Behavior of Graphene Aerogel of Different Morphologies

Julia A. Baimova [1,*] and Stepan A. Shcherbinin [2,3]

[1] Institute for Metals Superplasticity Problems, Russian Academy of Sciences, Ufa 450001, Russia
[2] Higher School of Theoretical Mechanics and Mathematical Physics, Peter the Great St. Petersburg Polytechnic University, Polytechnicheskaya 29, St. Petersburg 195251, Russia; stefanshcherbinin@gmail.com
[3] Laboratory "Discrete Models in Mechanics", Institute for Problems in Mechanical Engineering, Russian Academy of Sciences, St. Petersburg 199178, Russia
* Correspondence: julia.a.baimova@gmail.com

Abstract: Graphene aerogels are of high interest nowadays since they have ultralow density, rich porosity, high deformability, and good adsorption. In the present work, three different morphologies of graphene aerogels with a honeycomb-like structure are considered. The strength and deformation behavior of these graphene honeycomb structures are studied by molecular dynamics simulation. The effect of structural morphology on the stability of graphene aerogel is discussed. It is shown that structural changes significantly depend on the structural morphology and the loading direction. The deformation of the re-entrant honeycomb is similar to the deformation of a conventional honeycomb due to the opening of the honeycomb cells. At the first deformation stage, no stress increase is observed due to the structural transformation. Further, stress concentration on the junctions of the honeycomb structure and over the walls occurs. The addition of carbon nanotubes and graphene flakes into the cells of graphene aerogel does not result in a strength increase. The mechanisms of weakening are analyzed in detail. The obtained results further contribute to the understanding of the microscopic deformation mechanisms of graphene aerogels and their design for various applications.

Keywords: graphene; graphene aerogel; molecular dynamics; mechanical properties

---

Citation: Baimova, J.A.; Shcherbinin, S.A. Strength and Deformation Behavior of Graphene Aerogel of Different Morphologies. *Materials* **2023**, *16*, 7388. https://doi.org/10.3390/ma16237388

Academic Editor: Raul D. S. G. Campilho

Received: 7 November 2023
Revised: 23 November 2023
Accepted: 25 November 2023
Published: 27 November 2023

Copyright: © 2023 by the authors. Licensee MDPI, Basel, Switzerland. This article is an open access article distributed under the terms and conditions of the Creative Commons Attribution (CC BY) license (https://creativecommons.org/licenses/by/4.0/).

1. Introduction

The study of tree-dimensional (3D) porous graphene nanostructures is one of the hot issues in different industrial applications. Such structures as aerogels and crumpled graphene are lightweight and low-cost and have high porosity and high strength [1,2]. Among them are graphene aerogels (GA) or cellular honeycomb structures—a promising new class of materials, with properties that can be tuned by designing their morphology. Aerogels are highly porous solid foams that have an interconnected network of thin, solid walls. The details of the atomic arrangement and structural stability of GA were previously intensively studied [3–6]. Due to their outstanding properties, GA can be used for the fabrication of the composites [7], for electronic devises [8], and for energy harvesting [1,7,9]. The properties of such cellular structures are controlled by their cell geometry and intrinsic morphology [10,11].

One of the most promising methods is the fabrication of GA from graphene oxide, which is synthesized and dispersed in distilled water [12,13]. Graphene and carbon nanotubes have been used to synthesize aerogels with low density and high elasticity [14,15]. Recently, new nanoporous graphene aerogel was obtained by the nanoporous Ni (np-Ni)–based chemical vapor deposition method in which dealloyed np-Ni with 3D bicontinuous open na [16]. To date, the experimental investigation of such structures is commonly based on the understanding of compression behavior, which results in a better description of its main characteristic—high compressibility [17–19]. The main challenge is morphology control, for example, creating interactions among constituents or understanding the self-assembly mechanism, which can improve the mechanical and physical properties of GA [20,21]. From this point of view, it is very important to use simulation techniques to understand the basic principles of property control, of the increase in mechanical strength by morphology modifications. Molecular dynamics (MD) is especially effective for such studies and allows one to obtain a large number of new data with a full analysis of the atomic structure of such nanomaterials as GA. The MD method is very good at predicting the mechanical response of graphene, by calculating and summing the evolutions of all atoms. GA involved studies using both MD simulations [22–24] and experiments [25,26].

The mechanical properties of graphene aerogel are of high importance for stretchable electronics, wearable devices, and smart manufacturing. Some of the disadvantages of graphene aerogel are the volume shrinkage and the deformation of porous skeletons, which affect the mechanical properties [27–29]. GA also show brittle behavior under tension [26]. The poor mechanical strength because of the weak joints in their porous network is another negative characteristic [30]. At the same time, these structures are superelastic and highly compressible [31]. In [32], GA were studied under tension and nanoindentation. The hardness (50.9 GPa) and Young's modulus (461 ± 9 GPa) were defined from MD simulation. To enhance the mechanical properties of graphene aerogel and make it more stable to deformation, some supporting elements need to be added to the porous of graphene aerogels [33–35]. Graphene flakes, which appeared during the synthesis of graphene aerogel or metal nanoparticles, can be added as the reinforcing elements.

Despite the fact that numerous works have been conducted to date on the mechanical behavior of GA, there is still a lack of understanding of the effect of morphology on its strength [26–29,33–36]. The change in morphology gives rise to improved mechanical behavior and an increased surface area. Moreover, the majority of works are devoted to the study of compressive behavior in simulation as well as in experiments [37–39]. The reason for this is that compressible materials efficiently translate mechanical deformations into electrical signals that have various potential applications. It is desirable to search for high compressive strength and strength recovery, and superb compressibility, in combination with other properties. The systematic study of such factors as morphology, the presence of the reinforcing elements, temperature, and loading direction on the mechanical properties and deformation behavior of graphene aerogels is of high importance.

In the present work, the mechanical properties of three different morphologies of GA under tension are studied by molecular dynamics simulation. The high stretchability is studied and analyzed for tension along different lattice directions. The effect of high temperatures on the strain and deformation behavior is studied.

2. Simulation Details

Figure 1 shows three initial cellular structures (part of the simulation cell) under consideration. A part of the simulation cell in an enlarged scale is also presented in Figure 1b as the projection to the xz-plane and in perspective. All of the initial structures are generated by a homemade program, which is used to generate the graphene cellular structure with a different cell morphology. For comparison, graphene flakes and CNTs are added into each honeycomb cell because they are the most studied reinforcement elements for different aerogels [17,40–43]. Unless otherwise noted, the coordinate system for all structures is the same as the coordinate system shown in Figure 1. In other words, the tension parallel to bridges is equivalent to the tension along the z direction, while the tension along lamella is equivalent to the tension along the x direction.

The morphology, which will be addressed as a honeycomb is widespread and has been previously fabricated and studied in numerous works [17,26,39,44,45]. This honeycomb-like structure is a geometry that has proven to be highly beneficial to maximizing a bulk-specific elastic modulus [10,46]. Here, the size of the honeycomb wall is $l = 16$ Å, and the number of atoms is 2500.

The morphology, which will be addressed as a re-entrant honeycomb, has been studied in [47–49] as a structure with auxetic properties (negative Poisson's ratio). Here, the main parameters are $a = 20$ Å, $b = 11$ Å ($a/b = 0.5$), and the inclination angle $\theta = 60°$. These parameters were proposed in [47]. The number of atoms in the system is 2448.

One of the possible morphologies of GA, which will be addressed as an arrow honeycomb, is also studied for comparison. Here, the main parameters are $c = 20$ Å, $d = 11$ Å. The number of atoms in the system is 1836. This structure is very close to that discussed in [38,50]. The geometry of the considered structure consists of graphene bridges and lamellae. The bridges are positioned between the lamellae. In the present work, curved graphene nanoribbons (along z-axis) can also be called bridges, while long graphene nanoribbons (along x-axes) can be called lamellae. The difference with the structure considered in [38] is the curvature of the bridges.

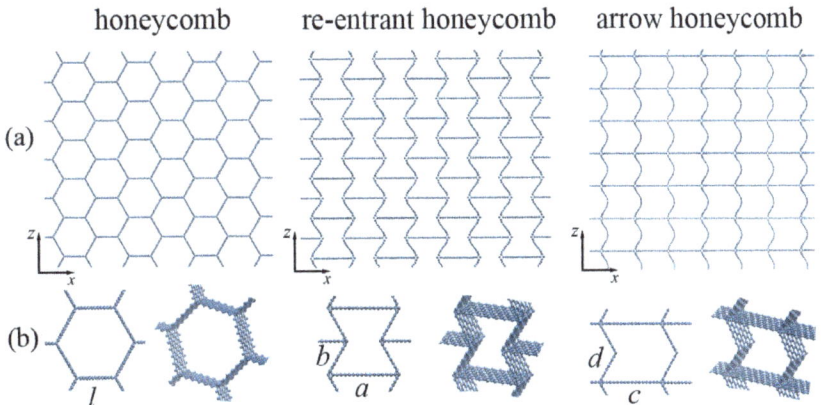

Figure 1. Three morphologies of honeycomb graphene aerogel: honeycomb, re-entrant honeycomb, and arrow-honeycomb. (**a**) Part of the simulation cell in projection to xz plane. (**b**) Part of the simulation cell on an enlarged scale as the projection to xz-plane and in perspective.

All calculations are performed using the LAMMPS [51–53] software package with the AIREBO interatomic potential [54], which has been successfully used to study various properties of a large number of carbon nanosystems, including graphene aerogels [24,55–59], and is proven to reliably reproduce experimentally-obtained mechanical results for graphene [60]. This potential can also be modified to AIREBO-Morse interatomic potential for the description of C-C interactions during compression [61]. Despite the fact that the method of MD is very popular for such studies, especially for the study of the mechanical and fracture behavior of graphene aerogels [55], there are some limitations. For example, under compression, the Lennard–Jones potential contribution to the Adaptive Intermolecular Reactive Empirical Bond Order (AIREBO) potential plays a critical role in predicting early densification. With the increase in the density of GA, the densification of the structure occurs faster with the dissipation energy rising [55]. Despite the fact that AIREBO can effectively describe the realistic mechanical behavior of graphene, it still faces some challenges while reproducing the C-C bond in hybrid structures. In [20,62], Tersoff potential was used as a more suitable option for the simulation of GA.

Despite the slight uncertainty of the AIREBO application for the study of mechanical properties of GA, there is plenty of work devoted to this issue, where simulation is conducted with AIREBO. Based on such works as [37–39,63,64], where the deformation behavior of GA was studied and considered in comparison with the experiment, we choose AIREBO interatomic potential for the present work. To avoid non-physical post-hardening behavior under large strains, we set the cut-off distances in the AIREBO potential to 1.92 Å, as suggested by previous studies [37,65,66]. To note, in [67] the deformation behavior of honeycomb graphene was studied with the modified parameters for AIREBO.

In the present work, the tensile behavior of GA is analyzed at two temperatures, 300 and 1000 K, and thus the limitation of the AIREBO potential for the simulation of such effects needs to be discussed. In [68], ReaxFF [69] and AIREBO have been utilized to understand the deformation and fracture mechanics of graphene at the atomic scale to compare the results from DFTB. The other important factor that affects the obtained results is the simulation technique for tension. The deformation under static loading usually shows much higher failure strains than the dynamic loading due to the imposed symmetry [68]. It was also shown that thermal fluctuation under dynamic loading disturbs symmetry and can predict similar failing strains under dynamic loading. Despite having a lot of limitations, the stress–strain curves and failure strain from AIREBO relatively work well.

Periodic boundary conditions are applied in all directions. A Nose–Hoover thermostat is used with a target temperature, a constant number of atoms, and standard velocity–Verlet time integration with a timestep of 1 fs. OVITO is used to visualize the atomistic simulation data. At first, the energy minimization at 0 K is conducted to reach an equilibrium state of the system.

To study the deformation behavior, uniaxial tensile stress is applied along the x or z axes, and the corresponding stress components are calculated as it is implemented in the LAMMPS package. Stress and energy per atom are also calculated using LAMMPS. At low temperatures, it has been previously shown that the strain rate has only a slight effect on the fracture strength and fracture strain for graphene [70]. Thus, the strain rate of 0.1 fs^{-1} was chosen. The length of the simulation cell for all structures is about $L_x = L_z = 60$ Å, $L_y = 12$ Å.

3. Results and Discussion

The structural changes during the relaxation were analyzed: the structural parameters were changed insignificantly. For example, sharp edges were smoothed for re-entrant and arrow honeycomb. For arrow honeycomb, cells transform to a more square shape.

The compressive behaviors of such honeycombs under in-plane uniaxial compression was previously studied [37,38]. It was shown that three deformation regimes can be distinguished: (1) stable and nearly uniform deformation and a resulting relatively high stiffness, (2) the coexistence of collapsed and uncollapsed deformation and a resulting essentially zero stiffness, and (3) densification with relatively uniform and stable collapsed deformation and a much stiffer response. Some of the deformation regimes for tension and compression are similar. It is revealed that GA exhibited elastic instability and inelastic collapse during compression.

3.1. Effect of Temperature

The effect of temperature on the fracture behavior of GA is studied for all three structures. The tensile behavior was analyzed at 300 K and 1000 K. The thermal stability has been confirmed by MD simulations at temperatures of 300 K and 1000 K [4]. Figures 2–4a show stress–strain curves under tension along x-axes for honeycomb (Figure 2); re-entrant honeycomb Figure 3; and arrow honeycomb (Figure 4). Figures 2–4b present the snapshots of GA as the projection on the xz-plane at critical points. The deformation behavior of GA considerably depends on their morphologies. Additional analysis of the stress per atom (Figures 2–4c) and potential energy per atom (Figures 2–4d) is conducted for each structure during tension.

Let us consider the deformation process at 300 K in detail for honeycomb GA (see Figure 2). The critical points on the stress–strain curve are labeled as 1–5. No stress changes up to $\varepsilon = 0.25$ occur, which is explained by the simple changes in the shape of honeycomb cells. As seen in Figure 2, initial tension causes the mild flexing of the structure, followed by a series of collapses of cells, leading to an eventual flattening. This can be seen from the snapshots of the structure at critical points (shown by green circles on the black curve). The walls of the honeycomb cells are almost not stressed. The distribution of stress per atom can also be seen from Figure 2c: walls, which are parallel to the tensile direction that became stressed only at $\varepsilon = 0.5$. From the distribution of the potential energy per atom (Figure 2d), the pre-critical strain $\varepsilon = 0.5$ can also be estimated. Up to this value, the potential energy of the system was not increased. The contact of cell walls, which is usually accompanied by sliding and irreversibility, increases the complexities of the constitutive behavior of the honeycomb structure.

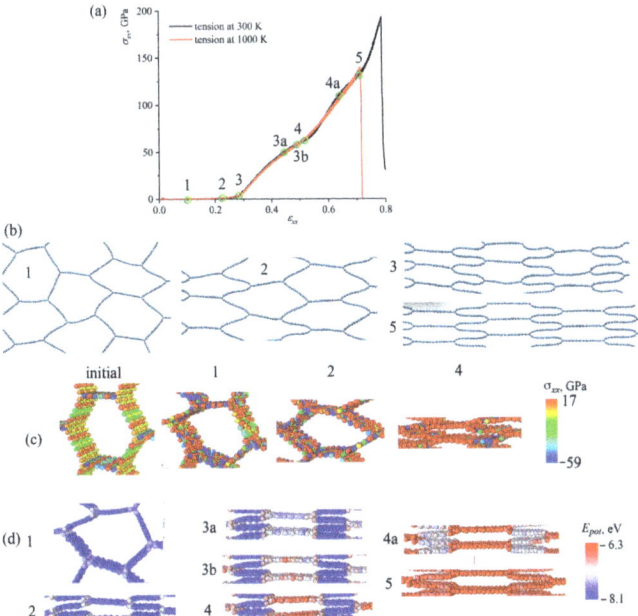

Figure 2. (a) Stress–strain curves as the function of strain during tension along x-axis for honeycomb GA. The critical points on the stress–strain curve are labeled as 1–5. (b) The snapshots of the structure as the projection on xz-plane at critical points for tension at 300 K. Part of the simulation cell is presented. (c,d) Stress σ_{xx} per atom (c) and potential energy per atom (d) during tension. Part of the simulation cell (one honeycomb cell) is presented.

Considerable structural changes occur at $\varepsilon = 0.3$, when honeycomb cells are flattened. This collapse mechanism is similar to the CNT collapse. It is well-known that CNT bundles can possess two main stable structural states: open CNTs and collapsed CNTs [71]. Thus, up to $\varepsilon = 0.25$, the structure is not stressed. Further, deformation is mainly defined by the stretching of the walls of honeycomb cells parallel to the tension direction. The initial length of the wall was 16 Å and was unchanged up to $\varepsilon = 0.25$. It was increased to 20 Å before the fracture occurred. The analysis of the interatomic bonds showed that on the GNR along the x-axis, all bonds a, b, c remained almost unchanged until $\varepsilon = 0.25$. Bonds a and b changed almost simultaneously (the difference took place due to the thermal fluctuations). These bonds are almost aligned with the loading direction and thus mostly strained. These bonds (and others of the same orientation) are continuously changing from 1.41 Å to 1.75 Å, while bond c changes from 1.41 Å to 1.55 Å. Very similar results were obtained previously [72].

As can be seen from Figure 2a, temperature also affects strength and failure strain for honeycomb GA. Previously, it was shown that tensile and compressive behavior considerably depend on the deformation temperature [63]; however, when the temperature is lower than 900 K, the temperature effect is slight, and the in-plane compressive stress–strain curve is similar to the response at room temperature. At higher temperatures, drastic change appears in the in-plane compressive stress–strain curve. The temperature effect can be explained by the larger chemical bond thermal fluctuations and bond length variation at higher temperatures.

Note, in [68] mechanical behavior was analyzed with AIREBO and DFTB for temperature range from 100 to 1000 K. It is found that the critical strain can reduce more than 20% in AIREBO at room temperature; DFTB shows less than a 10% decrease at room temperature. Also, the strength only drops less than 5% with DFTB, but AIREBO shows a 10% drop at room temperature. Such temperature sensitivity of AIREBO comes from the simple bond length criteria. Thus, the obtained results should be analyzed with this correction on the used potential.

For re-entrant honeycomb, just part of the stress–strain curves are presented with critical points labeled as 1–7. As can be seen from the structure snapshots during tension, re-entrant cells begin to open even at the first deformation steps. At $\varepsilon = 0.1$, the shape of the cells is different from the initial and further transforms to a square shape. Again, as for honeycomb, initial tension results in mild flexing of the walls of the structure. A very similar structure with square cells was obtained in [50] and further studied by simulation with a focus on the mechanical stabilizing mechanisms and properties of the deformed structure [38]. At $\varepsilon = 0.5$, honeycomb cells are formed, and the following deformation mechanisms are the same as for the simple honeycomb GA. Similar structural changes have been observed experimentally [44]. Up to $\varepsilon = 0.5$, no changes in the stress per atom and potential energy per atom occur; thus, only structural transformation is presented in Figure 3. Again, temperature slightly affects the stress–strain curves.

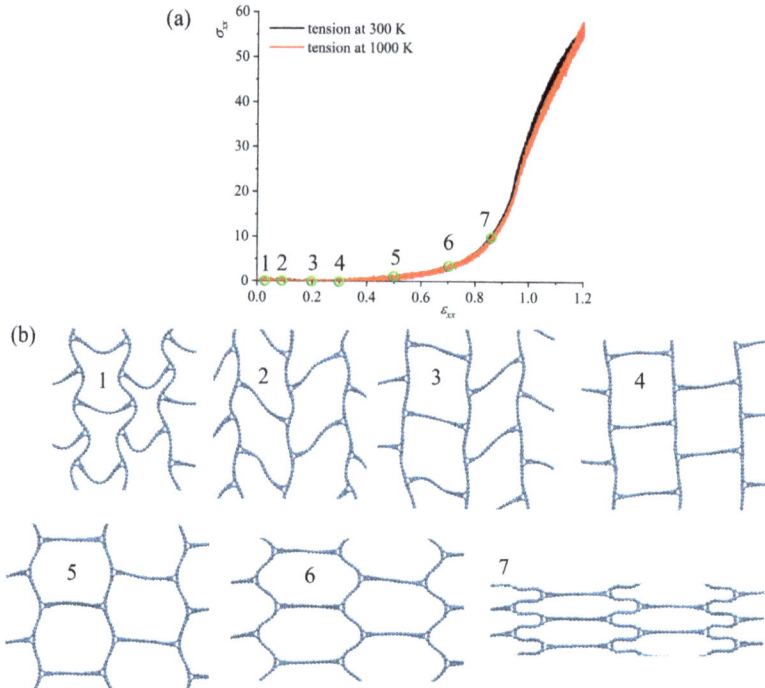

Figure 3. (**a**) Stress–strain curves as the function of strain during tension along x-axis for re-entrant honeycomb GA. The critical points on the stress–strain curve are labeled as 1–7. (**b**) The snapshots of the structure as the projection on xz-plane at critical points. Part of the simulation cell is presented.

As was shown in [38], for such a structure with square cells (called biomimetic), the length of the cell walls (bridges especially) plays a crucial role both in compression and tension: the longer the bridges, the lower the ultimate strength and strain. In the present work, the length of the bridges is equal to the lowest length of bridges from [38]. It was found that GA with short bridges have more bridge-lamellae bonds, so they may absorb more strain than specimens with longer bridges.

As can be seen from Figure 4, temperature slightly affects the tensile strength and failure strain of arrow honeycomb. The critical points on the stress–strain curve were labeled as 1–5. Before relaxation (see Figure 1), there was a sharp angle on the cells oriented along z, which was smoothed during relaxation. During tension, the walls of the cells can oscillate, and arrows transform into squares. Cell walls aligned along x are mostly stressed since they are oriented along the loading direction.

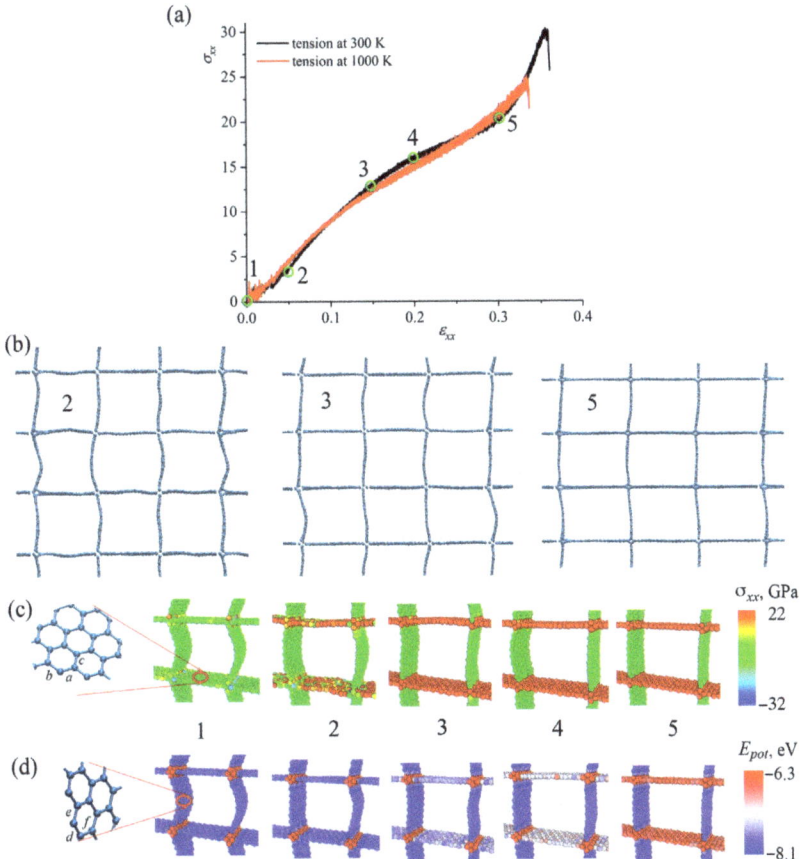

Figure 4. (a) Stress–strain curves during tension along x-axis for arrow-honeycomb. The critical points on the stress–strain curve were labeled as 1–5. (b) The snapshots of the structure as the projection on xz-plane at critical points. Part of the simulation cell is presented. (c,d) Stress σ_{xx} per atom (c) and potential energy per atom (d) during tension. Part of the simulation cell (one structural element) is presented.

At the initial state, junctions between graphene nanoribbons are strained a little with the highest concentrated energy (shown by red atoms in Figure 4d). Until $\varepsilon = 0.04$, deformation is determined by the elongation of cell walls oriented along z-axis. Until $\varepsilon = 0.035$, stress distribution per atom shows that the cell walls oriented along the z-axis are unstrained (Figure 4c), while cell walls oriented along the x-axis have higher energies. At the same time, no changes can be seen from energy distribution per atom (Figure 4d). At $\varepsilon = 0.05$ (point 2), all cell walls oriented along the x-axis were stressed. At $\varepsilon = 0.05$, the change of the slope on the stress–strain curve is found.

The other change in the slope of the stress–strain curve can be seen at $\varepsilon = 0.15$ (point 3), when all cell walls oriented along the x-axis are highly stressed and have much higher potential energy than walls oriented along the z-axis. Between $\varepsilon = 0.15$ and $\varepsilon = 0.2$ (point 4), this energy increases even higher, while no changes of stress per atom can be seen. At $\varepsilon = 0.3$ (point 5), there is a square net of GNRs, strained along the x-axis. Further, it can be seen that walls aligned along the loading direction have the highest stress and potential energy. The fracture occurs at $\varepsilon = 3.6$, near the junction between walls.

An analysis of the interatomic bonds showed that on the cell walls along the x-axis, all of the bonds a, b, c almost remained unhanged until $\varepsilon = 0.05$. Bonds a and b changed almost simultaneously (the difference took place due to the thermal fluctuations). These bonds are almost aligned with the loading direction and thus mostly strained. These bonds (and others of the same orientation) are continuously changing until $\varepsilon = 0.2$ with further abrupt change until $\varepsilon = 0.3$. At the same time, bond c almost does not change until $\varepsilon = 0.3$. Bonds d, e, f, which are analogous to a, b, c, are named differently to distinguish bonds on the cell walls that are differently oriented. These bonds are analyzed on the walls in the loading direction. They slightly change between 1.38 and 1.42 Å due to the thermal fluctuations. Both walls of the cell after relaxation are equal to about 19 Å; however, during tension the cell wall along the z-axis did not change, while the length of the cell wall along the x-axis increased to 26 Å (increased to 36%).

The deformation mechanisms at 300 K and 1000 K are the same. Strength is decreased just due to the thermal fluctuations.

3.2. Effect of Loading Direction

We also have studied the effect of the direction of the applied strain on the deformation behavior of the structure. The uniaxial strain was applied along the x- and z-axis. Figure 5 presents the stress–strain curves under tension along the x- and z-axes as the function of applied strain for (a) honeycomb; (b) re-entrant honeycomb; and (c) arrow honeycomb.

Figure 5a shows that for honeycomb GA and re-entrant honeycomb, a significant difference in deformation behavior during tension along armchair (x-axis) and zigzag (z-axis) was found. It was also previously shown for the same GA that when stretched along the armchair direction, mechanical strength and failure strain decrease with the increase in sidewall width [72]. Moreover, as the sidewall width increases, the strength and ductility gradually decreases. Here, the armchair coincides with the x-axis, while the zigzag coincides with the z-axis. Note that for re-entrant honeycomb, only part of the stress–strain curve for tension along the x-direction is presented, which is explained above. However, the continuation of the stress–strain curve for the re-entrant honeycomb is shown by a dashed line for reference.

For tension along z, both structures deformed similarly: the straightening of the walls of honeycomb cells took place with almost no stress concentration; then, stresses over the walls aligned with the tension direction began to increase, which was followed by fracture (analogous to the fracture of arrow honeycomb). For the honeycomb, the length of the wall is about 32 Å, and for the re-entrant honeycomb the length is 22 Å; thus, strength and failure strain for honeycomb GA are higher. Cells for both structures are presented in two structural states for comparison. The same deformation mode with the elongation of the honeycomb cell to the rectangular cell was shown in [63,67]. The obtained values of the fracture strain and strength are also in good agreement with the literature. Thus, the honeycomb GA can be effectively used for the verification of the model accuracy. Interestingly, it was shown in [64] that the resonant frequency of the zigzag honeycomb is lower than that of the armchair honeycomb, which means that the nonlinearity of the zigzag honeycomb is stronger than that of the armchair honeycomb.

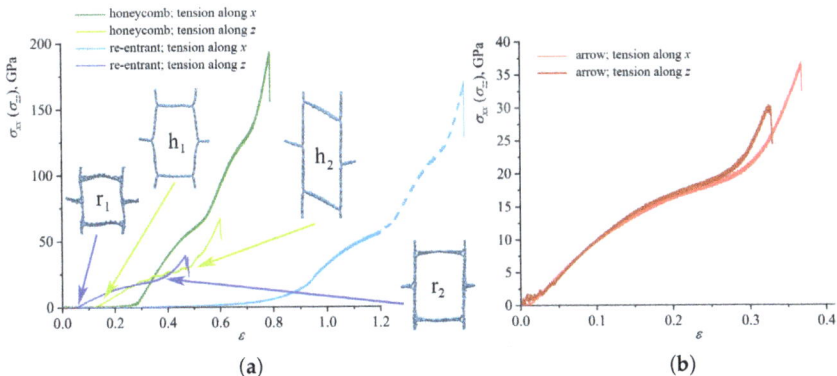

Figure 5. Stress–strain curves as the function of strain during tension along x- and z-axis for (**a**) honeycomb and re-entrant honeycomb; (**b**) for arrow honeycomb.

For arrow GA, the stress–strain curves for uniaxial tension along the x and z directions almost coincide. During relaxation, the cells of the arrow honeycomb change their shape into almost square cells. The size of the cell wall along x and z is 20 Å. Thus, the structure is isotropic in the xz plane. The lower failure strain and tensile strength for the arrow honeycomb stretched along x is explained solely by thermal fluctuations. The opened cells of arrow GA are very similar to the opened cells of re-entrant GA.

The addition of CNTs and graphene flakes in the present form does not result in an increase in strength or failure strain. Deformation occurs very differently due to the presence of CNTs and graphene: with these elements, honeycomb cells can transform into square, hexagonal, or collapsed shapes. However, additional studies are required to understand how to control morphology to obtain better mechanical properties of honeycomb. However, reinforcing elements increase the structure density, which results in a strength decrease. Table 1 presents the values of density ρ, tensile strength σ, and failure stress ε for all the considered structures.

Table 1. Density, tensile strength σ, and failure strain ε for all the considered GA.

	Honeycomb			Re-Entrant			Arrow		
	GA Matrix	Flakes	CNT	GA Matrix	Flakes	CNT	GA Matrix	Flakes	CNT
ρ, g/cm^3	0.58	0.86	0.92	1.01	1.31	1.58	0.69	1.21	1.0
σ, GPa	190	156	110	180	160	101	31	27	33
ε	0.79	0.76	0.72	0.8	0.76	0.7	0.36	0.37	0.38

As was shown in [16], the strength of the nanoporous graphene intrinsically depends on the density; however, this scaling law depends on the special morphology of the structure. In [16], it was shown that the higher the density, the higher the fracture strength, and this is very close to the behavior of crumpled graphene or paper [37,38]. However, this scaling law can be better used for nanoporous hollow graphene than for honeycomb graphene aerogel. For honeycomb and re-entrant honeycomb, the reverse situation is found: the higher the density, the lower the fracture strain and strength. This is because the key factor affecting the mechanical properties of GA is their microstructure. For GA materials, structural characteristics can vary greatly among the different works. Moreover, the density and mechanical properties of GA prepared by the same method can also vary considerably, while some GA have similar properties despite the use of different production methods [73]. In contrast to [16], the GA materials with lower density and higher mechanical properties were obtained in [74].

3.3. Elasticity

These morphologies of GA composed of cells not only possess high strength but also show good elasticity during tension. The elasticity of such structures is of significant interest for the materials used in applications where they are expected to deform periodically while still retaining most of their original strength and stiffness. Let us discuss the elastic behaviors of the considered GA in detail.

During tension in the x direction of the arrow honeycomb, the structure remains elastic until the first fracture starts. From the beginning of loading until fracture, there was no change in the atomic configuration, confirming the elasticity of GA. Before $\varepsilon = 0.15$, deformation can be explained by the straightening of the walls of cells. Following the point of initial fracture, the structure changes dramatically and the deformation becomes plastic. During tension in the z direction, deformation behavior is very close. Until the sharp increase in stress at $\varepsilon = 0.25$, elastic deformation took place owing to the lack of broken bonds or new close-range non-bonded interactions (see Figure 5). This increase in stress corresponds to pre-critical deformation when all covalent bonds inside the bridges are critically stressed. Interestingly, once the lamellae have buckled globally, new bonds can appear between the neighboring elements, which result in structure stability. Very similar behavior was observed for the other structural morphologies.

4. Conclusions

A detailed study has been performed to investigate the mechanical properties of graphene aerogels of a different morphology using MD simulations. The tensile loading of graphene aerogels is analyzed for two temperatures, 300 and 1000 K. Loading is conducted along two lattice directions to overview the structure anisotropy.

At the first deformation stages, the walls (bridges) of the honeycomb cells can easily bend or oscillate in the direction normal to the tension direction. With the strain increase, the stress concentration near the junctions of the honeycomb cell and over the walls oriented along tensile direction can be observed. The strength of GA is considerably dependent on the length of the walls of honeycomb cells and their density. The deformation of the re-entrant honeycomb is defined by the opening of the cells during the first deformation stages. Further, the re-entrant honeycomb transforms into the conventional honeycomb. Very similar deformation mechanisms were found for honeycomb under compression [57]. The new arrow morphology is the weakest since the walls of the cells almost cannot bend. Even at the beginning of the deformation process, the walls (lamellas) are straightened, and the arrows transform into square cells. For all cases, the temperature slightly decreases the strength and failure strain of the GA.

For comparison, aerogel carbon matrices were reinforced by graphene flakes and carbon nanotubes; however, this distribution of the reinforcing elements did not lead to an increase in the strength and failure strain. The search for the new morphologies with special distribution of the reinforcing elements is further required. The obtained results further contribute to the understanding of the microscopic deformation mechanisms of GA and their design for various applications.

Author Contributions: Conceptualization, J.A.B.; methodology, S.A.S.; software, S.A.S.; investigation, J.A.B.; resources, S.A.S.; and writing—original draft preparation, J.A.B. All authors have read and agreed to the published version of the manuscript.

Funding: The research is funded by the Ministry of Science and Higher Education of the Russian Federation as part of the World-class Research Center program: Advanced Digital Technologies (contract No. 075-15-2022-311 dated 20 April 2022.

Institutional Review Board Statement: Not applicable.

Informed Consent Statement: Not applicable.

Data Availability Statement: Data available on request due to restrictions eg privacy or ethical.

Acknowledgments: The authors wish to acknowledge Peter the Great Saint-Petersburg Polytechnic University Supercomputer Center "Polytechnic" for computational resources.

Conflicts of Interest: The authors declare no conflict of interest.

Abbreviations

The following abbreviations are used in this manuscript:

GA   graphene aerogel
MD   molecular dynamics
CNT   carbon nanotubes

References

1. Yu, C.; Song, Y.S. Analysis of Thermoelectric Energy Harvesting with Graphene Aerogel-Supported Form-Stable Phase Change Materials. *Nanomaterials* **2021**, *11*, 2192. [CrossRef] [PubMed]
2. Tafreshi, O.; Mosanenzadeh, S.; Karamikamkar, S.; Saadatnia, Z.; Park, C.; Naguib, H. A review on multifunctional aerogel fibers: Processing, fabrication, functionalization, and applications. *Mater. Today Chem.* **2022**, *23*, 100736. [CrossRef]
3. Zhu, L.; Wang, J.; Zhang, T.; Ma, L.; Lim, C.W.; Ding, F.; Zeng, X.C. Mechanically Robust Tri-Wing Graphene Nanoribbons with Tunable Electronic and Magnetic Properties. *Nano Lett.* **2010**, *10*, 494–498. [CrossRef] [PubMed]
4. Zhang, Z.; Kutana, A.; Yang, Y.; Krainyukova, N.V.; Penev, E.S.; Yakobson, B.I. Nanomechanics of carbon honeycomb cellular structures. *Carbon* **2017**, *113*, 26–32. [CrossRef]
5. Kawai, T.; Okada, S.; Miyamoto, Y.; Oshiyama, A. Carbon three-dimensional architecture formed by intersectional collision of graphene patches. *Phys. Rev. B* **2005**, *72*, 035428. [CrossRef]
6. Pang, Z.; Gu, X.; Wei, Y.; Yang, R.; Dresselhaus, M.S. Bottom-up Design of Three-Dimensional Carbon-Honeycomb with Superb Specific Strength and High Thermal Conductivity. *Nano Lett.* **2016**, *17*, 179–185. [CrossRef] [PubMed]
7. Yu, C.; Youn, J.R.; Song, Y.S. Reversible thermo-electric energy harvesting with phase change material (PCM) composites. *J. Polym. Res.* **2021**, *28*, 279. [CrossRef]
8. Wang, H.; Lu, W.; Di, J.; Li, D.; Zhang, X.; Li, M.; Zhang, Z.; Zheng, L.; Li, Q. Ultra-Lightweight and Highly Adaptive All-Carbon Elastic Conductors with Stable Electrical Resistance. *Adv. Funct. Mater.* **2017**, *27*, 1606220. [CrossRef]
9. Thakur, A. Graphene aerogel based energy storage materials—A review. *Mater. Today Proc.* **2022**, *65*, 3369–3376. [CrossRef]
10. Gibson, L.J.; Ashby, M.F. *Cellular Solids*; Cambridge University Press: Cambridge, UK, 1997. [CrossRef]
11. Schaedler, T.A.; Jacobsen, A.J.; Torrents, A.; Sorensen, A.E.; Lian, J.; Greer, J.R.; Valdevit, L.; Carter, W.B. Ultralight Metallic Microlattices. *Science* **2011**, *334*, 962–965. [CrossRef]
12. Yang, J.; Li, X.; Han, S.; Zhang, Y.; Min, P.; Koratkar, N.; Yu, Z.Z. Air-dried, high-density graphene hybrid aerogels for phase change composites with exceptional thermal conductivity and shape stability. *J. Mater. Chem. A* **2016**, *4*, 18067–18074. [CrossRef]
13. Jing, J.; Qian, X.; Si, Y.; Liu, G.; Shi, C. Recent Advances in the Synthesis and Application of Three-Dimensional Graphene-Based Aerogels. *Molecules* **2022**, *27*, 924. [CrossRef] [PubMed]
14. Hu, H.; Zhao, Z.; Wan, W.; Gogotsi, Y.; Qiu, J. Ultralight and Highly Compressible Graphene Aerogels. *Adv. Mater.* **2013**, *25*, 2219–2223. [CrossRef] [PubMed]
15. Nardecchia, S.; Carriazo, D.; Ferrer, M.L.; Gutiérrez, M.C.; del Monte, F. Three dimensional macroporous architectures and aerogels built of carbon nanotubes and/or graphene: Synthesis and applications. *Chem. Soc. Rev.* **2013**, *42*, 794–830. [CrossRef]
16. Kashani, H.; Ito, Y.; Han, J.; Liu, P.; Chen, M. Extraordinary tensile strength and ductility of scalable nanoporous graphene. *Sci. Adv.* **2019**, *5*, eaat6951. [CrossRef]
17. Afroze, J.D.; Tong, L.; Abden, M.J.; Yuan, Z.; Chen, Y. Hierarchical honeycomb graphene aerogels reinforced by carbon nanotubes with multifunctional mechanical and electrical properties. *Carbon* **2021**, *175*, 312–321. [CrossRef]
18. Peng, X.; Wu, K.; Hu, Y.; Zhuo, H.; Chen, Z.; Jing, S.; Liu, Q.; Liu, C.; Zhong, L. A mechanically strong and sensitive CNT/rGO–CNF carbon aerogel for piezoresistive sensors. *J. Mater. Chem. A* **2018**, *6*, 23550–23559. [CrossRef]
19. Wasalathilake, K.C.; Galpaya, D.G.; Ayoko, G.A.; Yan, C. Understanding the structure-property relationships in hydrothermally reduced graphene oxide hydrogels. *Carbon* **2018**, *137*, 282–290. [CrossRef]
20. Shang, J.J.; Yang, Q.S.; Liu, X. New Coarse-Grained Model and Its Implementation in Simulations of Graphene Assemblies. *J. Chem. Theory Comput.* **2017**, *13*, 3706–3714. [CrossRef]
21. Si, Y.; Wang, X.; Dou, L.; Yu, J.; Ding, B. Ultralight and fire-resistant ceramic nanofibrous aerogels with temperature-invariant superelasticity. *Sci. Adv.* **2018**, *4*, eaas8925. [CrossRef]
22. Fan, Z.; Gong, F.; Nguyen, S.T.; Duong, H.M. Advanced multifunctional graphene aerogel–Poly (methyl methacrylate) composites: Experiments and modeling. *Carbon* **2015**, *81*, 396–404. [CrossRef]
23. Zheng, B.; Liu, C.; Li, Z.; Carraro, C.; Maboudian, R.; Senesky, D.G.; Gu, G.X. Investigation of mechanical properties and structural integrity of graphene aerogels via molecular dynamics simulations. *Phys. Chem. Chem. Phys.* **2023**, *25*, 21897–21907. [CrossRef] [PubMed]

24. Lei, J.; Liu, Z. The structural and mechanical properties of graphene aerogels based on Schwarz-surface-like graphene models. *Carbon* **2018**, *130*, 741–748. [CrossRef]
25. Qin, Z.; Jung, G.S.; Kang, M.J.; Buehler, M.J. The mechanics and design of a lightweight three-dimensional graphene assembly. *Sci. Adv.* **2017**, *3*, e1601536. [CrossRef] [PubMed]
26. Xu, Z.; Zhang, Y.; Li, P.; Gao, C. Strong, Conductive, Lightweight, Neat Graphene Aerogel Fibers with Aligned Pores. *ACS Nano* **2012**, *6*, 7103–7113. [CrossRef] [PubMed]
27. Ha, H.; Shanmuganathan, K.; Ellison, C.J. Mechanically Stable Thermally Crosslinked Poly(acrylic acid)/Reduced Graphene Oxide Aerogels. *ACS Appl. Mater. Interfaces* **2015**, *7*, 6220–6229. [CrossRef]
28. Hong, J.Y.; Yun, S.; Wie, J.J.; Zhang, X.; Dresselhaus, M.S.; Kong, J.; Park, H.S. Cartilage-inspired superelastic ultradurable graphene aerogels prepared by the selective gluing of intersheet joints. *Nanoscale* **2016**, *8*, 12900–12909. [CrossRef]
29. Chen, Y.; Yang, Y.; Xiong, Y.; Zhang, L.; Xu, W.; Duan, G.; Mei, C.; Jiang, S.; Rui, Z.; Zhang, K. Porous aerogel and sponge composites: Assisted by novel nanomaterials for electromagnetic interference shielding. *Nano Today* **2021**, *38*, 101204. [CrossRef]
30. Ren, W.; Cheng, H.M. When two is better than one. *Nature* **2013**, *497*, 448–449. [CrossRef]
31. Guo, F.; Jiang, Y.; Xu, Z.; Xiao, Y.; Fang, B.; Liu, Y.; Gao, W.; Zhao, P.; Wang, H.; Gao, C. Highly stretchable carbon aerogels. *Nat. Commun.* **2018**, *9*, 881. [CrossRef]
32. Wang, H.; Cao, Q.; Peng, Q.; Liu, S. Atomistic Study of Mechanical Behaviors of Carbon Honeycombs. *Nanomaterials* **2019**, *9*, 109. [CrossRef] [PubMed]
33. Tong, H.; Chen, H.; Zhao, Y.; Liu, M.; Cheng, Y.; Lu, J.; Tao, Y.; Du, J.; Wang, H. Robust PDMS-based porous sponge with enhanced recyclability for selective separation of oil-water mixture. *Colloids Surf. A Physicochem. Eng. Asp.* **2022**, *648*, 129228. [CrossRef]
34. Yu, X.; Liang, X.; Zhao, T.; Zhu, P.; Sun, R.; Wong, C.P. Thermally welded honeycomb-like silver nanowires aerogel backfilled with polydimethylsiloxane for electromagnetic interference shielding. *Mater. Lett.* **2021**, *285*, 129065. [CrossRef]
35. Sun, H.; Xu, Z.; Gao, C. Multifunctional, Ultra-Flyweight, Synergistically Assembled Carbon Aerogels. *Adv. Mater.* **2013**, *25*, 2554–2560. [CrossRef] [PubMed]
36. Park, O.K.; Tiwary, C.S.; Yang, Y.; Bhowmick, S.; Vinod, S.; Zhang, Q.; Colvin, V.L.; Asif, S.A.S.; Vajtai, R.; Penev, E.S.; et al. Magnetic field controlled graphene oxide-based origami with enhanced surface area and mechanical properties. *Nanoscale* **2017**, *9*, 6991–6997. [CrossRef] [PubMed]
37. Cao, L.; Fan, F. Deformation and instability of three-dimensional graphene honeycombs under in-plane compression: Atomistic simulations. *Extrem. Mech. Lett.* **2020**, *39*, 100861. [CrossRef]
38. Morris, B.; Becton, M.; Wang, X. Mechanical abnormality in graphene-based lamellar superstructures. *Carbon* **2018**, *137*, 196–206. [CrossRef]
39. Meng, F.; Chen, C.; Hu, D.; Song, J. Deformation behaviors of three-dimensional graphene honeycombs under out-of-plane compression: Atomistic simulations and predictive modeling. *J. Mech. Phys. Solids* **2017**, *109*, 241–251. [CrossRef]
40. Shokrieh, M.M.; Rafiee, R. A review of the mechanical properties of isolated carbon nanotubes and carbon nanotube composites. *Mech. Compos. Mater.* **2010**, *46*, 155–172. [CrossRef]
41. Young, R.J.; Kinloch, I.A.; Gong, L.; Novoselov, K.S. The mechanics of graphene nanocomposites: A review. *Compos. Sci. Technol.* **2012**, *72*, 1459–1476. [CrossRef]
42. Yang, Y.; Shi, E.; Li, P.; Wu, D.; Wu, S.; Shang, Y.; Xu, W.; Cao, A.; Yuan, Q. A compressible mesoporous SiO2 sponge supported by a carbon nanotube network. *Nanoscale* **2014**, *6*, 3585. [CrossRef]
43. Fan, Z.; Tng, D.Z.Y.; Lim, C.X.T.; Liu, P.; Nguyen, S.T.; Xiao, P.; Marconnet, A.; Lim, C.Y.; Duong, H.M. Thermal and electrical properties of graphene/carbon nanotube aerogels. *Colloids Surf. A Physicochem. Eng. Asp.* **2014**, *445*, 48–53. [CrossRef]
44. Qiu, L.; Liu, J.Z.; Chang, S.L.; Wu, Y.; Li, D. Biomimetic superelastic graphene-based cellular monoliths. *Nat. Commun.* **2012**, *3*, 1241. [CrossRef]
45. Zhu, C.; Han, T.Y.J.; Duoss, E.B.; Golobic, A.M.; Kuntz, J.D.; Spadaccini, C.M.; Worsley, M.A. Highly compressible 3D periodic graphene aerogel microlattices. *Nat. Commun.* **2015**, *6*, 6962. [CrossRef] [PubMed]
46. Hyun, S.; Torquato, S. Effective elastic and transport properties of regular honeycombs for all densities. *J. Mater. Res.* **2000**, *15*, 1985–1993. [CrossRef]
47. Grima, J.N.; Oliveri, L.; Attard, D.; Ellul, B.; Gatt, R.; Cicala, G.; Recca, G. Hexagonal Honeycombs with Zero Poisson's Ratios and Enhanced Stiffness. *Adv. Eng. Mater.* **2010**, *12*, 855–862. [CrossRef]
48. Goldstein, R.V.; Gorodtsov, V.A.; Lisovenko, D.S. The elastic properties of hexagonal auxetics under pressure. *Phys. Status Solidi (B)* **2016**, *253*, 1261–1269. [CrossRef]
49. Goldstein, R.; Lisovenko, D.; Chentsov, A.; Lavrentyev, S. Experimental study of defects influence on auxetic behavior of cellular structure with curvilinear elements. *Lett. Mater.* **2017**, *7*, 355–358. [CrossRef]
50. Yang, M.; Zhao, N.; Cui, Y.; Gao, W.; Zhao, Q.; Gao, C.; Bai, H.; Xie, T. Biomimetic Architectured Graphene Aerogel with Exceptional Strength and Resilience. *ACS Nano* **2017**, *11*, 6817–6824. [CrossRef]
51. Available online: https://www.lammps.org (accessed on 24 November 2023).
52. Plimpton, S. Fast Parallel Algorithms for Short-Range Molecular Dynamics. *J. Comput. Phys.* **1995**, *117*, 1–19. [CrossRef]
53. Thompson, A.P.; Aktulga, H.M.; Berger, R.; Bolintineanu, D.S.; Brown, W.M.; Crozier, P.S.; in 't Veld, P.J.; Kohlmeyer, A.; Moore, S.G.; Nguyen, T.D.; et al. LAMMPS—A flexible simulation tool for particle-based materials modeling at the atomic, meso, and continuum scales. *Comput. Phys. Commun.* **2022**, *271*, 108171. [CrossRef]

54. Stuart, S.J.; Tutein, A.B.; Harrison, J.A. A reactive potential for hydrocarbons with intermolecular interactions. *J. Chem. Phys.* **2000**, *112*, 6472–6486. [CrossRef]
55. Patil, S.P.; Shendye, P.; Markert, B. Molecular Investigation of Mechanical Properties and Fracture Behavior of Graphene Aerogel. *J. Phys. Chem. B* **2020**, *124*, 6132–6139. [CrossRef]
56. Baimova, J.; Rysaeva, L.; Rudskoy, A. Deformation behavior of diamond-like phases: Molecular dynamics simulation. *Diam. Relat. Mater.* **2018**, *81*, 154–160. [CrossRef]
57. Shang, J.; Yang, Q.S.; Liu, X.; Wang, C. Compressive deformation mechanism of honeycomb-like graphene aerogels. *Carbon* **2018**, *134*, 398–410. [CrossRef]
58. Rysaeva, L.K.; Baimova, J.A.; Lisovenko, D.S.; Gorodtsov, V.A.; Dmitriev, S.V. Elastic Properties of Fullerites and Diamond-Like Phases. *Phys. Status Solidi (B)* **2018**, *256*, 1800049. [CrossRef]
59. Safina, L.; Baimova, J.; Krylova, K.; Murzaev, R.; Mulyukov, R. Simulation of metal-graphene composites by molecular dynamics: A review. *Lett. Mater.* **2020**, *10*, 351–360. [CrossRef]
60. Wei, Y.; Wu, J.; Yin, H.; Shi, X.; Yang, R.; Dresselhaus, M. The nature of strength enhancement and weakening by pentagon–heptagon defects in graphene. *Nat. Mater.* **2012**, *11*, 759–763. [CrossRef]
61. O'Connor, T.C.; Andzelm, J.; Robbins, M.O. AIREBO-M: A reactive model for hydrocarbons at extreme pressures. *J. Chem. Phys.* **2015**, *142*, 024903. [CrossRef]
62. Du, Y.; Zhou, J.; Ying, P.; Zhang, J. Effects of cell defects on the mechanical and thermal properties of carbon honeycombs. *Comput. Mater. Sci.* **2021**, *187*, 110125. [CrossRef]
63. Hu, J.; Zhou, J.; Zhang, A.; Yi, L.; Wang, J. Temperature dependent mechanical properties of graphene based carbon honeycombs under tension and compression. *Phys. Lett. A* **2021**, *391*, 127130. [CrossRef]
64. Li, B.; Wei, Y.; Meng, F.; Ou, P.; Chen, Y.; Che, L.; Chen, C.; Song, J. Atomistic simulations of vibration and damping in three-dimensional graphene honeycomb nanomechanical resonators. *Superlattices Microstruct.* **2020**, *139*, 106420. [CrossRef]
65. Zhang, P.; Ma, L.; Fan, F.; Zeng, Z.; Peng, C.; Loya, P.E.; Liu, Z.; Gong, Y.; Zhang, J.; Zhang, X.; et al. Fracture toughness of graphene. *Nat. Commun.* **2014**, *5*, 3782. [CrossRef] [PubMed]
66. Zhao, H.; Min, K.; Aluru, N.R. Size and Chirality Dependent Elastic Properties of Graphene Nanoribbons under Uniaxial Tension. *Nano Lett.* **2009**, *9*, 3012–3015. [CrossRef] [PubMed]
67. Liu, Y.; Liu, J.; Yue, S.; Zhao, J.; Ouyang, B.; Jing, Y. Atomistic Simulations on the Tensile Deformation Behaviors of Three-Dimensional Graphene. *Phys. Status Solidi (B)* **2018**, *255*, 1700680. [CrossRef]
68. Jung, G.S.; Irle, S.; Sumpter, B.G. Dynamic aspects of graphene deformation and fracture from approximate density functional theory. *Carbon* **2022**, *190*, 183–193. [CrossRef]
69. Srinivasan, S.G.; van Duin, A.C.T.; Ganesh, P. Development of a ReaxFF Potential for Carbon Condensed Phases and Its Application to the Thermal Fragmentation of a Large Fullerene. *J. Phys. Chem. A* **2015**, *119*, 571–580. [CrossRef]
70. Zhao, H.; Aluru, N.R. Temperature and strain-rate dependent fracture strength of graphene. *J. Appl. Phys.* **2010**, *108*, 064321. [CrossRef]
71. Magnin, Y.; Rondepierre, F.; Cui, W.; Dunstan, D.; San-Miguel, A. Collapse phase diagram of carbon nanotubes with arbitrary number of walls. Collapse modes and macroscopic analog. *Carbon* **2021**, *178*, 552–562. [CrossRef]
72. Gu, X.; Pang, Z.; Wei, Y.; Yang, R. On the influence of junction structures on the mechanical and thermal properties of carbon honeycombs. *Carbon* **2017**, *119*, 278–286. [CrossRef]
73. Qi, P.; Zhu, H.; Borodich, F.; Peng, Q. A Review of the Mechanical Properties of Graphene Aerogel Materials: Experimental Measurements and Computer Simulations. *Materials* **2023**, *16*, 1800. [CrossRef] [PubMed]
74. Qiu, L.; Huang, B.; He, Z.; Wang, Y.; Tian, Z.; Liu, J.Z.; Wang, K.; Song, J.; Gengenbach, T.R.; Li, D. Extremely Low Density and Super-Compressible Graphene Cellular Materials. *Adv. Mater.* **2017**, *29*, 1701553. [CrossRef] [PubMed]

Disclaimer/Publisher's Note: The statements, opinions and data contained in all publications are solely those of the individual author(s) and contributor(s) and not of MDPI and/or the editor(s). MDPI and/or the editor(s) disclaim responsibility for any injury to people or property resulting from any ideas, methods, instructions or products referred to in the content.

Article

Fatigue Life Prediction for Injection-Molded Carbon Fiber-Reinforced Polyamide-6 Considering Anisotropy and Temperature Effects

Joeun Choi [1], Yohanes Oscar Andrian [1,†], Hyungtak Lee [2], Hyungyil Lee [1] and Naksoo Kim [1,*]

1. Department of Mechanical Engineering, Sogang University, Seoul 04107, Republic of Korea; oscar998@sogang.ac.kr (Y.O.A.)
2. Polymer R&D Team, GS Caltex R&D Center, Daejeon 34122, Republic of Korea
* Correspondence: nskim@sogang.ac.kr; Tel.: +82-2-705-8635
† Current address: Department of Mechanical Engineering, Politeknik ATMI, Surakarta 57145, Indonesia; oscar.andrian@atmi.ac.id.

Abstract: The effects of anisotropy and temperature of short carbon fiber-reinforced polyamide-6 (CF-PA6) by the injection molding process were investigated to obtain the static and fatigue characteristics. Static and fatigue tests were conducted with uniaxial tensile and three-point bending specimens with various fiber orientations at temperatures of 40, 60, and 100 °C. The anisotropy caused by the fiber orientations along a polymer flow was calculated using three software connecting analysis sequences. The characteristics of tensile strength and fatigue life can be changed by temperature and anisotropy variations. A semi-empirical strain–stress fatigue life prediction model was proposed, considering cyclic and thermodynamic properties based on the Arrhenius equation. The developed model had a good agreement with an $R^2 = 0.9457$ correlation coefficient. The present fatigue life prediction of CF-PA6 can be adopted when designers make suitable decisions considering the effects of temperature and anisotropy.

Keywords: fiber-reinforced composite; injection molding process; numerical simulation; anisotropy; temperature effect; fatigue life prediction

1. Introduction

Fiber-reinforced plastics (FRPs) are widely utilized as structural materials in the aerospace and automotive industries due to their low density and high specific strength [1–6]. The injection molding process efficiently produces FRP material, facilitating the manufacturing of complex geometries and ultimately increasing the material production rate [7]. The manufacturing conditions and operating environments influence the failure behavior of injection-molded FRPs. The complex fiber orientation distribution, determined by the polymer flow path at different locations, substantiates the considerable anisotropic behavior of FRPs [8]. In addition, the tensile strength strongly differs at different temperatures. Consequently, the need to characterize the combined effects of anisotropy and temperature is a pressing theme in assessing the static and fatigue behavior of FRPs [9].

The influence of fiber orientation distribution, fiber length, and fiber volume fraction on mechanical strength and stiffness, mechanical elasticity, and anisotropy has been extensively investigated [10–19]. Brighenti et al. [13] evaluated the static and fatigue behavior of short fiber-reinforced plastics utilizing a micromechanical model based on a Gaussian-like distribution function using the average fiber orientation. Just et al. [14] developed a crack growth model depending on the degree of anisotropy and fiber orientations, predicting the crack growth path with good agreement.

Most of the fracture analysis for FRP has been proposed at room temperature, and fracture behavior considering the effect of various temperatures has been studied far less.

The bonding ability at the interface of FRP is weakened by increased temperature. As a result, the debonding length of the fibers is increased, and the tensile strength is partially reduced [20–25]. Li et al. [26] proposed a theoretical model to predict the longitudinal tensile strength of FRP under various temperatures. Kawai et al. [27] investigated the temperature dependence of static strength employing Arrhenius-type equations and clarified that the decrease in compressive strength with increasing temperature is smaller than that in tensile strength.

As concerns the fatigue life prediction models, theory-based formula models such as the Manson–Coffin model [28] and the SWT model [29] have been used for various composite materials, modified to consider the effects of anisotropy [30–35] and temperature [36–40]. Regarding FRP, Launay et al. [34] developed a predictive fatigue criterion based on the dissipated energy density per cycle of polyamide 66 base polymer filled with glass fibers, contemplating both anisotropic and nonlinear behavior. Fouchier [41] proposed an energetic fatigue criterion of injection molded short fiber-reinforced plastics at 100 °C.

Although the fracture behavior and fatigue life have been studied, none seem to deal with the fatigue fracture analysis combining the effects of anisotropy and temperature of injection-molded short fiber-reinforced plastics. The application of FRP, which combines temperature conditions with fiber orientation distribution, means that the influence of environmental temperature and random locations of load impact regions can be considered.

For these reasons, we investigated the static and cyclic fracture behavior of 20% volume fraction CF-PA6 material at three different temperatures along 0° (injection direction), 45°, and 90° direction. All specimens were machined on injection-molded plates to analyze the anisotropic behavior caused by fiber orientation distribution through polymer flow. Short fiber-reinforced plastic manufactured by the injection process has an arbitrary fiber orientation in the center of thickness and a fiber arrangement parallel to the polymer flow direction when it deviates from the center [42,43]. The direction of the specimen having a fiber orientation arranged parallel to the polymer flow direction was designated as 0° to evaluate the difference in mechanical properties due to the relationship between fiber orientation and principal stress, and the directions of 45° and 90° were determined in consideration of the angle during specimen processing.

The three experimental temperatures (40, 60, and 100 °C) adopted in static and fatigue tests were determined in consideration of the fact that the operating environment temperature of CFRP applied to various products such as automotive, aircraft, and solar panels exceeds room temperature and the glass transition temperature of the material, 64 °C [44]. Three commercial software (Moldflow Insight 2019, Autodesk Helius 2019, and Abaqus 2020) have been utilized to describe the fiber orientation distribution from a numerical point of view, showing a maximum difference between finite element analysis and experimental load–stroke curve integrals equal to 5.12%.

To enhance the mechanical reliability of CF-PA6, a semi-empirical fatigue life prediction model based on the strain–stress-based fatigue failure theory has been developed and verified using tensile and three-point bending test specimens. The proposed model can predict the failure cycle with high accuracy and ensures the reliability of injection-molded short fiber-reinforced plastic products under high temperatures and various load conditions. The developed semi-empirical fatigue life prediction model shows a high correlation coefficient of $R^2 = 0.9457$. The results provide the possibility of predicting the fracture behavior considering the anisotropic behavior and temperature effect of CF-PA6 with the proposed model.

2. Fatigue Life Prediction Procedure

Fatigue life prediction considering the anisotropy and temperature effect of 20% short carbon fiber volume fraction CF-PA6 material has been conducted through the procedure shown in Figure 1. The fiber volume fraction is calculated based on the Rule of Mixtures. The uniaxial and three-point bending static and cyclic tests were carried out to evaluate the mechanical properties. The specimens were cut from the injection-molded plate in three directions, and experiments were conducted at three different temperatures.

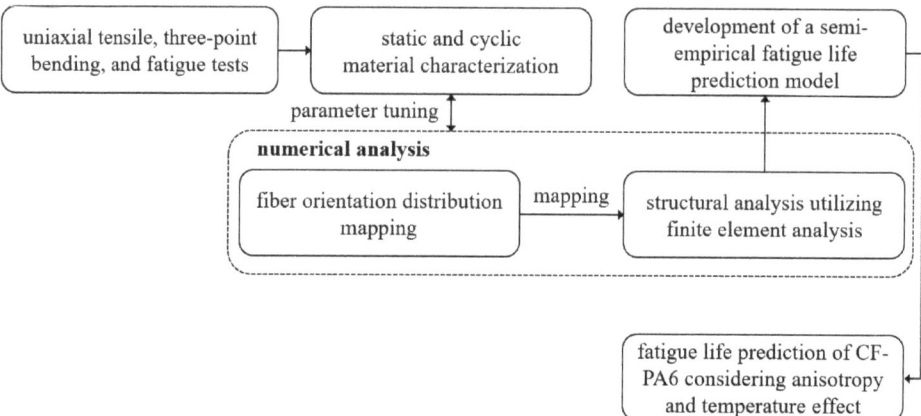

Figure 1. Fatigue life prediction procedure of CF-PA6 based on the semi-empirical model.

The test results obtained through the machined specimens and selected test conditions were used to characterize the mechanical properties of CF-PA6. In addition, the obtained load–displacement curve was adopted to adjust the material parameters used for numerical analysis, and the stress–strain curve was analyzed to confirm brittle properties and yield strength. Finally, cycle behavior was evaluated from fatigue tests under various load-controlled test conditions.

The numerical analysis models were reverse-engineered through static experiments to apply the semi-empirical model's stress, strain, and temperature values. The model was divided into two categories: a model for calculating fiber orientation distribution and a model for structural analysis. The entire numerical analysis procedure is constructed through three commercial software.

The semi-empirical model based on the life prediction model utilizing strain amplitude was developed and modified by formalizing an effect depending on temperatures and directions. The finite element method extracted the minimum and maximum stress and strain values in the stabilized hysteresis loop. The given values were used to input the proposed semi-empirical model and the fracture strain and stress in the uniaxial tensile experiments.

3. Material Characterization

3.1. Static Mechanical Properties Characterization

The pellets used to manufacture tensile and three-point bending specimens for mechanical characteristics of 20% volume fraction short CF-PA6 were produced utilizing a mixture of PA6 and mono-carbon fibers. From the four mm-thick injection-molded plates, the tensile test specimens were cut along 0 (injection flow direction), 45°, and 90° directions according to the ASTM-D 638-02a-TYPE IV specification, as shown in Figure 2. The ASTM-D 638 specimens were manufactured 1.2 times larger than the standard in the 14 mm diameter hole made for the fixture on the heat chamber jig. Therefore, the fracture could be well-achieved in the center of the specimen through dimensional adjustment. The gauge length of the 1.2 times-larger ASTM-D 638 specimen was 30 mm.

For checking the stress triaxiality on the material, the three-point bending test jig was designed to match the dimension of the room inside the heating chamber, which was 300 mm in height and 100 mm in diameter. In addition, the static and cyclic behavior under complex stress conditions were analyzed by machined three-point bending specimens, similarly cut in three directions, to investigate the anisotropy caused by fiber orientation distribution, as shown in Figure 3.

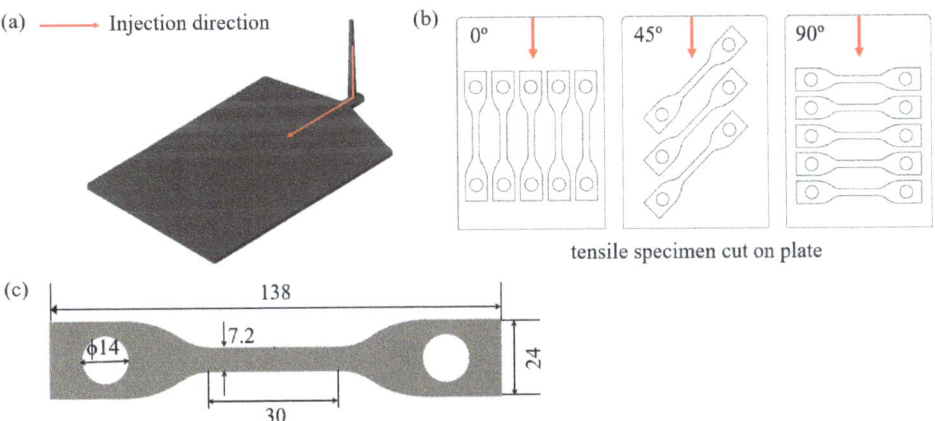

Figure 2. (**a**) The injection-molded plate with a runner. (**b**) 0°, 45°, and 90° directions-cut ASTM-D 638 specimens. (**c**) The dimensions of 1.2 times ASTM-D 638 specimen.

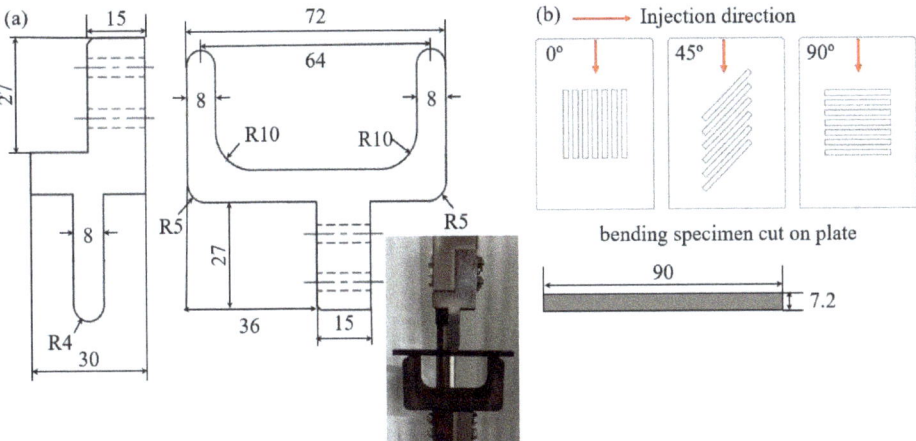

Figure 3. (**a**) Test jig dimensions for the upper part (**left**), lower part (**right**), and experimental setting for three-point bending test in an environmental chamber. (**b**) Schematic dimensions of injection plates for extracting three-point bending test specimens and specimen shape and specifications with ASTM-D 790.

The temperature expansion inside the heating chamber was investigated in the preliminary stage. Thermocouples were attached to the jig and the specimen to observe the thermal expansion by recording the temperature by heating time. As a result, target temperatures of 40, 60, and 100 °C were maintained stable with a range of ±5 °C after 30, 20, and 15 min, respectively. By temperature tracking, all experiences were carried out at stable temperatures by comparing temperature recordings between repeated tests.

Tensile experiments and three-point bending tests were conducted at a tensile and bending speed of 2 mm/min. All tests were performed with the heating chamber maintained at a constant operating temperature of 40, 60, and 100 °C. The specimens were dried at 100 °C for 1 h with 12 m^3/h heat flow in an oven dryer to remove the humidity before proceeding with experiments. All static experiments were repeated three times to increase experimental accuracy. The strain was obtained by means of a 20 mm gauge length extensometer in the case of the uniaxial tensile test.

3.2. Cyclic Mechanical Properties Characterization

Fatigue experiments were conducted based on load control with the same temperature conditions and specimens as the tensile and three-point bending tests. The mean loads of fatigue tests are summarized in Table 1, considering the total stress range. The same mean load was used for three maximum and minimum load combinations.

Table 1. Mean loads of fatigue tests.

Temperature [°C]	Selected Mean Load for Each Specimen Direction [kN]		
	0°	45°	90°
40	2.000	1.250	1.125
60	1.750	1.125	1.000
100	1.500	1.000	0.750

In addition, complete elasticity and complete plastic deformation region conditions were utilized. All fatigue tests were carried out with a frequency of 1 Hz. In addition, three-point bending fatigue experiments were conducted to consider complex stress states. Fifty-four fatigue test data of uniaxial tensile and three-point bending specimens were obtained for five and three conditions considering each temperature and specimen direction. For the reliability of the experimental results, all three repetitive tests were performed for the fatigue test of each load case, and the intermediate value was adopted and used as the result. The mean and amplitude loads were selected to consider both low-cycle and high-cycle fatigue failure. In addition, the cycle when the load decreases to 40% or less of the previous cycle due to damage to the specimen was selected as the failure cycle. The results of each condition are summarized in Tables 2 and 3.

In the initial stage of the fatigue test, the hysteresis loop shows instabilities due to experimental settings. For this reason, the stress and strain were calculated in the last stabilized cycle, where the maximum and minimum displacements differ by less than 5% from the fatigue fracture cycle's hysteresis loop. The periodic spectrum of fatigue life was considered from 10^2 to 10^6. The hysteresis loop moves in a positive direction of strain as the load amplitude increases, indicating that the degree of asymmetry intensifies. An example of a hysteresis loop of one low-cycle fatigue and one high-cycle fatigue is reported in Figure 4.

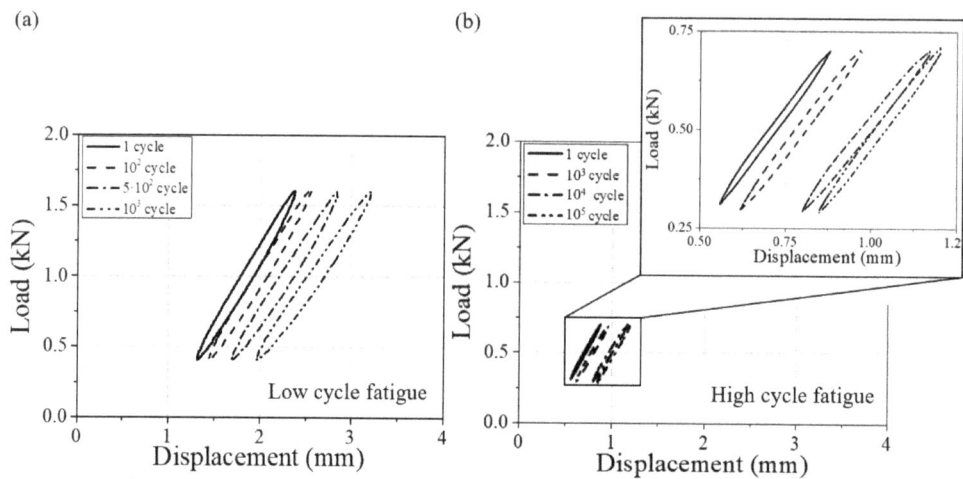

Figure 4. Test results of hysteresis loops for low- and high-cycle fatigues (90° and T = 60 °C). (**a**) fatigue failure at 3800 cycles (0.6 kN load amplitude and 1.0 mean load) and (**b**) fatigue failure at 346,900 cycles (0.2 kN load amplitude and 0.5 mean load).

Table 2. Summary of fatigue tests on the ASTM-D 638 fatigue test specimens.

Temperature [°C]	Specimen Angle [°]	Max. Load [kN]	Min. Load [kN]	Mean Load [kN]	Amplitude Load [kN]	N_f [Cycles]
40	0	3.20	0.80	2.00	1.20	1512
		3.40	0.60	2.00	1.40	980
		2.80	1.20	2.00	0.80	3400
		1.40	0.60	1.00	0.40	106,263
		2.40	2.20	2.30	0.10	674,785
	45	1.90	0.60	1.25	0.65	7500
		2.10	0.40	1.25	0.85	2420
		1.70	0.80	1.25	0.45	46,700
		0.90	0.40	0.65	0.25	268,410
		1.60	1.40	1.50	0.10	340,485
	90	1.75	0.50	1.12	0.62	1774
		1.95	0.30	1.12	0.82	1670
		1.55	0.70	1.12	0.42	9800
		0.90	0.30	0.60	0.30	76,000
		1.50	1.30	1.40	0.10	361,807
60	0	2.90	0.60	1.75	1.15	1820
		3.25	0.25	1.75	1.50	770
		2.60	0.90	1.75	0.85	8700
		1.30	0.50	0.90	0.40	79,805
		2.10	1.90	2.00	0.10	892,500
	45	1.75	0.50	1.12	0.62	5840
		1.95	0.30	1.12	0.82	642
		1.55	0.70	1.12	0.42	10,400
		0.80	0.30	0.55	0.25	89,600
		1.50	1.30	1.40	0.10	428,500
	90	1.50	0.50	1.00	0.50	11,650
		1.60	0.40	1.00	0.600	3800
		1.30	0.70	1.00	0.30	51,007
		0.70	0.30	0.50	0.20	346,900
		1.40	1.20	1.30	0.10	276,000
100	0	2.25	0.75	1.50	0.75	4753
		2.50	0.50	1.50	1.00	1670
		2.00	1.00	1.50	0.50	23,060
		1.00	0.40	0.70	0.30	660,650
		1.80	1.60	1.70	0.10	475,439
	45	1.50	0.50	1.00	0.50	4460
		1.60	0.40	1.00	0.60	1790
		1.30	0.70	1.00	0.30	29,800
		0.70	0.30	0.50	0.20	108,699
		1.40	1.20	1.30	0.10	90,335
	90	1.15	0.35	0.75	0.40	5700
		1.35	0.15	0.75	0.60	1213
		1.00	0.50	0.75	0.25	39,355
		0.50	0.20	0.35	0.15	270,883
		1.10	0.90	1.00	0.10	166,068

Table 3. Summary of fatigue experiments on the three-point bending specimens.

Temperature [°C]	Specimen Angle [°]	Max. Load [kN]	Min. Load [kN]	Mean Load [kN]	Amplitude Load [kN]	N_f [Cycles]
40	0	0.13	0.11	0.12	0.01	2109
	45	0.10	0.80	0.90	0.01	905
	90	0.08	0.06	0.07	0.01	2160
60	0	0.11	0.09	0.10	0.01	15,041
	45	0.08	0.06	0.07	0.01	9800
	90	0.08	0.06	0.07	0.01	5061
100	0	0.09	0.07	0.08	0.01	1200
	45	0.06	0.04	0.05	0.01	671
	90	0.05	0.03	0.04	0.01	2780

3.3. Numerical Analysis

A numerical analysis model was constructed to utilize the fatigue test results in the semi-empirical model and to investigate the cyclic mechanical properties of CF-PA6 under complex stress conditions. The three-point bending test to examine fracture behavior in complex stress states obtained true stress and strain through numerical analysis. Three commercial software were used to account for the influence of the fiber orientation. The injection molding process was simulated using Autodesk Moldflow Insight/Synergy (AMI), and the fiber orientation distribution was calculated at the end of the cooling phase.

The accuracy of the fiber orientation distribution from Moldflow was evaluated by comparing the fiber alignment from the 3D X-ray CT (XCT) results, as shown in Choi et al. [33]. The obtained optimal RSC parameter utilizing the XCT data in a previous study [33] was adapted to the Moldflow simulation to calculate the fiber orientation distribution accurately. The XCT data were taken with 140 kV voltage and 2 μm pixel size. One injection gate was used for the injection molding simulation. The injection and cooling times were 2 and 20 s for the analysis, and the mold and melt temperatures were 85 and 285 °C in the simulation. Each element's mechanical properties were mapped to the structural simulation mesh using Advanced Material Exchange Helius 2019 (AME) in the second step. Finally, the Abaqus input files with the inbuilt element-based mesh sets were created. This procedure is essential since the fiber orientation results affect the static and cyclic mechanical properties. The developed numerical analysis models for fiber orientation distribution calculation with AMI, AME, and structural analysis with Abaqus are shown in Figure 5a and b, respectively.

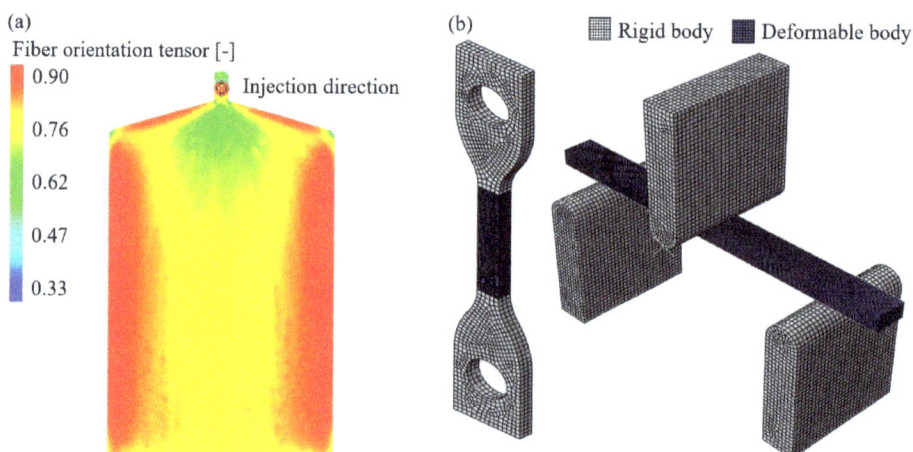

Figure 5. (**a**) Fiber orientation distribution from AME. (**b**) Computational domain for structural analysis for the uniaxial specimen (**left**) and the three-point bending specimen (**right**).

The fiber orientation tensor was calculated using the Folgar–Tucker orientation model in Moldflow simulation, as shown in Equation (1). a_{ij} is the fiber orientation tensor, $0.5\omega_{ij}$ is the vorticity tensor, $0.5\dot{\lambda}_{ij}$ is the deformation rate tensor, and C_I is the fiber interaction coefficient.

$$\frac{Da_{ij}}{Dt} = -\frac{1}{2}\left(\omega_{ik}a_{kj} - a_{ik}\omega_{kj}\right) + \frac{1}{2}\lambda\left(\dot{\gamma}_{ik}a_{kj} + a_{ik}\dot{\gamma}_{kj} - 2a_{ijkl}\dot{\gamma}_{kl}\right) + 2C_I\dot{\gamma}\left(\delta_{ij} - 3a_{ij}\right) \quad (1)$$

The Ramberg–Osgood flow stress model was combined with a modified Hill' 48 yield function to account for the influence of the fiber orientation on the mechanical properties of the CF-PA6. The equations of the Ramberg–Osgood flow stress model and a modified Hill' 48 yield function are shown in Equation (2). The definition of model parameters and the optimized Ramberg–Osgood model constants for the CF-PA6 are summarized in Tables 4 and 5.

$$\sigma = E^{1/n}(K)^{(n-1)/n}\left(\varepsilon_{p,eff}\right)^{1/n}$$

$$\sigma_{eff} = \sqrt{\frac{(\alpha\sigma_{11} - \beta\sigma_{22})^2 + (\beta\sigma_{22} - \beta\sigma_{33})^2 + (\beta\sigma_{33} - \alpha\sigma_{11})^2 + 6\left[(\sigma_{12})^2 + (\sigma_{23})^2 + (\sigma_{31})^2\right]}{2}} \quad (2)$$

$$\alpha(\lambda_I) = \theta + \left[\frac{(\alpha_m - \theta)}{(\lambda_{m,I} - 1/2)}\right](\lambda_I - 1/2), \beta(\lambda_I) = \theta + \left[\frac{(\beta_m - \theta)}{(\lambda_{m,I} - 1/2)}\right](\lambda_I - 1/2)$$

Table 4. Parameter and constant definitions for the Ramberg–Osgood model.

Symbol	Definition
$\varepsilon_{p,eff}$	Effective plastic strain
T	Temperature in the environmental chamber in °C
K	Strength coefficient
n	Hardening exponent
α_m	Weight factor for the fiber direction
β_m	Weight factor for the direction normal to the fibers
E_m	Polymer matrix elastic modulus
E_f	Fiber's elastic modulus
$\lambda_{m,I}$	The first eigenvalue of the fiber orientation matrix in the region with strong fiber alignment with the polymer flow

Table 5. Ramberg–Osgood model constants for each experimental temperature in injection molding process analysis by AME.

T [°C]	σ_0 [MPa]	n	α_m	β_m	E_m [GPa]	E_f [GPa]	$\lambda_{m,I}$
40	250.10	3.34	18.85	11.33	0.78	60.02	0.85
60	215.76	4.14	7.13	8.40	0.56	30.37	0.85
100	227.12	4.46	16.71	17.65	0.40	30.48	0.85

The mechanical properties according to fiber orientation were calculated considering anisotropy through the Ramberg–Osgood model on the Moldflow simulation. Each mechanical property was derived through seven constants ranging from σ_0 to $\lambda_{m,I}$ in Table 5. σ_0 is the stress level at which plastic strain becomes dominant. Using the experimental results by specimen direction and temperature, the constant with the highest agreement between the analysis and experimental result was derived through reverse engineering. Each constant was determined through the BFGS optimization technique to minimize the area difference between the load–displacement curve FEA result and the experimental result according to each coefficient combination. The orientation and elastic modulus of the fiber had a more significant influence on the calculation of the anisotropic behavior of the material, so the polymer matrix elastic modulus was derived relatively low.

In general, the elastic modulus of carbon fiber is higher than the value obtained through the optimization technique presented in Table 5. However, the Ramberg–Osgood model constants obtained in this study were not calculated by evaluating the elastic modulus of the matrix and the fiber through the experiment. They were derived by reverse engineering so that the difference between the experimental and numerical analysis load was the least in consideration of the influence of each constant on each other in one combination consisting of seven coefficients from σ_0 to $\lambda_{m,I}$. Therefore, the elastic modulus of fiber was obtained lower than the actual fiber value. Even though the elastic modulus was not directly evaluated and used, the coefficient obtained is meaningful because the stress–strain value utilized in predicting fatigue life can be derived through numerical analysis by minimizing the error between analysis and experiment. When an optimization technique is applied by experimentally deriving the elastic modulus of each matrix and the fiber, the constants other than the elastic modulus would have different values in Table 5, and a new combination constant would be derived.

In this study, the accuracy of numerical analysis was considered a significant factor in determining the input variables of the fatigue life prediction model because it is essential to accurately calculate the stress and strain field when subjecting the load. To match the experimental data with numerical analysis results, the reduced strain closer model parameter, ARD-RSC, was optimized with the BFGS optimization module in the Python program. The short fiber flow, which changes fiber orientation distribution, was controlled by tuning the ARD-RSC parameter, allowing proper consideration of complex stress states.

4. Fatigue Life Prediction Model

The fatigue life prediction model of CF-PA6 was developed to consider the effects of temperature and anisotropic behavior due to fiber orientation distribution, starting with the strain-based Manson–Coffin model (Equation (3)) [28]. The $\varepsilon_{p,\max}$ and $\varepsilon_{p,\min}$ are the maximum and minimum plastic strain values. The model correlates the plastic strain amplitude, $\Delta\varepsilon_p$, with the failure cycle, N_f, to predict fatigue life through the material coefficients A and c.

$$f(\varepsilon) = \frac{(\varepsilon_{p,\max} - \varepsilon_{p,\min})}{2} = \Delta\varepsilon_p = A\left(N_f\right)^c \qquad (3)$$

Efforts were made by Choi et al. [31] to express the anisotropic behavior of a directional material by adding a maximum von Mises stress ratio between the angle and 0° directions considered as in Equation (4). ε_f is the fracture strain of tensile experiment for each specimen angle. $\varepsilon_{\max}$ and $\varepsilon_{\min}$ are the total maximum and minimum strain of fatigue test conditions. $\sigma_{peak,\theta}$ is the maximum von Mises stress and the stress term's denominator is the value of 0° directions. For the stress term, the anisotropic effect was considered by setting the sine value added by one as an index.

$$f(\varepsilon,\theta) = \left(\frac{\varepsilon_{\max} - \varepsilon_{\min}}{2\varepsilon_f}\right)\left(\frac{\sigma_{peak,\theta}}{\sigma_{peak,\theta=0}}\right)^{(1+\sin\theta)} = A\left(N_f\right)^c \qquad (4)$$

In the case of CF-PA6, however, different fiber orientation distributions are shown despite minor position changes due to the injection molding process. If anisotropic behaviors are identified through a ratio between specific and reference directions, predicting critical fracture parts or fatigue life can be significantly reduced. Since the anisotropic behavior of CF-PA6 appears from the relationship between the fiber direction and the principal stress direction, the fatigue life is affected by fiber orientation distribution. Due to the difficulties of considering the fiber orientation distribution in the fatigue life prediction model, numerical analysis or SEM microscopic image photography must be accompanied to calculate the stress and strain. Therefore, the fatigue life prediction model was developed considering the relationship between principal stress and fiber orientation vector in terms of stress rather than the fiber orientation distribution. The model is presented in Equation (5).

$$f(\varepsilon,\sigma) = \left(\frac{\varepsilon_{max,i} - \varepsilon_{min,i}}{2\varepsilon_{f,i}}\right)\left(\frac{\sigma_{max,i} - \sigma_{min,i}}{2\sigma_{f,i}}\right)^{\ln\left(\frac{\sigma_{f,i}}{\sigma_{max,i}}\right)}\left(e^{-T_{ref}/T_{ope}}\right) = A\left(N_f\right)^c \quad (5)$$

The first term is that of the Manson–Coffin model, and the second term is anisotropic. $\varepsilon_{f,i}$ and $\sigma_{f,i}$ are the fracture strain and stress of tensile experiments for each specimen direction and temperature condition. $\varepsilon_{max,i}$, $\varepsilon_{min,i}$, $\sigma_{max,i}$, and $\sigma_{min,i}$ are the maximum and minimum total strain and von Mises stress. The anisotropic term includes the influence of FOD through stress values. The effect of anisotropy on fracture failure can be considered effectively in the developed model by combining fracture stress, $\sigma_{f,i}$, and maximum stress, $\sigma_{max,i}$, using a logarithmic function. The last term concerns the temperature effect with the reference temperature (T_{ref}) of CF-PA6, 20 °C, and the experimental temperature in Arrhenius law.

5. Results

The true and engineering stress–strain curves of the uniaxial tensile experiment are summarized in Figure 6. The static behavior of CF-PA6 shows a linear elastic stress–strain relationship until fracture. The tensile strengths of CF-PA6 from the uniaxial tensile test are summarized in Table 6 for each direction and temperature. As shown in the load–stroke curve of the uniaxial tensile test, the higher mechanical properties appear when the main load direction and the fiber orientation tensor coincide.

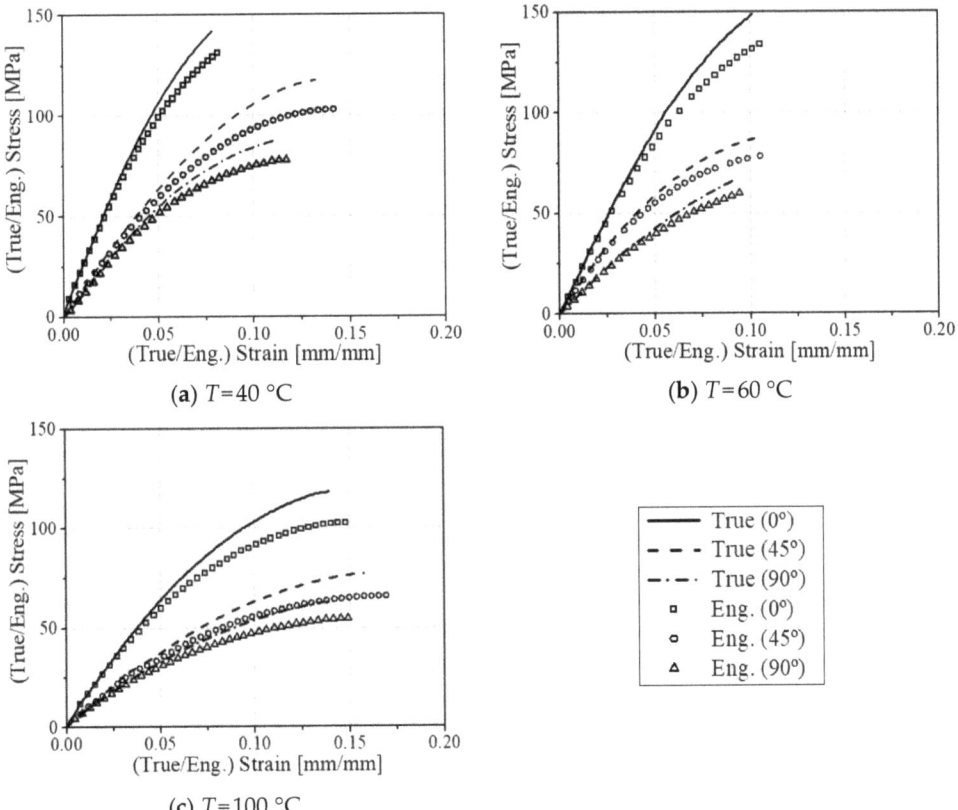

Figure 6. Experimental results of true and engineering (eng.) stress–strain curves for 0°, 45°, and 90° specimens in uniaxial tensile tests.

Table 6. Tensile strengths of CF-PA6 in the uniaxial tensile tests.

Temperature [°C]	Tensile Strength of Each Specimen Direction [MPa]		
	0°	45°	90°
40	130.3	101.3	77.2
60	122.6	79.6	60.8
100	98.8	62.8	51.2

The engineering strain is the changed displacement at each measurement moment divided by the initial gauge length, and the true strain is the changed displacement divided by the displacement immediately before. The engineering strain does not consider the changed length of the specimen, and the true strain takes into account the changed length of the specimen at every moment. The nominal stress is based on the cross-sectional area of the initial specimen when calculating the stress, and the true stress is based on the actual cross-sectional area that continues to change during the tensile test. Just before the specimen is broken, the cross-sectional area becomes very small, and the engineering stress does not consider the decrease in the cross-sectional area, so the closer to the fracture stress, the greater the true stress than the engineering stress, as shown in Figure 6.

In order to minimize the phenomenon that mechanical properties differ due to fiber orientation distribution depending on the location of the injection molded plate from which the specimen was machined, the stress–strain curves were compared by adopting the results of the specimen collected from the center to confirm the tendency by angle and temperature. The average difference in tensile strength due to the specimen location was measured as 10.4%, 8.2%, and 9.7% at 0°, 45°, and 90°, respectively. In order to consider the difference in mechanical properties, structural analysis was performed by mapping the fiber orientation at the location where each sample was machined.

The load–displacement curve comparison between experimental and FEA results on the uniaxial tensile specimen and three-point bending test is reported in Figures 7 and 8, respectively. The tensile displacements of all the experiments were obtained using the 20 mm extensometer to compare the experiment results and numerical analysis accurately. The numerical analysis results indicate that higher anisotropy is observed according to the test direction with the increase in test temperature. In addition, as the load decreases, the influence of the matrix increases. It can also be confirmed that the elasticity increases as the test direction coincides with the injection direction. For the three-point bending tests, the stress and strain values are calculated by optimizing the material parameters by utilizing the tensile and three-point bending test results together. Implemented numerical models and proposed structural equations were validated and used to explain the anisotropy and temperature effects of CF-PA6.

The correlation coefficient (R^2) is slightly lower when considering various temperatures simultaneously compared to a single temperature; however, the proposed semi-empirical model confirms that fatigue life expectancy is well predicted. Based on a validated numerical analysis model, as presented in the previous section of the paper, each combination of investigated temperature and specimen direction, the von Mises stress, and strain at the minimum and maximum load associated with the fatigue experiment were exported from the simulation. The temperature was selected as the reference, room, and experimental temperature inside the heating chamber. The constants of the developed model Equation (5) were calculated by $A = 5.6734$ and $c = -0.692$ based on a total set of 54 data, as shown in Figure 9, and the correlation coefficient is $R^2 = 0.9457$. Thus, the reliability of the proposed function of the anisotropic fatigue test data integration obtained under various stress states and temperatures is demonstrated.

In order to evaluate the performance of the developed fatigue life prediction model, the regression results of Equations (4) and (5) were compared. From a macroscopic point of view, the tensile fracture strength of 0° was adopted as the denominator of the stress term of Equation (4), and the maximum stress value of each fatigue test was used in the numerator. In addition, since Equation (4) does not consider the temperature effect, Equation (5) was calculated without the last term to compare the model performance in the same environment. As a result of the calculation, it was confirmed that A and c of Equation (4) and modified Equation (5) were $A = 2.4959$, $c = -0.549$, and $A = 14.138$, $c = -0.69$, respectively. The correlation coefficients were $R^2 = 0.9729$ and $R^2 = 0.6548$, respectively, which was further excellent in the performance of Equation (5). It was confirmed that the fatigue life prediction model developed using the principal stress worked well without considering the fiber orientation distribution, which is difficult to calculate, as an angle. The comparison of calculated results is shown in Figure 10.

Moreover, the fracture behavior of the matrix and fiber according to low- and high-cycle fatigue fracture by test temperature was analyzed through SEM analysis. The SEM image was obtained at a magnification at which the matrix and the fiber could be observed simultaneously, and the result of enlarging a single fiber for the analysis of the fracture surface of the fiber was also taken. Each result is reported in Figures 11 and 12.

As shown in Figure 11, high-cycle fatigue shows more significant irregularities in the base material compared to low-cycle fatigue. From these results, it can be confirmed that the polymer matrix causes more significant deformation as the fatigue failure cycle number increases. This result is related to the fact that the shorter the fatigue fracture cycle, the more the load is transmitted by the fiber. In addition, it can be seen that the fiber remains longer as the temperature increases, and the matrix is broken. Figure 12 shows that the higher the test temperature and the longer the fracture cycle, the more even the fracture surface of the fiber is.

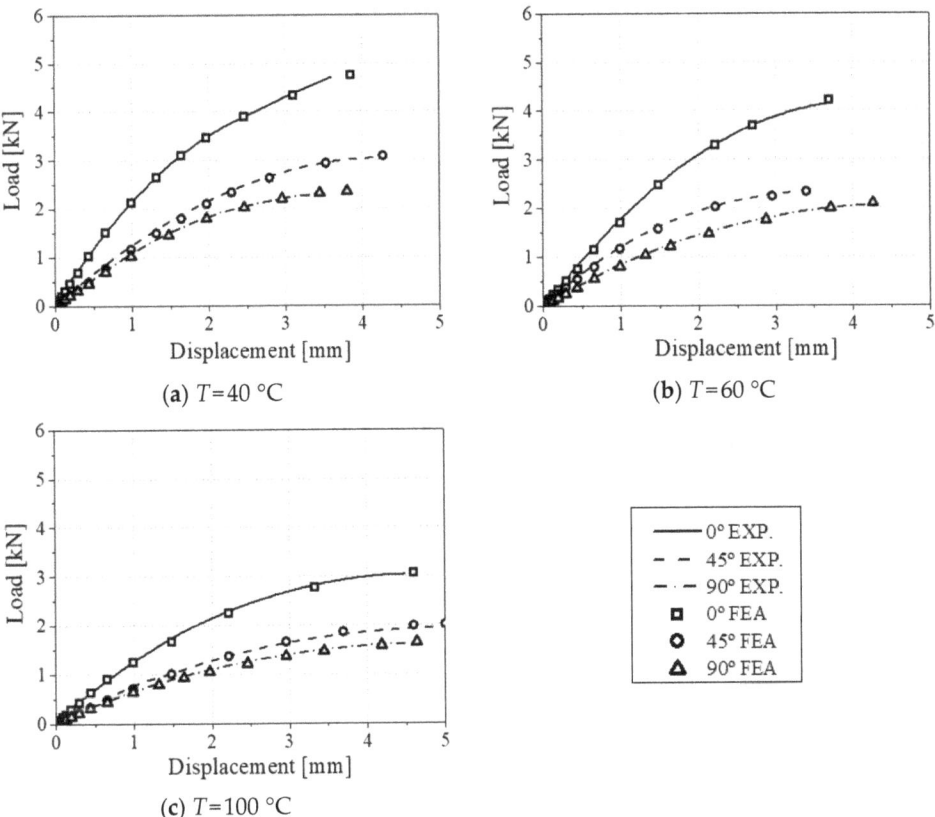

Figure 7. Load–displacement comparison between experimental and FEA results for 0°, 45°, and 90° specimens in uniaxial tensile test conditions.

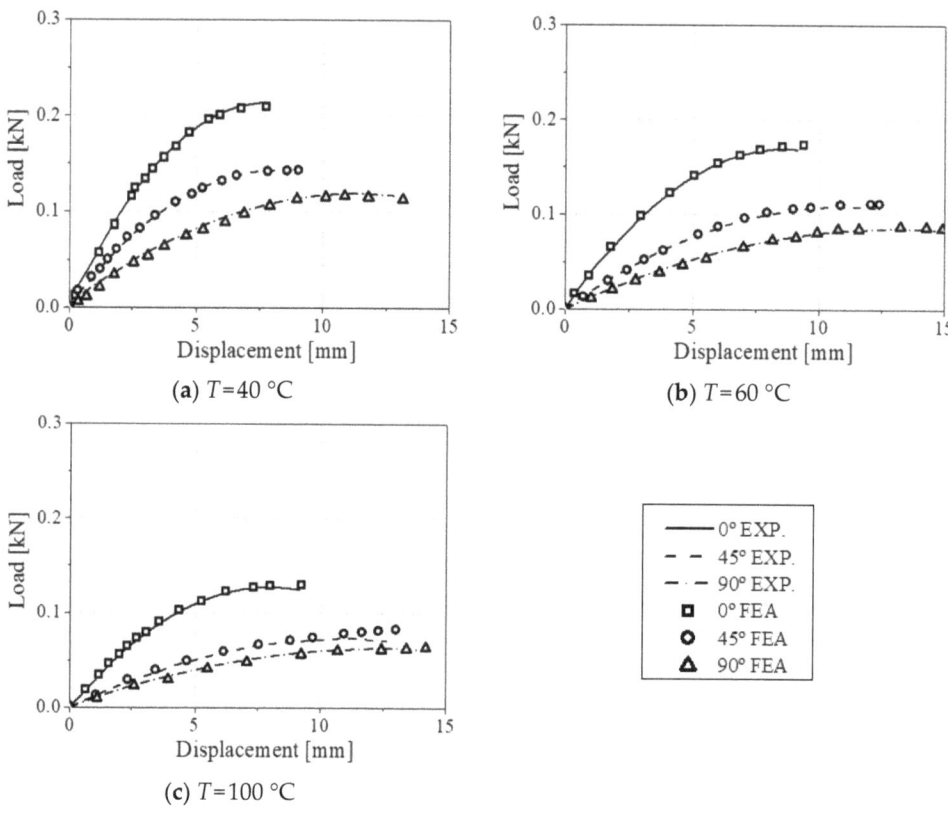

Figure 8. Load–displacement comparison between experimental and FEA results for 0°, 45°, and 90° specimens in three-point bending test conditions.

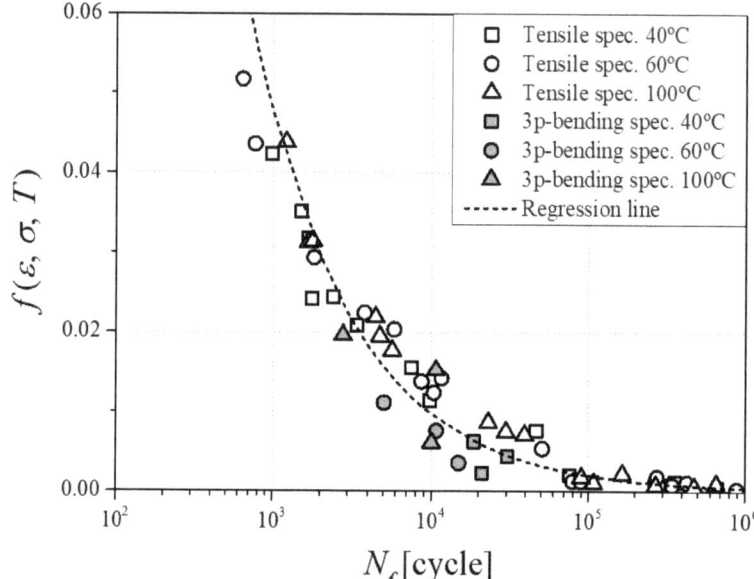

Figure 9. Discrete energy function values from various fatigue tests and their regression line.

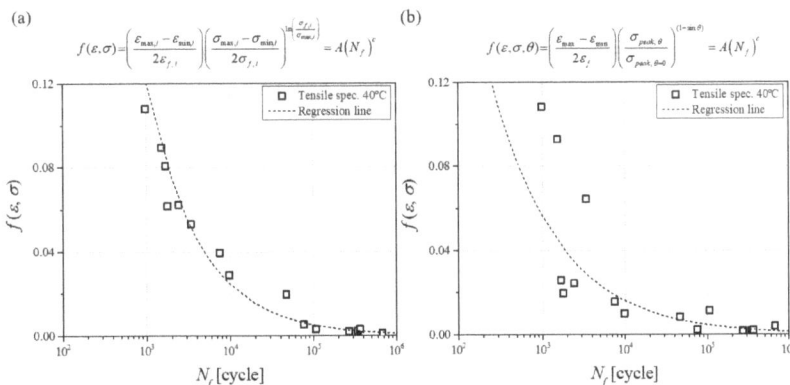

Figure 10. Fatigue experiments regression on (**a**) modified Equation (5) model and (**b**) Equation (4) model.

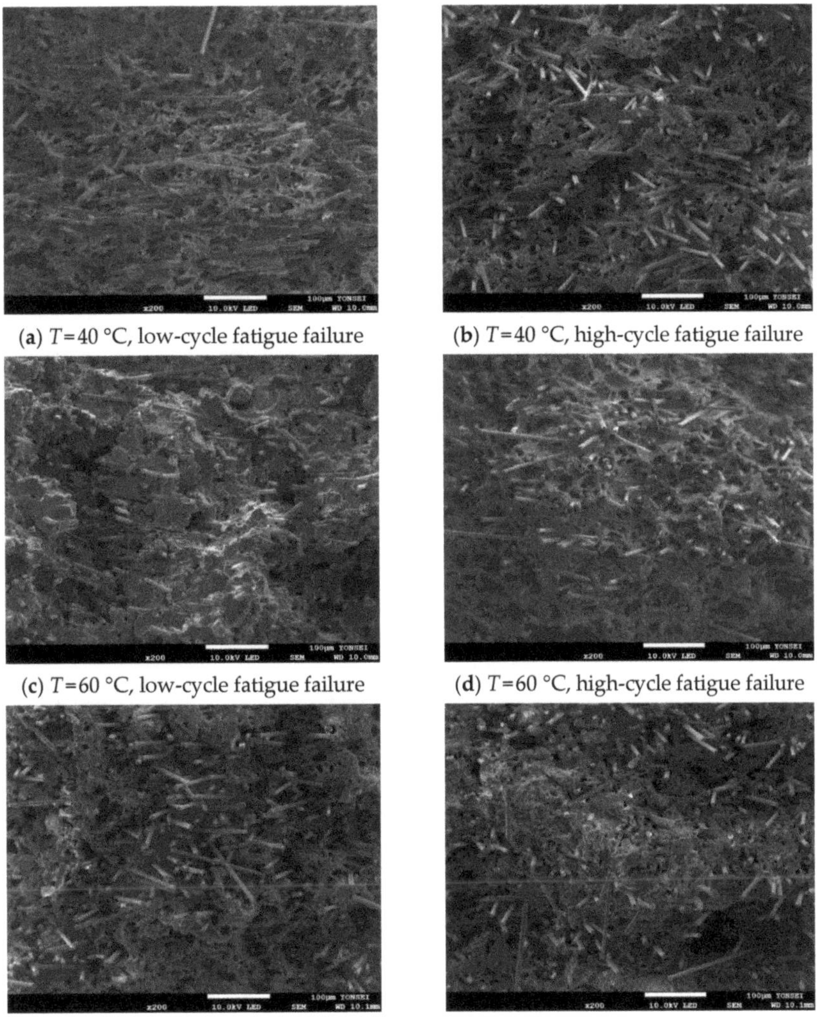

(**a**) $T=40\ °C$, low-cycle fatigue failure

(**b**) $T=40\ °C$, high-cycle fatigue failure

(**c**) $T=60\ °C$, low-cycle fatigue failure

(**d**) $T=60\ °C$, high-cycle fatigue failure

(**e**) $T=100\ °C$, low-cycle fatigue failure

(**f**) $T=100\ °C$, high-cycle fatigue failure

Figure 11. Matrix and fiber at the failure surface in case of low- and high-cycle fatigue.

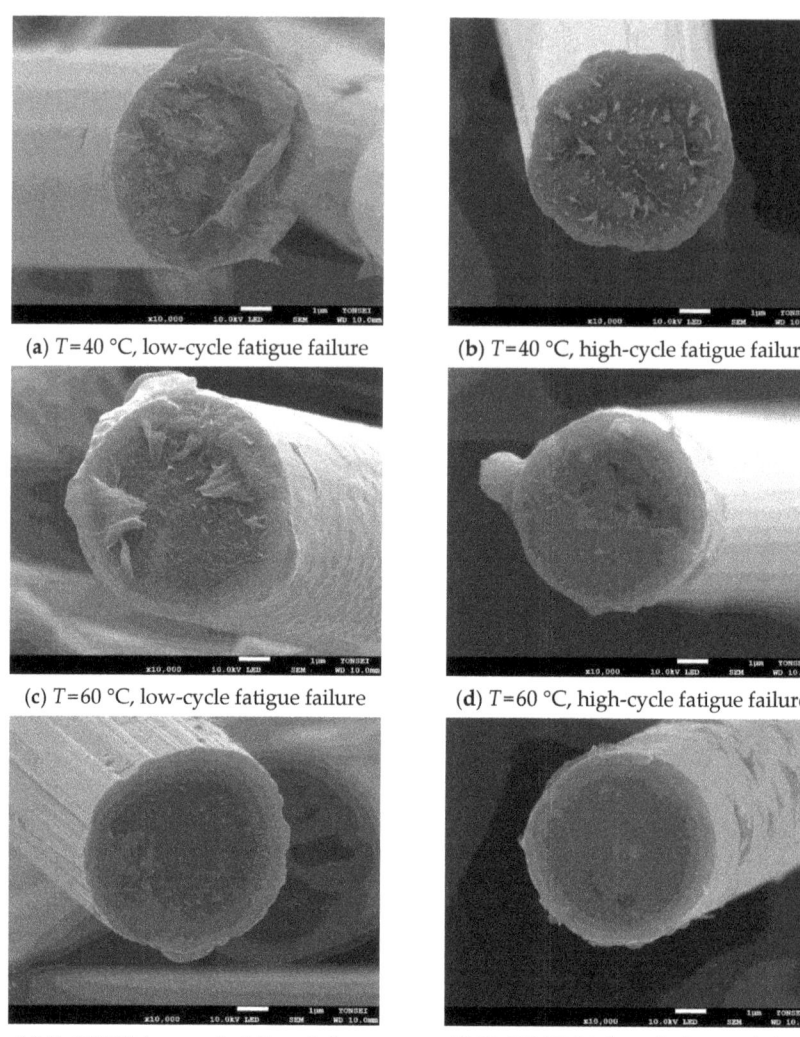

Figure 12. Detail of fiber's failure surface in low- and high-cycle fatigue.

6. Conclusions

- This study predicted the fatigue life expectancy of CF-PA6, a plastic reinforced with short fiber, through a strain-based semi-empirical model with a high correlation factor. A three-point bending test was performed to investigate various multi-axial stress states in actual components.
- A meaningful, intuitive fatigue life prediction model is proposed considering anisotropy as a stress term, which directly utilizes experimental results with a theoretical approach. It can be concluded that the fatigue life of materials with high temperature and anisotropy fiber orientation and polymers can be predicted with reasonable accuracy.
- SEM photography revealed that the higher the temperature and fatigue fracture cycle, the greater the deformation of the polymer matrix, and inversely, the more the deformation of the fiber. The higher the temperature, the more evenly the fiber's fracture cross-section is.

- The developed numerical model and structural equation are highly consistent between experiments and FEA results. Furthermore, they could accurately export stress and strain as inputs to a semi-empirical model.
- The usefulness of the results proposed in this paper can be outlined in two parts. First, the paper summarizes the static and fatigue behavior considering the anisotropy and temperature of short fiber-reinforced plastic materials, which are increasingly utilized exponentially in the industry. Secondly, it provides insight into the availability of the developed semi-empirical model to predict the fatigue life of CF-PA6.
- The use of FRP affected by temperature and fiber orientation is a remaining challenge for research on much colder temperatures and compressive forces below 0 °C. In addition, using compressive force in testing and investigating the mechanical properties of FRP can accurately describe the complex stress states in industries. Therefore, it can be a better solution to predict the fatigue life and composite use of FRP considering low temperature and compression stress states in the future.

Author Contributions: Conceptualization, N.K.; methodology, J.C.; software, H.L. (Hyungyil Lee); validation, J.C.; formal analysis, J.C.; investigation, Y.O.A.; resources, H.L. (Hyungyil Lee); data curation, J.C.; writing—original draft preparation, J.C.; writing—review and editing, Y.O.A.; visualization, J.C.; supervision, N.K.; project administration, N.K.; funding acquisition, H.L. (Hyungtak Lee). All authors have read and agreed to the published version of the manuscript.

Funding: This research was funded by the Carbon Industrial Cluster Development Program (10083609) funded by the Ministry of Trade, Industry & Energy (MOTIE, Korea), by the Material Component Technology Development Program (20004983) funded by the Ministry of Trade, Industry & Energy (MOTIE, Korea), by the Material Component Technology Development Program (20013060) funded by the Ministry of Trade, Industry & Energy (MOTIE, Korea), by the Technology Innovation Program (20016443) funded by the Ministry of Trade, Industry & Energy (MOTIE, Korea), and by the Sogang University Research Grant of 2023 (202312001.01).

Institutional Review Board Statement: Not applicable.

Informed Consent Statement: Not applicable.

Data Availability Statement: Data are contained within the article.

Conflicts of Interest: The authors declare no conflicts of interest.

References

1. Offringa, A.R. Thermoplastic Composites—Rapid Processing Applications. *Compos. Part A Appl. Sci. Manuf.* **1996**, *27*, 329–336. [CrossRef]
2. Pini, T.; Caimmi, F.; Briatico-Vangosa, F.; Frassine, R.; Rink, M. Fracture Initiation and Propagation in Unidirectional CF Composites Based on Thermoplastic Acrylic Resins. *Eng. Fract. Mech.* **2017**, *184*, 51–58. [CrossRef]
3. Li, L.; Xiao, N.; Guo, C.; Wang, F. A Study on Processing Defects and Parameter Optimization in Abrasive Suspension Jet Cutting of Carbon-Fiber-Reinforced Plastics. *Materials* **2023**, *16*, 7064. [CrossRef] [PubMed]
4. Yardim, Y.; Yilmaz, S.; Corradi, M.; Thanoon, W.A. Strengthening of Reinforced Concrete Non-Circular Columns with FRP. *Materials* **2023**, *16*, 6973. [CrossRef] [PubMed]
5. Tan, X.; Zhu, M.; Liu, W. Experimental Study and Numerical Analysis of the Seismic Performance of Glass-Fiber Reinforced Plastic Tube Ultra-High Performance Concrete Composite Columns. *Materials* **2023**, *16*, 6941. [CrossRef] [PubMed]
6. Pan, X.; Tian, A.; Zhang, L.; Zheng, Z. A Study on the Performance of Prestressed Concrete Containment with Carbon Fiber-Reinforced Polymer Tendons under Internal Pressure. *Materials* **2023**, *16*, 6883. [CrossRef] [PubMed]
7. Pastukhov, L.V.; Govaert, L.E. Crack-Growth Controlled Failure of Short Fibre Reinforced Thermoplastics: Influence of Fibre Orientation. *Int. J. Fatigue* **2021**, *143*, 105982. [CrossRef]
8. Lazzarin, P.; Molina, G.; Molinari, L.; Quaresimin, M. Numerical Simulation of SMC Component Moulding. *Key Eng. Mater.* **1998**, *144*, 191–200. [CrossRef]
9. Li, Y.; Li, W.; Shao, J.; Deng, Y.; Kou, H.; Ma, J.; Zhang, X.; Zhang, X.; Chen, L.; Qu, Z. Modeling the Effects of Interfacial Properties on the Temperature Dependent Tensile Strength of Fiber Reinforced Polymer Composites. *Compos. Sci. Technol.* **2019**, *172*, 74–80. [CrossRef]
10. Fu, S.-Y.; Lauke, B. Effects of Fiber Length and Fiber Orientation Distributions on the Tensile Strength of Short-Fiber-Reinforced Polymers. *Compos. Sci. Technol.* **1996**, *56*, 1179–1190. [CrossRef]

11. Launay, A.; Maitournam, M.H.; Marco, Y.; Raoult, I.; Szmytka, F. Cyclic Behaviour of Short Glass Fibre Reinforced Polyamide: Experimental Study and Constitutive Equations. *Int. J. Plast.* **2011**, *27*, 1267–1293. [CrossRef]
12. Arif, M.F.; Saintier, N.; Meraghni, F.; Fitoussi, J.; Chemisky, Y.; Robert, G. Multiscale Fatigue Damage Characterization in Short Glass Fiber Reinforced Polyamide-66. *Compos. Part B Eng.* **2014**, *61*, 55–65. [CrossRef]
13. Brighenti, R.; Carpinteri, A.; Scorza, D. Micromechanical Model for Preferentially-Oriented Short-Fibre-Reinforced Materials under Cyclic Loading. *Eng. Fract. Mech.* **2016**, *167*, 138–150. [CrossRef]
14. Judt, P.O.; Zarges, J.C.; Ricoeur, A.; Heim, H.P. Anisotropic Fracture Properties and Crack Path Prediction in Glass and Cellulose Fiber Reinforced Composites. *Eng. Fract. Mech.* **2018**, *188*, 344–360. [CrossRef]
15. Tanaka, K.; Kitano, T.; Egami, N. Effect of Fiber Orientation on Fatigue Crack Propagation in Short-Fiber Reinforced Plastics. *Eng. Fract. Mech.* **2014**, *123*, 44–58. [CrossRef]
16. Tanaka, K.; Oharada, K.; Yamada, D.; Shimizu, K. Fatigue Crack Propagation in Short-Carbon-Fiber Reinforced Plastics Evaluated Based on Anisotropic Fracture Mechanics. *Int. J. Fatigue* **2016**, *92*, 415–425. [CrossRef]
17. Russo, A.; Sellitto, A.; Curatolo, P.; Acanfora, V.; Saputo, S.; Riccio, A. A Robust Numerical Methodology for Fatigue Damage Evolution Simulation in Composites. *Materials* **2021**, *14*, 3348. [CrossRef]
18. Qureshi, H.J.; Saleem, M.U.; Khurram, N.; Ahmad, J.; Amin, M.N.; Khan, K.; Aslam, F.; Al Fuhaid, A.F.; Arifuzzaman, M. Investigation of CFRP Reinforcement Ratio on the Flexural Capacity and Failure Mode of Plain Concrete Prisms. *Materials* **2022**, *15*, 7248. [CrossRef]
19. Li, Y.; Deng, H.; Takamura, M.; Koyanagi, J. Durability Analysis of CFRP Adhesive Joints: A Study Based on Entropy Damage Modeling Using FEM. *Materials* **2023**, *16*, 6821. [CrossRef]
20. Aklilu, G.; Adali, S.; Bright, G. Tensile Behaviour of Hybrid and Non-Hybrid Polymer Composite Specimens at Elevated Temperatures. *Eng. Sci. Technol. Int. J.* **2020**, *23*, 732–743. [CrossRef]
21. Cheng, T. Ultra-High-Temperature Mechanical Behaviors of Two-Dimensional Carbon Fiber Reinforced Silicon Carbide Composites: Experiment and Modeling. *J. Eur. Ceram. Soc.* **2021**, *41*, 2335–2346. [CrossRef]
22. Meng, J.; Wang, Y.; Yang, H.; Wang, P.; Lei, Q.; Shi, H.; Lei, H.; Fang, D. Mechanical Properties and Internal Microdefects Evolution of Carbon Fiber Reinforced Polymer Composites: Cryogenic Temperature and Thermocycling Effects. *Compos. Sci. Technol.* **2020**, *191*, 108083. [CrossRef]
23. Jia, Z.; Li, T.; Chiang, F.P.; Wang, L. An Experimental Investigation of the Temperature Effect on the Mechanics of Carbon Fiber Reinforced Polymer Composites. *Compos. Sci. Technol.* **2018**, *154*, 53–63. [CrossRef]
24. Carpier, Y.; Vieille, B.; Coppalle, A.; Barbe, F. About the Tensile Mechanical Behaviour of Carbon Fibers Fabrics Reinforced Thermoplastic Composites under Very High Temperature Conditions. *Compos. Part B Eng.* **2020**, *181*, 107586. [CrossRef]
25. Jia, S.; Wang, F.; Zhou, J.; Jiang, Z.; Xu, B. Study on the Mechanical Performances of Carbon Fiber/Epoxy Composite Material Subjected to Dynamical Compression and High Temperature Loads. *Compos. Struct.* **2021**, *258*, 113421. [CrossRef]
26. Li, Y.; Li, W.; Tao, Y.; Shao, J.; Deng, Y.; Kou, H.; Zhang, X.; Chen, L. Theoretical Model for the Temperature Dependent Longitudinal Tensile Strength of Unidirectional Fiber Reinforced Polymer Composites. *Compos. Part B Eng.* **2019**, *161*, 121–127. [CrossRef]
27. Kawai, M.; Takeuchi, H.; Taketa, I.; Tsuchiya, A. Effects of Temperature and Stress Ratio on Fatigue Life of Injection Molded Short Carbon Fiber-Reinforced Polyamide Composite. *Compos. Part A Appl. Sci. Manuf.* **2017**, *98*, 9–24. [CrossRef]
28. Milella, P.P. *Fatigue and Corrosion in Metals*; Springer: Milan, Italy, 2013. ISBN 9788578110796.
29. Smith, P.; Topper, T.; Watson, P. A Stress-Strain Function for the Fatigue of Metals. *J. Mater.* **1970**, *5*, 767–778.
30. Miao, C.; Tippur, H.V. Fracture Behavior of Carbon Fiber Reinforced Polymer Composites: An Optical Study of Loading Rate Effects. *Eng. Fract. Mech.* **2019**, *207*, 203–221. [CrossRef]
31. Choi, J.; Quagliato, L.; Shin, J.; Kim, N. Investigation on the Static and Cyclic Anisotropic Mechanical Behavior of Polychloroprene Rubber (CR) Reinforced with Tungsten Nano-Particles. *Eng. Fract. Mech.* **2020**, *235*, 107183. [CrossRef]
32. Choi, J.; Quagliato, L.; Lee, S.; Shin, J.; Kim, N. Multiaxial Fatigue Life Prediction of Polychloroprene Rubber (CR) Reinforced with Tungsten Nano-Particles Based on Semi-Empirical and Machine Learning Models. *Int. J. Fatigue* **2021**, *145*, 106136. [CrossRef]
33. Choi, J.; Lee, H.; Lee, H.; Kim, N. A Methodology to Predict the Fatigue Life under Multi-Axial Loading of Carbon Fiber-Reinforced Polymer Composites Considering Anisotropic Mechanical Behavior. *Materials* **2023**, *16*, 1952. [CrossRef] [PubMed]
34. Launay, A.; Maitournam, M.H.; Marco, Y.; Raoult, I. Multiaxial Fatigue Models for Short Glass Fibre Reinforced Polyamide. Part II: Fatigue Life Estimation. *Int. J. Fatigue* **2013**, *47*, 390–406. [CrossRef]
35. Launay, A.; Maitournam, M.H.; Marco, Y.; Raoult, I. Multiaxial Fatigue Models for Short Glass Fiber Reinforced Polyamide—Part I: Nonlinear Anisotropic Constitutive Behavior for Cyclic Response. *Int. J. Fatigue* **2013**, *47*, 382–389. [CrossRef]
36. Choi, J.; Choi, J.; Lee, K.; Hur, N.; Kim, N. Fatigue Life Prediction Methodology of Hot Work Tool Steel Dies for High-Pressure Die Casting Based on Thermal Stress Analysis. *Metals* **2022**, *12*, 1744. [CrossRef]
37. Tang, H.C.; Nguyen, T.; Chuang, T.J.; Chin, J.; Wu, F.; Lesko, J. Temperature Effects on Fatigue of Polymer Composites. In Proceedings of the Composites Engineering, 7th Annual International Conference, ICCE/7, Denver, CO, USA, 2–8 July 2000.
38. Mivehchi, H.; Varvani-Farahani, A. The Effect of Temperature on Fatigue Strength and Cumulative Fatigue Damage of FRP Composites. *Procedia Eng.* **2010**, *2*, 2011–2020. [CrossRef]

39. Fouchier, N.; Nadot-Martin, C.; Conrado, E.; Bernasconi, A.; Castagnet, S. Fatigue Life Assessment of a Short Fibre Reinforced Thermoplastic at High Temperature Using a Through Process Modelling in a Viscoelastic Framework. *Int. J. Fatigue* **2019**, *124*, 236–244. [CrossRef]
40. Okayasu, M.; Tsuchiya, Y. Mechanical and Fatigue Properties of Long Carbon Fiber Reinforced Plastics at Low Temperature. *J. Sci. Adv. Mater. Devices* **2019**, *4*, 577–583. [CrossRef]
41. Yang, C.; Kim, Y.; Ryu, S.; Gu, G.X. Prediction of Composite Microstructure Stress-Strain Curves Using Convolutional Neural Networks. *Mater. Des.* **2020**, *189*, 108509. [CrossRef]
42. Mortazavian, S.; Fatemi, A. Fatigue Behavior and Modeling of Short Fiber Reinforced Polymer Composites Including Anisotropy and Temperature Effects. *Int. J. Fatigue* **2015**, *77*, 12–27. [CrossRef]
43. Murata, Y.; Kanno, R. Effects of Heating and Cooling of Injection Mold Cavity Surface and Melt Flow Control on Properties of Carbon Fiber Reinforced Semi-Aromatic Polyamide Molded Products. *Polymers* **2021**, *13*, 587. [CrossRef] [PubMed]
44. Karsli, N.G.; Aytac, A. Tensile and Thermomechanical Properties of Short Carbon Fiber Reinforced Polyamide 6 Composites. *Compos. Part B Eng.* **2013**, *51*, 270–275. [CrossRef]

Disclaimer/Publisher's Note: The statements, opinions and data contained in all publications are solely those of the individual author(s) and contributor(s) and not of MDPI and/or the editor(s). MDPI and/or the editor(s) disclaim responsibility for any injury to people or property resulting from any ideas, methods, instructions or products referred to in the content.

Article

Fracture Behavior of a Unidirectional Carbon Fiber-Reinforced Plastic under Biaxial Tensile Loads

Kosuke Sanai [1], Sho Nakasaki [1], Mikiyasu Hashimoto [1], Arnaud Macadre [2] and Koichi Goda [2,*]

1 Graduate School of Sciences and Technology for Innovation, Yamaguchi University, Ube 755-8611, Japan; sanaikousuke@gmail.com (K.S.); c005wcw@yamaguchi-u.ac.jp (M.H.)
2 Department of Mechanical Engineering, Yamaguchi University, Ube 755-8611, Japan; macadre@yamaguchi-u.ac.jp
* Correspondence: goda@yamaguchi-u.ac.jp

Abstract: In order to clarify the fracture behavior of a unidirectional CFRP under proportional loading along the fiber (0°) and fiber vertical (90°) directions, a biaxial tensile test was carried out using a cruciform specimen with two symmetric flat indentations in the thickness direction. Three fracture modes were observed in the specimens after the test. The first mode was a transverse crack (TC), and the second was fiber breakage (FB). The third mode was a mixture mode of TC and FB (TC&FB). According to the measured fracture strains, regardless of the magnitude of the normal strain in the 0° direction, TC and TC&FB modes occurred when the normal strain in the 90° direction, ε_y, ranged from 0.08% to 1.26% (positive values), and the FB mode occurred when ε_y ranged from -0.19% to -0.79% (negative values). The TC&FB mode is a unique mode that does not appear as a failure mode under uniaxial tension; it only occurs under biaxial tensile loading. Biaxial tensile tests were also conducted under non-proportional loading. The result showed three fracture modes similarly to the proportional loading case, each of which was also determined by the positive or negative value of ε_y. Thus, this study reveals that the occurrence of each fracture mode in a unidirectional CFRP is characterized by only one parameter, namely ε_y.

Keywords: carbon fibers; epoxy resin; unidirectional laminate; biaxial test; fracture mode

Citation: Sanai, K.; Nakasaki, S.; Hashimoto, M.; Macadre, A.; Goda, K. Fracture Behavior of a Unidirectional Carbon Fiber-Reinforced Plastic under Biaxial Tensile Loads. *Materials* **2024**, *17*, 1387. https://doi.org/10.3390/ma17061387

Academic Editor: Raul D. S. G. Campilho

Received: 14 February 2024
Revised: 13 March 2024
Accepted: 14 March 2024
Published: 18 March 2024

Copyright: © 2024 by the authors. Licensee MDPI, Basel, Switzerland. This article is an open access article distributed under the terms and conditions of the Creative Commons Attribution (CC BY) license (https://creativecommons.org/licenses/by/4.0/).

1. Introduction

Carbon fiber-reinforced plastics (CFRPs) are a composite material with excellent specific strength and stiffness, superior to those of other synthetic fiber-reinforced plastics; they are used in various industrial fields, such as aerospace, automobiles, construction materials, and so on. Needless to say, materials in practical use as structural members are inevitably subjected to multi-axial loading, and CFRPs are no exception. The failure phenomena of CFRPs under multiaxial stresses are complicated because the mechanical properties vary greatly depending on the direction of loading due to its strong orthogonal anisotropy and include many interactive damage modes, such as fiber breaks, matrix fracture, interfacial debonding, delamination, and fiber kinking [1–7]. The above modes interact in a complex manner, eventually leading to fractures in the CFRPs.

Although CFRPs are generally applied in a laminated form, representative data on their mechanical properties are published as results obtained from uniaxial tensile tests of a unidirectional form [8]. As is widely known, when unidirectional CFRPs are subjected to tensile loading in the direction of the fiber axis (0° direction), individual fiber breaks accumulate and eventually lead to material failure. In contrast, tensile loading in the fiber vertical axis (90° direction) results in transverse cracking, thereby bisecting the material. The question then arises: What kind of fracture behavior is observed when tensile loading occurs simultaneously in the 0° and 90° directions? Such fracture behavior should be of interest, but for some reason has not been discussed. The reason, the authors presume, is that although the biaxial tensile loading test has been successfully conducted on laminated

CFRPs, it is difficult to perform for unidirectional CFRPs. According to the authors' experience, the CFRP does not fracture in the region working on biaxial tensile stresses, but it does in the uniaxial loading portion along the 90° direction when the specimen shape is a cruciform, as shown schematically in Figure 1. In fact, there are very few papers that address biaxial tensile loading tests of unidirectional CFRPs in the 0° and 90° directions. On the other hand, various failure criteria [9–13] have been proposed for the fracture of unidirectional fiber composites, and many computational simulations [14,15] as well as continuum damage mechanics models [16–19] have been applied. Their validity has been confirmed by comparing experimental data, such as off-axial tests [9,10,12,13], axial tension/transverse compression biaxial loading [13], tension or compression/shear biaxial loading [13,19], transverse tension [20] or compression [21], axial tension and transverse compression [22], in-plane [23] or out-of-plane shear [24], and so on. However, there are only a few comparisons using biaxial tensile loading tests [13,19], and those are based on data from Soden et al. [25] for unidirectional E-glass/epoxy, with no comparison to CFRP. As with the aforementioned fracture behavior, there is currently insufficient information on stress and strain at fracture under biaxial tensile loading.

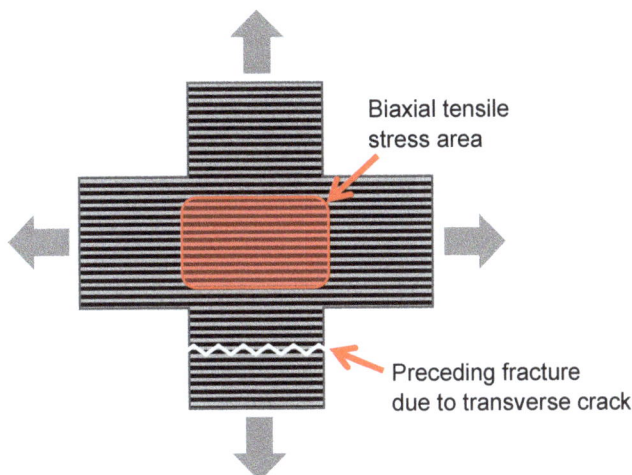

Figure 1. Schematic of fracture mode under T-T loading in a unidirectional CFRP (the fracture does not occur in the biaxial tensile stress area).

Cruciform specimens are often used to determine the mechanical properties of CFRPs under biaxial tensile loading. However, most of the papers have been conducted on CFRP laminates, such as cross-ply and angle-ply [26–30], and there are few papers that apply the cruciform specimens to unidirectional CFRPs. The only reported paper is the case where the specimen was tensile-loaded from two transverse directions perpendicular to the fiber axis [31]. Meanwhile, the shape of the biaxial test specimen has been studied according to the loading conditions as follows. Goto et al. [32] investigated the tensile strength of a unidirectional CFRP laminate in the 0° direction subjected to compressive loading in the 90° direction. The specimen used was dumbbell-shaped, but the two surfaces on both sides subjected to transverse compressive loading were simple planes. Rev et al. [33] proposed a thin-ply specimen composed of angle-ply and unidirectional CFRP, where an in-plane biaxial stress state of longitudinal tension and transverse compression is induced in the central 0° layers by the scissoring deformation of the angle-plies. The mechanism of transverse compressive loading differs from that of Goto et al., but the specimen also has a long strip shape in the 0° direction. Potter et al. [34] and Kang et al. [35] conducted a biaxial compression test on CFRP laminates. In this test, square-shaped specimens were used to apply compressive loads simultaneously from four directions. Thus, the specimen shape used for biaxial tests differs between tension–compressive loading (T-C loading) and compression–compressive loading (C-C loading). From such a background, the authors developed a unique unidirectional CFRP specimen for tension–tension loading (T-T loading) and succeeded in fracturing it in its biaxial tensile loading area [36].

The purpose of this study is to characterize the fracture modes of a unidirectional CFRP under proportional and non-proportional T-T loading, using the developed cruciform specimens. This study also aims to investigate the relationship between the fracture behavior and the strains in the 0° and 90° directions by measuring them at fracture. The results showed that three different fracture modes, namely transverse crack (TC), fiber breakage (FB), and TC including FB (TC&FB), were observed in both T-T loading tests. In addition, it was found that TC and TC&FB modes occurred when the strain in the 90° direction was positive, regardless of the difference in the two loading tests. However, the strains in the 90° direction (denoted as ε_y) causing the TC&FB mode, were clearly lower than those of the TC mode. This fact implies that the maximum strain criterion [37] is not applicable for the unidirectional CFRP under T-T loading. The strain range for the FB mode occurrence was also clarified, such that this mode occurred only when ε_y was negative, despite the fact that the FB and TC&FB modes were comparable in strains in the 0° direction (denoted as ε_x). Thus, this study reports that the occurrence of each fracture mode is characterized by only one parameter ε_y.

2. Materials and Methods

2.1. Materials

The reinforcing material used was PAN-based carbon fiber tows (TR50S 12L and 15L supplied by Mitsubishi Chemical Corporation, Tokyo, Japan). The carbon fibers are hereafter abbreviated as CFs. For the matrix resin, epoxy EP-4901 (ADEKA Co., Ltd., Tokyo, Japan) was used as the resin base, and Jeffarmin T-403 (Huntsman International LLC, The Woodlands, TX, USA) was used as the hardener. The epoxy resin base and hardener were mixed at the manufacturer's recommended mixing ratio of 100:46. The obtained mixture was degassed using a vacuum dryer and injected into the mold by hand, as described in Section 2.2. The properties of the CF tows and epoxy resin are shown in Table 1. The properties of the epoxy resin were obtained by tensile test in the authors' laboratory.

Table 1. CF tow [38] and matrix resin properties.

Constituent Material	Density [Mg/m^3]	Filament Diameter [mm]	Tensile Strength [MPa]	Young's Modulus [GPa]	Fracture Strain [%]	$\sigma_{0.2}$ [MPa]	H' [MPa]
CF tow	1.82	7	4900	235	-	-	-
Epoxy resin	1.16	-	65.2	3.2	5.0	48.0	447

$\sigma_{0.2}$: 0.2% proof stress, H': Linear plastic work hardening coefficient.

2.2. Preparation Method of Cruciform Specimen

A cruciform specimen has been used for T-T loading tests on such laminates as cross-ply and angle-ply [26–30]. The present study applies this specimen to a unidirectional CFRP. According to these papers [26–30], conventional cruciform specimen shapes have a circular or square indentation referred to as 'gauge area' in the center. The authors also attempted to make this indentation and to fracture it at the center where the biaxial tensile load occurs, but it often broke prematurely at the arm. Through trial and error, we found that by embedding the CFRP laminae in the center of the resin outside the gage area and making it longer in the fiber direction, all fracture modes shown later occurred within the gauge area without arm fracture. Thus, the shape of the gauge area differs from that of the above papers.

In the design of the cruciform specimen, the thickness dimension of the gauge area was determined such that the specimen is fractured in the 0° direction within the load capacity of the biaxial testing machine used. In other words, since the number of CF tows is adjustable according to the capacity, the dimension can simply be calculated. The numbers of CF tows used were three for the 12L CF tow and two for the 15L CF tow, respectively. To achieve specimen preparation with such a gauge area, a pair of convex aluminum parts with flat top and bottom surfaces were used, as shown in Figure 2. The shape and aspect ratio of the convex parts are empirically optimal. In the gap between the two parts, resin pre-impregnated CF tows are aligned with physical restraints, as mentioned later. The gap distances between the two convex parts are designed to be 0.45 mm for the 12L CF tow and 0.3 mm for the 15L CF tow, respectively.

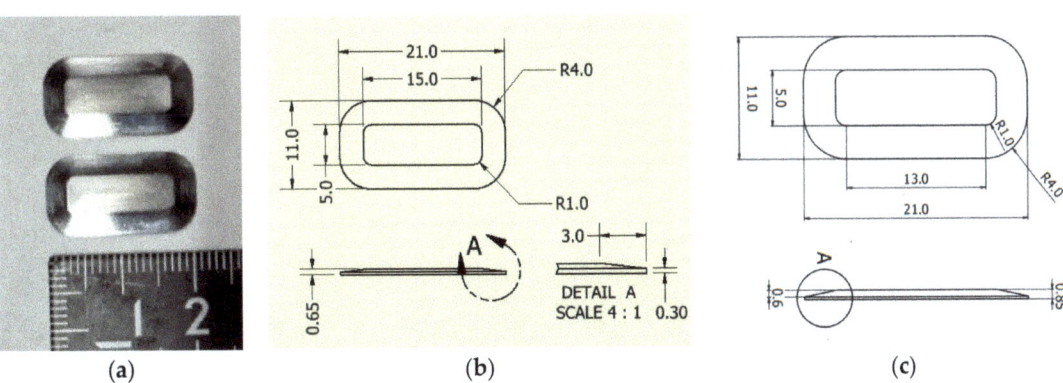

Figure 2. Convex parts to form gauge area. (**a**) Convex parts [36]; (**b**) shape and dimension of the convex part for 12L CF tow (unit: mm); (**c**) shape and dimension of the convex part for 15L CF tow (unit: mm). Note: The detail of A in (**c**) is the same as in (**b**).

The preparation method of the cruciform specimen is described in detail in the previous report [36], but this paper provides an outline. First, a convex part was glued to the center of the glass plate, which was placed into a stainless-steel frame with L-shaped metal fittings (Figure 3a). Then, a Teflon-die fixing frame was slid into them, and a Teflon-die cut out in a cruciform shape was placed into the frame, as shown in Figure 3b,c. Next, the prepared epoxy resin was impregnated to the CF tows. In order to place the CF tows in the center of the thickness direction, a 9.8 N weight was suspended at both ends to tension the CF tow, as shown in Figure 3d. This weight was adopted through several attempts in order to achieve a tension that did not loosen the CF tows and which was negligible in fracture load. During this process, air bubbles were generated, so the CF tows were degassed with a vacuum pump in the state shown in Figure 3d. After that, the epoxy resin was again poured into the Teflon-die (Figure 3e), and another glass plate, with the convex part of the counterpart shown in Figure 3f, was placed in the stainless-steel frame to avoid air bubbles. The entire apparatus, including the material before curing, was placed in a thermostatic dryer and cured for 24 h at 25 °C and post-cured for 6.5 h at 60 °C. The appearance of the resultant cruciform specimen is shown in Figure 4a, and its dimensions and shape are shown in Figure 4b. Figure 5 shows a typical fiber vertical cross-section of CFRP filled in the 'gap' described above. Although there is some variation in the fiber dispersion, it can be confirmed that large-scale resin-rich areas occurring between laminae are not found on the specimen surface due to the constraint by the convex parts.

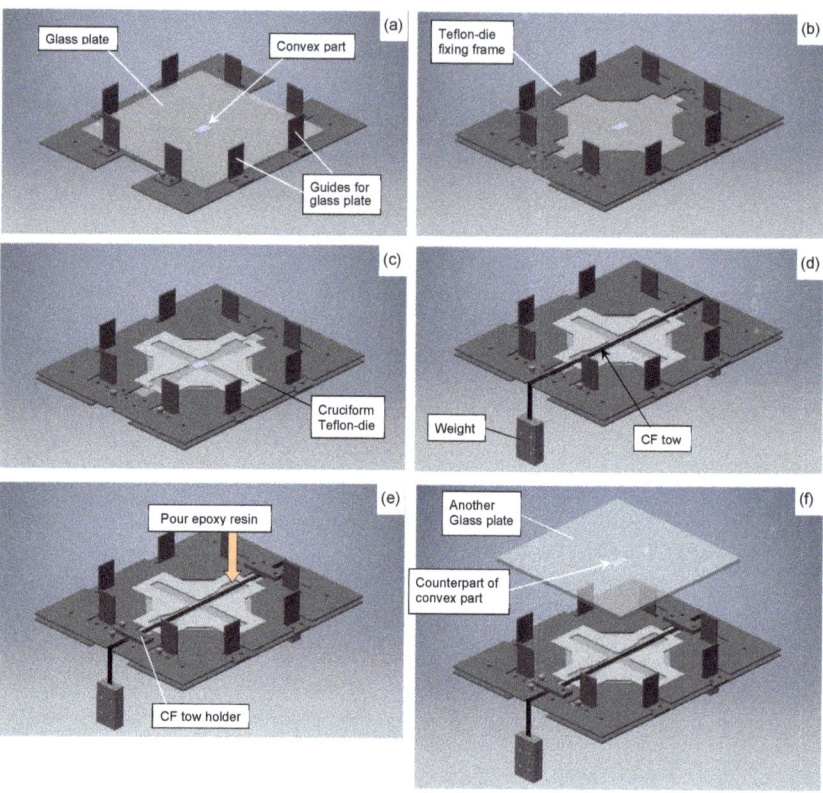

Figure 3. Schematic of cruciform specimen preparation procedure. (**a**) Set glass plate and position convex part in the center. (**b**) Set Teflon-die fixing frame. (**c**) Set cruciform Teflon-die. (**d**) Set pre-impregnated CF tows and fixation with weights. (**e**) Pour resin. (**f**) Set another glass plate. In (**f**), the counterpart convex part is attached to the back side of the glass plate.

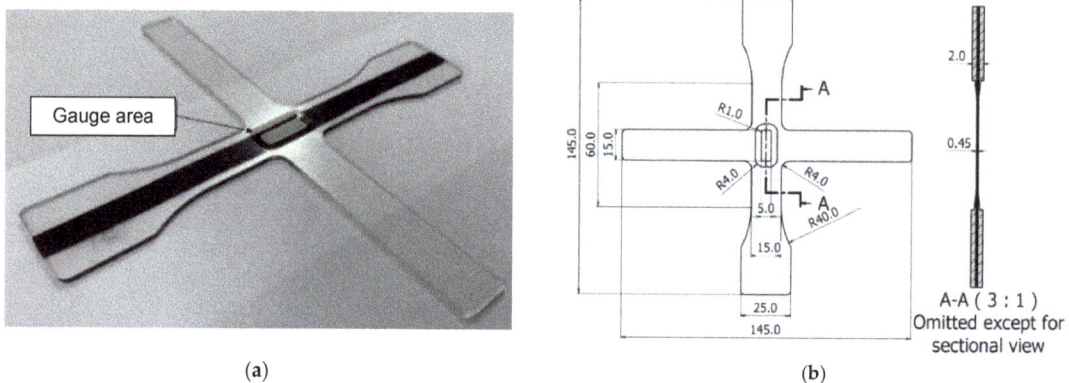

Figure 4. Cruciform specimen developed for the T-T loading test [36]. (**a**) Completed cruciform specimen; (**b**) shape and dimension used for the 12L CF tow (unit: mm). A-A bold line means cross section.

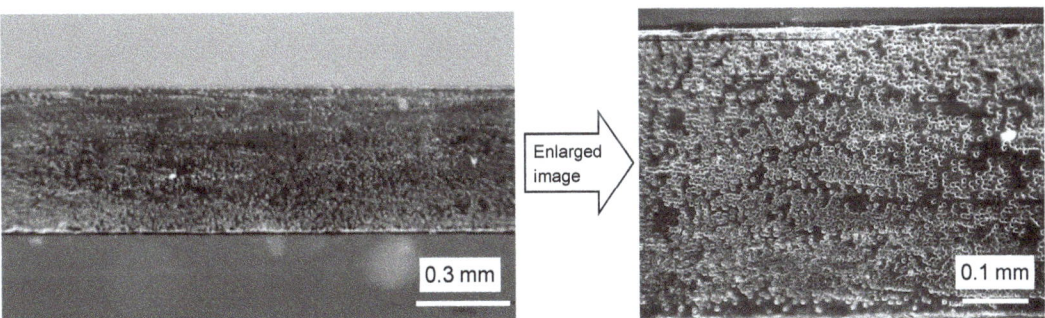

Figure 5. Transverse cross-section in gauge area of the cruciform specimen using the 12L CF tow.

Since the thickness of the CFRP cross-section is considered to be approximately equal to the gap distance described above, and the widths are both 7 mm, the fiber volume fractions are, respectively, calculated to be 44.0% and 55.0% for the 12L and 15L CF tows.

2.3. Stress Distribution

It was also demonstrated in reference [36] that the stress distributions in the fiber and fiber vertical directions at the gauge area are nearly uniform when using general-purpose finite element analysis software (ANSYS, ver. 18.0). The simulation results are summarized below. Figure 6 shows the element mesh of the gauge area and its vicinity used in the analysis, where the model shape is a 1/8 model in consideration of the symmetry of the specimen. The element type used is a linear tetrahedral element. The model consists of two areas, namely CFRP and epoxy areas, which constitute the cruciform specimen. The numbers of nodes and elements are 55,740 and 24,780, respectively. The material properties of CFRP were set as an orthotropic elastic material, and those of epoxy were chosen as an isotropic elastoplastic material. The material properties of the epoxy were based on the stress–strain diagram of the epoxy resin used for the specimens, and the multilinear approximate isotropic hardening law was applied for the plastic region. The interfacial contact between the CFRP and epoxy was assumed to be fully bonded, and the deformation perpendicular to the symmetry plane was constrained. The load boundary condition was applied at the ends of the specimen along x- and y-axes. The material constants used in the simulation are shown in Table 2.

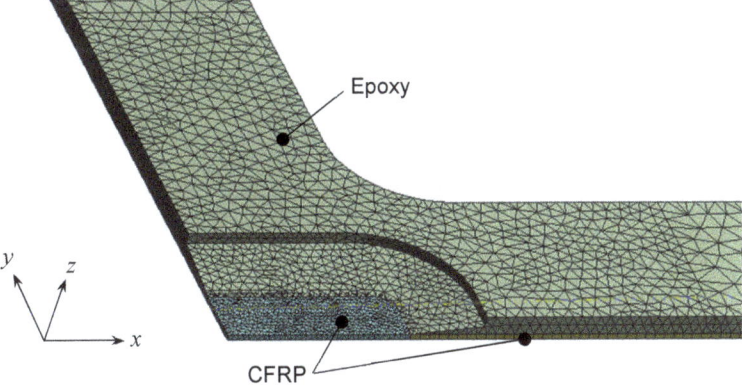

Figure 6. Finite element mesh applied for the cruciform specimen (CFRP: peacock blue area and bottom area under the epoxy. Epoxy: lime green area).

Table 2. Material properties of CFRP used for the 12L CF tow.

E_1 = 112.8 [GPa]	ν_{12} = 0.38	G_{12} = 6.0 [GPa]
E_2 = 6.6 [GPa]	ν_{13} = 0.38	G_{13} = 6.0 [GPa]
E_3 = 6.6 [GPa]	ν_{23} = 0.49	G_{23} = 2.2 [GPa]

E: Young's modulus, ν: Poisson's ratio, G: shear elastic modulus. Subscripts 1, 2, and 3 correspond to the x-, y-, and z-axes, respectively.

Figure 7a,b show the stress distributions, σ_x and σ_x, along the x- and y-axes, respectively. In this simulation, the loads, 1000 N and 100 N, were, respectively, applied at the ends of the x- and y-axes as boundary conditions. Although some stress concentration occurred at the border between the flat CFRP and the inclined epoxy areas, the stress distributions in the gauge area were stable without extreme variation, as shown in both figures. Table 3 shows the ranges in the gauge area of the simulated maximum and minimum stresses under various boundary conditions. It is found from Table 3 that the ranges do not have a large variation. Thus, the cruciform specimen shown in Figure 4 was used.

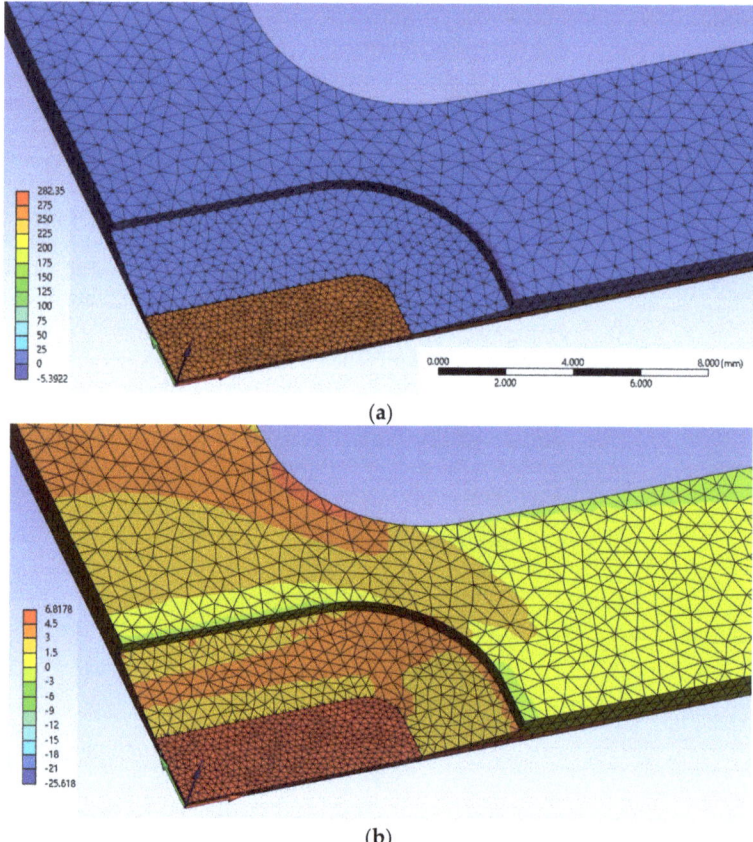

Figure 7. Simulation results of stress distributions, σ_x and σ_y (the unit of the numbers next to the color display is MPa, and the top and bottom are the maximum and minimum values, respectively). (**a**) Stress distribution of σ_x; (**b**) stress distribution of σ_y.

Table 3. Ranges in the gauge area of the computed maximum and minimum stresses.

Applied Loads		Simulated Stresses	
F_x [N]	F_y [N]	σ_x [MPa]	σ_y [MPa]
50	100	3.7–6.8	6.6–10.4
100	100	17.7–20.5	6.6–10.1
200	100	45.5–48.1	6.6–9.6
400	100	101–103	6.4–8.5
800	100	210–212	6.1–6.6
1000	100	267–279	5.2–6.2
1400	100	377–380	3.5–5.8
1000	-	285–286	0.06–0.52

2.4. Biaxial Tensile Test

For the biaxial tensile test, a biaxial tension and compression testing machine (manufactured by Kisan Chemical Engineering & Works Co., Ltd., Yamaguchi Shunan, Japan, with a capacity of 10 kN in each axis) shown in Figure 8a was used. The machine has independent structures on the x-axis and y-axis, and is equipped with a motor and a load cell at the bases of both the x-axis and y-axis. The power of each motor is transmitted to each shaft through gears, and the y-axis base moves along the rail when the tensile or compressive load is applied to the x-axis, and vice versa. The load conditions were set by changing the displacement speed ratios of the x- and y-axes cross-heads. Table 4 shows the loading conditions applied by changing the cross-head speed. The U10 condition is a uniaxial tensile test using a cruciform specimen with the y-axis arm of the 90° direction cut off. On the other hand, it is known that the fiber end-face effect often appears in uniaxial tests in the fiber vertical direction. Hence, the loading condition on the y-axis that results in zero stress in the 90° direction was estimated in advance by the finite element analysis software, and the cruciform specimen was used as it is. B114 corresponds to this condition.

Table 4. Proportional biaxial tensile loading conditions.

Loading Condition	Cross-Head Speed		Cross-Head Speed Ratio
	x-Axis [mm/s]	y-Axis [mm/s]	
B11	0.01	0.01	1:1
B21	0.02	0.01	2:1
B31	0.03	0.01	3:1
B41	0.04	0.01	4:1
B51	0.05	0.01	5:1
B12	0.01	0.02	1:2
B114	0.01	0.14	1:14
U10	0.01	-	-

x-axis: 0° direction, y-axis: 90° direction.

Non-proportional tests were also conducted in this study, in which several arbitrary loads were applied in the x-axis first; the loads were then fixed, and the y-axis loading started and continued up to the specimen's fracture. This condition is denoted as X1Y2. Another condition, denoted as X2Y1, is the case when the order of loading is the reverse of X1Y2. Both of the cross-head speeds were set as 0.01mm/s.

To measure strains along the 0° and 90° directions, a biaxial superimposed strain gauge (KFGS-1-120- D16-11, Kyowa Electronic Instruments Co., Ltd., Tokyo, Japan) was attached to the center of the gauge area, as shown in Figure 8b. The measurement was carried out until the specimen was fractured.

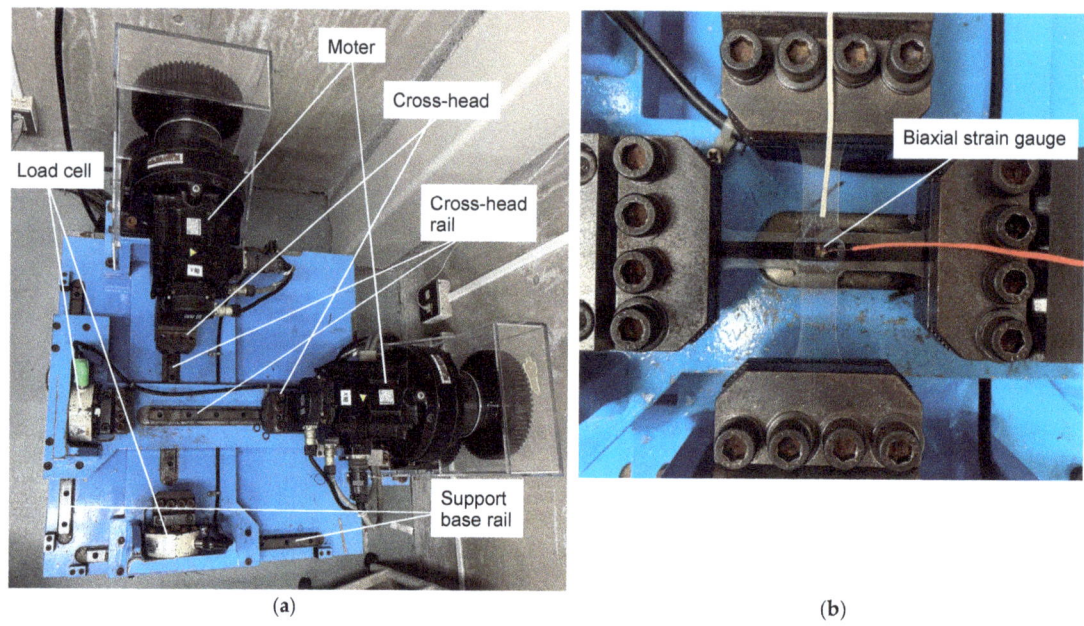

Figure 8. Biaxial tensile and compressive testing machine. (**a**) Top view of the biaxial tensile and compressive testing machine; (**b**) Top view after mounting the cruciform specimen attached with the biaxial strain gauge. This product consists of two uniaxial strain gages, which are placed perpendicularly to each other.

3. Results and Discussion

3.1. Fracture Modes

After the proportional and non-proportional T-T loading tests, the gauge areas of the fractured specimens were observed visually. It was found that there were three types of fracture modes. The first is the transverse crack (referred to as TC), in which a single crack runs along the fiber direction, as shown in Figure 9a,b. This fracture mode was observed in all specimens under B11, B12, B114 conditions, and in several X1Y1 conditions. In B11, there were several specimens in which the main crack branched and secondary cracks occurred in the resin, as shown by the arrows in Figure 9a. The second mode appeared such that the TC occurred within the gauge area but did not extend so much to the outside. In addition, cracks perpendicular to the fibers are observed at two locations just outside the gauge area, as shown in Figure 9c,d. This is undoubtedly a fracture mode due to fiber breakage (referred to as FB). This mode is referred to as TC&FB, which was observed in all specimens under B21 and B31 conditions, and under several X1Y2 conditions. The TC&FB mode is a unique mode that does not appear under uniaxial tension; instead, it only appears under biaxial tensile loading. The third one is only the FB mode, as shown in Figure 9e,f, which was observed in most of the specimens under B41 and B51 conditions and in all the specimens under U10 and X2Y1 conditions.

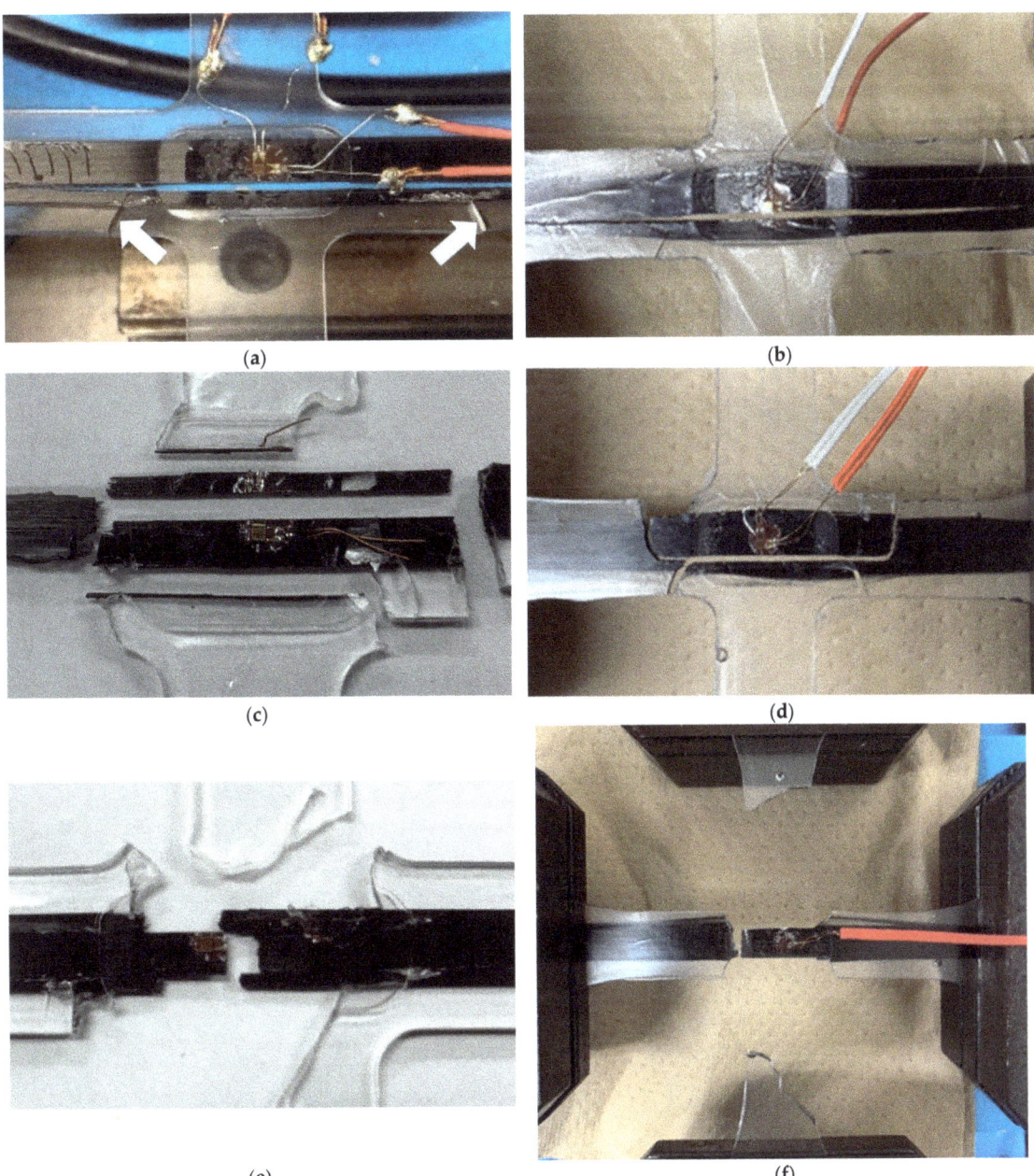

Figure 9. Various fracture modes under biaxial tensile loads of the cruciform specimens. (**a**) TC mode (B11) [36]; (**b**) TC mode (X1Y2); (**c**) TC&FB mode (B31) [36]; (**d**) TC&FB mode (X1Y2); (**e**) FB mode (B51) [36]; (**f**) FB mode (X2Y1). The arrows in (**a**) indicate resin cracks.

Table 5 shows the average fracture loads F_x in the 0° direction and F_y in the 90° direction for proportional loading conditions along with fracture modes. F_y values of B41 and B51 showing the FB mode were lower than those in the other loading conditions, which suggests that the FB mode was mainly caused by F_x. On the other hand, the F_x of

B31 was close to the F_x of B41, and the F_y of B31 was also close to the F_y of B11 and B12. Such results imply that it is not clear which mode (FB or TC) occurs first in the FB&TC mode. Due to the capacity of the instrument used in this study, the strain measurement interval was set to 5.0 μs. Unexpectedly, at the time of fracture of the specimens, both strains on the x- and y-axes changed at the same time within the above interval. This fact means that it is difficult to distinguish which mode occurred first. Nevertheless, we believe that in the TC&FB mode, TC occurs first. As mentioned above, the FB in the TC&FB mode occurred at two outside locations where the thickness was larger than the gauge area, but there was no case where two FB modes appeared individually without the TC mode. Also, there was no FB mode at only one location outside the gauge area. If FB mode(s) occur first on the outside, such specimens without the TC mode should exist. However, the TC was always present in the gauge area. To rephrase the above, it is reasonable to conclude that in TC&FB mode, TC occurs prior to FB and the occurrence of TC triggers FB a moment later.

Table 5. Average fracture loads along x- and y-axes under proportional loading.

Loading Condition	Number of Specimens	Fracture Loads [N]		Fracture Mode
		F_x	F_y	
B11	3	3292	835	TC
B21	3	5728	818	TC&FB
B31	3	7457	773	TC&FB
B41	3	7415	595	FB *
B51	3	8058	567	FB
B12	3	1608	825	TC
B114	3	235	685	TC
U10	3	7318	0	FB

CF tows used are all 12L; * one of the specimens is unknown.

Thus, we suggest the formation process of the TC&FB mode, as shown in Figure 10. Since B21 and B31 conditions provide large loads in the 0° direction, local single fiber-breaks are able to occur, and subsequent fiber–matrix interfacial debonding easily occurs due to another loading in the 90° direction. As a result, although the fracture approaches the FB mode, such multiple localized debonding can also cause the TC mode even at smaller strain levels than B12 or B11. Once TC occurs, it propagates instantaneously, as shown in Figure 10a,b. After that, the crack propagates outside the gauge area at high speed, as shown in Figure 10c. In general, cracks of rigid polymers are known to branch not only under uniaxial tensile loading but also under biaxial tensile loading [39,40]. In this case, the crack also branches at the epoxy area. Finally, the tensile stress along the 0° direction greatly increases at the two branching locations, as shown in Figure 10d, where two FB modes occur instantaneously.

Table 6 shows the fixed loads and the average fracture loads for non-proportional loading conditions along with fracture modes. The fracture load F_x of X2Y1* showing the FB mode is almost at the same level as those of the proportional condition, but the F_x values of X2Y1 are lower than those in the proportional loading conditions, irrespective of the fixed load F_y. This is because the number of tows is reduced from three to two in the specimen using the 15L CF tow. Since the ratio of the total numbers of monofilaments is 6:5 for the 12L CF tow to 15L, 6605 N and 6535 N of F_x in the X2Y1 conditions are, respectively, estimated as 7926 N and 7842 N at the same number of monofilaments. Thus, we can say that these values correspond to fracture loads of the FB modes measured at the proportional loading.

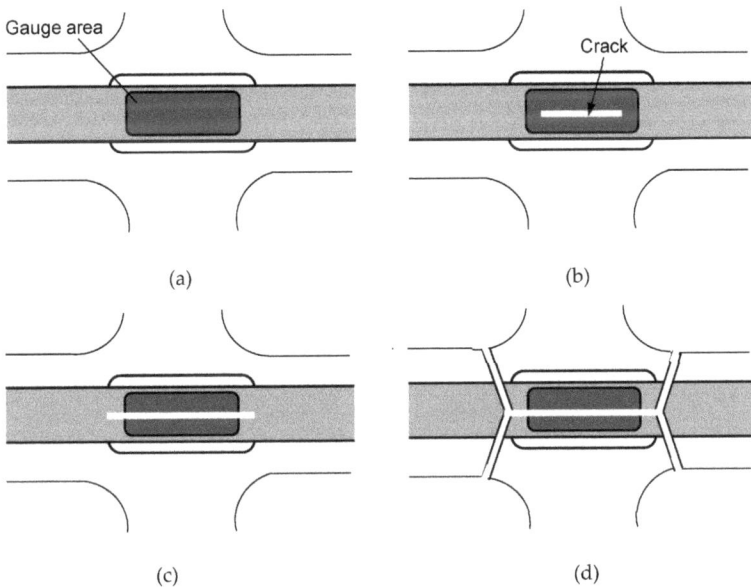

Figure 10. Schematic of the TC&FB formation process. (**a**) Initial state; (**b**) TC occurrence in the gauge area; (**c**) TC propagation outside the gauge area; (**d**) TC branching followed by two FB modes.

Table 6. Average fracture loads along the x- and y-axes under non-proportional loading.

Loading Condition	Number of Specimens	Fixed Load [N]		Fracture Load [N]		Fracture Mode
		F_x	F_y	F_x	F_y	
X2Y1 *	3	-	460	7203	-	FB
X2Y1	1	-	610	6605	-	FB
X2Y1	1	-	420	6535	-	FB
X1Y2	1	5515	-	-	890	TC&FB
X1Y2	3	4500	-	-	807	TC&FB
X1Y2	2	2265	-	-	813	TC
X1Y2	7	2200	-	-	853	TC
X1Y2	2	500	-	-	790	TC

CF tow used is 12L for X2Y1 * only; it was15 L for all others.

TC&FB modes were obtained under several X1Y2 conditions, at which the fracture loads of F_y are close to the fracture loads obtained under B21 condition. TC modes were also obtained under several X1Y2 conditions, at which the fracture loads of F_y are also similar to the fracture loads under the B11 and B12 conditions. According to Reuss' model and other models, Young' modulus along the fiber vertical direction of a unidirectional lamina is insensitive to changes in the fiber volume fraction up to about 60% [41,42]. Therefore, the fracture loads showing TC and TC&FB modes may be at a similar level to those of proportional loading despite the different volume fraction, if the fracture mechanism of the two CFRP specimens is the same. In any case, it is concluded from the above results that the same fracture modes appear at similar fracture loads, regardless of the difference in the loading history.

3.2. Strain Histories and Fracture Strains

The results of the strain measurements of the unidirectional CFRP cruciform specimens under proportional T-T loading are shown in Figure 11. The horizontal and vertical axes, ε_x

and ε_y, in the figure are, respectively, strains along the 0° and 90° directions measured up to fracture. As can be seen from the $\varepsilon_x - \varepsilon_y$ curves of the B21, B31, B41, and B51 conditions, the specimens initially behave with a linear deformation and then change slightly to the positive side of ε_y as they approach the final fracture. On the other hand, B141, B12, and U10 behave almost linearly up to fracture. The fracture modes were the TC mode for the B141, B12, and B11 conditions, and the TC&FB mode for the B21 and B31 conditions. The FB mode appeared with the B41, B51, and U10 conditions. Despite these differences in fracture modes, the slight changes in ε_y in the B21, B31, B41, and B51 conditions are considered to be attributed to the nonlinearity of the matrix resin caused by tensile loading on the y-axis. This means that these are the conditions under which the matrix is greatly deformed plastically. In fact, when the neat epoxy resin used for the matrix was tensile-tested, plastic deformation was observed. Since the plastic work is stored in the specimen during deformation, it is inferred that a greater release energy is required to fracture the specimen, resulting in a complex fracture mode, such as TC&FB. It should be noted here that a positive ε_y at the final fracture always shows the TC or TC&FB modes. That is satisfied even if the specimen exhibits a partially negative strain during the deformation process, as can be seen in B31. On the other hand, if the change in ε_y remains negative, the fracture modes appeared as FB.

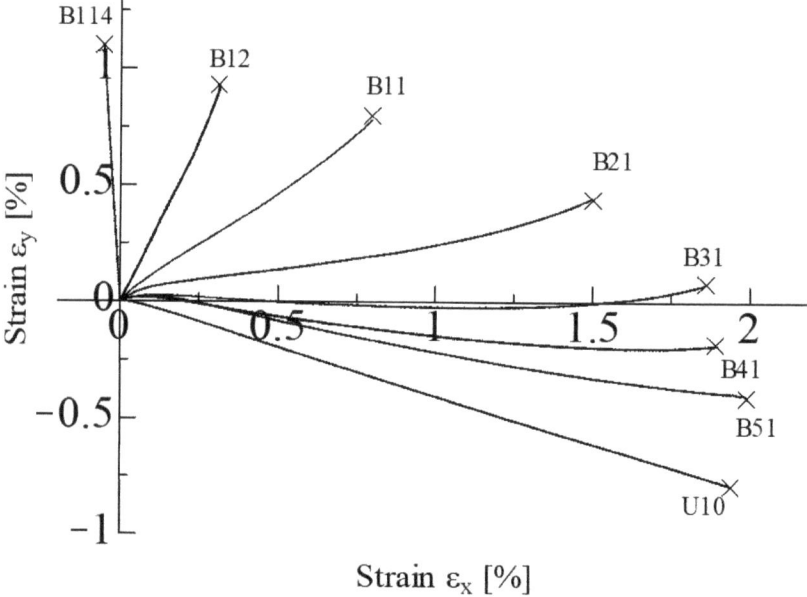

Figure 11. Typical $\varepsilon_x - \varepsilon_y$ diagrams under proportional biaxial loading. (The "×" in the figure indicates a fracture).

Typical $\varepsilon_x - \varepsilon_y$ diagrams under non-proportional T-T loading are shown in Figure 12a,b for the X1Y2 and X2Y1 conditions, respectively. The former condition caused the TC mode, while the latter resulted in the FB mode. In each figure, the initial deformation occurs in two directions due to the Poisson effect despite the fact that the load is first applied along a single axis. It is found from these figures and the observation results that the fracture mode is determined by the positive or negative value of ε_y, as in the proportional loading test.

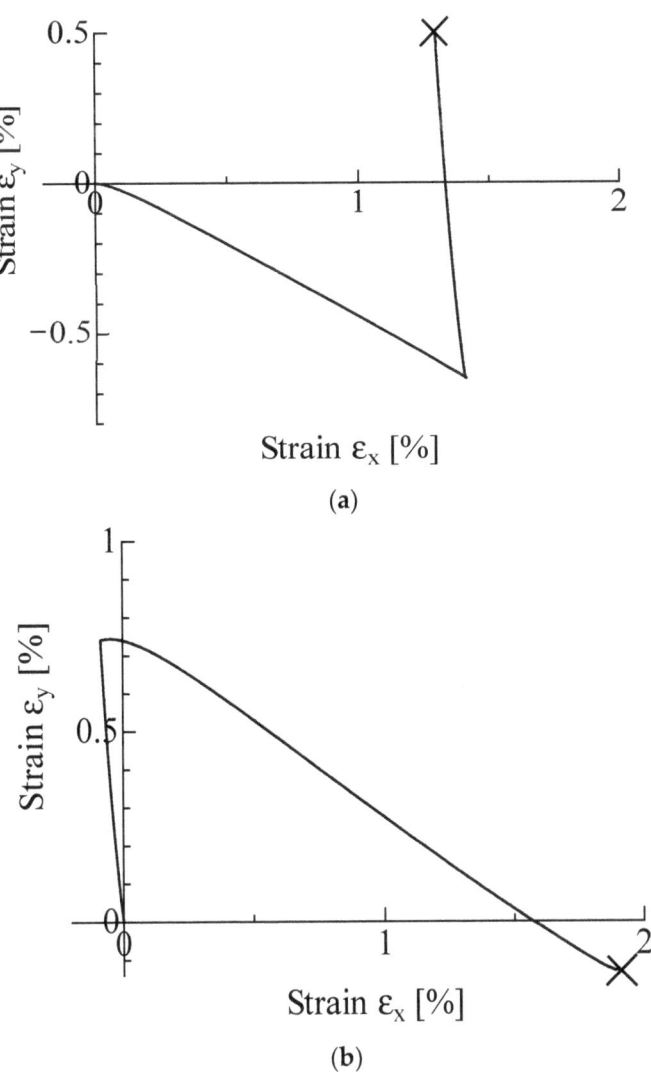

Figure 12. Typical $\varepsilon_x - \varepsilon_y$ diagrams under non-proportional biaxial loading. (**a**) X1Y2 condition; (**b**) X2Y1 condition. (The "×" in the figure indicates a fracture).

We next discuss the transition point from the TC&FB to FB modes through the fracture strain data. Figure 13 shows the fracture strains of the cruciform specimens under proportional T-T loading with filled symbols. It is confirmed that when the fracture strain ε_y is in the range 0.60 to 1.26%, the fracture mode is the TC mode. The strains ε_y causing the TC&FB mode are in the range of 0.08 to 0.50%. Although these strains are all positive, those of the TC&FB mode are given with smaller ε_y. This difference signifies that the maximum strain criterion [37] is not satisfied. On the other hand, when the strain ε_y is in the range of −0.19 to −0.79%, the fracture mode is FB. All specimens of U10 showed the strain ε_y as being less than −0.50%. The results of fracture strains derived from the non-proportional loading test are added to Figure 13 with open symbols. It can be seen that the strains are distributed in a similar range to the proportional loading case. It is also found that FB modes appear when ε_y is negative, and TC and TC&FB modes occur when ε_y is positive.

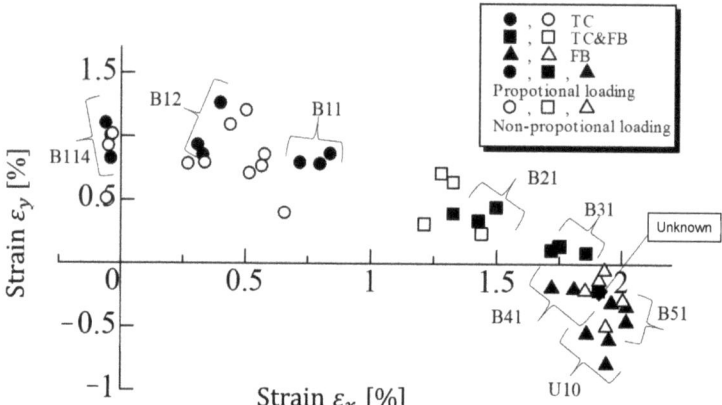

Figure 13. Fracture strains of cruciform specimens under proportional and non-proportional loading. Note: The number overlapping an open circle near the vertical axis is 0.5.

In general, the fiber breakage condition is significantly important, because it determines the load-bearing capacity of structural members. Although Hashin's or the maximum stress failure criterion has been used as the condition in many theoretical models and numerical simulations [7,11–16], the above results newly show that ε_y is the key point when the members are under biaxial tensile loading. In other words, even if the simulated stresses reach the fiber breakage condition, attention should be paid to the phenomenon that a positive ε_y does not generate isolated FB modes, but rather TC&FB first.

Thus, the fracture modes depend on the positive or negative value of ε_y, irrespective of the loading conditions. This leads to the conclusion that the transition of the fracture modes occurs when ε_y is 0%. This is achieved without any difference in ε_x, as seen in the comparison of B31 and B41. Unlike C-T and T-C loading, T-T loading is a combined loading condition that mutually constrains deformation. Therefore, when high tensile loads are applied in the x-axis as in the B41 and B51 conditions, compressive deformation due to the Poisson effect is dominant in the 90° direction rather than tension. The ε_y increases further in the negative direction as the fiber–axis tension continues; hence, the TC mode no longer occurs. The tensile load then continues to be applied, and eventually the FB mode occurs.

The evaluation of fracture stress in the gauge area was not carried out in this study, but this is an issue to be covered in the near future, along with the proposal of the fracture criterion of unidirectional CFRPs under T-T loading.

4. Conclusions

The failure behavior of unidirectional CFRPs subjected to biaxial tensile loading in the 0° and 90° directions has not been experimentally clarified prior to this study. Therefore, the fracture behavior of the specimens was investigated under various proportional and non-proportional T-T loading conditions using a new cruciform specimen for tension–tension loading (T-T loading) developed for a unidirectional CFRP. The results obtained are summarized as follows:

(1) Biaxial tensile fracture of the unidirectional CFRP is enabled by a cruciform specimen with a gauge area longer in the fiber direction than in the fiber vertical direction at the center. This is applicable regardless of proportional or non-proportional loading.

(2) There were three fracture modes in the specimens: a transverse crack (TC), fiber breakage (FB), and both modes (TC&FB) occurring simultaneously. The TC&FB mode is a unique mode that does not appear in the fracture modes under uniaxial tension (FB and TC), but only under biaxial tensile loading.

(3) Under several conditions of proportional loading, the failure loads in the 0° direction, indicating the FB and TC&FB modes, were almost identical. In another condition, the

fracture loads in the 90° direction, indicating the TC and TC&FB modes, were close to each other. From the aspect of fractured specimens, it was finally inferred that the TC occurred before FB, and the occurrence of the TC triggered FB instantaneously.

(4) The TC and TC&FB modes occurred when the strain in the 90° direction, ε_y, was positive. On the other hand, the FB mode occurred when ε_y was negative despite the fact that FB and TC&FB modes showed almost the same strains in the 0° direction. It was concluded that the occurrence of each fracture mode is characterized by only one parameter, namely ε_y.

Author Contributions: Conceptualization, K.G.; methodology, K.S., S.N. and K.G.; validation, K.S. and K.G.; resources, K.G.; data curation, K.S., S.N., M.H. and K.G.; writing—original draft preparation, K.S., M.H., A.M. and K.G.; writing—review and editing, K.S. and K.G.; visualization, K.S. and K.G.; supervision, K.G.; project administration, A.M. and K.G.; funding acquisition, A.M. and K.G. All authors have read and agreed to the published version of the manuscript.

Funding: The biaxial testing machine used in this research was funded by JSPS KAKENHI, grant number 22360051. (JSPS: Japan Society for the Promotion of Science, KAKENHI: Grants-in-Aid for Scientific Research).

Institutional Review Board Statement: Not applicable.

Informed Consent Statement: Not applicable.

Data Availability Statement: Data are contained within the article.

Acknowledgments: The Mitsubishi Chemical Group Corporation (MCG) provided the carbon fiber tows (TR 50S12L and 15L) for this research. The authors would like to express their gratitude to the MCG.

Conflicts of Interest: The authors declare no conflicts of interest.

References

1. Kaddour, A.S.; Hinton, M.J.; Soden, P.D. A comparison of the predictive capabilities of current failure theories for composite laminates: Additional contributions. *Compos. Sci. Technol.* **2004**, *64*, 449–476. [CrossRef]
2. Hinton, M.J.; Kaddour, A.S.; Soden, P.D. A further assessment of the predictive capabilities of current failure theories for composite laminates comparison with experimental evidence. *Compos. Sci. Technol.* **2004**, *64*, 549–588. [CrossRef]
3. Christensen, R.M. Failure criteria for fiber composite materials, the astonishing sixty year search, definitive usable results. *Compos. Sci. Technol.* **2019**, *182*, 107718. [CrossRef]
4. Daniel, I.M. Yield and failure criteria for composite materials under static and dynamic loading. *Prog. Aerosp. Sci.* **2016**, *81*, 18–25. [CrossRef]
5. Kaddour, A.S.; Hinton, M.J.; Smith, P.A.; Li, S. A comparison between the predictive capability of matrix cracking, damage and failure criteria for fibre reinforced composite laminates: Part A of the third world-wide failure exercise. *J. Compos. Mater.* **2013**, *47*, 2749–2779. [CrossRef]
6. Puck, A.; Schürmann, H. Failure analysis of FRP laminates by means of physically based phenomenological models. *Compos. Sci. Technol.* **2002**, *62*, 1633–1662. [CrossRef]
7. Pinho, S.T.; Iannucci, L.; Robinson, P. Physically based failure models and criteria for laminated fibre-reinforced composites with emphasis on fibre kinking. Part II: FE implementation. *Compos. Part A* **2006**, *37*, 766–777. [CrossRef]
8. Clyne, T.W.; Hull, D. Chapter 1: General Introduction. In *An Introduction to Composite Materials*, 3rd ed.; Cambridge University Press: Cambridge, UK, 2019; pp. 1–8.
9. Tsai, S.W.; Wu, E.M. A general theory of strength for anisotropic materials. *J. Compos. Mater.* **1971**, *5*, 58–80. [CrossRef]
10. Hashin, Z. Failure Criteria for unidirectional fiber composites. *J. Appl. Mech.* **1980**, *47*, 329–334. [CrossRef]
11. Christensen, R.M. Stress based yield/failure criteria for fiber composites. *Int. J. Solids Struct.* **1997**, *34*, 529–543. [CrossRef]
12. Pinho, S.T.; Iannucci, L.; Robinson, P. Physically-based failure models and criteria for laminated fibre-reinforced composites: Part I: Development. *Compos. Part A* **2006**, *37*, 63–73. [CrossRef]
13. Gu, J.; Chen, P. Some modifications of Hashin's failure criteria for unidirectional composite materials. *Compos. Struct.* **2017**, *182*, 143–152. [CrossRef]
14. Wan, L.; Ismail, Y.; Sheng, Y.; Ye, J.; Yang, D. A review on micromechanical modelling of progressive failure in unidirectional fibre-reinforced composites. *Compos. Part C Open Access* **2023**, *10*, 100348. [CrossRef]
15. Wan, L.; Ullah, Z.; Yang, D.; Falzon, B.G. Probability embedded failure prediction of unidirectional composites under biaxial loadings combining machine learning and micromechanical modelling. *Compos. Struct.* **2023**, *312*, 116837. [CrossRef]
16. Matzenmiller, A.; Lubliner, J.; Taylor, R.L. A constitutive model for anisotropic damage in fiber-composites. *Mech. Mater.* **1995**, *20*, 125–152. [CrossRef]

17. Maimí, P.; Camanho, P.P.; Mayugo, J.A.; Dávila, C.G. A continuum damage model for composite laminates Part I—Constitutive model. *Mech. Mater.* **2007**, *39*, 897–908. [CrossRef]
18. Liu, P.F.; Zheng, J.Y. Progressive failure analysis of carbon fiber epoxy composite laminates using continuum damage mechanics. *Mater. Sci. Eng. A* **2008**, *485*, 711–717. [CrossRef]
19. Shahabi, E.; Forouzan, M.R. A damage mechanics based failure criterion for fiber reinforced polymers. *Compos. Sci. Technol.* **2017**, *140*, 23–29. [CrossRef]
20. Ismail, Y.; Sheng, Y.; Yang, D.; Ye, J. Discrete element modelling of unidirectional fibre-reinforced polymers, under transverse tension. *Compos. Part B* **2015**, *73*, 118–125. [CrossRef]
21. González, C.; Llorca, J. Mechanical behavior of unidirectional fiber-reinforced polymers under transverse compression: Microscopic mechanisms and modeling. *Compos. Sci. Technol.* **2007**, *67*, 2795–2806. [CrossRef]
22. Romanowicz, M. A numerical approach for predicting the failure locus of fiber reinforced composites under combined transverse compression and axial tension. *Comput. Mater. Sci.* **2012**, *51*, 7–12. [CrossRef]
23. Totry, E.; Molina-Aldareguía, J.M.; González, C.; Llorca, J. Effect of fiber, matrix and interface properties on the in-plane shear deformation of carbon-fiber reinforced composites. *Compos. Sci. Technol.* **2010**, *70*, 970–980. [CrossRef]
24. Tory, E.; González, C.; Llorca, J. Failure locus of fiber-reinforced composites under transverse compression and out-of-plane shear. *Compos. Sci. Technol.* **2008**, *68*, 829–839. [CrossRef]
25. Soden, P.D.; Hinton, M.J.; Kaddour, A.S. Biaxial test results for strength and deformation of a range of E-glass and carbon fibre reinforced composite laminates: Failure exercise benchmark data. *Compos. Sci. Technol.* **2002**, *62*, 1489–1514. [CrossRef]
26. Youssef, Y.; Labonte, S.; Roy, C.; Lefebvre, D. An Effective Flat Cruciform-Shaped Specimen for Biaxial Testing of CFRP laminates. *Sci. Eng. Compos. Mater.* **1994**, *3*, 259–267. [CrossRef]
27. Kumazawa, H.; Hayashi, H.; Susuki, I.; Utsunomiya, T. Damage and permeability evolution in CFRP cross-ply laminates. *Compos. Struct.* **2006**, *76*, 73–81. [CrossRef]
28. Gower, M.R.L.; Shaw, R.M. Towards a Planar Cruciform Specimen for Biaxial Characterisation of Polymer Matrix Composites. *Appl. Mech. Mater.* **2010**, *24–25*, 115–120.
29. Gutiérrez, J.C.; Lozano, A.; Manzano, A.; Flores, M.S. Numerical and Experimental Analysis for Shape Improvement of a Cruciform Composite Laminates Specimen. *Fibres Text. East. Eur.* **2016**, *24*, 89–94. [CrossRef]
30. Zhang, X.; Zhu, H.; Lv, Z.; Zhao, X.; Wang, J.; Wang, Q. Investigation of Biaxial Properties of CFRP with the Novel-Designed Cruciform Specimens. *Materials* **2022**, *15*, 7034. [CrossRef]
31. Correa, E.; Barroso, A.; Pérez, M.D.; París, F. Design for a cruciform coupon used for tensile biaxial transverse tests on composite materials. *Compos. Sci. Technol.* **2017**, *145*, 138–148. [CrossRef]
32. Goto, K.; Arai, M.; Nishimura, M.; Dohi, K. Strength evaluation of unidirectional carbon fiber reinforced plastic laminates based on tension-compression biaxial stress tests. *J. Jpn. Soc. Compos. Mater.* **2017**, *43*, 48–57. [CrossRef]
33. Rev, T.; Wisnom, M.R.; Xu, X.; Czél, G. The effect of transverse compressive stresses on tensile failure of carbon fibre/epoxy composites. *Compos. Part A* **2022**, *156*, 106894. [CrossRef]
34. Potter, D.; Gupta, V.; Chen, X.; Tian, J. Mechanisms-based failure laws for AS4/3502 graphite/epoxy laminates under in-plane biaxial compression. *Compos. Sci. Technol.* **2005**, *65*, 2105–2117. [CrossRef]
35. Kang, H.; Liang, J.; Li, Y.; Cui, H.; Li, Y. Dynamic biaxial compression of CFRP laminates using electromagnetic loading. *Acta Mech. Solida Sin.* **2022**, *35*, 891–900. [CrossRef]
36. Nakasaki, S.; Nakamura, S.; Kataoka, Y.; Macadre, A.; Goda, K. Fracture Characteristics of Unidirectional CFRP Composites under Biaxial Tensile Load. *J. Jpn. Soc. Compos. Mater.* **2022**, *48*, 77–85. [CrossRef]
37. Hart-Smith, L.J. Predictions of the original and truncated maximum-strain failure models for certain fibrous composite laminates. *Compos. Sci. Technol.* **1998**, *58*, 1151–1178. [CrossRef]
38. MITSUBISHI CHEMICAL Corp. Available online: https://www.m-chemical.co.jp/carbon-fiber/en/product/tow/ (accessed on 10 February 2024).
39. Fineberg, J.; Marder, M. Instability in dynamic fractur. *Phys. Rep.* **1999**, *313*, 1–108. [CrossRef]
40. Lee, J.; Hong, J.W. Dynamic crack branching and curving in brittle polymers. *Int. J. Solids Struct.* **2016**, *100–101*, 332–340. [CrossRef]
41. Hull, D. Chapter 5: Elastic properties. In *An Introduction to Composite Materials*, 1st ed.; Cambridge University Press: Cambridge, UK, 1981; pp. 81–101.
42. Clyne, T.W.; Hull, D. Chapter 3: Elastic Deformation of Long Fiber Composites. In *An Introduction to Composite Materials*, 3rd ed.; Cambridge University Press: Cambridge, UK, 2019; pp. 31–42.

Disclaimer/Publisher's Note: The statements, opinions and data contained in all publications are solely those of the individual author(s) and contributor(s) and not of MDPI and/or the editor(s). MDPI and/or the editor(s) disclaim responsibility for any injury to people or property resulting from any ideas, methods, instructions or products referred to in the content.

Graphene/Heterojunction Composite Prepared by Carbon Thermal Reduction as a Sulfur Host for Lithium-Sulfur Batteries

Jiahao Li, Bo Gao *, Zeyuan Shi, Jiayang Chen, Haiyang Fu and Zhuang Liu

Key Laboratory for Ecological Metallurgy of Multimetallic Mineral, Ministry of Education, Northeastern University, Shenyang 110819, China; surflijh@163.com (J.L.); surfshizy@163.com (Z.S.); neuchenjy@163.com (J.C.); surffuhy@163.com (H.F.); surfliuz@163.com (Z.L.)
* Correspondence: gaob@smm.neu.edu.cn

Abstract: An interlayer nanocomposite (CC@rGO) consisting of a graphene heterojunction with CoO and Co_9S_8 was prepared using a simple and low-cost hydrothermal calcination method, which was tested as a cathode sulfur carrier for lithium-sulfur batteries. The CC@rGO composite comprises a spherical heterostructure uniformly distributed between graphene sheet layers, preventing stacking the graphene sheet layer. After the introduction of cobalt heterojunction on a graphene substrate, the Co element content increases the reactive sites of the composite and improves its electrochemical properties to some extent. The composite exhibited good cycling performance with an initial discharge capacity of 847.51 mAh/g at 0.5 C and a capacity decay rate of 0.0448% after 500 cycles, which also kept 452.91 mAh/g at 1 C and in the rate test from 3 C back to 0.1 C maintained 993.27 mAh/g. This article provides insight into the design of cathode materials for lithium-sulfur batteries.

Keywords: graphene; nanocomposite; heterojunction; lithium-sulfur battery

1. Introduction

Lithium-ion batteries have been studied extremely extensively over the past few decades and are used in portable and mobile electronic devices [1,2]. However, the theoretical energy density of 300 Wh/kg cannot meet the requirements of the growing new energy storage field. Therefore, lithium-sulfur batteries are considered one of the most competitive energy storage devices for the next generation due to their high theoretical specific capacity (1675 mAh/g), energy density (2600 Wh/kg), and the significant advantages of their active substance sulfur such as non-toxicity, low cost and wide source [3]. Despite the above advantages, lithium-sulfur batteries are still hampered by the intrinsic disadvantages of sulfur in the application process. The sulfur's extremely poor intrinsic conductivity, a large volume change rate during charging and discharging, and the "shuttle effect" caused by soluble polysulfides all cause unstable electrode structure, short cycle life, and low Coulomb efficiency [4–6].

Various strategies have been proposed to overcome these challenges, including well-designed cathodes, modified separators, and new-developed electrolytes [7]. In the cathode material design, the use of conductive carbon materials, the design of nanostructured sulfur cathodes, and the incorporation of polar metal materials are the most common and widely used methods. Conductive carbon materials, such as carbon nanotubes [8–12] and graphene [12–16], can encapsulate polysulfides in the internal pores by physical adsorption [17]. However, the interaction between non-polar carbon materials and polar polysulfides is weak and cannot inhibit the diffusion of polysulfides in the long cycle process [18,19]. Although heteroatom doping can effectively polarize the surface of carbon materials [20–22], the concentration of existing doping methods is too low to play a role in sulfur fixation. As for polar materials such as metal oxides, they can effectively trap and transform polysulfides due to the strong bonding between them and sulfur, but there are disadvantages such as their poor electrical conductivity and too-strong bonding with polysulfides [19,23,24]. In contrast, polar materials such as metals, metal sulfides, and metal phosphides have high conductivity, however, they have poor adsorption capacity for polysulfides compared to metal oxides [25–27], which limits the conversion of polysulfides for multiple uses. Therefore combining metal oxides and metal sulfides and forming heterojunctions is a promising strategy [28]. Different strategies for synthesizing heterojunctions have been continuously proposed, including SnO_2-SnS_2 [29], Mn_3O_4-MnS [30], and Co_3O_4-CoP [31]. For instance, Wang et al. synthesized SnO_2-SnS_2 nanosheet heterojunctions to determine for the first time the interfacial effect in lithium-sulfur batteries, which improved the diffusion efficiency of ions and significantly accelerated the redox reaction [29]. Qin et al. designed metal–metal three-layer hollow spheres as sulfur carriers and utilized the separated spatial constraints of the hollow multishell structure to fully utilize the active sites and the built-in electric field, and the assembled cells had remarkable cycling performance and rate performance [30]. Zhang et al. synthesized CC on carbon nanotubes, combining the strong adsorption of oxides with the conversion of phosphides to prepare a composite material with superior electrochemical properties [31]. Most of these studies focus on the structural design and material selection of heterojunctions, while simple and low-cost preparation of heterojunction–carbon materials has been rarely investigated. Therefore, it is instructive to develop heterojunction-carbon materials with reasonable structures and study them for the development of cathode materials for lithium-sulfur batteries.

This study presents a hydrothermal–calcination method for synthesizing rGO-CoO/Co_9S_8 heterojunction composite nanomaterials (CC@rGO). $CoSO_4$, graphene, and chitosan were used as the Co source, carbon source, and supplementary carbon source, respectively. The Co^{2+} ions were assembled with negatively charged functional groups on graphene oxide through electrostatic adsorption, followed by a hydrothermal-high temperature reduction to generate CoO/Co_9S_8 heterojunctions uniformly distributed between graphene lamellae. The CC@rGO electrode exhibited excellent cycling stability and rate performance, thanks to graphene's large surface area and numerous ion-electron transport channels, the built-in accelerating electric field of CoO/Co_9S_8, and the ability to trap and transform polysulfides. This study provides valuable insights into the potential application of carbon material–metal compound composites in lithium-sulfur battery cathode materials, which may lead to the development of more efficient and sustainable energy storage systems.

2. Experimental

2.1. Preparation of GO

Graphene oxide was prepared by modified Hummer's method. Two grams g scaled graphite powder were added to 250 mL of a mixed acid solution of sulfuric acid-nitric acid (volume 9:2). A total of 12 g of potassium permanganate was added to it and the temperature was increased to 50 °C (40 min), 60 °C (7 h), 90 °C (30 min), and finally, 30 mL of H_2O_2 was added to it to obtain a bright yellow graphene oxide solution. After being cooled, the graphene oxide was left to separate, centrifuged to a pH of 7, and finally freeze-dried, as previously reported by our subject group [32,33].

2.2. Preparation of Co_9S_8/CoO/rGO

A suspension of graphene oxide was prepared (1.5 mg/mL, solution A), then 2 g of cobalt sulfate and 0.3 g of chitosan were added to 50 mL of deionized water (solution B), and A and B were homogeneously mixed, and transferred to a PTFE-lined reactor, hydrothermally heated at 180 °C for 12 h. The precursor powder was obtained by freeze-drying. The black powder (Co_9S_8/CoO/rGO, abbreviated as CC@rGO) was obtained by calcination at 500 °C for 2 h in the Ar atmosphere. rGO samples were made by the same procedure without the addition of cobalt sulfate and chitosan.

2.3. Preparation of Battery Cathode

The monomeric sulfur was mixed with the sample at a mass ratio of 7:3 and then molten at 155 °C for 12 h. The active material, conductive agent (Super-P), and binder

(PVDF) were ground at 7:2:1 and added to a certain amount of NMP, and stirred for 12 h to make a slurry, coated on carbon-coated aluminum foil, dried at 60 °C for 12 h, and then stamped into a 12 mm × 12 mm circular positive electrode by a press.

2.4. Material Characterizations

The microscopic morphology of the samples was observed by scanning electron microscopy (SEM, TESCAN MIRA LMS, Brno, Czech Republic) as well as elemental analysis (EDS, TESCAN MIRA LMS, Czech Republic). X-ray diffraction analysis (XRD, ultima IV, Rigaku, Japan) was used to characterize the sample lattice structure in the interval 10°–90° at 5° per minute. Transmission electron microscopy images (TEM, FEI Tecnai G2F 20, Hillsboro, OR, USA) were used to obtain the internal microstructure. X-ray photoelectron spectroscopy images were acquired by K-Alpha (XPS, Thermo Scientific, Waltham, MA, USA). To obtain Raman spectra, a Renishaw micro confocal laser Raman spectrometer (633 nm) by HR800 (Raman, HORIBA JobinYvon, Palaiseau, France) was used. Thermal gravimetric analysis (TG, TA TGA 550, New Castle, DE, USA) was used to analyze sample sulfur loading. The specific surface area and pore size distribution before and after modification were analyzed using N_2 adsorption and desorption experiments (BET, Micromeritics 3FLEX, Norcross, GA, USA).

2.5. Electrochemical Measurements

A coin cell (type CR2032) was used for electrochemical testing of the rGO as well as the CC@rGO. A 12 mm × 12 mm circular positive electrode sheet was assembled into a cell in a glove box, and the electrolyte used consisted of 1.0 M LiTFSI -DOL: DME with a 1:1 volume ratio and 2% $LiNO_3$ (dodochemicals.com, accessed on 11 November 2022). Lithium sheets (Φ15.6 mm, 0.45 mm, China Energy Lithium Co., Tianjin, China) were used for the negative electrode of the half-cells. Coin cell constant current charge/discharge test was performed in the voltage window range of 1.7–2.8 V. Electrochemical AC impedance (EIS) testing, as well as cyclic voltammetry (CV) testing was performed at the Princeton Electrochemical Workstation (VersaSTAT3, Oak Ridge, TN, USA), with EIS frequencies ranging from 1×10^{-2} Hz to 10^6 Hz. Li_2S_6 adsorption experiments were used to verify the adsorption capacity of the material (0.05 M Li_2S_6 dissolved in 1:1 DME: DOL).

3. Results and Discussion

As shown in Figure 1, Co^{2+} combines with the negatively charged oxygen-containing functional group on GO under the effect of electrostatic adsorption, and after the removal of the oxygen-containing functional group by hydrothermal heat, Co^{2+} fills the oxygen vacancies generated after the removal of the oxygen-containing functional group and achieves the uniform distribution of Co^{2+} among the graphene sheets. After calcination at 500 °C, $CoSO_4$ was calcined and reduced to CoO and Co_9S_8 heterogeneous spheres, which were uniformly distributed between the graphene lamellae, forming a typical wrapping structure. Graphene and heterojunction (composed of cobalt oxide and nine cobalt octa sulfide) composites can not only physically limit the dissolution of polysulfides into the electrolyte through the porous structure of graphene, but also trap and transform polysulfides through the heterojunction, inhibiting the shuttle effect that occurs when polysulfides shuttle through the diaphragm to the surface of the lithium sheet and react to cause a decrease in capacity and lifetime.

To prepare CoO/Co_9S_8@rGO composites, $CoSO_4$/C@GO precursors were prepared by hydrothermal mixing and calcined to generate CoO/Co_9S_8@rGO composites (labeled as CC@rGO).

The compositional analysis of rGO and CC@rGO was carried out using XRD, as shown in Figure 2a. The diffraction peaks of CC@rGO at 36.49°, 42.38°, 61.49°, 73.67°, and 77.53° mainly correspond to (111), (200), (220), (311), (222) crystallographic planes, respectively, with the standard card PDF#48-1719 matches (CoO), while the diffraction peaks at 29.38°, 31.29°, 52.09° correspond to (311), (200), (440) of Co_9S_8, respectively, proving the generation of CoO and Co_9S_8. rGO exhibits a distinct (002) interface at 26°, which has the amorphous broad peak that is amorphous carbon at 26°. The composites, on the other hand, did not show a clear broad peak at 26°, which indicates a clear amorphization trend of graphene during the carbon thermal reduction process and the disappearance of the broad peak [34].

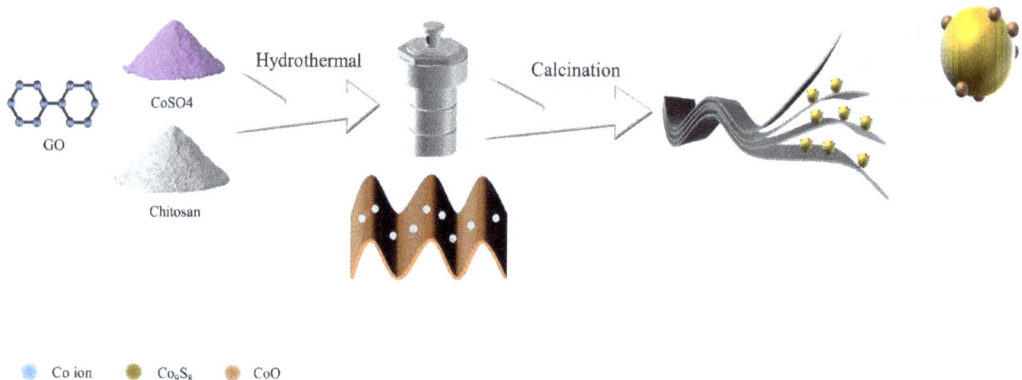

Figure 1. Flow chart of CC@rGO synthesis.

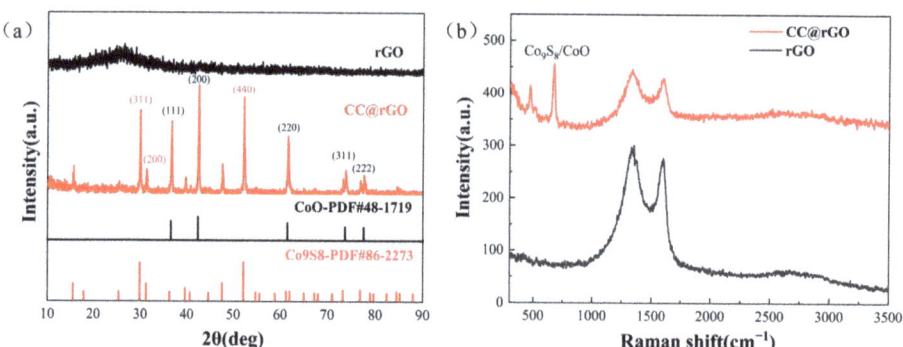

Figure 2. XRD (**a**) and Raman plots (**b**) of rGO and CC@rGO.

Raman spectroscopy was performed for CC@rGO and rGO to measure the disorder of different samples, and the results are shown in Figure 2b. The two broad bands of CC@rGO appear at 1359 cm^{-1} and 1600 cm^{-1}, corresponding to the D and G peaks [12,35], respectively, while the D and G peaks of CC@rGO are shifted to the right at 1335 cm^{-1} and 1590 cm^{-1}, respectively, which was believed to be due to the addition of heterojunctions that make the material structure vibrate and shift. The double peaks of CC@rGO at 466 cm^{-1} and 676 cm^{-1} were considered to be strong interactions between CoO and Co9S8 [36,37]. Compared to rGO, the I_D/I_G of CC@rGO is reduced from 1.1 to 1.035 and the defects of the material are reduced [38]. The addition of Co oxide particles, encapsulated by graphene, decreases the D peak, which is caused by the lamellar encapsulation property of graphene on transition metals, and the D peak includes oxygen-containing functional groups and defects in the material itself [39]. Meanwhile, the 2D peaks of CC@rGO did not change significantly compared to the 2D peaks of rGO, indicating that the incorporation of heterojunctions did not lead to the stacking of graphene lamellae.

The microstructure and elemental distribution of rGO and CC@rGO were observed by scanning electron microscopy and elemental energy spectroscopy, respectively. As shown in Figure 3e,f, rGO shows a distinct muslin lamellar shape after hydrothermal calcination, which is consistent with the morphology of reduced graphene oxide prepared in the literature, while irregular heterogeneous ellipses or squares formed by CoO and Co_9S_8 have uniformly distributed between graphene lamellar layers as seen in Figure 3a–d. This helps to suppress the stacking of graphene lamellar layers due to van der Waals forces, increase ion and electron transport channels, and enhance the sulfur-carrying capacity. Meanwhile, the large specific surface area of rGO provides abundant space for sulfur loading, which can give full play to the adsorption and conversion ability of polysulfides by heterojunctions and effectively improve the electrical conductivity of the composites, and form sulfur cathode composites with stable structures. An elemental analysis of CC@rGO is shown in Figure 3g, which further demonstrates the uniform distribution of Co, S, and O in the sample. The homogeneous distribution of Co elements also demonstrates the binding and homogeneous distribution of Co^{2+} with oxygen-containing functional groups on graphene during the hydrothermal process.

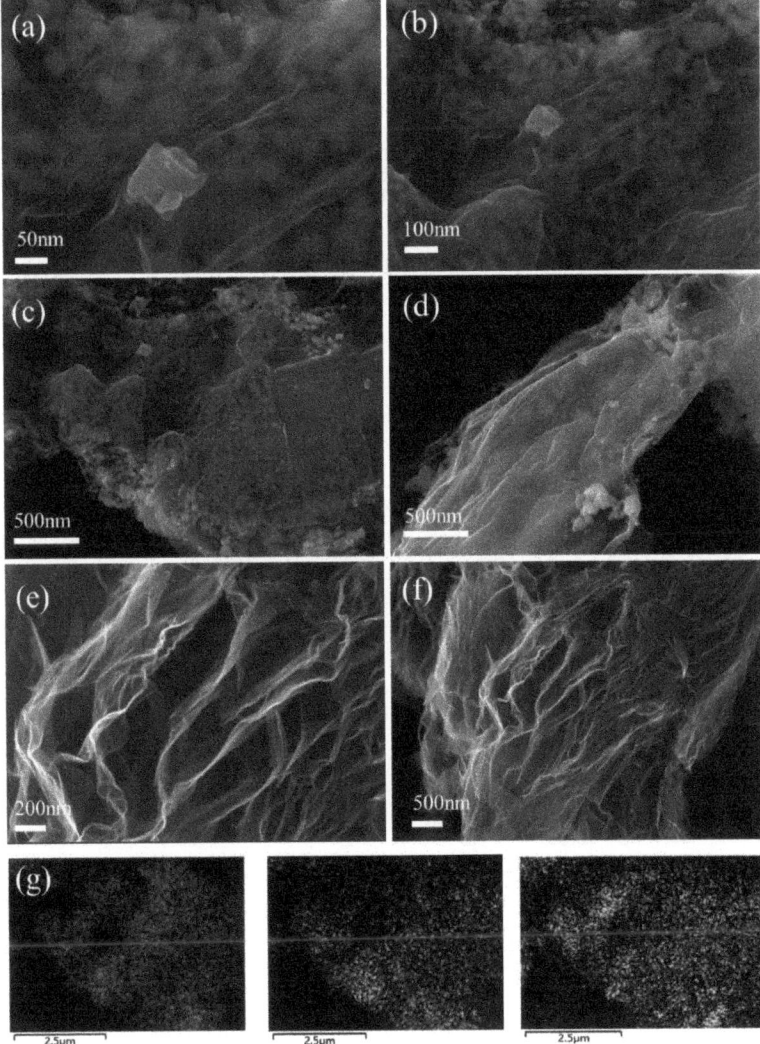

Figure 3. SEM images of CC@rGO (**a–d**) and rGO (**e,f**) and elemental energy spectrum of Co, O, S (**g**).

As shown in Figure 4a,b, the heterojunctions in CC@rGO show irregular ellipses or squares. The 0.213 nm lattice stripes correspond to the (200) crystal plane of CoO, while 0.124 nm and 0.175 nm lattice stripes correspond to the (800) and (440) crystal planes of Co_9S_8, respectively (Figure 4c–e). This is consistent with the XRD results, which demonstrate the generation of CoO and Co_9S_8 and the presence of heterogeneous interfaces. Due to the interaction between CoO and Co_9S_8, abundant inhomogeneous interfaces are formed between their particles, and the built-in electric field induced through the heterogeneous interfaces helps ions migrate in the lithium-sulfur cell, while graphene as the substrate provides abundant transport channels and provide support for good electrochemical performance.

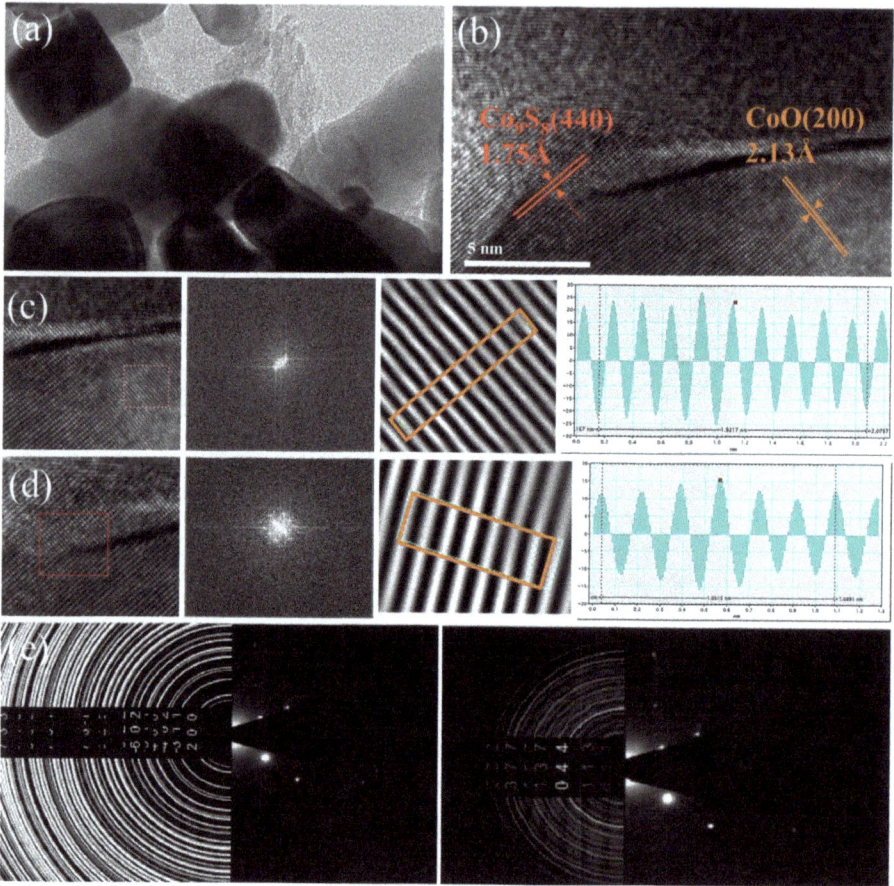

Figure 4. TEM images of CC@rGO (**a**), HRTEM images of CC@rGO (**b**), lattice streak analysis, and SAED diagram of CC@rGO (**c–e**).

The chemical bonding of the composites was analyzed by XPS. As shown in Figure 5c, an XPS analysis of CC@rGO samples existed for C, O, S, and Co. Figure 5d shows the high-resolution peak fitting curves for C1s, having 284.2 eV, 285.58 eV, and 287.83 eV corresponding to the C=C bond, C-O-C bond, and O-C=O bond, respectively [40,41]. Referring to the high-resolution XPS spectra of Co2p shown, the two peaks at 779 and 795 eV correspond to $2p_{3/2}$ and $2p_{1/2}$ of Co^{3+}, the peaks at 781.2 and 797 eV correspond to $2p_{3/2}$ and $2p_{1/2}$ of Co^{2+}, while the two peaks at 785.8 and 802 eV correspond to the satellite peaks of Co [35,42].

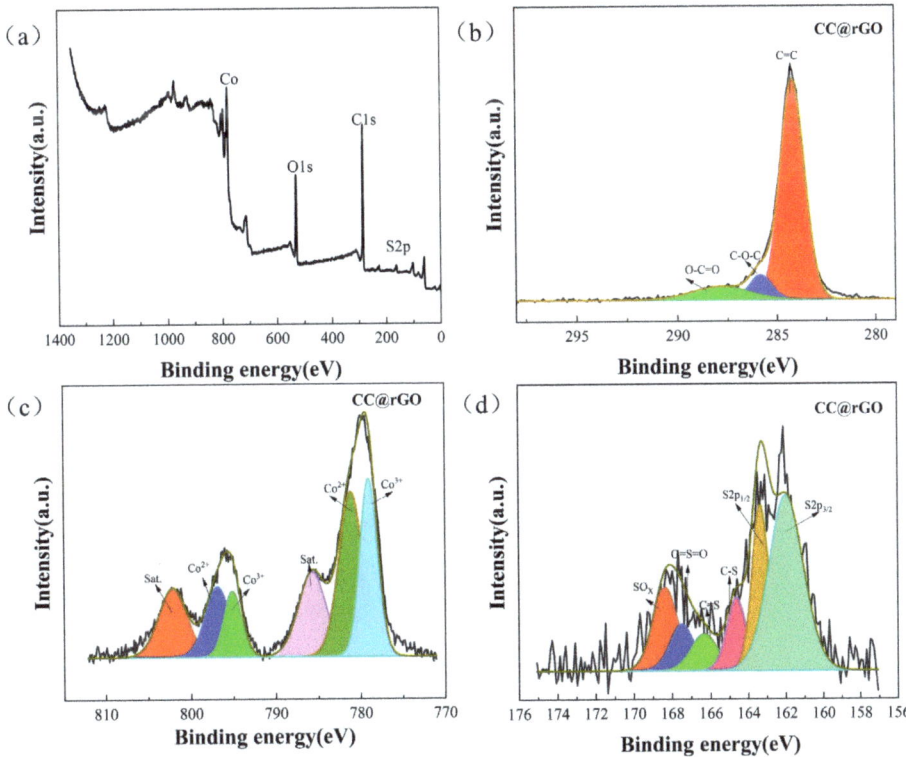

Figure 5. (**a**) Elemental spectrum, (**b**) C spectra, (**c**) Co spectra, (**d**) S spectra of CC@rGO.

There are also two sets of satellite peaks (802.08 eV, 785.83 eV) in the S2p high-resolution spectrum, with peaks at S2p$_{3/2}$ (161.8 eV) and S2p$_{1/2}$ (162.8 eV) attributed to metal-S bonds [35,43]. The two peaks at 163.7 and 165.3 eV were attributed to C-S-C bonding and C-S bonding, respectively. The remaining peaks at 168.8 and 170 eV represent the C-SO$_x$-C species [16,41,44]. The multivalent form of Co aids in accelerating the LiPSs power conversion process during battery discharge.

As shown in Figure 6, both rGO and CC@rGO exhibit type IV desorption curves [45,46], indicating that both are mesoporous materials and in sheet form. Compared with rGO, the CC@rGO specific surface area decreased from 49.73 m^2/g to 46.112 m^2/g after compounding the heterojunctions due to the incorporation of heterogeneous spheres with a smaller specific surface area. The pore size (calculation: 4 V/A by BET), on the other hand, decreased from 8.16 nm to 4.46 nm. The smaller pore size will melt the sulfur-loaded process to encapsulate the sulfur well in the pore channel and maintain sufficient stability. At the same time, during the melting process, due to the good wettability between carbon and sulfur, sulfur will penetrate inward through the capillary of carbon, and when the pore size becomes larger, the capillary force is not enough to let sulfur enter completely. In this way, the small pore size will achieve a better encapsulation of sulfur as well as a better inhibition of polysulfide diffusion. In addition, the small pore size will make the sulfur layer on the material surface thinner, improve the sulfur utilization, and accelerate the electron and ion transfer efficiency; meanwhile, increasing the contact area of the electrolyte will promote the electrolyte to infiltrate the material, reduce the interfacial impedance, and finally reduce the electrochemical polarization [47–49].

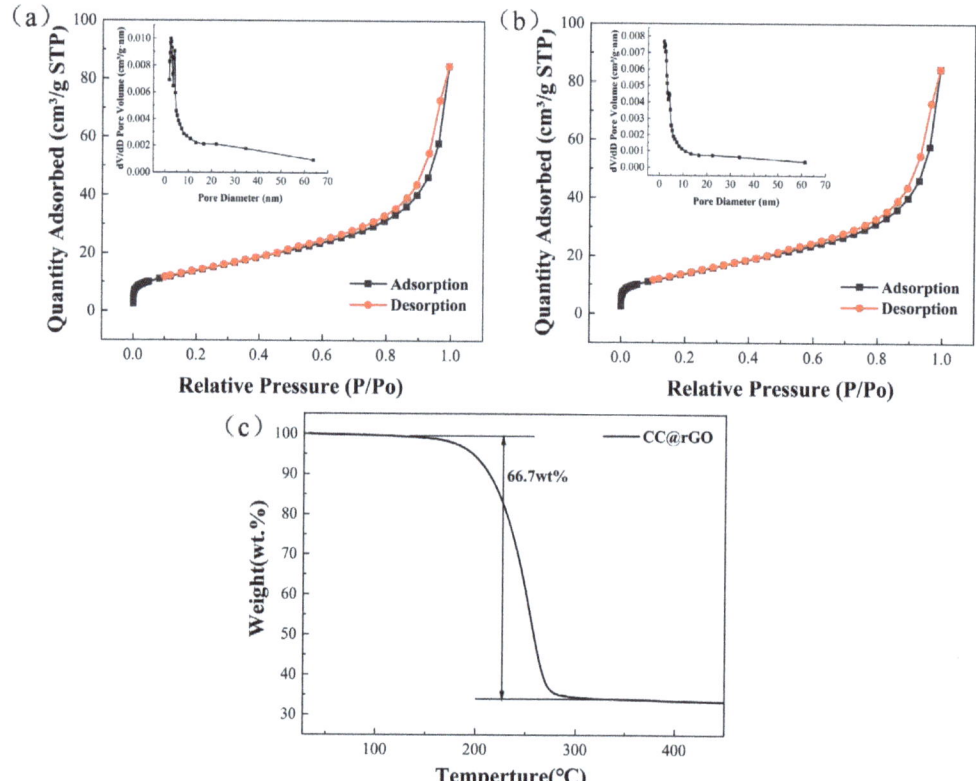

Figure 6. Desorption curve and pore size distribution of (**a**) rGO (**b**) CC@rGO, (**c**) TG of CC@rGO.

The composites' large surface area and thin mesoporous structure will make it easier for the cell's electrolyte to penetrate the material, which will lower interface impedance, increase charge exchange efficiency, hasten the conversion of sulfur intermediate species, and enhance the cell's electrochemical performance. As shown in Figure 6c, the thermogravimetric curves of CC@rGO in argon atmosphere, the sample starts to lose weight around 160 °C and stops losing weight around 280 °C. The results indicate a sulfur content of 66.7%, which is in general agreement with the sulfur loading of the experimental part.

To characterize the electrochemical performance of the composites with wrapping structure, a series of electrochemical characterizations were performed for the cells assembled with rGO and CC@rGO, respectively. To characterize the electrochemical properties of the composites, CC@rGO and rGO were assembled into CR2032 button cells, respectively. As shown in Figure 7a, compared with rGO with 437.81 mAh/g in the first cycle, CC@rGO has a high discharge capacity of 847.51 mAh/g in the first cycle, and the decay rate is only 0.0448% per cycle after 500 cycles, which is much lower than that of rGO; this indicates that CC@rGO has effectively improved the conductivity and structural stability of the material after the introduction of CoO/Co_9S_8 heterostructure. It can effectively activate more sulfur-active material while maintaining good cycling stability and promoting the kinetic conversion of polysulfides. The cycling test of CC@rGO using 1C cycling current shows that the first turn discharge capacity is 452.91 mAh/g and the decay rate is 0.0639% after 500 cycles, which further demonstrates the good structural stability and electrical conductivity of the composite material.

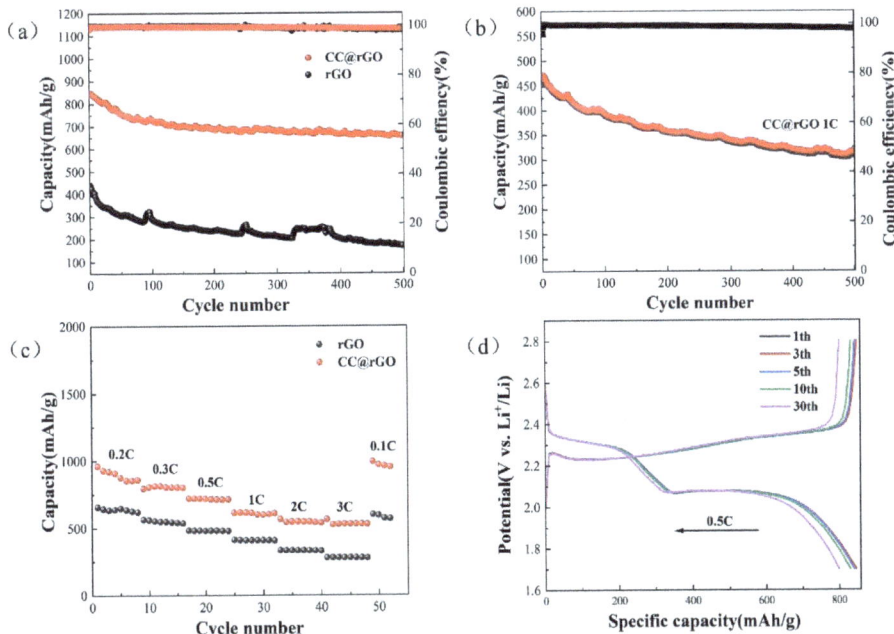

Figure 7. (**a**) 0.5 C cycle plots of CC@rGO and rGO, (**b**) 1 C cycle plots of CC@rGO, (**c**) rate performance, (**d**) charge/discharge curves of CC@rGO.

The rate electrochemical characterization of CC@rGO was performed and the results are shown in Figure 7c. In the button cell, the discharge capacities of rGO and CC@rGO in 0.2 C, 0.3 C, 0.5 C, 1 C, 2 C, 3 C, and 0.1 C were 657.79 mAh/g, 564.07 mAh/g, 483.37 mAh/g, 413.28 mAh/g, 334.37 mAh/g, 277.64 mAh/g, 600.13 mAh/g and 962.96 mAh/g, 807.01 mAh/g, 722.08 mAh/g, 614.81 mAh/g, 568.04 mAh/g, 529.53 mAh/g, and 993.27 mAh/g, respectively. The rate performance proves that CC@rGO has good multiplicative properties, which further supports the structural stability and cycling performance of the composite under high currents. The cyclic charge–discharge plateau curves in Figure 7d shows that the lithium-sulfur battery prepared by CC@rGO exhibits a typical dual charge–discharge voltage plateau (phase I, 2.4–2.1 V; phase II, 2.1–1.7 V) [13]. After 30 cycles, the typical charge–discharge plateau can still be maintained and the △E is unchanged, which indicates that CC@rGO as a sulfur carrier has good electrode structural stability and electrode kinetic performance, corroborating that the composite material can achieve excellent long-cycle performance. In terms of electrochemical performance, the introduction of polar CoO/Co_9S_8 heterojunctions between nonpolar graphene sheets provides abundant polysulfide action sites, effectively suppressing the loss of active material sulfur caused by the "shuttle effect" and reducing the appearance of dead sulfur during the cycling of lithium-sulfur batteries.

The electrochemical performance of rGO and CC@rGO cells was compared at a scan rate of 0.1 mv/s and a voltage window of 1.7–2.8 V. The two typical reduction peaks (2.25 V, 2.01 V) characterized by CC@rGO correspond to the reduction of S to soluble polysulfide (Li_2S_x, $4 \leq x \leq 8$) and the reduction of soluble polysulfide to insoluble polysulfide (Li_2S_2, Li_2S) during discharge, respectively [50]. Compared with rGO, the CC@rGO has a smaller electrochemical polarization with a higher peak after the introduction of the heterostructure, implying that the composite has better kinetic catalytic performance as well as charge and discharge capacity [51]. This indicates that the introduction of the heterojunctions results in a stronger catalytic effect and faster reaction kinetics of the cell [52]. In Figure 8b, the electrochemical performance of the CC@rGO cell is investigated at different sweep rates, and the cell polarization increases with increasing sweep rate, but the curve shape does not change much and the multi-turn curves at uniform sweep rates are in good agreement, which proves that CC@rGO has good structural stability and is consistent with the results of charge/discharge plateau curve and rate curve.

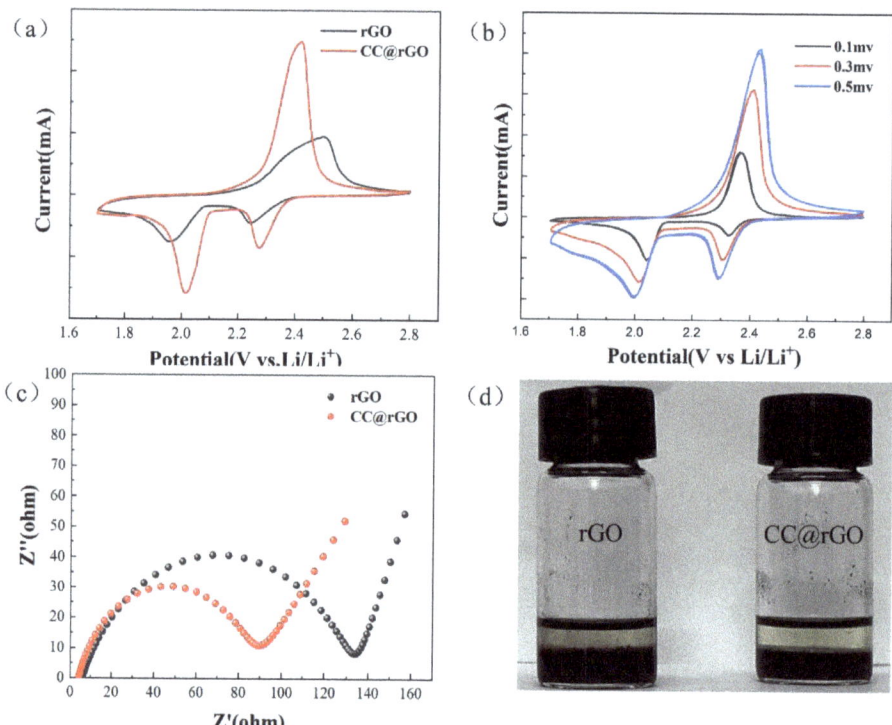

Figure 8. (a) CV plots of CC@rGO and rGO at 0.1 mv/s sweep speed, (b) CV plots of CC@rGO at different sweep speeds, (c) EIS plots of CC@rGO and rGO, (d) Li_2S_6 adsorption test chart.

The electrical conductivity of the cells assembled from the composites was further tested, and the Nyquist curves were shown in Figure 8c. Compared with rGO, the R_{ct} (charge transfer resistance) of CC@rGO is reduced from 133 Ω to 88 Ω. The improved tilt of the Warburg curve in the low-frequency region is attributed to the increased conductivity of the composite material after the addition of the heterostructure, which can better perform the adsorption-transformation of polysulfides [7]; meanwhile, the electrostatically adsorbed heterojunction effectively prevents the stacking of graphene and increases the reactive sites as well as ion transport channels, which enhances the ion transfer diffusion in the cell.

As shown in Figure 8d, Li_2S_6 adsorption experiments were carried out to further verify the adsorption capacity of the composites. After 12 h adsorption in 0.05 M Li_2S_6 solution, the supernatant of CC@rGO on the right side was clarified, while the rGO on the left side still showed a pale yellow color, indicating that CC@rGO has excellent static adsorption ability, which is consistent with the previous characterization results. This indicates that the composites have a better ability to trap polysulfides, which is attributed to the excellent adsorption of polysulfides by metal oxides and the porous structure of graphene, both of which synergistically enhance the performance of the composites.

In Table 1, the work is compared with the previous graphene-based composite lithium-sulfur battery cathode, and although the initial discharge capacity is slightly lower at 847.51 mAh/g, the material shows excellent capacity retention over long cycles, maintaining a high capacity retention and low decay rate (0.0448%) over 500 cycles, demonstrating that the material has good structural stability and can achieve longer cycle life and better Coulomb efficiency.

Table 1. Comparison of the results of this work with previous studies.

Sample	Electrochemical Performance	Decay Rate	
ZIF-8@rGO/S	1544 mAh/g at 0.2 C	0.33% after 200 cycles	[53]
Co_9S_8@S/GO	1057 mAh/g at 0.1 C	0.033% after 100 cycles	[54]
TiO_2 NTs/GO hybrid	850.7 mAh/g at 0.1 C (after 100 cycles)	0.409% after 100 cycles	[55]
$Ni_3(HITP)_2$@GO/S	959.3 mAh/g at 0.5 C	0.1325% after 400 cycles	[56]
NCF-G@S	923.8 mAh/g at 0.5 C	0.212% after 150 cycles	[21]
CC@rGO	847.51 mAh/g at 0.5 C	0.0448% after 500 cycles	This work

4. Conclusions

In summary, we propose the preparation of CoO/Co_9S_8 heterojunction-rGO composites by electrostatic adsorption and calcination. CoO/Co_9S_8 heterojunction acts as a polysulfide adsorption transformation, while rGO acts as a sulfur-carrying carbon carrier and provides ion and electron transport channels. Due to the wrapping structure, it can play a synergistic-catalytic role well, suppressing the "shuttle effect" of LiPSs and accelerating the electro-kinetic reaction. The results show that the CC@rGO sulfur cathode has good electrochemical performance, good rate performance, and cycling stability at different current densities, and the Coulomb efficiency is always maintained above 97%. Therefore, this electrode material design can provide new insights into the design and development of sulfur carriers for lithium-sulfur battery cathodes.

Author Contributions: Software, J.C., H.F., and Z.L.; Investigation, Z.S.; Data curation, B.G.; Writing—original draft, J.L. All authors have read and agreed to the published version of the manuscript.

Funding: This work was supported by the National Natural Science Foundation of China (51671052, 51750110513, and 52250610222), the Fundamental Research Funds for the Central Universities (N182502042, N2025001), and the Liao Ning Revitalization Talents Program (XLYC1902105).

Data Availability Statement: The data presented in this study are available on request from the corresponding author. The data are not publicly available due to personal privacy.

Conflicts of Interest: The authors declare no conflict of interest.

References

1. Tian, J.; Yang, Z.; Yin, Z.; Ye, Z.; Wang, J.; Cui, C.; Qian, W. Perspective to the Potential Use of Graphene in Li-Ion Battery and Supercapacitor. *Chem. Rec.* **2019**, *19*, 1256–1262. [CrossRef] [PubMed]
2. Tang, Y.; Wang, X.; Chen, J.; Wang, X.; Wang, D.; Mao, Z. High-Level Pyridinic-N-Doped Carbon Nanosheets with Promising Performances Severed as Li-Ion Battery Anodes. *Energy Technol.* **2020**, *8*, 2000361. [CrossRef]
3. Pang, Q.; Kundu, D.; Nazar, L.F. A graphene-like metallic cathode host for long-life and high-loading lithium-sulfur batteries. *Mater. Horiz.* **2016**, *3*, 130–136. [CrossRef]
4. Chen, T.; Ma, L.; Cheng, B.; Chen, R.; Hu, Y.; Zhu, G.; Wang, Y.; Liang, J.; Tie, Z.; Liu, J.; et al. Metallic and polar Co9S8 inlaid carbon hollow nanopolyhedra as efficient polysulfide mediator for lithium-sulfur batteries. *Nano Energy* **2017**, *38*, 239–248. [CrossRef]
5. Tan, Z.; Ni, K.; Chen, G.; Zeng, W.; Tao, Z.; Ikram, M.; Zhang, Q.; Wang, H.; Sun, L.; Zhu, X.; et al. Incorporating Pyrrolic and Pyridinic Nitrogen into a Porous Carbon made from C60 Molecules to Obtain Superior Energy Storage. *Adv. Mater.* **2017**, *29*, 1603414. [CrossRef]
6. Wu, C.; Zhang, Y.; Dong, D.; Xie, H.; Li, J. Co9S8 nanoparticles anchored on nitrogen and sulfur dual-doped carbon nanosheets as highly efficient bifunctional electrocatalyst for oxygen evolution and reduction reactions. *Nanoscale* **2017**, *9*, 12432–12440. [CrossRef]
7. Li, F.; Wu, Y.; Lin, Y.; Li, J.; Sun, Y.; Nan, H.; Wu, M.; Dong, H.; Shi, K.; Liu, Q. Achieving job-synergistic polysulfides adsorption-conversion within hollow structured MoS2/Co4S3/C heterojunction host for long-life lithium–sulfur batteries. *J. Colloid Interface Sci.* **2022**, *626*, 535–543. [CrossRef]
8. Zhang, J.; Yu, L.; Lou, X.W.D. Embedding CoS2 nanoparticles in N-doped carbon nanotube hollow frameworks for enhanced lithium storage properties. *Nano Res.* **2017**, *10*, 4298–4304. [CrossRef]

9. Guo, J.; Xu, Y.; Wang, C. Sulfur-Impregnated Disordered Carbon Nanotubes Cathode for Lithium–Sulfur Batteries. *Nano Lett.* **2011**, *11*, 4288–4294. [CrossRef]
10. Jeong, S.S.; Choi, Y.J.; Kim, K.W. Effects of Multiwalled Carbon Nanotubes on the Cycle Performance of Sulfur Electrode for Li/S Secondary Battery. *Mater. Sci. Forum*, 2006; 510–511, 1106–1109.
11. Huang, P.; Li, C.; Wang, Y.; Dang, B. The preparation and electrochemical performance study of the carbon nanofibers/sulfur composites. *Ionics* **2021**, *27*, 2609–2613. [CrossRef]
12. Zhou, G.; Pei, S.; Li, L.; Wang, D.-W.; Wang, S.; Huang, K.; Yin, L.-C.; Li, F.; Cheng, H.-M. A Graphene–Pure-Sulfur Sandwich Structure for Ultrafast, Long-Life Lithium–Sulfur Batteries. *Adv. Mater.* **2014**, *26*, 625–631. [CrossRef]
13. Li, L.; Ruan, G.; Peng, Z.; Yang, Y.; Fei, H.; Raji, A.-R.O.; Samuel, E.L.G.; Tour, J.M. Enhanced Cycling Stability of Lithium Sulfur Batteries Using Sulfur–Polyaniline–Graphene Nanoribbon Composite Cathodes. *ACS Appl. Mater. Interfaces* **2014**, *6*, 15033–15039. [CrossRef]
14. Zhang, K.; Qin, F.; Lai, Y.; Li, J.; Lei, X.; Wang, M.; Lu, H.; Fang, J. Efficient Fabrication of Hierarchically Porous Graphene-Derived Aerogel and Its Application in Lithium Sulfur Battery. *ACS Appl. Mater. Interfaces* **2016**, *8*, 6072–6081. [CrossRef]
15. Huang, J.-Q.; Zhuang, T.-Z.; Zhang, Q.; Peng, H.-J.; Chen, C.-M.; Wei, F. Permselective Graphene Oxide Membrane for Highly Stable and Anti-Self-Discharge Lithium–Sulfur Batteries. *ACS Nano* **2015**, *9*, 3002–3011. [CrossRef]
16. Zhang, L.; Liang, P.; Man, X.-l.; Wang, D.; Huang, J.; Shu, H.-b.; Liu, Z.-g.; Wang, L. Fe, N co-doped graphene as a multi-functional anchor material for lithium-sulfur battery. *J. Phys. Chem. Solids* **2019**, *126*, 280–286. [CrossRef]
17. Bao, W.; Zhang, Z.; Chen, W.; Zhou, C.; Lai, Y.; Li, J. Facile synthesis of graphene oxide @ mesoporous carbon hybrid nanocomposites for lithium sulfur battery. *Electrochim. Acta* **2014**, *127*, 342–348. [CrossRef]
18. Sun, Z.; Zhang, J.; Yin, L.; Hu, G.; Fang, R.; Cheng, H.-M.; Li, F. Conductive porous vanadium nitride/graphene composite as chemical anchor of polysulfides for lithium-sulfur batteries. *Nat. Commun.* **2017**, *8*, 14627. [CrossRef]
19. Cheng, R.; Xian, X.; Manasa, P.; Liu, J.; Xia, Y.; Guan, Y.; Wei, S.; Li, Z.; Li, B.; Xu, F.; et al. Carbon Coated Metal-Based Composite Electrode Materials for Lithium Sulfur Batteries: A Review. *Chem. Rec.* **2022**, *22*, e202200168. [CrossRef]
20. Zhang, L.; Liang, P.; Shu, H.B.; Man, X.L.; Du, X.Q.; Chao, D.L.; Liu, Z.G.; Sun, Y.P.; Wan, H.Z.; Wang, H. Design rules of heteroatom-doped graphene to achieve high performance lithium–sulfur batteries: Both strong anchoring and catalysing based on first principles calculation. *J. Colloid Interface Sci.* **2018**, *529*, 426–431. [CrossRef]
21. Lee, J.; Park, S.-K.; Piao, Y. N-doped Carbon Framework/Reduced Graphene Oxide Nanocomposite as a Sulfur Reservoir for Lithium-Sulfur Batteries. *Electrochim. Acta* **2016**, *222*, 1345–1353. [CrossRef]
22. Zha, C.; Liu, S.; Zhou, L.; Li, K.; Zhang, T. One-Pot Pyrolysis to Nitrogen-Doped Hierarchically Porous Carbon Nanosheets as Sulfur-Host in Lithium–Sulfur Batteries. *J. Electrochem. Energy Convers. Storage* **2022**, *19*, 021018. [CrossRef]
23. Zhu, S.; Li, Y. Carbon-metal oxide nanocomposites as lithium-sulfur battery cathodes. *Funct. Mater. Lett.* **2018**, *11*, 1830007. [CrossRef]
24. Yang, W.; Li, X.; Li, Y.; Zhu, R.; Pang, H. Applications of Metal-Organic-Framework-Derived Carbon Materials. *Adv. Mater.* **2019**, *31*, 1804740. [CrossRef] [PubMed]
25. Liu, X.; Huang, J.Q.; Zhang, Q.; Mai, L. Nanostructured Metal Oxides and Sulfides for Lithium-Sulfur Batteries. *Adv. Mater.* **2017**, *29*, 1601759. [CrossRef] [PubMed]
26. Chen, L.; Li, X.; Xu, Y. Recent advances of polar transition-metal sulfides host materials for advanced lithium–sulfur batteries. *Funct. Mater. Lett.* **2018**, *11*, 1840010. [CrossRef]
27. Wu, J.; Ye, T.; Wang, Y.; Yang, P.; Wang, Q.; Kuang, W.; Chen, X.; Duan, G.; Yu, L.; Jin, Z.; et al. Understanding the Catalytic Kinetics of Polysulfide Redox Reactions on Transition Metal Compounds in Li-S Batteries. *ACS Nano* **2022**, *16*, 15734–15759. [CrossRef]
28. Liu, H.; Yang, X.; Jin, B.; Cui, M.; Li, Y.; Li, Q.; Li, L.; Sheng, Q.; Lang, X.; Jin, E.; et al. Coordinated Immobilization and Rapid Conversion of Polysulfide Enabled by a Hollow Metal Oxide/Sulfide/Nitrogen-Doped Carbon Heterostructure for Long-Cycle-Life Lithium-Sulfur Batteries. *Small* **2023**, 2300950. [CrossRef]
29. Wang, M.; Fan, L.; Wu, X.; Qiu, Y.; Wang, Y.; Zhang, N.; Sun, K. SnS_2/SnO_2 Heterostructures towards Enhanced Electrochemical Performance of Lithium–Sulfur Batteries. *Chem. A Eur. J.* **2019**, *25*, 5416–5421. [CrossRef]
30. Qin, B.; Wang, Q.; Yao, W.; Cai, Y.; Chen, Y.; Wang, P.; Zou, Y.; Zheng, X.; Cao, J.; Qi, J.; et al. Heterostructured Mn_3O_4-MnS Multi-Shelled Hollow Spheres for Enhanced Polysulfide Regulation in Lithium–Sulfur Batteries. *Energy Environ. Mater.* **2022**, e12475. [CrossRef]
31. Zhang, X.; Yu, Z.; Wang, C.; Gong, Y.; Ai, B.; Zhang, L.; Wang, J. NC@CoP–Co_3O_4 composite as sulfur cathode for high-energy lithium–sulfur batteries. *J. Mater. Sci.* **2021**, *56*, 10030–10040. [CrossRef]
32. Liu, Z.; Fu, H.; Gao, B.; Wang, Y.; Li, K.; Sun, Y.; Yin, J.; Kan, J. In-situ synthesis of Fe_2O_3/rGO using different hydrothermal methods as anode materials for lithium-ion batteries. *Rev. Adv. Mater. Sci.* **2020**, *59*, 477–486. [CrossRef]
33. Fu, H.; Gao, B.; Hu, C.; Liu, Z.; Hu, L.; Kan, J.; Feng, Z.; Xing, P. 3D nitrogen-doped graphene created by the secondary intercalation of ethanol with enhanced specific capacity. *Nanotechnology* **2022**, *33*, 075703. [CrossRef]
34. Tu, C.; Peng, A.; Zhang, Z.; Qi, X.; Zhang, D.; Wang, M.; Huang, Y.; Yang, Z. Surface-seeding secondary growth for CoO@Co_9S_8 P-N heterojunction hollow nanocube encapsulated into graphene as superior anode toward lithium ion storage. *Chem. Eng. J.* **2021**, *425*, 130648. [CrossRef]

35. Wang, T.; Li, C.; Liao, X.; Li, Q.; Hu, W.; Chen, Y.; Yuan, W.; Lin, H. Fe-doped Co$_9$S$_8$@CoO aerogel with core-shell nanostructures for boosted oxygen evolution reaction. *Int. J. Hydrogen Energy* **2022**, *47*, 21182–21190. [CrossRef]
36. Sun, L.; Liu, Y.; Xie, J.; Fan, L.; Wu, J.; Jiang, R.; Jin, Z. Polar Co9S8 anchored on Pyrrole-Modified graphene with in situ growth of CNTs as multifunctional Self-Supporting medium for efficient Lithium-Sulfur batteries. *Chem. Eng. J.* **2023**, *451*, 138370. [CrossRef]
37. Pachfule, P.; Shinde, D.; Majumder, M.; Xu, Q. Fabrication of carbon nanorods and graphene nanoribbons from a metal–organic framework. *Nat. Chem.* **2016**, *8*, 718–724. [CrossRef]
38. Shi, Z.; Shi, Z.; Gao, B.; Yin, J.; Liu, Z.; Wang, L. Preparation of Co-nanocluster graphene composite by asymmetric domain-limited electrochemical exfoliation for functionalized lithium-sulfur battery separator applications. *J. Alloys Compd.* **2023**, *960*, 170827. [CrossRef]
39. Jiang, Y.; Wang, J.; Liu, B.; Jiang, W.; Zhou, T.; Ma, Y.; Che, G.; Liu, C. Superhydrophilic N, S, O-doped Co/CoO/Co$_9$S$_8$@carbon derived from metal-organic framework for activating peroxymonosulfate to degrade sulfamethoxazole: Performance, mechanism insight and large-scale application. *Chem. Eng. J.* **2022**, *446*, 137361. [CrossRef]
40. Xu, Y.; Long, J.; Tu, L.; Dai, W.; Yang, L.; Zou, J.; Luo, X.; Luo, S. CoO engineered Co$_9$S$_8$ catalyst for CO$_2$ photoreduction with accelerated electron transfer endowed by the built-in electric field. *Chem. Eng. J.* **2021**, *426*, 131849. [CrossRef]
41. Wang, N.; Chen, B.; Qin, K.; Liu, E.; Shi, C.; He, C.; Zhao, N. Rational design of Co$_9$S$_8$/CoO heterostructures with well-defined interfaces for lithium sulfur batteries: A study of synergistic adsorption-electrocatalysis function. *Nano Energy* **2019**, *60*, 332–339. [CrossRef]
42. Song, H.; Li, T.; He, T.; Wang, Z.; Fang, D.; Wang, Y.; Li, X.L.; Zhang, D.; Hu, J.; Huang, S. Cooperative catalytic Mo-S-Co heterojunctions with sulfur vacancies for kinetically boosted lithium-sulfur battery. *Chem. Eng. J.* **2022**, *450*, 138115. [CrossRef]
43. Wang, Y.; Zhu, T.; Zhang, Y.; Kong, X.; Liang, S.; Cao, G.; Pan, A. Rational design of multi-shelled CoO/Co$_9$S$_8$ hollow microspheres for high-performance hybrid supercapacitors. *J. Mater. Chem. A* **2017**, *5*, 18448–18456. [CrossRef]
44. Kaneko, K.; Otsuka, H. New IUPAC recommendation and characterization of nanoporous materials with physical adsorption. *Acc Mater. Surf. Res.* **2020**, *5*, 25–32.
45. Sing, K.S.W.; Williams, R.T. Physisorption Hysteresis Loops and the Characterization of Nanoporous Materials. *Adsorpt. Sci. Technol.* **2004**, *22*, 773–782. [CrossRef]
46. Deng, W.; Zhou, X.; Fang, Q.; Liu, Z. Graphene/Sulfur Composites with a Foam-Like Porous Architecture and Controllable Pore Size for High Performance Lithium–Sulfur Batteries. *ChemNanoMat* **2016**, *2*, 952–958. [CrossRef]
47. Zhou, W.; Wang, C.; Zhang, Q.; Abruña, H.D.; He, Y.; Wang, J.; Mao, S.X.; Xiao, X. Tailoring Pore Size of Nitrogen-Doped Hollow Carbon Nanospheres for Confining Sulfur in Lithium–Sulfur Batteries. *Adv. Energy Mater.* **2015**, *5*, 1401752. [CrossRef]
48. Hu, L.; Lu, Y.; Zhang, T.; Huang, T.; Zhu, Y.; Qian, Y. Ultramicroporous Carbon through an Activation-Free Approach for Li–S and Na–S Batteries in Carbonate-Based Electrolyte. *ACS Appl. Mater. Interfaces* **2017**, *9*, 13813–13818. [CrossRef]
49. Qi, X.; Huang, L.; Luo, Y.; Chen, Q.; Chen, Y. Ni3Sn2/nitrogen-doped graphene composite with chemisorption and electrocatalysis as advanced separator modifying material for lithium sulfur batteries. *J. Colloid Interface Sci.* **2022**, *628*, 896–910. [CrossRef]
50. Wei, J.; Qiao, X.; Ye, X.; Chen, B.; Zhang, Y.; Wang, T.; Hui, J. The Metal–Organic Frameworks Derived Co$_3$O$_4$/TiO$_2$ Heterojunction as a High-Efficiency Sulfur Carrier for Lithium–Sulfur Batteries. *Energy Technol.* **2023**, *11*, 2300092. [CrossRef]
51. Li, H.-J.; Song, Y.-H.; Xi, K.; Wang, W.; Liu, S.; Li, G.-R.; Gao, X.-P. Sulfur vacancies in Co$_9$S$_{8-x}$/N-doped graphene enhancing the electrochemical kinetics for high-performance lithium-sulfur batteries. *J. Mater. Chem. A* **2021**, *9*, 10704–10713. [CrossRef]
52. Wang, P.; Zhang, Z.; Hong, B.; Zhang, K.; Li, J.; Lai, Y. Multifunctional porous VN nanowires interlayer as polysulfides barrier for high performance lithium sulfur batteries. *J. Electroanal. Chem.* **2019**, *832*, 475–479. [CrossRef]
53. Wang, J.; Gao, L.; Zhao, J.; Zheng, J.; Wang, J.; Huang, J. A facile in situ synthesis of ZIF-8 nanoparticles anchored on reduced graphene oxide as a sulfur host for Li-S batteries. *Mater. Res. Bull.* **2021**, *133*, 111061. [CrossRef]
54. Wei, J.; Chen, B.; Su, H.; Jiang, C.; Li, X.; Qiao, S.; Zhang, H. Co9S8 nanotube wrapped with graphene oxide as sulfur hosts with ultra-high sulfur content for lithium-sulfur battery. *Ceram. Int.* **2021**, *47*, 2686–2693. [CrossRef]
55. Song, H.; Zuo, C.; Xu, X.; Wan, Y.; Wang, L.; Zhou, D.; Chen, Z. A thin TiO$_2$ NTs/GO hybrid membrane applied as an interlayer for lithium–sulfur batteries. *RSC Adv.* **2018**, *8*, 429–434. [CrossRef]
56. Zhang, T.; Wu, Y.; Yin, Y.; Chen, H.; Gao, C.; Xiao, Y.; Zhang, X.; Wu, J.; Zheng, B.; Li, S. Rational design of Ni$_3$(HITP)$_2$@GO composite for lithium-sulfur cathode. *Appl. Surf. Sci.* **2021**, *572*, 151479. [CrossRef]

Disclaimer/Publisher's Note: The statements, opinions and data contained in all publications are solely those of the individual author(s) and contributor(s) and not of MDPI and/or the editor(s). MDPI and/or the editor(s) disclaim responsibility for any injury to people or property resulting from any ideas, methods, instructions or products referred to in the content.

Article

Enhancing Epoxy Composite Performance with Carbon Nanofillers: A Solution for Moisture Resistance and Extended Durability in Wind Turbine Blade Structures

Angelos Ntaflos [1], Georgios Foteinidis [1], Theodora Liangou [2], Elias Bilalis [2], Konstantinos Anyfantis [2], Nicholas Tsouvalis [2], Thomais Tyriakidi [3], Kosmas Tyriakidis [3], Nikolaos Tyriakidis [3] and Alkiviadis S. Paipetis [1,*]

[1] CSMLab, Department of Materials Science & Engineering, University of Ioannina, 45110 Ioannina, Greece; a.ntaflos@uoi.gr (A.N.); g.foteinidis@uoi.gr (G.F.)

[2] Shipbuilding Technology Laboratory, Department of Naval Architecture & Marine Engineering, National Technical University of Athens, 15780 Zografos, Greece; doraliangou@yahoo.com (T.L.); ebilalis@mail.ntua.gr (E.B.); kanyf@naval.ntua.gr (K.A.); tsouv@mail.ntua.gr (N.T.)

[3] B&T Composites, Agrokthma Florina AA 1834, 53100 Florina, Greece; thomai@btcomposites.gr (T.T.); kosmas@btcomposites.gr (K.T.); nikos@btcomposites.gr (N.T.)

* Correspondence: paipetis@uoi.gr

Citation: Ntaflos, A.; Foteinidis, G.; Liangou, T.; Bilalis, E.; Anyfantis, K.; Tsouvalis, N.; Tyriakidi, T.; Tyriakidis, K.; Tyriakidis, N.; Paipetis, A.S. Enhancing Epoxy Composite Performance with Carbon Nanofillers: A Solution for Moisture Resistance and Extended Durability in Wind Turbine Blade Structures. *Materials* **2024**, *17*, 524. https://doi.org/10.3390/ma17020524

Academic Editors: Andrea Sorrentino and Raul D. S. G. Campilho

Received: 27 November 2023
Revised: 10 January 2024
Accepted: 17 January 2024
Published: 22 January 2024

Copyright: © 2024 by the authors. Licensee MDPI, Basel, Switzerland. This article is an open access article distributed under the terms and conditions of the Creative Commons Attribution (CC BY) license (https:// creativecommons.org/licenses/by/ 4.0/).

Abstract: The increasing prominence of glass-fibre-reinforced plastics (GFRPs) in the wind energy industry, due to their exceptional combination of strength, low weight, and resistance to corrosion, makes them an ideal candidate for enhancing the performance and durability of wind turbine blades. The unique properties of GFRPs not only contribute to reduced energy costs through improved aerodynamic efficiency but also extend the operational lifespan of wind turbines. By modifying the epoxy resin with carbon nanofillers, an even higher degree of performance can be achieved. In this work, graphene nanoplatelet (GNP)-enhanced GFRPs are produced through industrial methods (filament winding) and coupons are extracted and tested for their mechanical performance after harsh environmental aging in high temperature and moisture. GNPs enhance the in-plane shear strength of GFRP by 200%, while reducing their water uptake by as much as 40%.

Keywords: wind turbine blades; glass-fibre-reinforced plastics (GFRPs); graphene nanoplatelets (GNPs); filament winding; environmental aging

1. Introduction

The global shift from fossil fuels to renewable energy sources, primarily wind energy, represents a pivotal moment in the ongoing battle against climate change. This transition, driven by mounting environmental concerns and the recognition of finite fossil fuel resources, signifies a commitment to a more sustainable and cleaner future. Wind energy offers an abundant and renewable source of power, significantly reducing greenhouse gasses [1]. GFRP's lightweight yet strong nature is pivotal in ensuring the blades' efficient rotation and optimizing energy production [2]. Furthermore, its resistance to corrosion is vital for withstanding harsh outdoor conditions where wind turbines are typically situated [3].

Epoxy resins represent a class of exceptionally versatile polymers utilized across diverse high-performance industries owing to their exceptional amalgamation of mechanical strength, chemical stability, and physical properties. They exhibit a broad range of applications, serving as vital components for structural elements [4]. Their inherent compatibility with a wide array of reinforcing fibres, minimal shrinkage during the curing process, low mass, and cost-effectiveness makes epoxy resins and fibre-reinforced epoxy composites attractive alternatives to traditional materials [5]. Epoxy-based composites offer unique advantages, particularly in high-stress settings characterized by exposure to moisture and elevated temperatures known to expedite degradation processes, ultimately leading to premature component failure. The presence of moisture initiates various adverse effects, such as swelling, plasticization, and overall material degradation. The water molecules permeate the epoxy resin matrix, effectively diminishing its mechanical properties through the establishment of hydrogen bonds with the hydrophilic functional groups within the epoxy structure. This interaction triggers a swelling effect, infiltrating the material's "free volume" and substantially compromising its long-term durability [6].

The Integration of carbon nanofillers into epoxy composites has gained substantial attention within the realm of scientific investigation, primarily due to the immense potential for advanced applications [7]. These advanced epoxy/carbon nanocomposites are promising for mitigating the adverse impact of moisture, simultaneously enhancing their mechanical properties [8], and thus extending their operational lifespan. Extensive research efforts into exploring the influence of these diverse carbon fillers on the moisture absorption characteristics of epoxy resins have revealed notable enhancements in the resilience of the matrix when subjected to environmental exposure [9]. This improvement is intricately linked to the geometry and dimensions of the incorporated nanofillers [10]. These findings underscore the promising potential of carbon nanofillers in fortifying epoxy resins against moisture-induced degradation and advancing their overall performance characteristics, fostering their suitability for a range of advanced applications [11].

Graphene nanoplatelets (GNPs) are commonly commended for their ability to enhance the barrier properties of epoxy composites [12]. GNPs are stacks of graphene sheets with thicknesses from a few nm up to 100 nm [13]. The advantages of GNPs are their low manufacturing cost and capabilities for mass production. These characteristics make GNPs ideal for industrial applications. Enhancing epoxy GFRP with GNPs can result in epoxy nanocomposites with improved barrier, electrical, and mechanical properties. When a network of GNPs is formed in the matrix, it can significantly decrease the permeation of erosive substances like moisture by creating a tortuous path, forcing the water molecules to follow a complicated pathway [14]. Research has established a correlation between the geometry of GNPs, including high specific surface area and aspect ratio, and their efficacy as barriers [15].

In this study, physical and mechanical characterization through international standards was performed in pristine and aged unmodified (neat) and GNP-reinforced GFRP post environmental degradation in harsh environments of elevated moisture and temperature. The GFRPs were produced in an industrial environment using filament winding, which is deemed a sustainable manufacturing methodology for large structures such as wind turbine blades. For better evaluation of the results, dynamic mechanical analysis was performed to examine the effect of water ingress on the properties of the GFRP. This work encompasses all evaluation methodologies, including industrial manufacturing, to prove the viability of the nanomodification in the relevant industrial environment with the direct transfer beneficial effect of the nanomodification on real structures.

2. Materials and Methods

2.1. Materials

XGNPs C-300 graphene nanoplatelets, provided by XGSciences, Lansing, MI, USA, were used as the carbon nanofiller of choice for the improvement of the GFRP. The GNPs had a thickness of a few nanometres, lateral size smaller than 2 μm, and 300 g/m^2 surface area with a Raman ID/IG ratio of 0.85 [16–18]. A commercial-grade epoxy resin, diglycidyl ether of bisphenol A (DGEBA) Epikote 828, was provided by Hexion, Columbus, OH, USA, along with complementary Epikure curing agent 866 and Epicure Catalyst 101 in a 10.8.3.0.15 mixing ratio. The epoxy viscosity at room temperature was 10.000 mPa s. E6-CR 386T by Jushi E-glass fibres designed for filament winding applications were applied for the manufacturing of the GFRP structures.

2.2. Dispersion and Manufacturing

High shear mixing was selected for dispersing the nanofillers in the polymer matrix. The dispersion protocol was performed using a laboratory dissolver device (Dispermat AE by Gentzman, Reichshof, Germany) supplied with a double wall vacuum container in combination with a thermostatic bath by GRANT capable of temperature control within ±1 °C accuracy. The conditions of the dispersion protocols were rotary speed of 3000 rounds per minute (rpm) and temperature of 25 °C. The selected nanofiller weight content was selected in a previously unpublished work to be 1% wt. GNP. Introducing 1% wt. GNP in the epoxy system showcased the optimum overall performance, including reduced water absorption and increased mechanical properties, in lab-scale manufactured GFRP.

SEM spectroscopy was performed on a Phenom Pharos Desktop SEM by Thermo Fisher Scientific, Waltham, MA, USA. Epoxy matrix specimens were tested to evaluate the dispersion state. In Figure 1, two SEM images are presented: Figure 1a corresponds to the GNP-enhanced matrix specimen post single-edged notched beam (SENB) testing, while Figure 1b corresponds to the neat matrix post SENB. After examination, the dispersion of GNPs in the matrix was homogonous, while the fracture mechanisms presented were in line with the literature. GNPs introduced additional fracture mechanisms, increasing the fracture surface roughness [19].

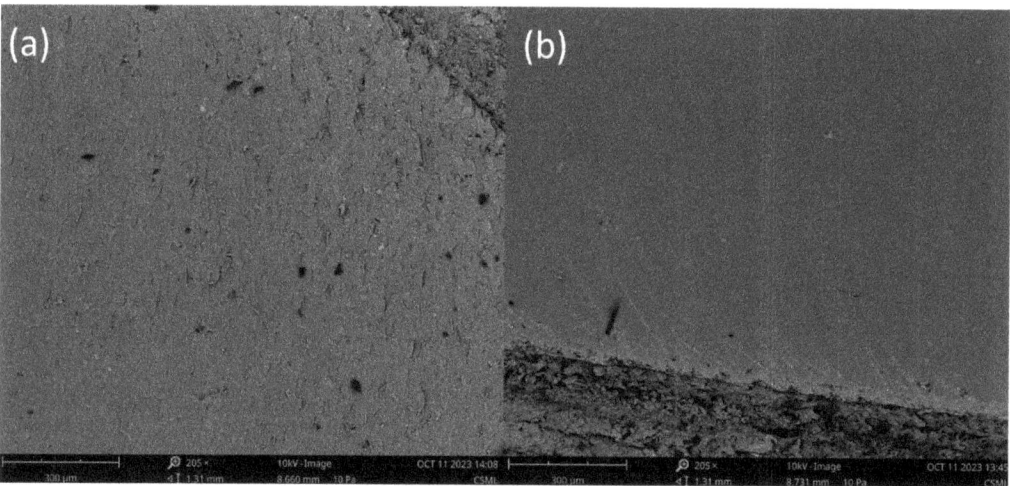

Figure 1. SEM images of (**a**) GNP-enhanced epoxy resin and (**b**) Neat epoxy resin post single-edged notched beam testing.

B&T Composites in Florina, Greece, produced two sets of large-scale industrial GFRP structures purely via filament winding (Figure 2), one set utilized a conventional resin, while the other employed resin modified with 1% wt. GNPs. Each set consisted of three composite variants with varying fibre orientations: 0°, 90°, and a biaxial orientation of approximately ±45°.

Coupons from the configurations above were collected and subjected to physical characterization based on the ISO 1172:1996 [20] standard to assess fibre content (calcination Method A) and ISO 1183-1:2004 [21] standard to determine coupon density (Method A, immersion method). The coupons' dimensions were defined according to the ASTM D3039 [22] and ASTM D3518 [23] standards, used for the measurement of the tensile and the in-plane shear properties, respectively. Both types of coupons were plane and orthogonal, having length equal to 160 mm, width equal to 15.3 mm with coefficient of variation (CoV) = 1.8% for the neat resin coupons and equal to 15.7 mm with CoV = 2.3% for the modified resin ones, and thickness equal to 8.8 mm with CoV = 5.0% for the neat resin coupons and equal to 9.0 mm with CoV = 5.6% for the modified resin ones. Figure 3 depicts typical GFRP coupons. Both types of coupons were plane and orthogonal, having length equal to 160 mm, width equal to 15.3 mm with coefficient of variation (CoV) = 1.8% for the neat resin coupons and equal to 15.7 mm with CoV = 2.3% for the modified resin ones, and thickness equal to 8.8 mm with CoV = 5.0% for the neat resin coupons and equal to 9.0 mm with CoV = 5.6% for the modified resin ones.

Figure 2. Filament wound GFRP structures for material mechanical characterization (neat resin above, modified resin below).

Figure 3. Typical GFRP coupons (neat resin above, modified resin below).

Subsequently, these coupons were categorized into six groups as outlined in Table 1:

Table 1. Nomenclature and physical properties of each group of coupons.

Name	Resin Type	Fibre Orientation (°)	Fibre Content (%)	Density (kg/m^3)
N_0	Neat	0	72	1874
N_90	Neat	90	71	1793
N_45	Neat	±45	69	1787
MD_0	1% wt. GNP	0	64	1781
MD_90	1% wt. GNP	90	64	1774
MD_45	1% wt. GNP	±45	68	1820

2.3. Mechanical Testing

Mechanical testing was carried out in a 250 kN capacity hydraulic testing machine. All tests were displacement controlled with an imposed displacement rate equal to 2 mm/min for the 0° coupons, 0.5 mm/min for the 90° coupons, and 1 mm/min for the ±45° ones. Strains were measured with the aid of an extensometer in the case of the tensile tests of 0° and 90° coupons (gauge length equal to 50 mm), and with the aid of a 5 mm gauge length, 0/90 strain gauge rosette, in the case of the ±45° coupons for measuring in-plane shear properties (Figure 4). The reaction force of the testing machine was also measured for each test, which, by dividing it by the respective cross section area of each coupon, resulted in the applied stress. Therefore, Young's modulus and tensile strength were measured from the 0° and 90° coupons and shear modulus and shear strength from the ±45° ones.

Figure 4. Tensile test of a 0° neat resin coupon (**left**) and of a 0° modified resin coupon (**right**).

2.4. Hydrothermal Aging

Hydrothermal exposure was performed on 5 coupons from each category of Table 1. The specimens were exposed to 70 °C temperature and 85% relative humidity in an environmental chamber. For all configurations, the coupons were sealed with commercial high-temperature-resistant silicon and aged for 90 days. The coupons were periodically weighed to determine their water absorption. Post degradation mechanical evaluation was performed.

2.5. Dynamic Mechanical Analysis

Dynamic mechanical analysis was performed in DMA Q850 (TA Instruments, New Castle, DE, USA) in 3-point bending configuration. The coupons tested were extracted from the 90° sample. The testing parameters were:

- Amplitude: 20.0 µm
- Frequency: 1.0 Hz
- Temperature scan: from 40 °C to 180 °C
- Heating rate: 3.0 °C/min

The DMA results were used to calculate the molecular weight between crosslinks according to the basic equation of rubber elasticity:

$$E_R = 3\,(d/M_c)\,RT \qquad (1)$$

where E_R is the storage modulus at the rubbery plateau, d is the density of the composite, R is the universal gas constant, and T is the temperature at the rubbery plateau. For this research the T at the rubbery plateau was set as Tg + 30 °C [24].

3. Results and Discussion

3.1. Water Uptake

Before starting the moisture absorption tests, all coupons were dried in an oven at 60 °C until significant change in mass was not observed (0.1 mg). The water absorption curves correspond to the average water uptake of the neat structures compared to the 1% wt. GNP-enhanced structures (Figure 5). In all structures, despite the fibre orientation, the GNP-enhanced composites outperformed the neat composites. All coupons exhibited near-identical water uptake curves, with the fibre orientation that presented the highest water absorption being the 90° coupons, despite the resin type, as presented in Table 2. The detrimental effects of water absorption were less evident in the case of the GNP composites, leading to higher retention of mechanical properties compared to their neat counterparts.

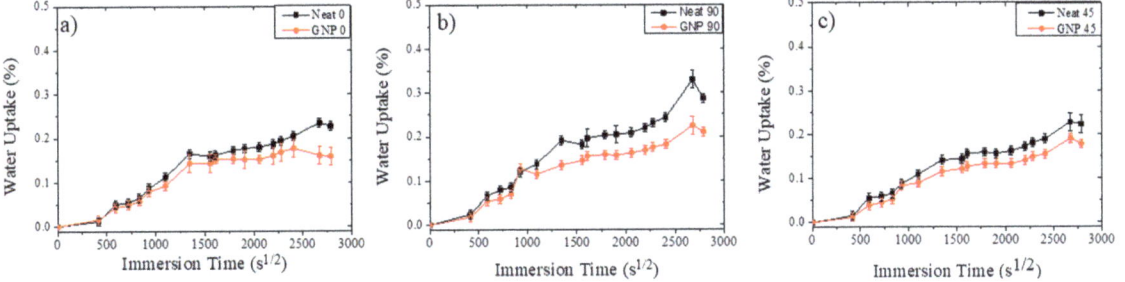

Figure 5. Average water uptake (%) per square root of time ($s^{1/2}$) of neat and nano-modified (GNP) coupons. (**a**) 0 oriented coupons, (**b**) 90 oriented coupons, (**c**) 45 oriented coupons.

Table 2. Water uptake of each group at day 90.

	Water Uptake (%)		
Orientation	0°	90°	±45°
Neat	0.22 ± 0.01	0.28 ± 0.01	0.22 ± 0.02
GNP	0.16 ± 0.02	0.20 ± 0.01	0.17 ± 0.01

Due to the natural affinity of epoxies to absorb moisture, considerable research efforts have been made to reduce that effect. Epoxies absorb water within the voids of their polymeric network. Generally, two types of water can be identified when water is absorbed by epoxies:

- Unbound free water (Type-I water), which occupies nano-voids within the epoxy without inducing any significant swelling.
- Hydrogen-bonded water (Type-II water), which is responsible for causing swelling in the epoxy due to the formation of multiple hydrogen bonds with unreacted epoxy groups [25,26].

In the initial absorption stage, water ingress increases linearly with time until it reaches a saturation point. This saturation point is linked to the free volume within the polymer. After reaching the saturation point, stage-2 absorption begins and the rate of water ingress decreases. During this stage, Type-II water molecules form hydrogen bonds with unreacted epoxy groups, leading to swelling of the polymer. This, in turn, results in the degradation of the material's performance [11]. In GFRP, water absorption can occur with various mechanisms either at the locations of micro-cracks in the matrix or at the interface between the fibres and the matrix. The latter mechanism involves the diffusion of water into the surrounding polymer network through unreacted polymeric chains. In highly cross-linked epoxy systems, the increased free volume tends to result in higher absorption of Type-I water during the early stages of exposure [27]. Consequently, when GNPs are effectively dispersed within the epoxy network, the reduction of available free space within the material can lead to notable enhancements in both the sorption (absorption) properties and the mechanical characteristics of the composite system. The sorption curves for both groups followed the same trend. The GNP-enhanced GFRP had an aggregate water absorption reduction of 12%, which is a significant improvement considering the already low absorption of the neat coupons. As separate groups, the GNP modification of the GFRP improved the water absorption by as much as 40% in the case of the 90° coupons. The improvement is even higher if the difference in fibre content is considered due to glass fibres' higher hydrophobicity compared to the epoxy resin. In Figure 5 it is observed that despite fibre orientation the water uptake curves have similar trends, with most of the water intake taking place in the first $\sec^{1/2}$ of exposure. These trends can be attributed to coupon similarities since the core materials are the same, manufactured with the same method, and weighed on the same day during exposure in the hydrothermal chamber. Their only difference is their edges, which were sealed with commercial silicon.

3.2. Physical and Mechanical Characterization

The coupons that incorporated nano-modifications with fibre orientation parallel to the loading direction displayed an ultimate tensile strength that was 7% lower than that of the unmodified coupons and near-identical Young's modulus as seen in Figure 6. This observed behaviour can be primarily attributed to the reduction in fibre content compared to the unmodified GFRP. The tensile strength of 0 oriented composite materials is mostly affected by the fibre content. The ultimate tensile strength of the fibres is significantly higher than the tensile strength of the matrix and, as a result, the longitudinal tensile strength is mostly affected by the volume of fibres in the system. In composite coupons oriented at $\theta = 90°$, failure is primarily due to the occurrence of transverse matrix cracking. While one might intuitively assume that in these coupons where the fibres are not under tensile stress the material would exhibit the characteristics of a pure polymer, the presence of transverse fibres has an adverse impact on the tensile strength of the coupons caused by debonding between the fibre/matrix interface. This debonding phenomenon leads to a reduction in tensile strength perpendicular to the orientation of the fibres [28]. Compared to the neat coupons, nanomodified coupons showcased an increase in tensile strength perpendicular to the fibre orientation of over 10%, while the modulus of elasticity was relatively the same. An impressive increase due to the nano-modification of the resin was exhibited by the coupons subjected to in-plane shear according to ASTM—D3518. The nano-modified coupons exhibited ultimate shear strength over 200% compared to the neat GFRP. Carbon nanofillers show an improvement in the shear strength of composites. Pinto et al. observed an increase in shear strength of more than 50% with incorporation of 0.1 wt.% GNPs [25]. The effective reinforcement of the epoxy matrix can be attributed to the 3D orientation of GNPs. GNPs introduce additional reinforcement mechanisms, such as bridging effects, enhancing mechanical interactions between the fibres and the matrix and, as a result, increasing shear strength [29].

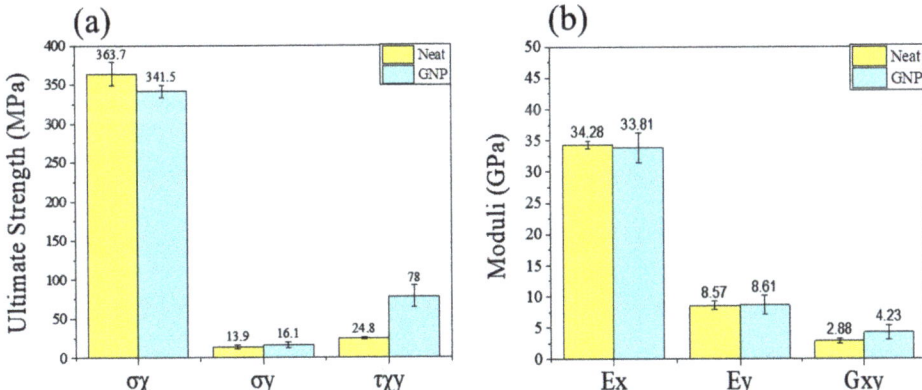

Figure 6. Pristine coupons (**a**) strength and (**b**) moduli. Ex and σx are the Young's modulus and ultimate strength of the 0° coupons; Ey and σy for the 90° coupons. Gxy and τxy are the in-plane shear moduli and strengths of the ±45° coupons.

Environmental aging had a notable impact on the mechanical properties of GFRP coupons as observed in Figure 7. In the case of coupons oriented at 0°, both systems showed an increase in their Young's modulus. Neat GFRP exhibited a 3% increase in Young's modulus after aging, while GNP-GFRP exhibited a significant increase of 10%. However, the neat coupons experienced a 20% drop in ultimate tensile strength (UTS), whereas GNP-GFRP demonstrated better retention of mechanical properties with a 10% decrease, showcasing higher UTS post environmental degradation despite the lower fibre content.

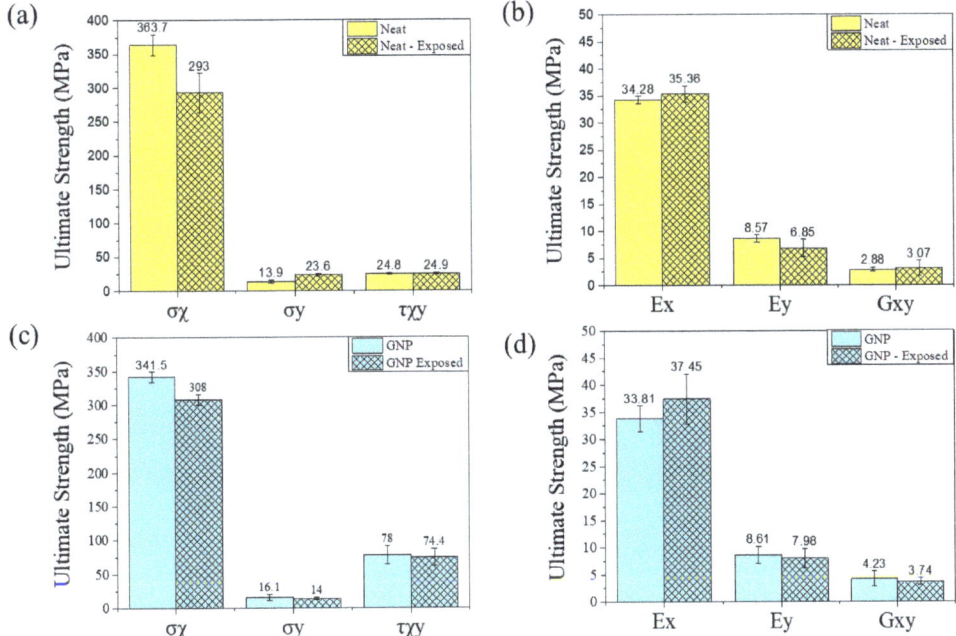

Figure 7. Moduli and strength comparison of pristine and aged (exposed) coupons. (**a**) Strength of neat coupons, (**b**) moduli of neat coupons, (**c**) strength of GNP coupons, (**d**) moduli of GNP coupons.

For the 90° coupons, neat coupons appeared to show an almost 70% increase in UTS. This apparent increase can be attributed to curing reactions occurring during environmental exposure, often referred to as "pseudo cross-linking", which results in improved properties [29]. Despite the increase in UTS, the Young's modulus decreased by 20%. Regarding the GNP 90° coupons, both the Young's modulus and the UTS presented a drop of 7% and 13%, respectively. For the ±45° coupons, the in-plane shear strength remained relatively stable and did not exhibit significant changes.

3.3. Dynamic Mechanical Analysis

In pristine coupons (those without exposure to hydrothermal conditions), the incorporation of GNPs into the epoxy increased its storage modulus by nearly 10% compared to the pure epoxy (Figure 8, Table 3). This enhancement can be attributed to the stiff nature of GNPs, which restrict the movement of polymer chains, despite the smaller crosslink density indicated by the molecular weight between crosslinks' (Mc) values. Similarly, the inclusion of GNPs led to an increase in the glass transition temperature (Tg) of the material. This rise in Tg is due to the strong interaction between the stiff GNPs and the epoxy matrix, reducing both the mobility of the polymeric chains and the free volume [30]. Despite the reduction in the crosslink density of the material mentioned above, the GNP sample experiences higher thermal stability compared to the neat resin, reaching the rubbery state at higher temperatures [31]. As expected, the addition of GNPs in the material also led to a decrease in the height of the tan (d) curve, indicating higher energy dissipation compared to its internal losses. The tan (d) curve of the neat coupons peaked at 0.77 compared to the 0.71 of the GNP coupons. The decrease in the half-width of the tan (d) curve is noticeable, which can be interpreted as a decrease in the heterogeneity of the polymeric network and lower distribution of the relaxation times of the polymer chains compared to the neat GFRP [32].

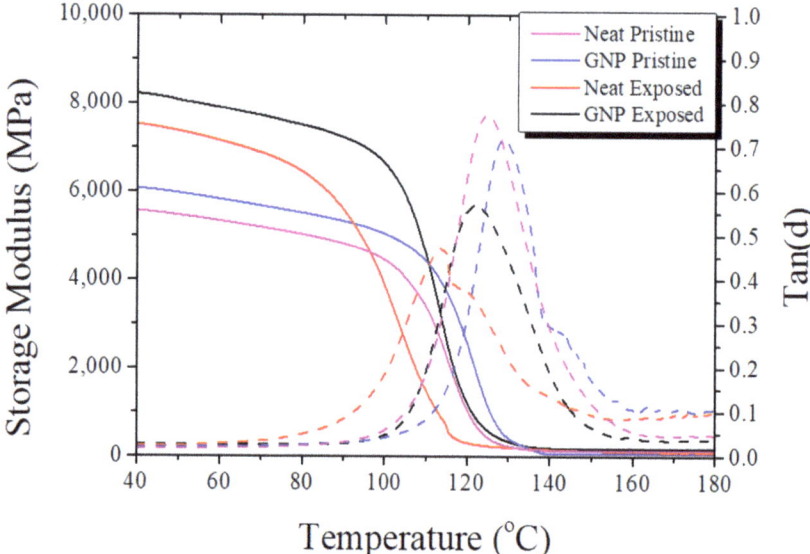

Figure 8. Storage modulus (MPa) and tan (d) of the measured pristine and environmentally aged (exposed) coupons.

Table 3. DMA values for pristine and aged composites.

	Neat Pristine	Neat Exposed	GNP Pristine	GNP Exposed
T_g (from $\tan(\delta)$ peak)	124 ± 0.55	113 ± 1.8	128 ± 0.05	121 ± 0.50
M_c (g/mol)	159 ± 70	146 ± 59	321 ± 30.5	112 ± 22.5
$\tan(\delta)$ peak	0.77 ± 0.05	0.47 ± 0.01	0.71 ± 0.01	0.56 ± 0.01
Width $\tan(\delta)$ at half peak maximum	22.7 ± 0.93	26.8 ± 3.05	19.0 ± 1.30	22.8 ± 1.05

When comparing the pristine coupons to those that have undergone hydrothermal aging, a significant increase in the storage modulus is evident: both the pure epoxy and GNP-enhanced resins showed an increase of more than 20%. The storage modulus of the pure resin increased from 5564 MPa to 7522 MPa at 40 °C, while the GNP-enhanced resin increased from 6074 MPa to 8222 MPa. During hydrothermal aging, different mechanisms come into play. The prolonged exposure to elevated temperatures and moisture can lead to secondary post-curing mechanisms, increasing the crosslink density of the coupons. It has been reported that when epoxies are immersed in water, pseudo cross-linking phenomena can occur. Water molecules, both in the free volume of the resin and the water bound to the polymer chains, restrict the mobility of polymer chains, leading to increased stiffness and binding with unreacted polymer chains [11]. This notion is supported by the lower Mc values of the aged coupons, with the neat coupons' Mc decreasing from 156 to 146 g/mol and the GNP-enhanced coupons experiencing a massive decrease of 321 to 112 g/mol.

Hydrothermal aging also significantly affected the behaviour of the coupons. Both types of GFRP showcased a lower $\tan(\delta)$ peak, consistent with the increased stiffness mentioned earlier. The pure resin, however, exhibited a deterioration in its $\tan(\delta)$ profile, with a broader peak, indicating increased network heterogeneity. Furthermore, the aged pure resin displayed a distinct leftward shift in its $\tan(\delta)$ curve and the appearance of a double peak at 113 °C and 119 °C. This corresponds to a drop in Tg of 11 °C compared to the GNP resin's drop of 7 °C. The GNP coupon's smaller change in its $\tan(\delta)$ profile, if connected with the results of the absorption curve, can be attributed to the reduced hydrolytic degradation process in the polymer due to the lower amount of water absorbed into the body of the resin [33]. The GNP-enhanced coupons showcased the lowest $\tan(\delta)$ peak between all groups due to their enhanced behaviour.

4. Conclusions

This study highlights the considerable potential of GNP modifications in reducing moisture absorption and enhancing the mechanical and dynamic properties of composite materials for large structures. The findings revealed that GNP-enhanced composites consistently outperformed their neat counterparts in terms of moisture absorption. The improvement was particularly significant, with up to a 40% reduction in water absorption observed, in the 90° GNP-enhanced coupons.

Moving on to the physical and mechanical characterization of the composite materials, this study revealed some intriguing insights. Coupons with GNP modifications and fibres oriented parallel to the loading direction exhibited a slightly lower ultimate tensile strength (UTS) compared to the neat composites, despite the 5% reduction in fibre content in the GNP-modified coupons. In contrast, GNP-modified coupons displayed a remarkable increase in shear strength, surpassing their neat counterparts by 200%. This study also conducted dynamic mechanical analysis on the pristine and hydrothermally aged coupons, revealing that the incorporation of GNPs into the epoxy resulted in an approximately 10% increase in storage modulus, indicating greater stiffness. Comparing pristine and aged coupons, both the pure GFRP and GNP-enhanced GFRP demonstrated a significant increase of more than 20% in storage modulus post environmental degradation. This increase was attributed to the impact of prolonged exposure to elevated temperatures and moisture, leading to secondary post-curing mechanisms and an increased crosslink density. As a result, the higher retention of mechanical properties and decreased water absorption, with capabilities for mass production, can lead to improved GFRP composites for wind turbine applications by increasing their lifetime with a marginal increase in their cost.

Author Contributions: Conceptualization, A.S.P.; Methodology, A.N., G.F. and A.S.P.; Software, T.L., E.B. and K.A.; Validation, A.N., T.L., E.B. and K.A.; Formal analysis, A.N., T.L. and N.T. (Nicholas Tsouvalis); Investigation, E.B., T.T., K.T. and N.T. (Nikolaos Tyriakidis); Resources, N.T. (Nicholas Tsouvalis), T.T., K.T. and N.T. (Nikolaos Tyriakidis); Data curation, A.N.; Writing—review & editing, A.S.P.; Supervision, N.T. (Nicholas Tsouvalis) and A.S.P.; Project administration, A.S.P. All authors have read and agreed to the published version of the manuscript.

Funding: This research has been co-financed by the European Union and Greek national funds through the Operational Program Competitiveness, Entrepreneurship, and Innovation, under the call RESEARCH—CREATE—INNOVATE (project code: AIOLOS T2EDK-02971).

Data Availability Statement: Data are contained within the article.

Conflicts of Interest: Authors Thomais Tyriakidi, Kosmas Tyriakidis, Nikolaos Tyriakidis were employed by the company B&T Composites. The remaining authors declare that the research was conducted in the absence of any commercial or financial relationships that could be construed as a potential conflict of interest.

References

1. Torres, J.F.; Petrakopoulou, F. A Closer Look at the Environmental Impact of Solar and Wind Energy. *Glob. Chall.* **2022**, *6*, 2200016. [CrossRef] [PubMed]
2. Zhu, J.; Cai, X.; Gu, R. Multi-objective aerodynamic and structural optimization of horizontal-axis wind turbine blades. *Energies* **2017**, *10*, 101. [CrossRef]
3. Brijder, R.; Hagen, C.H.M.; Cortés, A.; Irizar, A.; Thibbotuwa, U.C.; Helsen, S.; Vásquez, S.; Ompusunggu, A.P. Review of corrosion monitoring and prognostics in offshore wind turbine structures: Current status and feasible approaches. *Front. Energy Res.* **2022**, *10*, 1433. [CrossRef]
4. Baltzis, D.; Bekas, D.; Tzachristas, G.; Parlamas, A.; Karabela, M.; Zafeiropoulos, N.; Paipetis, A. Multi-scaled carbon reinforcements in ternary epoxy composite materials: Dispersion and electrical impedance study. *Compos. Sci. Technol.* **2017**, *153*, 7–17. [CrossRef]
5. Hossain, M.E. The current and future trends of composite materials: An experimental study. *J. Compos. Mater.* **2011**, *45*, 2133–2144. [CrossRef]
6. Rocha, I.B.C.M.; Raijmaekers, S.; Nijssen, R.P.L.; van der Meer, F.P.; Sluys, L.J. Hygrothermal ageing behaviour of a glass/epoxy composite used in wind turbine blades. *Compos. Struct.* **2017**, *174*, 110–122. [CrossRef]
7. Prolongo, S.G.; Jiménez-Suárez, A.; Moriche, R.; Ureña, A. Influence of thickness and lateral size of graphene nanoplatelets on water uptake in epoxy/graphene nanocomposites. *Appl. Sci.* **2018**, *8*, 1150. [CrossRef]
8. Vavouliotis, A.; Karapappas, P.; Loutas, T.; Voyatzi, T.; Paipetis, A.; Kostopoulos, V. Multistage fatigue life monitoring on carbon fibre reinforced polymers enhanced with multiwall carbon nanotubes. *Plastics Rubber Compos.* **2009**, *38*, 124–130. [CrossRef]
9. Barkoula, N.M.; Paipetis, A.; Matikas, T.; Vavouliotis, A.; Karapappas, P.; Kostopoulos, V. Environmental degradation of carbon nanotube-modified composite laminates: A study of electrical resistivity. *Mech. Compos. Mater.* **2009**, *45*, 21–32. [CrossRef]
10. Gkikas, G.; Douka, D.D.; Barkoula, N.M.; Paipetis, A.S. Nano-enhanced composite materials under thermal shock and environmental degradation: A durability study. *Compos. Part B Eng.* **2015**, *70*, 206–214. [CrossRef]
11. Baltzis, D.; Bekas, D.; Tsirka, K.; Parlamas, A.; Ntaflos, A.; Zafeiropoulos, N.; Lekatou, A.G.; Paipetis, A.S. Multi-scaled carbon epoxy composites underwater immersion: A durability study. *Compos. Sci. Technol.* **2020**, *199*, 108373. [CrossRef]
12. Cui, Y.; Kundalwal, S.I.; Kumar, S. Gas barrier performance of graphene/polymer nanocomposites. *Carbon N. Y.* **2016**, *98*, 313–333. [CrossRef]
13. Cataldi, P.; Athanassiou, A.; Bayer, I.S. Graphene nanoplatelets-based advanced materials and recent progress in sustainable applications. *Appl. Sci.* **2018**, *8*, 1438. [CrossRef]
14. Tan, B.; Thomas, N.L. A review of the water barrier properties of polymer/clay and polymer/graphene nanocomposites. *J. Memb. Sci.* **2016**, *514*, 595–612. [CrossRef]
15. Wang, Y.; Zhu, W.; Zhang, X.; Cai, G.; Wan, B. Influence of thickness on water absorption and tensile strength of BFRP laminates in water or alkaline solution and a thickness-dependent accelerated ageing method for BFRP laminates. *Appl. Sci.* **2020**, *10*, 3618. [CrossRef]
16. Pullicino, E.; Zou, W.; Gresil, M.; Soutis, C. The Effect of Shear Mixing Speed and Time on the Mechanical Properties of GNP/Epoxy Composites. *Appl. Compos. Mater.* **2017**, *24*, 301–311. [CrossRef]

17. Kim, S.-G.; Park, O.-K.; Lee, J.H.; Ku, B.-C. Layer-by-layer assembled graphene oxide films and barrier properties of thermally reduced graphene oxide membranes. *Carbon Lett.* **2013**, *14*, 247–250. [CrossRef]
18. Maulana, A.; Nugraheni, A.Y.; Jayanti, D.N.; Mustofa, S.; Baqiya, M.A. Defect and Magnetic Properties of Reduced Graphene Oxide Prepared from Old Coconut Shell. *IOP Conf. Ser. Mater. Sci. Eng.* **2017**, *196*, 012021. [CrossRef]
19. Shirodkar, N.; Cheng, S.; Seidel, G.D. Enhancement of Mode I fracture toughness properties of epoxy reinforced with graphene nanoplatelets and carbon nanotubes. *Compos. Part B Eng.* **2021**, *224*, 109177. [CrossRef]
20. ISO-1172:1996; Textile-Glass-Reinforced Plastics Prepregs, Moulding Compounds and Laminates Determination of the Textile-glass and Mineral-Filler Content Using Calcination Methods. International Organization for Standardization: Geneva, Switzerland, 2005.
21. *Dansk Standard DS/EN ISO 1183-1*; Plast—Metoder til Bestemmelse af Densiteten af Ikke-Celleplast—Del 1: Nedsænkningsmetode, Flydende Pyknometermetode og Titreringsmetode Plastics—Methods for Determining the Density Deskriptorer. Dansk Standar–Eftertryk uden Tilladelse Forbudt: Geneva, Switzerland, 2011.
22. D3039/D3039M; Standard Test Method for Tensile Properties of Polymer Matrix Composite Materials. Annu B ASTM Stand.: West Conshohocken, PA, USA, 2017; p. 15.
23. D 3518; Standard Test Method for In-Plane Shear Response of Polymer Matrix Composite Materials by Tensile Test of a 645° Laminate 1. Annu B ASTM Stand.: West Conshohocken, PA, USA, 2007; pp. 1–7.
24. Karger-Kocsis, J.; Friedrich, K. Microstructure-related fracture toughness and fatigue crack growth behaviour in toughened, anhydride-cured epoxy resins. *Compos. Sci. Technol.* **1993**, *48*, 263–272. [CrossRef]
25. Xu, K.; Chen, W.; Zhu, X.; Liu, L.; Zhao, G. Chemical, mechanical and morphological investigation on the hygrothermal aging mechanism of a toughened epoxy. *Polym. Test.* **2022**, *110*, 107548. [CrossRef]
26. Glaskova-Kuzmina, T.; Aniskevich, A.; Martone, A.; Giordano, M.; Zarrelli, M. Effect of moisture on elastic and viscoelastic properties of epoxy and epoxy-based carbon fibre reinforced plastic filled with multiwall carbon nanotubes. *Compos. Part A Appl. Sci. Manuf.* **2016**, *90*, 522–527. [CrossRef]
27. Oun, A.; Manalo, A.; Alajarmeh, O.; Abousnina, R.; Gerdes, A. Long-Term Water Absorption of Hybrid Flax Fibre-Reinforced Epoxy Composites with Graphene and Its Influence on Mechanical Properties. *Polymers* **2022**, *14*, 3679. [CrossRef]
28. Brunbauer, J.; Pinter, G. Effects of mean stress and fibre volume content on the fatigue-induced damage mechanisms in CFRP. *Int. J. Fatigue* **2015**, *75*, 28–38. [CrossRef]
29. Alessi, S.; Pitarresi, G.; Spadaro, G. Effect of hydrothermal ageing on the thermal and delamination fracture behaviour of CFRP composites. *Compos. Part B Eng.* **2014**, *67*, 145–153. [CrossRef]
30. Hu, B.; Cong, Y.H.; Zhang, B.Y.; Zhang, L.; Shen, Y.; Huang, H.Z. Enhancement of thermal and mechanical performances of epoxy nanocomposite materials based on graphene oxide grafted by liquid crystalline monomer with Schiff base. *J. Mater. Sci.* **2020**, *55*, 3712–3727. [CrossRef]
31. Wijerathne, D.; Gong, Y.; Afroj, S.; Karim, N.; Abeykoon, C. Mechanical and thermal properties of graphene nanoplatelets-reinforced recycled polycarbonate composites. *Int. J. Light. Mater. Manuf.* **2023**, *6*, 117–128. [CrossRef]
32. Bocchini, S.; Fornasieri, G.; Rozes, L.; Trabelsi, S.; Galy, J.; Zafeiropoulos, N.E.; Stamm, M.; Gérard, J.-F.; Sanchez, C. New hybrid organic-inorganic nanocomposites based on functional [$Ti_{16}O_{16}(OEt)_{24}(OEMA)_8$] nano-fillers. *Chem. Commun.* **2005**, *16*, 2600–2602. [CrossRef]
33. Oliveira, M.S.; da Luz, F.S.; Pereira, A.C.; Costa, U.O.; Bezerra, W.B.A.; Cunha, J.d.S.C.d.; Lopera, H.A.C.; Monteiro, S.N. Water Immersion Aging of Epoxy Resin and Fique Fabric Composites: Dynamic–Mechanical and Morphological Analysis. *Polymers* **2022**, *14*, 3650. [CrossRef]

Disclaimer/Publisher's Note: The statements, opinions and data contained in all publications are solely those of the individual author(s) and contributor(s) and not of MDPI and/or the editor(s). MDPI and/or the editor(s) disclaim responsibility for any injury to people or property resulting from any ideas, methods, instructions or products referred to in the content.

Article

Machinability Measurements in Milling and Recurrence Analysis of Thin-Walled Elements Made of Polymer Composites

Krzysztof Ciecieląg

Department of Production Engineering, Faculty of Mechanical Engineering, Lublin University of Technology, 36 Nadbystrzycka, 20-618 Lublin, Poland; k.ciecielag@pollub.pl

Abstract: The milling of polymer composites is a process that ensures dimensional and shape accuracy and appropriate surface quality. The shaping of thin-walled elements is a challenge owing to their deformation. This article presents the results of milling polymer composites made of glass and carbon fibers saturated with epoxy resin. The milling of each material was conducted using different tools (tools with polycrystalline diamond inserts, physically coated carbide inserts with titanium nitride and uncoated carbide inserts) to show differences in feed force and deformation after the machining of individual thin-walled samples. In addition, the study used recurrence analysis to determine the most appropriate quantifications sensitive to changes occurring in milling different materials with the use of different tools. The study showed that the highest forces occurred in milling thin-walled carbon-fiber-reinforced plastics using uncoated tools and the highest feeds per revolution and cutting speeds. The use of a high feed per revolution (0.8 mm/rev) in carbon-fiber-reinforced plastics machining by uncoated tools resulted in a maximum feed force of 1185 N. A cutting speed of 400 m/min resulted in a force of 754 N. The largest permanent deformation occurred in the milling of glass-fiber-reinforced composite samples with uncoated tools. The permanent deformation value of this material was 0.88 mm. Low feed per revolution (0.1 mm/rev) resulted in permanent deformations of less than 0.30 mm for both types of materials. A change in feed per revolution had the most significant effect on the deformations of thin-walled polymer composites. The analysis of forces and deformation made it possible to conclude that high feed per revolution were not recommended in composite milling. In addition to the analysis of machining thin-walled composites, the novelty of this study was also the use of recurrence methods. Recurrence methods were used to determine the most appropriate quantifications. Determinism, averaged diagonal length and entropy have been shown to be suitable quantifications for determining the type of machined material and the tools used.

Keywords: milling; glass-fiber-reinforced plastics; carbon-fiber-reinforced plastics; thin-walled elements; feed force; deformation; recurrence method

Citation: Ciecielag, K. Machinability Measurements in Milling and Recurrence Analysis of Thin-Walled Elements Made of Polymer Composites. *Materials* **2023**, *16*, 4825. https://doi.org/10.3390/ma16134825

Academic Editor: Raul D. S. G. Campilho

Received: 4 June 2023
Revised: 28 June 2023
Accepted: 3 July 2023
Published: 4 July 2023

Copyright: © 2023 by the author. Licensee MDPI, Basel, Switzerland. This article is an open access article distributed under the terms and conditions of the Creative Commons Attribution (CC BY) license (https://creativecommons.org/licenses/by/4.0/).

1. Introduction

Composite materials that are an alternative to aluminum, titanium and magnesium alloys have been used for several decades, and their share in shaping processes is significantly increasing compared to other materials [1–3]. The demand for polymer composites and their processing is increasing due to their specific and beneficial properties. In particular, the properties of low specific weight and high strength mean they are widely used in elements where material costs are not the main production criterion. In an early period of use of composite materials, the focus was on the needs of the space, defense and aviation industries, which meant that the development of these materials was oriented at achieving specified properties, and not reducing material costs. The relatively high price of composite materials and their machining is balanced by their advantages, namely, in addition to low specific weight and high strength, they have high corrosion resistance, vibration damping ability, ease of forming any shape and good electroinsulation properties [1]. Unfortunately, composites also have a shortcoming, which is their temperature resistance. The long-term use of composites at high temperatures is unfavorable for them. The upper approximate continuous working temperature of composites depending on the type of matrix (which is a factor determining the thermal properties of a composite) is 400 °C for the polymer matrix, 580 °C for the metallic matrix and 1000 °C for the ceramic matrix [4].

Given a growing demand for understanding the phenomena related to composites, many studies have been conducted on the machining of these materials in recent years. These are mainly studies investigating the influence of cutting parameters, type of tool and composite material on cutting forces, roughness parameters and surface topography.

A literature review shows that the technological parameters of milling polymer composites with glass and carbon fibers are within the following limits: depth of cut 0.1–4 mm [5,6], cutting speed 20–250 m/min [7], and feed per tooth 0.01–0.5 mm/tooth [7–10]. According to some studies, cutting speeds of up to 500 m/min are also used [11]. The ranges of the parameters affect the effects of milling. By increasing the cutting speed in machining glass-fiber-reinforced plastics (GFRP), tool wear is significantly increased [12], but at the same time, surface roughness is improved [5,13]. In turn, an increase in feed causes the deterioration of surface quality [5,14]. The machining of carbon-fiber-reinforced plastics (CFRPs) is characterized by similar relationships to those observed in the machining of glass-fiber-reinforced plastics. GFRP machining causes surface roughness to increase as the cutting speed decreases while the feed speed increases [15,16]. An increase in feed also has a negative effect on cutting forces, causing them to increase [17,18]. Researchers have also shown that an increase in cutting speed, which is beneficial for surface quality, unfortunately also causes damage in the form of delamination and leads to higher processing temperature and cutting forces [19].

In addition to the technological parameters of milling, another important aspect is the appropriate selection of tools. Among the tools used for milling polymer composites, there are tools with a polycrystalline diamond (PCD) insert, tools with a chemically applied thin coating (CVD), tools with a physically applied coating (PVD) and uncoated carbide tools [1,20]. This selection depends on the type of composite material, expected machining results and tool costs. In order to achieve a surface roughness characterized by low parameters, tools with polycrystalline diamond inserts are most often used [21]. The cost of the tools with polycrystalline diamond inserts is significantly higher than that of the tools with a physically or chemically applied coating or even uncoated carbide tools. Due to the high cost of PVD, CVD and uncoated tools are also successfully used with satisfactory results [22].

Previous research works devoted to the milling of polymer composites also focus on the type of composite material. Composites are materials with a heterogeneous structure, in which the type and arrangement of fibers determine the properties and workability of the entire material. The arrangement of fibers mainly affects the direction of the load transfer. In terms of obtaining low roughness values, an arrangement of 0° to 30° to the direction of machining is preferred. Exceeding the 30° angle causes the bending of the fibers and, consequently, deterioration of the surface quality [20,23]. The machining process conducted at an angle of 90° to the direction of the fiber arrangement additionally causes their shear [20]. When cutting carbon fiber with a 90° fiber orientation, in addition to shearing, the matrix breaks and the resin bond is damaged [20].

Due to the fact that composite materials have also been used as thin-walled elements, this aspect is the subject of this research. They are a very good alternative to thin-walled structures made of aluminum and titanium alloys [24,25], where it is necessary to ensure high stiffness and strength with a relatively low specific weight. Previous simulations of milling thin-walled composite structures based on FEM analysis enabled the analysis of the milling process plan, taking into account the impact of machining parameters [26]. Numerical analyses and experimental studies were also conducted to investigate the load capacity of thin-walled elements made of polymer composites. Based on their results, it was possible to determine the critical load of the structure [27]. Numerical calculations are a tool that also allows you to determine the stability of thin-walled structures [28]. Research has also been conducted on thin-walled structure machining, but it largely concerned the machining of aluminum alloys. The research in this field focuses on the determination of technological parameters of machining. The most significantly negative impact on deformation and surface roughness was observed by an increase in feed [29], and the second most important parameter was cutting speed. A significant influence of the machining strategy and wall thickness of the processed material on its dimensional accuracy and surface roughness was also found in [30,31]. The machining of thin-walled elements made of aluminum alloys was also studied in [32], where deformation was determined. Deformation is directly related to the resulting stresses generated during machining [33]. In terms of deformation research, it was also shown that even a two-fold reduction in the thickness of a thin-walled element made of aluminum alloy could lead to a double increase in deformation after machining [30]. For machining thin-walled elements made of aluminum alloys, tools with polycrystalline diamond inserts or tools made of sintered carbides are used [34].

In the analysis of economic, industrial and machining phenomena, recurrence plots (RPs) and recurrence quantifications (RQA) are used as machining analysis methods in research on various types of materials [35,36]. The recurrence plot is based on the method of delayed coordinates and is created based on the analysis of the signal which can be a force recorded during machining [37]. The reconstructed lag vector x obtained by the lag method is described by Formula (1):

$$x = (x_i, x_{i+d}, x_{i+2d}, \ldots, x_{i+(m-1)d}). \tag{1}$$

Vector reconstruction involves selecting an embedding delay d, embedding dimension m and threshold parameter ε. In the formula, x_i denotes the i-th coordinate in a given time course [38]. The reconstructed vector is used to create a recurrence plot. However, due to the fact that the plots only provide qualitative information, recurrence quantifications have been introduced [39–41]. They are an advanced tool for the analysis of non-linear signals [42]. Among the many recurrence quantifications, one can distinguish the recurrence rate (RR, percentage of darkened points), determinism (DET, percentage of recurrence points which create diagonal lines), laminarity (LAM, percentage rate of recurrence point which create vertical lines), trapping time (TT, average length of the vertical lines), averaged diagonal length (L, average length of the diagonal line), longest vertical line (V_{max}, longest vertical line of the recurrence structure), length of longest diagonal line (L_{max}, longest diagonal line of the recurrence structure), recurrence time 1st, 2nd (T1, T2, time distances of the recurrence points in the vertical direction), entropy (ENTR, probability distribution of the diagonal line lengths), recurrence period density entropy (RPDE, measure quantifying the extent of recurrences) and clustering coefficient (CC, the probability that two recurrence states are close) [41,43–45]. Recurrence quantifications can be analyzed using two main methods. The first one is based on the assumption that the threshold parameter ε is constant. The other method is that the recurrence rate (RR) is constant and that the threshold parameter ε is changed to ensure a constant value of RR [39].

The use of recurrence methods is associated with recording a signal that serves for further analyses. This method has been successfully employed in medicine [46–48] and materials engineering research [49,50]. Recurrence methods have also been used to assess rotor cracks [51] and to analyze engine operation [52]. In machining, recurrence methods have also been applied. Studies have shown that the RQA technique has great potential in detecting surface wear in cutters. RQA parameters such as entropy (ENTR), trapping time (TT) and laminarity (LAM) can be used to detect insert wear in face milling [53]. In terms of machining, recurrence methods have been used to detect defects during drilling and milling of polymer composites. Studies [43,54–56] used cutting force as an input signal. Recurrence analyses allowed us to demonstrate that there were recurrence quantifications enabling the identification of defects. It was shown that the use of recurrence methods in drilling and milling made it possible not only to detect a defect, but also to determine its size and location. Previous studies also showed that it was possible to determine recurrence quantifications such as laminarity, entropy and recurrence time for detecting defects in polymer composites [43,55,56]. Recurrence analysis was used to study the milling of thin-walled structures made of aluminum alloys. The input signal for analysis was the cutting force, thanks to which the dynamics of the milling process was determined [57].

This article presents the research on milling thin-walled composite structures, a problem which has not been widely studied and is still new. Two different thin-walled composite materials consisting of glass- and carbon-reinforced fibers were used for the study, and they were machined using three types of cutting tools. The novelty of this work is the use of recurrence analysis to describe the cutting process of thin-walled polymer composites. The aim of the research was to determine the recurrence quantification that would be the most sensitive to the type of machined material and tool. In addition, the deformation of composite thin-walled structures was examined depending on the composite material, cutting tool and technological parameters of milling. Previous research on the use of recurrence analysis for defect detection in machining produced satisfactory results.

2. Materials and Methods

The milling experiments were carried out on the AVIA-VMC 800 HS vertical machining center. The center is controlled by Heidenhain iTNC 530. A 3D Kistler dynamometer (type 9257B) was placed in the working space of the machine tool. A vice was mounted at the top of the dynamometer. Test samples were fixed in the jaws of the vice. Each sample was clamped in the vice in the same way. The clamp of the vice for each of the samples ensured a stable fixation. In order to ensure the accuracy and repeatability of the results, the length and location of the milled horizontal surface relative to the mounting was identical for each research. The signal measured with the dynamometer was the feed force in accordance with the X axis of the working space and the dynamometer. The signal was processed by the Kistler charge amplifier (type 5070), the data acquisition card Dynoware (type 5697A) and the Dynoware software (type 2825A). A measuring probe was used to measure deformations of thin-walled elements made of polymer composites after milling. Measurements of the maximum feed force values and deformation were carried out seven times. The measurement of deformation is a permanent change of the shape of the workpiece expressed in mm. The deformation after milling marked on the chart is a deviation from the "ideal" sample size. The study also involved using recurrence methods. A research methodology scheme is shown in Figure 1.

The samples were rectangular plates with the dimensions of 10 mm × 10 mm × 100 mm. The width of the 10 mm sample was selected to ensure reliable milling with the entire diameter of the cutter, which was 12 mm. The thickness of the sample was selected in order to obtain results from four trials of milling composites of different thicknesses, i.e., 10 mm, 8 mm, 6 mm and 4 mm. The length of the sample of 100 mm resulted from the need to fully insert the tool into the material (30 mm) and to ensure stable clamping in the vice. The samples were made of glass and carbon-fiber-reinforced plastics saturated with epoxy resin. The first type of material was a glass-fiber-reinforced plastics (GFRP) called HexPly 916G-7781. It consists of 42 layers of pre-impregnated fabrics with a thickness of 0.24 mm. The material consisting of glass fibers and epoxy resin contains 47.68% carbon, 7.55% silicon, 3.89% nitrogen, 30.87% oxygen, 4.19% aluminum and 4.36% calcium. Other amounts of elements do not exceed 1%. The other type of material was a carbon-fiber-reinforced plastics (CFRP) called HexPly AG193PW-3501, consisting of 33 layers of pre-impregnated fabrics with a thickness of 0.3 mm. The material consisting of carbon fibers and epoxy resin contains 81.85% carbon, 5.71% nitrogen, 10.52% oxygen and 1.23% sulfur. Other amounts of elements do not exceed 1%. In each material, individual prepregs were arranged alternately (0–90° arrangement). After assembling and placing them in a vacuum package, the samples were subjected to polymerization. This process took place in an autoclave for 3 h, at a pressure of 0.6 MPa. The process was carried out at 120 °C for GFRP and at 180 °C for CFRP.

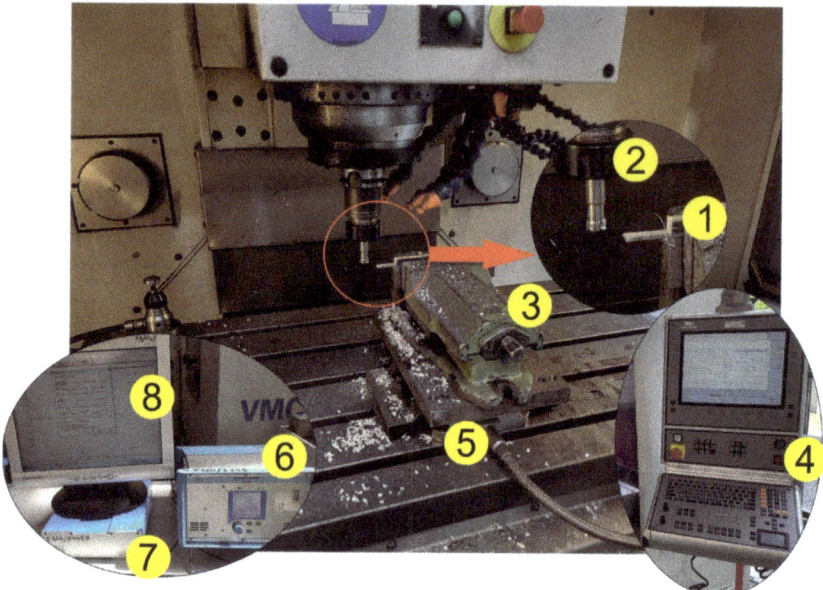

Figure 1. General scheme of the research methodology (1—sample; 2—tool; 3—vice; 4—control panel; 5—dynamometer; 6—charge amplifier; 7—data acquisition card; 8—software).

The machining of thin-walled polymer composites was carried out at a constant depth of 2 mm and variable feed per revolution and cutting speeds, the values of which are listed in Table 1. Technological parameters of milling were selected on the basis of a literature analysis. Each subsequent value of feed per revolution and cutting speed is twice as large as the previous one. Milling was conducted on composite samples with a thickness of 10 mm, 8 mm, 6 mm and 4 mm, over a length of 30 mm.

Table 1. Machining parameters.

No.	Feed per Revolution [mm/rev]	Cutting Speed v_c [m/min]	Depth of Cut a_p [mm]
1	0.1	100	2
2	0.2	100	2
3	0.4	100	2
4	0.8	100	2
5	0.2	50	2
6	0.2	100	2
7	0.2	200	2
8	0.2	400	2

The milling process was carried out using three different cutting tools that are used in the processing of polymer composites. Cutting tools are selected on the basis of a literature analysis and own research. Diamond-coated tools, coated and uncoated carbide tools are used for the machining of polymer composites. The tools were folding milling cutters with a diameter of 12 mm, equipped with a body of type R217.69-1212.0-06-2AN on which were mounted two polycrystalline diamond inserts with the symbol XOEX060204FR PCD05 (denoted by PCD), two physically coated carbide inserts with titanium nitride TiN XOEX060204FR-E03 F40M (denoted by F40M) and two uncoated carbide inserts XOEX060204FR-E03 H15 (denoted by H15). Detailed information about the cutting tools is presented in Table 2.

Table 2. Information about cutting tools (SECO).

Description	XOEX060204FR PCD05	XOEX060204FR-E03 F40M	XOEX060204FR-E03 H15
Gradetype	PCD	Carbide PVD	Carbide Uncoated
Clearance angle major	15°	15°	15°
Corner radius	0.4 mm	0.4 mm	0.4 mm
Cutting edge effective length	2.5 mm	6.0 mm	6.0 mm
Wiper edge length	1.1 mm	0.9 mm	0.9 mm
Insert thickness	2.45 mm	2.45 mm	2.45 mm

3. Results and Discussion

The research on the milling of thin-walled polymer composites made it possible to measure the feed force in order to present its value depending on the variable technological parameters, processed material and tool type. Standard deviations are marked in the plots in Figures 2 and 3, Figures 4a,b and 5a,b. Small values of the standard deviation indicate that the results are close to the average value. Figure 2 shows the effect of feed per revolution on the maximum values of the feed force F in milling thin-walled polymer composites with a thickness of 4 mm, made of glass-fiber-reinforced (GFRP) and carbon-fiber-reinforced (CFRP) plastics, using three types of tools (PCD, F40M and H15).

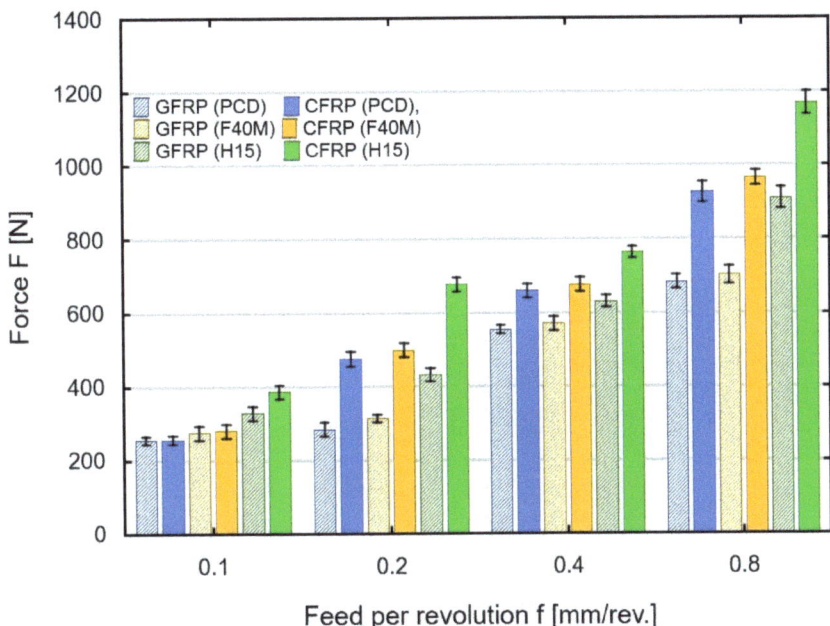

Figure 2. Feed per revolution versus maximum feed force in milling thin-walled polymer composites.

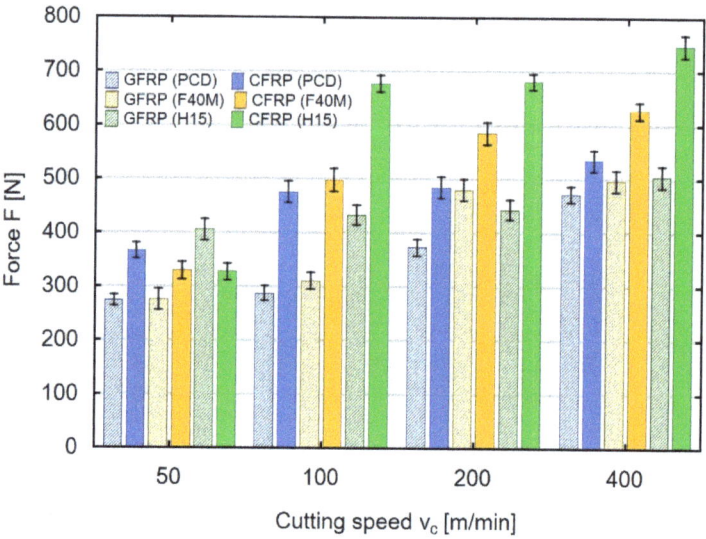

Figure 3. Cutting speed versus maximum feed force during milling of thin-walled polymer composites.

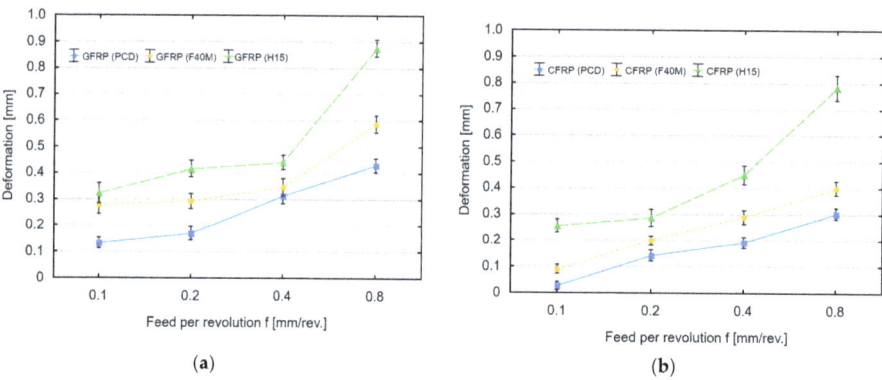

Figure 4. Feed per revolution versus permanent deformation after milling (**a**) GFRP and (**b**) CFRP.

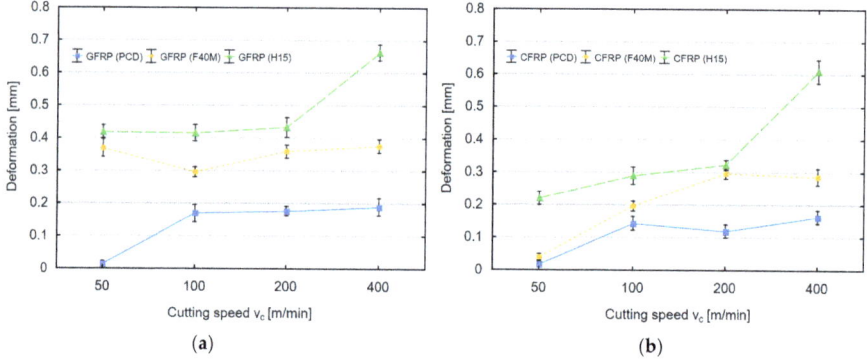

Figure 5. Cutting speed versus permanent deformation after milling for (**a**) GFRP and (**b**) CFRP.

The results presented in Figure 2 show that for each case of feed per revolution, higher maximum feed forces were obtained in machining thin-walled CFRP composites. At low feed per revolution values, the differences between the maximum force values for individual materials using the same tools are small. For the smallest value of the tested feed per revolution, maximum force values in the range of 260–390 N were obtained. Smaller force values (260–265 N) refer to the milling of GFRP and CFRP using PCD tools, and the largest (390 N) were obtained during the milling of CFRP using tools with uncoated inserts. At higher feeds per revolution, the differences begin to increase when comparing the maximum values of the feed force obtained for individual types of materials and tools. The most noticeable change is when milling CFRP with uncoated tools. For a feed per revolution of 0.2 mm/rev, a value of 685 N was obtained, and for a feed per revolution of 0.4 mm/rev, this value has already increased to 769 N. The highest values of the feed force were obtained as a result of machining thin-walled carbon-fiber-reinforced plastics using the uncoated carbide tools at a feed per revolution of 0.8 mm/rev. For the CFRP material, the use of uncoated tools has a negative effect on the force which reaches almost 1200 N. For comparison, the use of the PCD or F40M (PVD) tools at a feed per revolution of 0.8 mm/rev resulted in a reduction in the maximum feed force value by 20%. The increase in feed per revolution is significant in machining thin-walled polymer composites because for 0.8 mm/rev, when compared to 0.1 mm/rev, there was a three-fold increase in forces when the machining process was conducted with the uncoated tools. To sum up, the use of the highest feed per revolution value resulted in force values of 692 N, 711 N and 920 N, respectively, during GFRP milling for PCD, F40M and H15 tools. During CFRP machining, 935 N (for PCD), 974 N (for F40M) and 1185N (for H15) were obtained. A moving tool with a higher feed per revolution must remove the same amount of workpiece in less time. The use of a higher feed per revolution causes an increase in the tool load, and thus an increase in the maximum feed force values. Higher maximum feed force was created when milling CFRP because it is a material with greater strength.

Milling was also performed with variable cutting speed. Figure 3 shows the impact of cutting speed on the force values in milling glass- and carbon-fiber-reinforced plastics. The milling process was conducted with three different tools on the sample with a thickness of 4 mm.

An analysis of the cutting speeds showed that in the machining of thin-walled polymer composites, the maximum values of the feed force increase. The clearest increase was observed in the machining of carbon-fiber-reinforced plastics. Compared to the cutting speed of 50 m/min, the use of a cutting speed above 100 m/min caused a clear increase in the feed force in CFRP machining. The highest feed force values were obtained when the machining of both types of materials was conducted with the uncoated carbide tools. For the lowest cutting speed of 50 m/min, the maximum feed force values of 280 N were obtained when milling GFRP using PCD and F40M tools. The use of H15 tools, i.e., uncoated sintered carbides for GFRP machining, resulted in an increase in force to 414 N. Low cutting speeds during CFRP machining resulted in the values of 372 N (for PCD), 332 N (for F40M and 334 N (for H15). The increase in cutting speed meant that in the case of CFRP machining, the values of the maximum feed force increased significantly. For the cutting speed of 100 m/min, the use of PCD tools for CFRP machining resulted in a force of 483 N, for F40M a force of 498 N, and for H15 the force was already 682 N. A further increase in the cutting speed to 200 m/min and 400 m/min caused the maximum values of the feed force to continue to increase. At the highest cutting speed tested during CFRP machining, the forces obtained were 543 N (for PCD), 636 N (for F40M) and 754 N (for H15). The increase in the maximum value of the feed force when machining thin-walled GFRP and CFRP materials with the increase in cutting speed can be explained by the high resistance of the workpiece. One explanation is that despite the increasing cutting speed, and thus the higher rotational speed of the tool, composite materials are difficult to machine.

In addition to cutting force, the deformation of thin-walled polymer composites was also analyzed. Figure 4a,b show the influence of feed per revolution and tool type on the maximum permanent deformation of the GFRP and CFRP samples with a thickness of 4 mm after milling.

Figure 4a shows the maximum deformation obtained with increasing feed per revolution in milling the GFRP samples. For the smallest feed per revolution of 0.1 mm/rev, the deformation when machining GFRP with PCD tools is 0.14 mm, and for F40M and H15 tools, the deformation is about 0.30 mm. An increase in feed per revolution causes an increase in deformations, which for the highest feed per revolution value are 0.44 mm (for PCD), 0.58 mm (for F40M) and 0.88 mm (for H15). The highest deformations occurred in milling with the uncoated tools. A clear increase in deformations can be observed for a feed per revolution value of 0.8 mm/rev. Figure 4b shows the effect of feed per revolution on permanent deformation after CFRP milling. In the case of deformation analysis after CFRP machining, the values of 0.04 mm (for PCD), 0.10 mm (for F40M) and 0.25 mm (for H15) were obtained for the smallest value of the tested feed per revolution. Increasing the feed per revolution value caused further permanent deformations after CFRP machining and led to dimensional changes of 0.31 mm (for PCD), 0.41 mm (for F40M) and 0.78 mm (for H15) for the highest feed per revolution value. The results clearly show an increase in permanent deformation with increasing feed per revolution, with the largest deformation observed for the machining process conducted with the uncoated tools. A comparison of the two types of tested materials shows that higher deformations occurred in the sample made of GFRP. The average value of permanent deformation for the GFRP samples is 10–30% higher than that obtained for the CFRP material under the same processing conditions. The feed per revolution parameter is important when machining thin-walled polymer composites. Greater deformations during milling of composite materials along with increasing the feed per revolution are caused by the resistance that the tool must resist. As a result of increasing feed per revolution, the material is partially bent, leaving a permanent deformation after machining.

The plots in Figure 5a,b show the effect of cutting speed and tool type on the maximum permanent deformation of the GFRP and CFRP materials after milling. The thickness of the sample was 4 mm, and the milling depth was 2 mm.

Based on the plots illustrating the impact of cutting speed on permanent deformation, it can be concluded that cutting speed is a less important parameter than feed per revolution. The use of the polycrystalline diamond insert tools and the tools with a physically applied coating for different cutting parameters did not show any clear dependencies. For both types of materials, the use of these tools did not cause significant differences in deformation. Low cutting speed values caused GFRP deformation at the level of 0.02 mm for PCD and about 0.40 mm for F40M and H15. For the lowest cutting speed tested, the permanent deformation of CFRP was 0.02 mm (for PCD), 0.03 mm (for F40M) and 0.23 mm (for H15). Only the use of the uncoated tools for the cutting speed of 400 m/min resulted in a significant increase in deformation for both materials. Smaller values of the feed force and greater deformations of GFRP can be explained by the fact that this material is characterized by lower strength.

Another new aspect of the work is the introduction of recurrence analysis to the study of thin-walled polymer composites. Figure 6a–k show the values of recurrence quantifications depending on the tool type for the samples with a thickness of 4 mm, made of glass and carbon-fiber-reinforced plastics. The analysis was carried out with constant values of the milling parameters: f = 0.2 mm/rev and cutting speed v_c = 100 m/min. The recurrence analysis was carried out with a constant recurrence rate RR. To ensure a constant RR value, the threshold parameter was set to ε = 0.1. In the analysis, the embedding dimension was equal to 5 and the embedding delay was equal to 2.

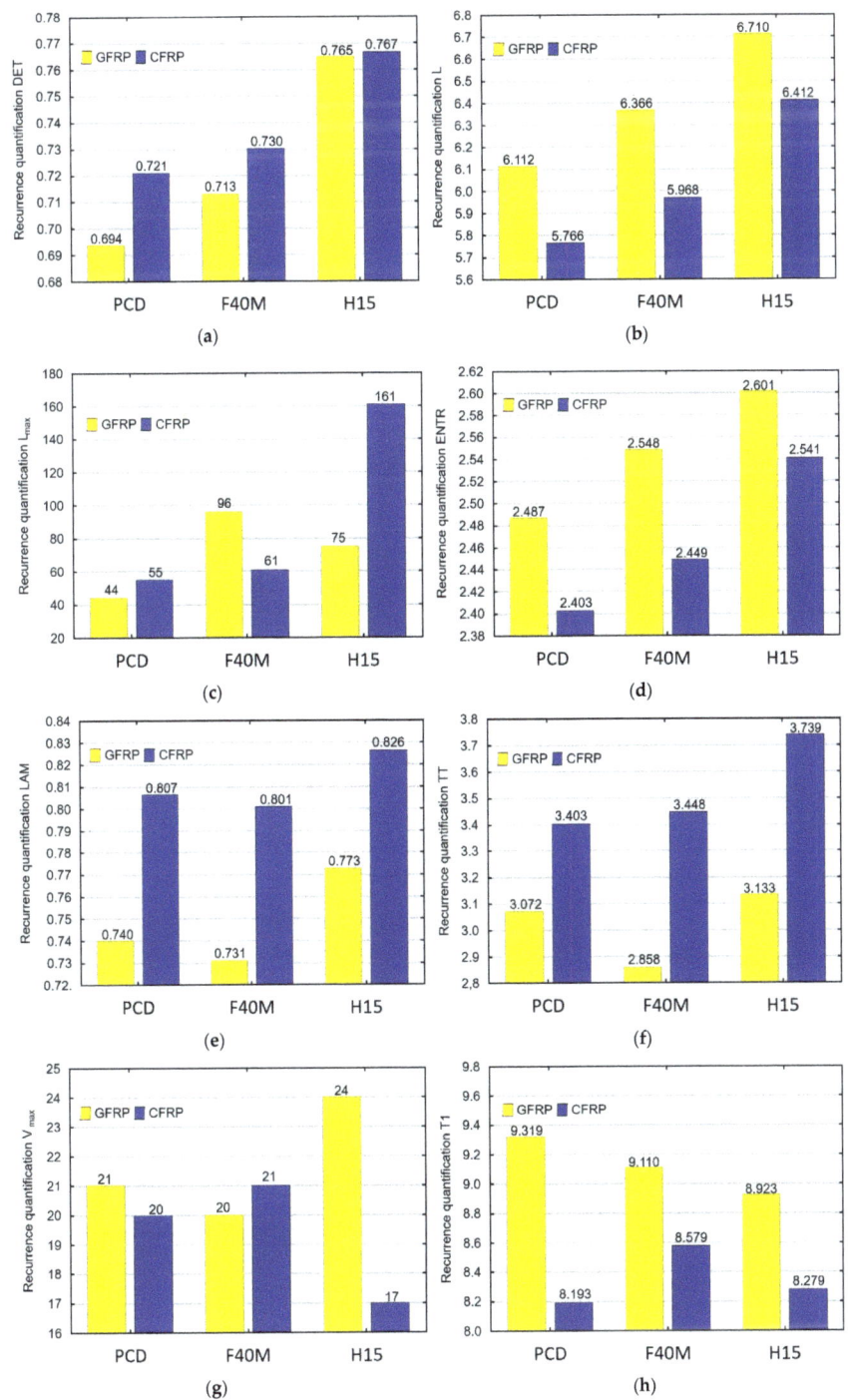

Figure 6. Cont.

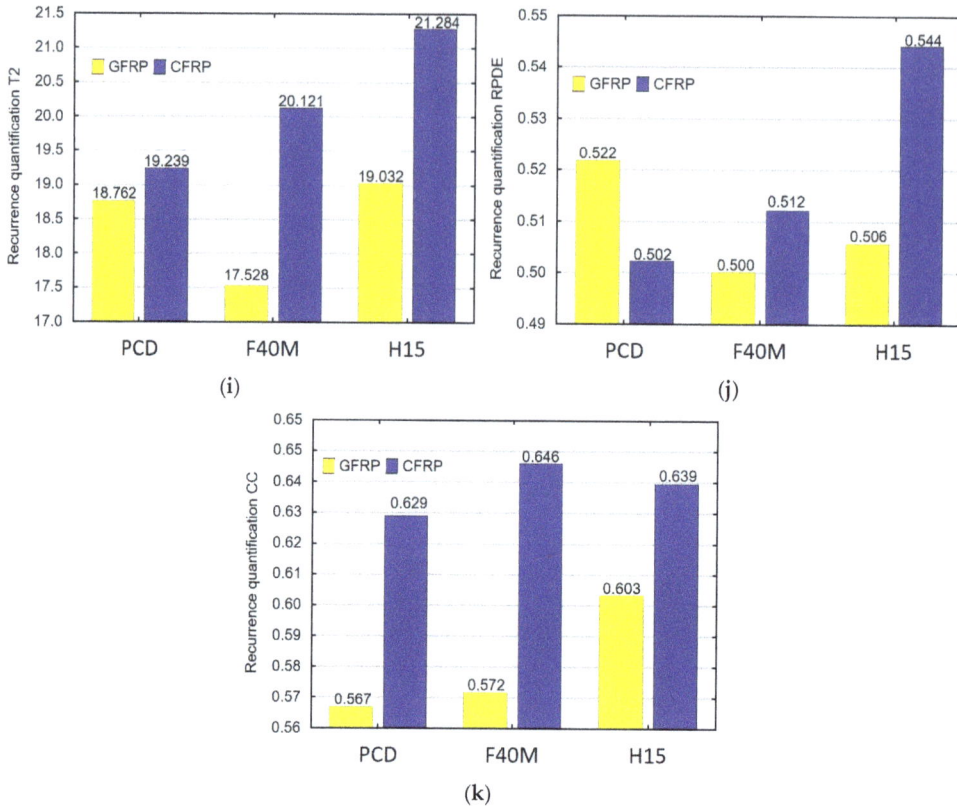

Figure 6. Type of tool and material versus recurrence quantifications: (**a**) DET, (**b**) L, (**c**) L_{max}, (**d**) ENTR, (**e**) LAM, (**f**) TT, (**g**) V_{max}, (**h**) T1, (**i**) T2, (**j**) RPDE and (**k**) CC.

The recurrence quantifications such as DET, L, ENTR, LAM, TT, T1, T2 and CC can be used to identify the type of material. The choice of these indicators is associated with a clear difference between their values for a given type of used tool.

Based on the above recurrence quantifications plots, it can be concluded that the indicators suitable for identifying the type of tool in GFRP milling are DET, L, ENTR, T1 and CC. This conclusion is supported by a clear change in the values of these quantifications in the plots. For the CFRP machining analysis, DET, L, L_{max}, ENTR, TT, T2 and RPDE can be used to identify the tool type. DET, L and ENTR can be used to identify the type of tool in machining two types of materials.

4. Conclusions

This study investigated the machinability of thin-walled polymer composites. The milling of two types of composite materials reinforced with glass and carbon fibers was conducted using three types of tools. These were as follows: milling cutters with polycrystalline diamond inserts, physically coated carbide inserts with titanium nitride and uncoated carbide inserts. Machining was carried out with variable feed per revolution and cutting speed parameters. The results showed that the use of high feeds per revolution and high cutting speeds in machining thin-walled polymer composites had a negative effect on the maximum values of the feed force. For each tool, the milling of CFRP produced higher maximum cutting forces than those observed for GFRP. Due to the force values, the feed per revolution had the greatest impact on the machining of thin-walled polymer composites. A three-fold increase in the maximum feed force was obtained by comparing feeds per revolution of 0.8 mm/rev and 0.1 mm/rev. Changing the cutting speed also increased the maximum feed force. However, the cutting speed was the second most important milling parameter. The research results obtained in this article are consistent with the results of the work of other researchers whose works were quoted in the introduction. The tendency is confirmed that increasing the feed per revolution and cutting speed negatively affects the cutting force. The analysis of deformations after milling showed that the feed per revolution affected the machining effects for both types of materials and for each of the tools used. An increase in feed per revolution caused deformations after milling. The analysis of the impact of the cutting speed did not show significant dependencies, but the use of the uncoated carbide tools was found to significantly increase deformations. The research results for samples with a thickness of 10 mm, 8 mm, 6 mm and 4 mm are analogous. This article shows the results for the smallest thickness of the tested sample.

The novelty of this study is that it used recurrence analysis to select appropriate quantifications for the identification of the type of workpiece and tool. The recurrence methods made it possible to determine that the most popular indicators of tool and material identification were DET, L and ENTR. The plots generated for these indicators showed a change in their values depending on the type of tool and workpiece. This change was determined by comparing the maximum values of the feed force. An analysis of the maximum feed force showed that the high values resulted from the use of the uncoated tools for machining CFRP. This relationship can also be seen in the plots of recurrence quantifications. The DET values increase with the use of a tool that causes higher maximum feed force and large deformations. The DET quantification is characterized by higher values when machining CFRP depending on the tool type. The L indicator also increases when the inferior tools are used. However, in the machining of CFRP, these values were lower than the values obtained for GFRP. A similar relationship to the L indicator can be seen when analyzing changes in the ENTR indicator. The course of the indicators in the plots is similar to the analysis of the maximum values of the feed force.

Future research will study the geometric structure of thin-walled composite material structure after milling. Also, indicators dependent on technological parameters of milling will be determined via recurrence analysis. The proposed methodology for the research of thin-walled composite structures can be successfully used in an industrial environment. Cutting force measurements require the use of a dynamometer whose placement in the working space does not interfere with industrial conditions.

Funding: The activities of the Polish Metrological Union are financed from the funds of the Ministry of Education and Science as part of a targeted subsidy for the implementation of the task titled "Establishment and Coordination of the activities of the Polish Metrological Union (PMU)" under contract No. MEiN/2021/DPI/179. This research was partially supported by the Mechanical Engineering Discipline Fund of Lublin University of Technology (Grant No. FD-20/IM-5/016).

Institutional Review Board Statement: Not applicable.

Informed Consent Statement: Not applicable.

Data Availability Statement: Not applicable.

Conflicts of Interest: The author declares no conflict of interest.

References

1. Teti, R. Machining of Composite Materials. *CIRP Ann.* **2002**, *51*, 611–634. [CrossRef]
2. Matuszak, J. Effect of Ceramic Brush Treatment on the Surface Quality and Edge Condition of Aluminium Alloy after Abrasive Waterjet Machining. *Adv. Sci. Technol. Res. J.* **2021**, *15*, 254–263. [CrossRef]
3. Kuczmaszewski, J.; Zaleski, K.; Matuszak, J.; Mądry, J. Testing Geometric Precision and Surface Roughness of Titanium Alloy Thin-Walled Elements Processed with Milling. In *Advances in Manufacturing II*; Diering, M., Wieczorowski, M., Brown, C.A., Eds.; Springer International Publishing: Cham, Switzerland, 2019; pp. 95–106. ISBN 978-3-030-18681-4.
4. Papakonstantinou, C.G.; Balaguru, P.; Lyon, R.E. Comparative Study of High Temperature Composites. *Compos. Part B Eng.* **2001**, *32*, 637–649. [CrossRef]

5. Azmi, A.I.; Lin, R.J.T.; Bhattacharyya, D. Machinability Study of Glass Fibre-Reinforced Polymer Composites during end Milling. *Int. J. Adv. Manuf. Technol.* **2013**, *64*, 247–261. [CrossRef]
6. Hintze, W.; Hartmann, D. Modeling of Delamination During Milling of Unidirectional CFRP. *Procedia CIRP* **2013**, *8*, 444–449. [CrossRef]
7. Ghidossi, P.; El Mansori, M.; Pierron, F. Edge Machining Effects on the Failure of Polymer Matrix Composite Coupons. *Compos. Part A Appl. Sci. Manuf.* **2004**, *35*, 989–999. [CrossRef]
8. Hosokawa, A.; Hirose, N.; Ueda, T.; Furumoto, T. High-Quality Machining of CFRP with High Helix end Mill. *CIRP Ann.* **2014**, *63*, 89–92. [CrossRef]
9. Karpat, Y.; Polat, N. Mechanistic Force Modeling for Milling of Carbon Fiber Reinforced Polymers with Double Helix Tools. *CIRP Ann.* **2013**, *62*, 95–98. [CrossRef]
10. Yuanyushkin, A.S.; Rychkov, D.A.; Lobanov, D.V. Surface Quality of the Fiberglass Composite Material after Milling. *Appl. Mech. Mater.* **2014**, *682*, 183–187. [CrossRef]
11. Teicher, U.; Rosenbaum, T.; Nestler, A.; Brosius, A. Characterization of the Surface Roughness of Milled Carbon Fiber Reinforced Plastic Structures. *Procedia CIRP* **2017**, *66*, 199–203. [CrossRef]
12. Azmi, A.I.; Lin, R.J.T.; Bhattacharyya, D. Experimental Study of Machinability of GFRP Composites by end Milling. *Mater. Manuf. Process.* **2012**, *27*, 1045–1050. [CrossRef]
13. Razfar, M.R.; Zadeh, M.R.Z. Optimum Damage and Surface Roughness Prediction in end Milling Glass Fibre-Reinforced Plastics, Using Neural Network and Genetic Algorithm. *Proc. Inst. Mech. Eng. Part B J. Eng. Manuf.* **2009**, *223*, 653–664. [CrossRef]
14. Jenarthanan, M.P.; Karthikeyan, M.; Kumar, K.P. Experimental Investigation of Surface Roughness and Delamination Using Artificial Intelligence. Antofagasta, Chile. 2023, p. 020014. Available online: https://pubs.aip.org/aip/acp/article-abstract/2715/1/020014/2890547/Experimental-investigation-of-surface-roughness?redirectedFrom=fulltext (accessed on 4 June 2023).
15. Davim, J.P.; Reis, P. Damage and Dimensional Precision on Milling Carbon Fiber-Reinforced Plastics Using Design Experiments. *J. Mater. Process. Technol.* **2005**, *160*, 160–167. [CrossRef]
16. Kiliçkap, E.; Yardimeden, A.; Çelik, Y.H. Investigation of Experimental Study of End Milling of CFRP Composite. *Sci. Eng. Compos. Mater.* **2015**, *22*, 89–95. [CrossRef]
17. Rusinek, R. Cutting Process of Composite Materials: An Experimental Study. *Int. J. Non-Linear Mech.* **2010**, *45*, 458–462. [CrossRef]
18. Bayraktar, S.; Turgut, Y. Investigation of the Cutting Forces and Surface Roughness in Milling Carbon-Fiber-Reinforced Polymer Composite Material. *Mater. Tehnol.* **2016**, *50*, 591–600. [CrossRef]
19. Chibane, H.; Serra, R.; Leroy, R. Optimal Milling Conditions of Aeronautical Composite Material under Temperature, Forces and Vibration Parameters. *J. Compos. Mater.* **2017**, *51*, 3453–3463. [CrossRef]
20. Wei, Y.; An, Q.; Cai, X.; Chen, M.; Ming, W. Influence of Fiber Orientation on Single-Point Cutting Fracture Behavior of Carbon-Fiber/Epoxy Prepreg Sheets. *Materials* **2015**, *8*, 6738–6751. [CrossRef]
21. Nurhaniza, M.; Ariffin, M.K.A.M.; Mustapha, F.; Baharudin, B.T.H.T. Analyzing the Effect of Machining Parameters Setting to the Surface Roughness during end Milling of CFRP-Aluminium Composite Laminates. *Int. J. Manuf. Eng.* **2016**, *2016*, 4680380. [CrossRef]
22. Saglam, H.; Unsacar, F.; Yaldiz, S. Investigation of the Effect of Rake Angle and Approaching Angle on Main Cutting Force and Tool Tip Temperature. *Int. J. Mach. Tools Manuf.* **2006**, *46*, 132–141. [CrossRef]
23. Palanikumar, K.; Karunamoorthy, L.; Karthikeyan, R. Assessment of Factors Influencing Surface Roughness on the Machining of Glass Fiber-Reinforced Polymer Composites. *Mater. Des.* **2006**, *27*, 862–871. [CrossRef]
24. Kurpiel, S.; Zagórski, K.; Cieślik, J.; Skrzypkowski, K. Investigation of Selected Surface Topography Parameters and Deformation during Milling of Vertical Thin-Walled Structures from Titanium Alloy Ti6Al4V. *Materials* **2023**, *16*, 3182. [CrossRef] [PubMed]
25. Ciecielag, K.; Zaleski, K. Milling of Three Types of Thin-Walled Elements Made of Polymer Composite and Titanium and Aluminum Alloys Used in the Aviation Industry. *Materials* **2022**, *15*, 5949. [CrossRef]
26. Rai, J.K.; Xirouchakis, P. Finite Element Method Based Machining Simulation Environment for Analyzing Part Errors Induced during Milling of Thin-Walled Components. *Int. J. Mach. Tools Manuf.* **2008**, *48*, 629–643. [CrossRef]
27. Rozylo, P.; Debski, H.; Wysmulski, P.; Falkowicz, K. Numerical and Experimental Failure Analysis of Thin-Walled Composite Columns with a Top-Hat Cross Section under Axial Compression. *Compos. Struct.* **2018**, *204*, 207–216. [CrossRef]
28. Rozylo, P.; Falkowicz, K. Stability and Failure Analysis of Compressed Thin-Walled Composite Structures with Central Cut-out, Using Three Advanced Independent Damage Models. *Compos. Struct.* **2021**, *273*, 114298. [CrossRef]
29. Ramanaiah, B.V.; Manikanta, B.; Ravi Sankar, M.; Malhotra, M.; Gajrani, K.K. Experimental Study of Deflection and Surface Roughness in Thin Wall Machining of Aluminum Alloy. *Mater. Today Proc.* **2018**, *5*, 3745–3754. [CrossRef]
30. Borojevic, S.; Lukic, D.; Milosevic, M.; Vukman, J.; Kramar, D. Optimization of Process Parameters for Machining of Al 7075 Thin-Walled Structures. *Adv. Prod. Eng. Manag.* **2018**, *13*, 125–135. [CrossRef]
31. Bałon, P.; Rejman, E.; Świątoniowski, A.; Kiełbasa, B.; Smusz, R.; Szostak, J.; Cieślik, J.; Kowalski, Ł. Thin-Walled Integral Constructions in Aircraft Industry. *Procedia Manuf.* **2020**, *47*, 498–504. [CrossRef]
32. Zawada-Michałowska, M.; Pieśko, P. Post-Machining Deformations of Thin-Walled Elements Made of EN AW-2024 T351 Aluminum Alloy as Regards the Mechanical Properties of the Applied, Rolled Semi-Finished Products. *Materials* **2021**, *14*, 7591. [CrossRef]

33. Singh, A.; Agrawal, A. Investigation of Surface Residual Stress Distribution in Deformation Machining Process for Aluminum Alloy. *J. Mater. Process. Technol.* **2015**, *225*, 195–202. [CrossRef]
34. Masoudi, S.; Amini, S.; Saeidi, E.; Eslami-Chalander, H. Effect of Machining-Induced Residual Stress on the Distortion of Thin-Walled Parts. *Int. J. Adv. Manuf. Technol.* **2015**, *76*, 597–608. [CrossRef]
35. Wiertel, M.; Zaleski, K.; Gorgol, M.; Skoczylas, A.; Zaleski, R. Impact of Impulse Shot Peening Parameters on Properties of Stainless Steel Surface. *Acta Phys. Pol. A* **2017**, *132*, 1611–1616. [CrossRef]
36. Skoczylas, A.; Zaleski, K.; Zaleski, R.; Gorgol, M. Analysis of Surface Properties of Nickel Alloy Elements Exposed to Impulse Shot Peening with the Use of Positron Annihilation. *Materials* **2021**, *14*, 7328. [CrossRef] [PubMed]
37. Eckmann, J.-P.; Kamphorst, S.O.; Ruelle, D. Recurrence Plots of Dynamical Systems. *Europhys. Lett.* **1987**, *4*, 973–977. [CrossRef]
38. Fraser, A.M.; Swinney, H.L. Independent Coordinates for Strange Attractors from Mutual Information. *Phys. Rev. A* **1986**, *33*, 1134–1140. [CrossRef]
39. Marwan, N.; Carmenromano, M.; Thiel, M.; Kurths, J. Recurrence Plots for the Analysis of Complex Systems. *Phys. Rep.* **2007**, *438*, 237–329. [CrossRef]
40. Webber, C.L.; Zbilut, J.P. Dynamical Assessment of Physiological Systems and States Using Recurrence Plot Strategies. *J. Appl. Physiol.* **1994**, *76*, 965–973. [CrossRef]
41. Zbilut, J.P.; Webber, C.L. Embeddings and Delays as Derived from Quantification of Recurrence Plots. *Phys. Lett. A* **1992**, *171*, 199–203. [CrossRef]
42. Almeida-Ñauñay, A.F.; Benito, R.M.; Quemada, M.; Losada, J.C.; Tarquis, A.M. Recurrence Plots for Quantifying the Vegetation Indices Dynamics in a Semi-Arid Grassland. *Geoderma* **2022**, *406*, 115488. [CrossRef]
43. Kecik, K.; Ciecielag, K.; Zaleski, K. Damage Detection by Recurrence and Entropy Methods on the Basis of Time Series Measured during Composite Milling. *Int. J. Adv. Manuf. Technol.* **2020**, *111*, 549–563. [CrossRef]
44. Schinkel, S.; Dimigen, O.; Marwan, N. Selection of Recurrence Threshold for Signal Detection. *Eur. Phys. J. Spec. Top.* **2008**, *164*, 45–53. [CrossRef]
45. Gao, J.; Cai, H. On the Structures and Quantification of Recurrence Plots. *Phys. Lett. A* **2000**, *270*, 75–87. [CrossRef]
46. Marwan, N.; Wessel, N.; Meyerfeldt, U.; Schirdewan, A.; Kurths, J. Recurrence-Plot-Based Measures of Complexity and Their Application to Heart-Rate-Variability Data. *Phys. Rev. E* **2002**, *66*, 026702. [CrossRef]
47. Thomasson, N.; Hoeppner, T.J.; Webber, C.L.; Zbilut, J.P. Recurrence Quantification in Epileptic EEGs. *Phys. Lett. A* **2001**, *279*, 94–101. [CrossRef]
48. Iwaniec, J.; Iwaniec, M. Application of Recurrence-Based Methods to Heart Work Analysis. In *Advances in Technical Diagnostics*; Timofiejczuk, A., Łazarz, B.E., Chaari, F., Burdzik, R., Eds.; Springer International Publishing: Cham, Switzerland, 2018; Volume 10, pp. 343–352. ISBN 978-3-319-62041-1.
49. Iwaniec, J.; Kurowski, P. Experimental Verification of Selected Methods Sensitivity to Damage Size and Location. *J. Vib. Control* **2017**, *23*, 1133–1151. [CrossRef]
50. Iwaniec, J.; Uhl, T.; Staszewski, W.J.; Klepka, A. Detection of Changes in Cracked Aluminium Plate Determinism by Recurrence Analysis. *Nonlinear Dyn.* **2012**, *70*, 125–140. [CrossRef]
51. Litak, G.; Sawicki, J.T.; Kasperek, R. Cracked Rotor Detection by Recurrence Plots. *Nondestruct. Test. Eval.* **2009**, *24*, 347–351. [CrossRef]
52. Ilie, C.O.; Alexa, O.; Lespezeanu, I.; Marinescu, M.; Grosu, D. Recurrence Plot Analysis to Study Parameters of a Gasoline Engine. *Appl. Mech. Mater.* **2016**, *823*, 323–328. [CrossRef]
53. Mhalsekar, S.D.; Rao, S.S.; Gangadharan, K.V. Investigation on Feasibility of Recurrence Quantification Analysis for Detecting Flank Wear in Face Milling. *Int. J. Eng. Sci. Technol.* **2010**, *2*, 23–38. [CrossRef]
54. Kecik, K.; Ciecielag, K.; Zaleski, K. Damage Detection of Composite Milling Process by Recurrence Plots and Quantifications Analysis. *Int. J. Adv. Manuf. Technol.* **2017**, *89*, 133–144. [CrossRef]
55. Ciecielag, K.; Kecik, K.; Zaleski, K. Defects Detection from Time Series of Cutting Force in Composite Milling Process by Recurrence Analysis. *J. Reinf. Plast. Compos.* **2020**, *39*, 890–901. [CrossRef]
56. Ciecielag, K.; Skoczylas, A.; Matuszak, J.; Zaleski, K.; Kęcik, K. Defect Detection and Localization in Polymer Composites Based on Drilling Force Signal by Recurrence Analysis. *Measurement* **2021**, *186*, 110126. [CrossRef]
57. Rusinek, R.; Zaleski, K. Dynamics of Thin-Walled Element Milling Expressed by Recurrence Analysis. *Meccanica* **2016**, *51*, 1275–1286. [CrossRef]

Disclaimer/Publisher's Note: The statements, opinions and data contained in all publications are solely those of the individual author(s) and contributor(s) and not of MDPI and/or the editor(s). MDPI and/or the editor(s) disclaim responsibility for any injury to people or property resulting from any ideas, methods, instructions or products referred to in the content.

Article

The Effect of Clearance Angle on Tool Life, Cutting Forces, Surface Roughness, and Delamination during Carbon-Fiber-Reinforced Plastic Milling

Tomáš Knápek *, Štěpánka Dvořáčková and Martin Váňa

Assembly and Engineering Metrology, Department of Machining, Faculty of Mechanical Engineering, Technical University of Liberec, 461 17 Liberec, Czech Republic
* Correspondence: tomas.knapek@tul.cz

Abstract: This study aimed to investigate the effect of the clearance angle of the milling tool on wear, cutting forces, machined edge roughness, and delamination during non-contiguous milling of carbon-fiber-reinforced plastic (CFRP) composite panels with a twill weave and 90° fiber orientation. To achieve the objective of the study, it was first necessary to design suitable tools (6 mm diameter sintered carbide shank milling cutters) with a variety of clearance angles (8.4°, 12.4°, and 16.4°) and all the machinery and measuring equipment for the research to be carried out. Furthermore, measurement and evaluation methods for cutting tool wear, cutting forces, machined edge roughness, and delamination were developed. Last but not least, the results obtained during the research were summarized and evaluated. From the experiments conducted in this study, it was found that the tool clearance angle has a significant effect on tool wear, roughness of the machined surface, and delamination of the carbon fiber composite board. The tool with a clearance angle of 8.4° wore faster than the tool with a clearance angle of 16.4°. The same trend was observed for cutting force, machined surface roughness, and delamination. In this context, it was also shown that the cutting force increased as the tool wear increased, which in turn increased surface roughness and delamination. These results are of practical significance, not only in terms of the quality of the machined surface but also in terms of time, cost, and energy savings when machining CFRP composite materials.

Keywords: milling; CFRP; tool wear; tool parameters; delamination

Citation: Knápek, T.; Dvořáčková, Š.; Váňa, M. The Effect of Clearance Angle on Tool Life, Cutting Forces, Surface Roughness, and Delamination during Carbon-Fiber-Reinforced Plastic Milling. *Materials* 2023, *16*, 5002. https://doi.org/10.3390/ma16145002

Academic Editors: Szymon Wojciechowski and Raul D. S. G. Campilho

Received: 26 May 2023
Revised: 14 June 2023
Accepted: 5 July 2023
Published: 14 July 2023

Copyright: © 2023 by the authors. Licensee MDPI, Basel, Switzerland. This article is an open access article distributed under the terms and conditions of the Creative Commons Attribution (CC BY) license (https://creativecommons.org/licenses/by/4.0/).

1. Introduction

The machining of composite materials is highly desirable due to their widespread use in various industries. However, machining these composite materials poses several difficulties due to their properties, which differ significantly from those of traditional metallic materials [1,2].

Key parameters that affect the cutting process include cutting conditions, such as axial and radial depth of cut, engagement angle, feed, and feed rate profile. These parameters, along with their variations along the tool path, have a direct impact on cutting speed, tool wear, and cutting force [3–5].

The damage forms and failure modes of carbon-fiber-reinforced plastic (CFRP) materials are strongly influenced by changes in their microstructures and interface properties. These properties include fiber orientation, fiber and matrix volume fraction, as well as the presence of voids and cracks. Since CFRP composites consist of a reinforced phase (carbon fiber) and a continuous phase (epoxy resin), the machining process for CFRP is more complex compared to that of homogeneous materials [6,7].

When machining CFRP, it is recommended not to exceed the glass transition temperature (Tg) of the material matrix. The cutting temperature significantly influences the subsurface damage of the hole wall, thereby influencing the tensile performance of the material [8,9].

The direction of the CFRP fiber has a significant effect on the quality of the machined layer and the wear of the cutting tool. The mechanical interaction between CFRP and the cutter is primarily influenced by the anisotropy of the material and the angle of the cutter. In CFRP, the fibers bear the load while the resin facilitates force transfer between the fibers. Therefore, the failure modes of the material can be analyzed primarily based on the stresses experienced by the fibers [10,11].

During milling in a direction parallel to the fiber direction (0°), the tool moves at a specific speed, and the main cutting force is generated by the interaction between the tool face and the material. This cutting force can be divided into components that are parallel and perpendicular to the fiber direction. The perpendicular force generates compressive stress on the fibers, causing them to be pushed and bent (buckling stress). Cracks formed in the fibers at the points of tensile stress lead to fiber breakage. The brittle polymer matrix surrounding the fibers is also subjected to compression, resulting in cracks and fragmentation. The direction of deformation caused by buckling and the location of cracks are primarily influenced by the tool angle [12–14].

Figure 1 illustrates the material removal mechanism for carbon-fiber-reinforced polymer composites machined in a direction parallel to the fiber direction (0°).

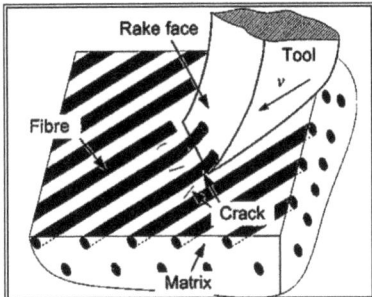

Figure 1. CFRP material removal mechanism at 0° fiber direction [12].

When machined in the direction of the fibers at 90°, the primary cutting force acts perpendicular to the fiber axis and can be divided into cutting forces within the fiber plane and those perpendicular to it. The fibers experience compressive stress at the point of contact with the tool, which can lead to crack formation. On the opposite side of the fiber, tensile stress occurs, which causes fiber breakage. The polymer matrix undergoes compressive stress in front of the cutter, resulting in crack formation and fragmentation of the brittle matrix into small particles [12,15–17].

Figure 2 illustrates the material removal mechanism of CFRP when machining at a fiber angle of 90°.

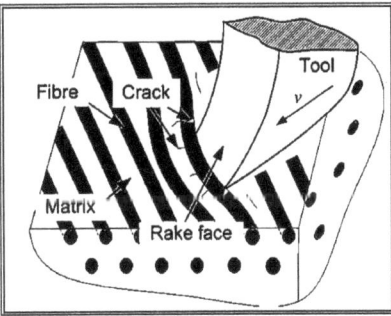

Figure 2. CFRP material removal mechanism at 90° fiber direction [1].

The completely different mechanical parameters of the reinforcing fibers and the matrix used are reflected in the machining in terms of the quality of the machined surface by the so-called delamination. Delamination, which is the most problematic, occurs because it exceeds the forces that hold the composite layers together. Therefore, it is determined by the cohesion of the individual layers of the composite applied, and also by the strength of the bond between the fibers and their binder. According to the literature [18–20], delamination can be described as follows.

During machining, the tool penetrates the composite material. Under pressure, the reinforcing fibers bend, leading to the formation of tensile cracks at the top of the fibers. The resulting cracks initiate the breakage of the fibers. On the bottom side, the fibers are pushed into the resin. The resin beneath the fibers is compressed and crushed. As the fibers in the upper layer of the machined specimen break, they become dislodged when the tool penetrates the material. Consequently, the upper layer of the material lacks fibers. This type of delamination is commonly referred to as type I delamination. Type I delamination is frequently observed when machining in the direction of the fibers at 45° and 90° [21].

The tool applies pressure to the reinforcing fibers, and the fibers bend or bend away from the path of the moving tool. The bending or avoidance of the fibers causes them to be uncut, which then return to their original position in the composite material. This results in protruding fibers on the machined surface. This is type II delamination. Type II delamination occurs most commonly when machining in the direction of the fibers at an angle of 135° [22,23].

Type III delamination describes loose fibers that partially adhere to the machined surface and lie parallel to the tool feed direction. Both Type I and Type III delaminations generate loose fibers "attached" to the machined surface and cause a poor quality "unsharp" machined surface very reminiscent of the burrs known from machining metallic materials. The types of delamination are shown in Figure 3 [24].

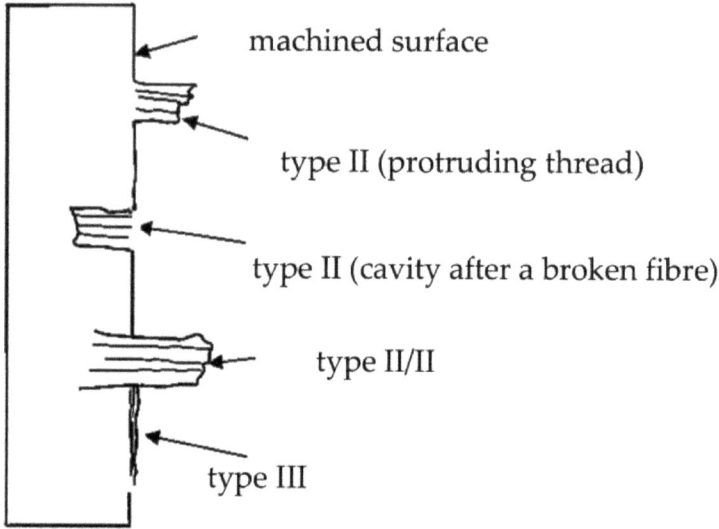

Figure 3. Types of delamination during milling [8].

In recent years, numerous studies have been conducted to address the issue of delamination in machining composite materials, with the aim of developing phenomenological and empirical models to predict its occurrence and emergence. These models serve as valuable tools for implementing delamination as a process monitoring criterion.

Hintze et al. [25] studied the occurrence and propagation of delamination and fiber protrusion during circumferential milling of unidirectional carbon fiber composite material (hereafter referred to as CFRP) and reported that delamination is closely related to the cutting edge (blade) wear and fiber orientation of the upper layer of the composite material. The occurrence of delamination was found to be more frequent in cutting tools with large tooltip radii and when machining composite materials with 90° and 180° fiber orientation. Subsequently, an analytical model was later derived to predict the length of protruding filaments [21]. A model for the analysis of the protrusions was developed by Hosokawa et al. [26], who investigated the effect of the helix angle of a shank cutter on tool wear and delamination during edge trimming of a multidirectional CFRP laminate. It was shown that machining with a large helix angle tool resulted in a smoother surface and less tool edge wear. In addition, tool wear and delamination were reduced when the milling direction was tilted so that the direction of the resultant force was parallel to the feed direction. This was subsequently confirmed by Qingliang et al. [19], who showed that minimal delamination damage was achieved when the tilt angle was equal to the helix angle of the cutter. They also showed that the type of delamination and its frequency of occurrence depended on the inclination angle and fiber orientation. When the inclination angle of the fibers was large, type I/II was the prominent type of delamination. At a low inclination angle, type I delamination occurred more frequently at 45° and 90° fiber orientations, and type II delamination occurred more frequently at 135° fiber orientation [27,28].

Despite the widespread use of composite materials reinforced with more than just carbon fiber, very few studies have attempted to systematically characterize the size and amount of burrs in CFRP machining to determine tool life, mainly due to the complexity of the geometry.

Therefore, the present study focuses on tool wear as a function of tool geometry and fiber orientation, and the analysis of the effect of tool wear on the resulting delamination in CFRP milling.

2. Materials and Methods

A laminated 3K CFRP board with a thickness of 4.3 mm was chosen as the machined material. The board was manufactured using the vacuum infusion method. An LG 120 epoxy resin (GRM Systems, Olomouc, Czech Republic) with an HG 356 hardener (GRM Systems, Olomouc, Czech Republic) was used as the matrix. For reinforcement, a CCH600 fabric (Kordkarbon, Strážnice, Czech Republic) with a weight of 600 g/cm^2 and a twill weave of 2 × 2 cm was used. The surface of the laminate consisted of a thin layer of pure resin with a thickness of 10–15 μm. The properties of the machined materials are shown in Table 1.

Table 1. Properties of the machined materials.

Resin Type	Epoxy
Fiber	Hyosung Tansome 12K H2550
Weave	Twill 2/2
Areal weight	600 g/m^2 ± 3%
Number of filaments per roving	3K
Ply thickness in laminate	4.3 mm

The plates used in this study had dimensions of 600 × 250. The choice of 250 mm width dimension was based on the design of a fixture that was specifically designed to accommodate a maximum plate width of 250 mm (see Figure 4). To effectively extract the resulting chips in the form of dust, a powerful NEDERMAN extraction system (Nederman Holding AB, Helsingborg, Sweden) was used, along with a specially designed nozzle created via 3D printing. These methods were selected based on scientific articles [27,28]. The most common form of machining of CFRP materials is contour cutting of laminated parts; therefore, this study focused on the problems that arise in this method of machining.

Figure 4. Clamping fixture with a suction device.

For the research, a specific type of monolithic 6 mm diameter uncoated sintered carbide cutter manufactured by UniCut s.r.o (Holoubkov, Czech Republic) was utilized. This cutter was specifically designed for the contour cutting of composite materials. The tool was securely clamped in a thermal chuck, which is known for its strong clamping force and low runout value of 0.003 mm (see Figure 5). The influence of the clearance angle on the observed accompanying and subsequent phenomena was investigated for the cutters. Three different clearance angles were chosen (8.4°, 12.4°, and 16.4°), and the rest of the tool geometry remained unchanged. The manufacturer of these tools was consulted, who recommended the clearance angle values. The clearance angle plays a significant role in machining composite materials. A larger clearance angle makes it more challenging for delaminated fibers to rub against the tool back and cause tool abrasion. However, a larger clearance angle also results in a lower blade angle, which reduces the stability of the blade. During the milling process, the generated forces and the following phenomena were monitored: tool wear, surface roughness of the milled edge of the workpiece, and delamination on the edge of the workpiece.

Figure 5. Clamped tool.

Cutting conditions for milling were chosen according to the tool manufacturer's recommendations, as shown in Table 2.

Table 2. Cutting conditions.

Parameter	Value
Diameter of the tool	6
Cutting speed v_c	220 m/min
Feed rate vf	1167 mm/min
Sidestep a_e	1 mm

Machining was performed on a three-axis milling center DMG MORI CMX 600 (DMG Mori Seiki, Nagoya, Japan) with a spindle power of 13 kW and a maximum speed of 12,000 rpm. Non-contiguous machining was selected to suppress delamination.

All selected parameters were measured after a certain time, as shown in Table 3.

Table 3. Effect of tool wear on milling time.

Tool	8.4°	12.4°	16.4°
time t [min]	wear VB [µm] ± measurement uncertainty U [µm]		
10	75.77 ± 0.89	73.43 ± 0.88	64.57 ± 0.99
20	118.09 ± 0.88	106.01 ± 0.87	90.91 ± 0.95
35	197.83 ± 0.95	157.09 ± 0.89	124.8 ± 0.98
50	-	-	165.88 ± 0.99

The displacement component of the cutting force F_y was measured using a KISTLER 9265 B piezoelectric dynamometer (Kistler Instrument Corp, Amherst, NY, USA). The surface roughness of the machined edge of the material was measured using a MITU-TOYO SV-2000N2 SURFTEST touch profile profilometer (Mitutoyo, Kanagawa, Japan), the evaluation of the measured profile was performed using Surfpak software (v.12.2, 2004, Mitutoyo, Kanagawa, Japan), and the evaluation was performed only for the parameter Ra, which is the most telling for machining composites. The wear, tooling, and size of the delaminated material fibers were investigated using a KEYENCE VK-X1100 3D laser confocal microscope (Keyence, Itasca, IL, USA). The microscope was used to evaluate the amount of wear on the back of the VB tool, and the size and type of the delaminated fibers were examined.

Statistical data processing involves calculating the arithmetic mean, denoted as 'x', from the measured data. Afterward, the measurement uncertainty was calculated.

The measurement uncertainty was calculated according to the valid document Guide to the Expression of Uncertainty in Measurement (document EA 4/02). First, the A-type uncertainty was determined, and then the B-type uncertainty, the combined standard uncertainty u(s), and the resulting expanded measurement uncertainty U were calculated. In this case, the standard uncertainties were the sample standard deviation of the mean based on the calculation. The procedure for determining the type B standard uncertainty was based on determining the uncertainty by means other than the statistical analysis of a series of observations. In this specific case, the standard uncertainty determination involved several factors. First, it relied on the information obtained from the calibration sheet of the measuring instrument. Additionally, the uncertainties associated with various influences acting on the measurement process were taken into account. These influences included, among others, the temperature coefficient of the length expansion of the cutting tool, temperature variations in the measuring room, and the influence of the operator handling the measuring instrument. The combined standard uncertainty was calculated as the geometric sum of the A-type uncertainty and the B-type uncertainty. This expanded uncertainty of measurement U was multiplied by the standard uncertainty u(y) of the estimate of the output quantity y by the expansion factor k: $U = k \times u(y)$. For the calculation, the normal (Gaussian) distribution of the measurand was determined (the standard uncertainty of the estimate of the output quantity was determined with sufficient reliability); therefore, the standard expansion coefficient k = 2 was used. The resulting expanded uncertainty corresponded to a coverage probability of approximately 95%.

3. Results

To obtain the necessary results, it was necessary to evaluate the magnitudes of the cutting forces, tool wear, the surface roughness of the machined material, and the size of the delaminated layer on the edge of the machined material.

3.1. Tool Wear Assessment

Wear measurements were taken at five selected locations around the circumference of the tool. At each selected location, the wear on the tool back was measured five times for each type of tool. The critical life of the VB_{krit} tool was determined according to the recommendations provided by the tool manufacturer. The dependence between the tool wear and milling time was investigated. Based on the measured results shown in Table 3 and graphically presented in Figure 6, the following conclusions can be drawn:

(1) Tool wear increased with increasing milling time.
(2) Tool wear increased with a lower tool clearance angle.

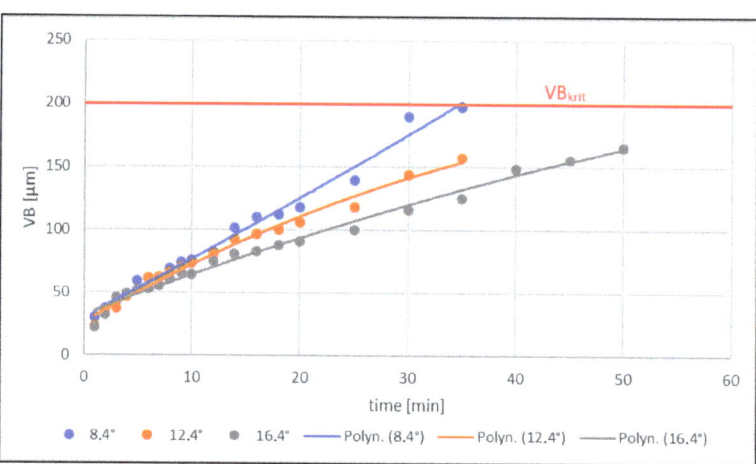

Figure 6. Effect of clearance angle value on tool wear VB [µm].

The VB wear [µm] of all three tools with different clearance angles (8.4°; 12.4°; 16.4°) increased with increasing milling time. The tool with a clearance angle of 8.4° showed faster back wear (the critical tool life of VB_{krit} = 200 µm was reached within t = 35 min) than the tool with a clearance angle of 16.4°, where the increase was rather gradual, and the critical tool life of VB_{krit} = 200 µm was not even reached in 50 min of milling. The wear on all three instruments increased with decreasing (smaller) clearance angle. The tool with a clearance angle of 8.4° had the highest wear (VB = 197.83 µm at the observed time t = 35 min), whereas the tool with the highest clearance angle of 16.4° had the lowest wear (VB = 124.80 µm at the observed time).

3.2. Roughness Evaluation of the Machined Edge/Surface

The dependence between the roughness of the machined edge and the tool wear was investigated. Ra [μm] was selected as the parameter for the study of this dependence. The surface roughness was measured using a MITUTOYO SV-2000 contact profilometer with a 5 μm radius diamond tip. The measured profile was evaluated according to the applicable standard, ČSN ISO 21920 [29]. The measured results are presented in Figure 7.

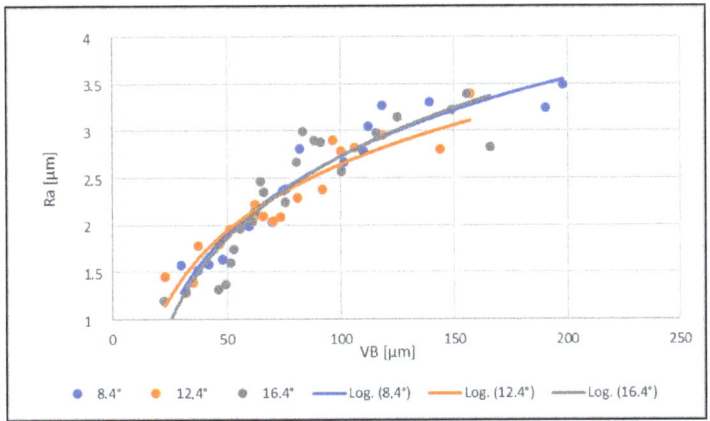

Figure 7. Effect of the machined edge roughness Ra [μm] on tool wear VB [μm].

From the measured results graphically presented in Figure 7, the following conclusions can be drawn:

(1) Tool wear VB [μm] affected the surface roughness for all three tools with different clearance angles (8.4°; 12.4°; 16.4°). The roughness Ra of the machined edge increased with increasing tool wear.
(2) The roughness of the machined edge and the tool wear did not increase or decrease significantly with a greater or lesser angle of the tool back.

3.3. Cutting Force Rating

Furthermore, the relationship between the cutting force (sliding cutting force F_y) and the tool wear was investigated. The measured results are shown in Figure 8.

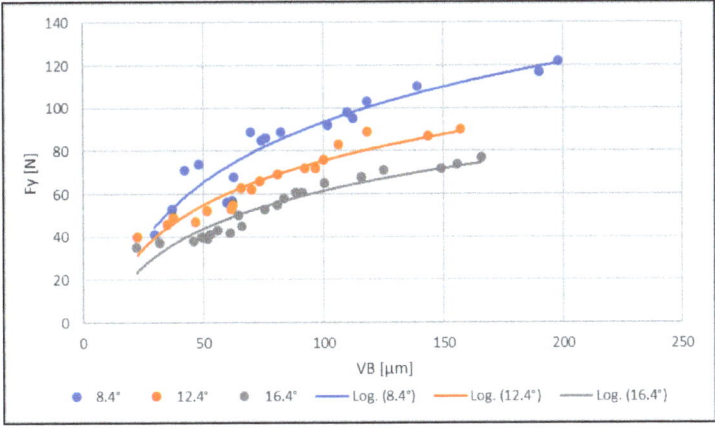

Figure 8. Effect of cutting force F_y [N] on tool wear VB [μm].

From the measured results graphically presented in Figure 8, the following conclusions can be drawn:
(1) The cutting force (sliding) F_y [N] increased with increasing tool wear for all three tools with different clearance angles (8.4°; 12.4°; 16.4°).
(2) The cutting force for all three instruments increased with a lower clearance angle.

The relationship between the cutting force (sliding cutting force F_y) and the roughness of the machined edge was also investigated. The measured results are shown in Figure 9.

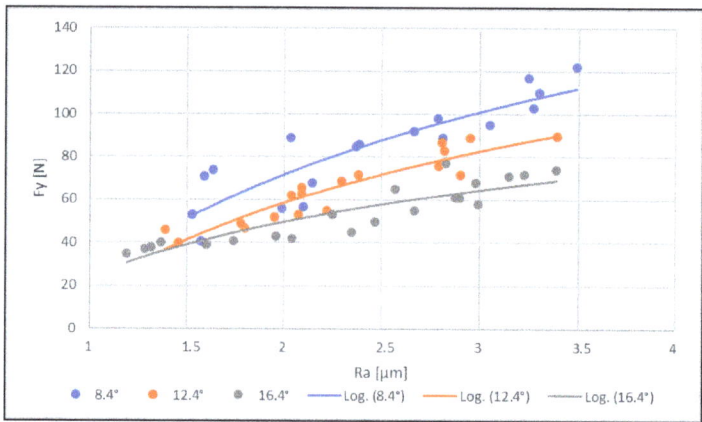

Figure 9. Effect of cutting force F [N] on the roughness of the machined edge Ra [µm].

3.4. Evaluation of Delamination

The dependence of the clearance angle and tool wear on the size of delamination was investigated. The measured results are graphically presented in Figures 10 and 11. The clearance angle influences the formation and length of the delaminated fibers. A larger clearance angle resulted in a smaller size of delaminated fibers. Additionally, the size of the delaminated fibers increased with increasing wear. Furthermore, the radius of curvature of the blade increased with more wear on the tool back (unworn blade, Figure 12a; worn blade, Figure 12b). This indicates that the blade became blunted and lost its ability to cut the individual carbon fibers of the top layer of the CFRP material.

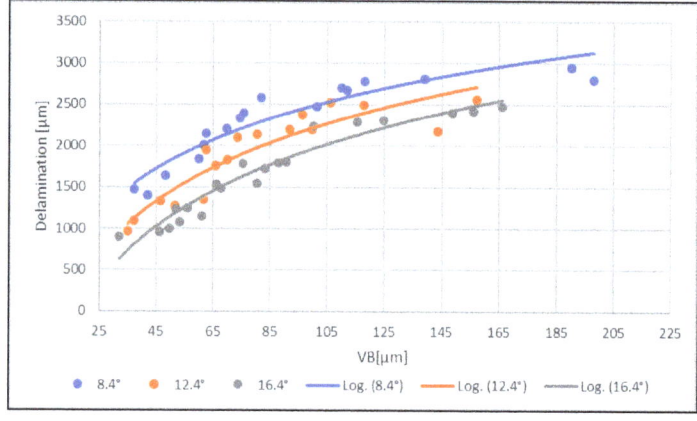

Figure 10. Effect of delamination [µm] on the top edge of the machined surface on tool wear VB [µm].

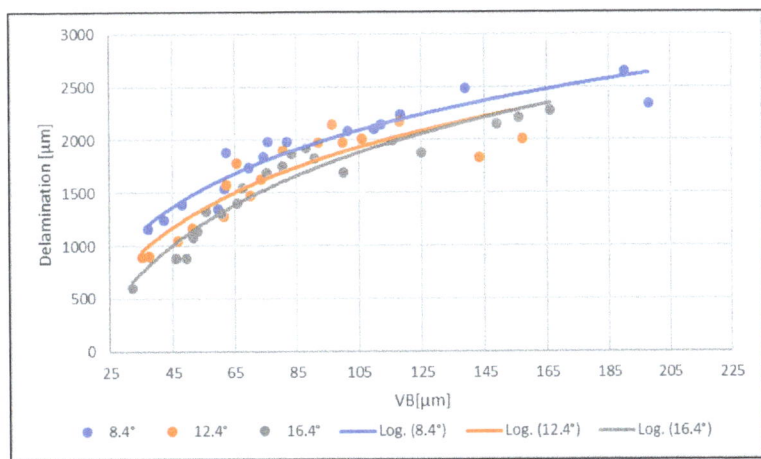

Figure 11. Effect of delamination [μm] on the bottom edge of the machined surface on tool wear VB [μm].

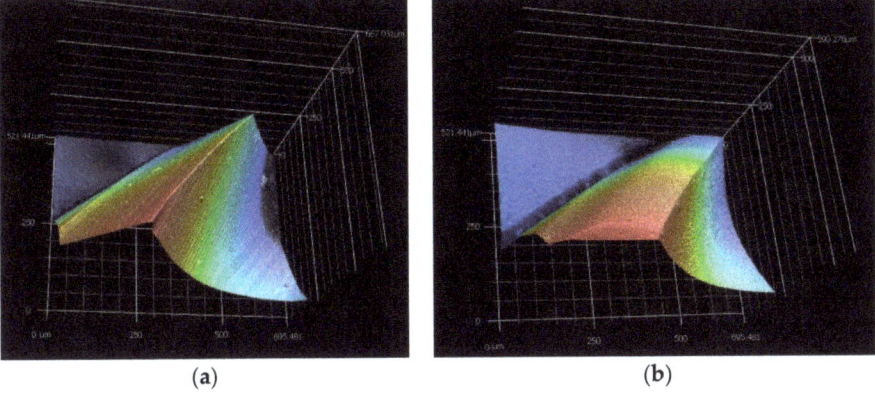

Figure 12. Detail of the tool edge: (**a**) new blade; (**b**) worn blade.

(1) Delamination [μm] increased with increasing tool wear for all three tools with different clearance angles (8.4°; 12.4°; 16.4°).
(2) Delamination and tool wear for all three tools increased with a lower clearance angle.

The delamination [μm] increased with increasing tool wear for all three tools with different clearance angles (8.4°; 12.4°; 16.4°).

Within the conducted measurements, attention was also given to the type of delamination that primarily occurred on the upper side of the machined CFRP edge.

Two types of delaminations were observed as follows: type I/II and type III. The delamination of type I/II was most noticeable at the lower machined edge and was characterized by a significant presence of uncut fibers that protruded and were either bent by the advancing tool in the cutting direction or only slightly trimmed (see Figure 13a–c). The poorest quality machined edge, indicated by uncut protruding fibers, was observed after a machining time of t = 35 min, demonstrating the significant influence of tool wear on delamination (Figure 13g–l). It was observed that a sharp tool resulted in clean fiber cuts with no delamination, whereas a tool with average wear already showed visible uncut protruding fibers. As the wear increased, pronounced delamination occurred, including uncut protruding fibers and fiber breakage within the individual layers of the milled CFRP edge. Additionally, clusters of crushed epoxy resin were observed on the surface of the machined edge, forming a fine crust on the uncut fibers and along the machined edge.

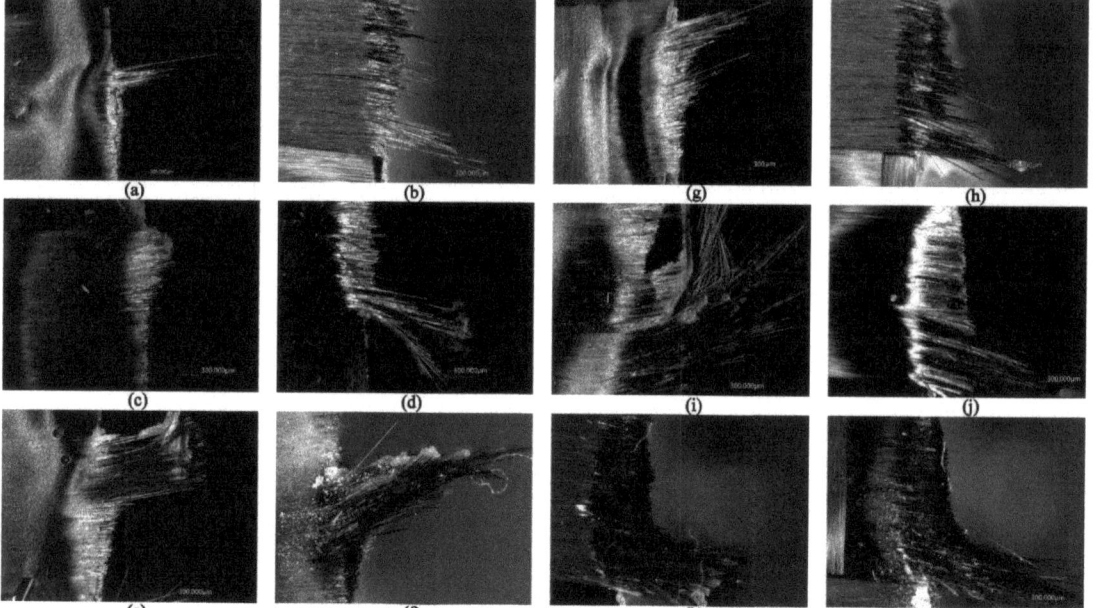

Figure 13. Delamination after milling time (**a,b**) 8.4° (10 min), top edge of the machined surface; (**c,d**) 12.4° (10 min), top edge of the machined surface; (**e,f**) 16.4° (10 min), top edge of the machined surface; (**g,h**) 8.4° (35 min), bottom edge of the machined surface; (**i,j**) 12.4° (35 min), bottom edge of the machined surface; (**k,l**) 16.4° (35 min), bottom edge of the machined surface.

Occasionally, type III delamination was also observed. Type III delamination was presented as loose fibers that partially adhered to the machined surface and were oriented parallel to the tool feed direction. These loose fibers resulted in a poor quality 'unsharp' machined surface, reminiscent of the burrs commonly encountered in machining metallic materials. This type of delamination primarily occurred at the upper edge of the machined surface.

4. Discussion

This study investigated the effect of the clearance angle of a milling tool on the wear, forces, roughness of the machined edge, and delamination of a non-consistently milled CFRP board with a twill weave and 90° fiber orientation.

Tool wear VB [μm] increased with increasing milling time for all three tools with different clearance angles (8.4°; 12.4°; 16.4°). Furthermore, tool wear increased with a lower clearance angle (8.4°).

It was also observed that the smooth and shiny surface in the tool wear area indicated the presence of abrasion wear. This abrasion wear was caused by the high abrasiveness of the carbon fibers. The chips that rubbed against the tool's back during the machining process acted as a polishing mechanism, resulting in a shiny and polished area in the tool wear zone. However, unlike the wave-like wear observed in [27,28], the chips did not produce wave-shaped wear patterns. This difference could be attributed to the distinct geometry of the examined tools, which featured numerous small teeth with opposite helix angle orientations.

The overall measurements show that the tool geometry, or the clearance angle, significantly influenced the tool wear during the milling process. The larger the tool clearance angle (16.4°), the less the wear; conversely, the smaller the tool clearance angle (8.4°), the greater the wear.

Ra [µm] was chosen as a parameter for the dependence study. The roughness of the machined edge increased with increasing tool wear for all three tools with different clearance angles (8.4°; 12.4°; 16.4°). Further, the roughness of the machined edge and tool wear did not increase or decrease significantly with smaller or larger tool clearance angles. The observed effect was not demonstrated in any way. Also, in [30], changing the clearance angle did not affect the amount of surface roughness.

The overall measurements showed that the tool geometry, or the clearance angle, did not significantly affect the worn tool's effect on the machined edge's roughness.

The cutting force F_y [N] increased with increasing tool wear for all three tools with different clearance angles (8.4°; 12.4°; 16.4°). Further, the cutting force and tool wear increased with a lower tool clearance angle (8.4°).

The overall measurement subsequently showed that the cutting (sliding) force increased significantly with increasing wear. Both the cutting force and tool wear decreased with increasing tool clearance angle. With a low tool clearance angle (8.4°), both the cutting force and tool wear increased. A low cutting force prevented tool wear.

The small magnitude of the cutting forces resulted in low tool wear, and thus low energy consumption.

The cutting force F_y [N] for all three tools with different clearance angles (8.4°; 12.4°; 16.4°) increased with increasing roughness of the machined edge. Furthermore, the cutting force and roughness of the machined edge increased with a lower tool clearance angle (8.4°).

The overall measurements showed that the cutting (sliding) force increased significantly with increasing surface roughness of the machined edge. Both the cutting force and surface roughness decreased with increasing tool clearance angle (16.4°). With a low tool clearance angle (8.4°), both the cutting force and surface roughness increased. The overall low cutting force helped to reduce surface roughness and prevent tool wear.

Delamination [µm] increased with increasing roughness of the machined edge, both on the upper and lower sides of the edge, for all three tools with different clearance angles (8.4°; 12.4°; 16.4°). Furthermore, the delamination and roughness of the machined edge increased with a lower tool clearance angle (8.4°). The resulting delamination, in the form of fibers protruding beyond the edge of the material, was the main type of delamination observed, in contrast to [6], where the delamination was mainly in the form of deep abrasion of the overlying material

The overall measurements subsequently showed that delamination increased significantly with an increase in the surface roughness of the machined edge. Both the delamination and surface roughness decreased with increasing tool clearance angle (16.4°). With a low tool clearance angle (8.4°), both delamination and surface roughness increased.

It was observed that delamination always occurred at the point of initial contact between the tool and the CFRP board fiber where the initial cutting occurred. The fibers were under compressive stress at the point of contact with the tool, and cracks started to form, with tension and subsequent fracture of the fibers on the opposite side. The polymer matrix was stressed by the pressure in front of the tool, cracks formed, and the brittle matrix was crushed into small particles.

Delamination [µm] increased with increasing tool wear for all three tools with different clearance angles (8.4°; 12.4°; 16.4°), both on the upper and lower sides of the machined edge. Furthermore, delamination and tool wear increased with a lower tool clearance angle (8.4°).

The above results show that delamination increased with increasing tool wear. Both delamination and tool wear decreased with increasing tool clearance angle (16.4°). With a low tool clearance angle (8.4°), both delamination and tool wear increased.

Delamination [μm] increased with increasing cutting force for all three tools with different clearance angles (8.4°; 12.4°; 16.4°), both on the upper and lower sides of the machined edge. Increasing the cutting force increased the degree of delamination. Furthermore, delamination and cutting force increased with a lower tool clearance angle (8.4°).

The above results show that delamination increased with increasing cutting force. Both delamination and cutting force decreased with increasing tool clearance angle (16.4°). With a low clearance angle (8.4°) of the tool, both delamination and cutting force increased. The formation of delamination is consistent with the data in [31].

5. Conclusions

Tool geometry has a significant influence when machining CFRP materials. This study focused on the investigation of the effect of the tool clearance angle on wear, cutting forces, machined edge roughness, and delamination in the non-contiguous milling of CFRP laminates.

The following conclusions were drawn from the measurements:

- The angle of the tool back has a significant effect on tool wear, the roughness of the machined edge, and the delamination of CFRP.
- The tool with a clearance angle of 8.4° wore out faster than the tool with a clearance angle of 16,4°. The same was true for the cutting force, the observed roughness of the machined edge, and delamination. In this context, it was shown that the cutting force increased with increasing tool wear, which in turn increased surface roughness and delamination.
- From the tool surface topography images, it was observed that abrasion wear occurred within the tool wear.
- Type I/II and III delaminations were observed in the machined upper and lower edges. Type I/II delamination was the most pronounced and increased with increasing tool wear, especially on the lower machined edge.

The results of this research hold significant practical importance, not only in terms of the quality of the machined surface but also in relation to time, financial costs, and energy savings when machining CFRP composite materials. Another possible direction for future research could be to investigate other aspects of tool geometry, such as face angle, helix pitch angle, etc., and their influence on the parameters investigated in the present research.

Author Contributions: Conceptualization, T.K. and Š.D.; methodology, Š.D.; validation, Š.D.; writing—original draft preparation, T.K.; resources, T.K. and Š.D.; data curation; M.V.; writing—review and editing, T.K. and M.V.; supervision, Š.D. All authors have read and agreed to the published version of the manuscript.

Funding: Research was funded by the institutional funding of science and research at the Technical University of Liberec and by the Student Grant Competition of the Technical University of Liberec under the project number SGS-2022-5043—"Research and development in the field of machining of metal and composite materials using new knowledge for industrial practice".

Institutional Review Board Statement: Not applicable.

Informed Consent Statement: Not applicable.

Data Availability Statement: Data sharing does not apply to this article.

Conflicts of Interest: The authors declare no conflict of interest. The funders had no role in the design of the study; in the collection, analyses, or interpretation of data; in the writing of the manuscript; or in the decision to publish the results.

References

1. Diefendorf, R.J.; Tokarsky, E. High-Performance Carbon Fibers. *Polym. Eng. Sci.* **1975**, *15*, 150–159. [CrossRef]
2. Qiu, J.; Zhang, S.; Li, B.; Li, Y.; Wang, L. Research on Tool Wear and Surface Integrity of CFRPs with Mild Milling Parameters. *Coatings* **2023**, *13*, 207. [CrossRef]
3. Petrovic, A.; Lukic, L.; Ivanovic, S.; Pavlovic, A. Optimisation of Tool Path for Wood Machining on CNC Machines. *Proc. Inst. Mech. Eng. Part C J. Mech. Eng. Sci.* **2016**, *231*, 72–87. [CrossRef]
4. Mughal, K.; Mughal, M.P.; Farooq, M.U.; Anwar, S.; Ammarullah, M.I. Using Nano-Fluids Minimum Quantity Lubrication (NF-MQL) to Improve Tool Wear Characteristics for Efficient Machining of CFRP/Ti6Al4V Aeronautical Structural Composite. *Processes* **2023**, *11*, 1540. [CrossRef]
5. Doluk, E.; Rudawska, A.; Kuczmaszewski, J.; Miturska-Barańska, I. Surface Roughness after Milling of the Al/CFRP Stacks with a Diamond Tool. *Materials* **2021**, *14*, 6835. [CrossRef]
6. Ozkan, D.; Panjan, P.; Gok, M.S.; Karaoglanli, A.C. Experimental Study on Tool Wear and Delamination in Milling CFRPs with TiAlN- and TiN-Coated Tools. *Coatings* **2020**, *10*, 623. [CrossRef]
7. Xu, Z.; Wang, Y. Study on Milling Force and Surface Quality during Slot Milling of Plain-Woven CFRP with PCD Tools. *Materials* **2022**, *15*, 3862. [CrossRef]
8. Wang, B.; Wang, Y.; Zhao, H.; Wang, M.; Sun, L. Mechanisms and Evaluation of the Influence of Cutting Temperature on the Damage of CFRP by Helical Milling. *Int. J. Adv. Manuf. Technol.* **2021**, *113*, 1887–1897. [CrossRef]
9. Salem, B.; Mkaddem, A.; Ghazali, S.; Habak, M.; Felemban, B.F.; Jarraya, A. Towards an Advanced Modeling of Hybrid Composite Cutting: Heat Discontinuity at Interface Region. *Polymers* **2023**, *15*, 1955. [CrossRef]
10. Wu, M.Y.; Tong, M.L.; Wang, Y.W.; Ji, W.; Wang, Y. Study on Carbon Fiber Composite Materials Cutting Tools. *Appl. Mech. Mater.* **2013**, *401–403*, 721–727. [CrossRef]
11. Elgnemi, T.; Songmene, V.; Kouam, J.; Jun, M.B.G.; Samuel, A.M. Experimental Investigation on Dry Routing of CFRP Composite: Temperature, Forces, Tool Wear, and Fine Dust Emission. *Materials* **2021**, *14*, 5697. [CrossRef]
12. Shyha, I.; Huo, D. (Eds.) *Advances in Machining of Composite Materials: Conventional and Non-Conventional Processes*; Engineering Materials; Springer International Publishing: Cham, Switzerland, 2021; ISBN 978-3-030-71437-6.
13. Knápek, T.; Kroisová, D.; Dvorackova, Š.; Knap, A. Destruction of Fibrous Structures during Machining of Carbon Fiber Composites. In Proceedings of the 14th International Conference on Nanomaterials—Research & Application, OREA Congress Hotel, Brno, Czech Republic, 19–21 October 2022; pp. 242–248.
14. Qi, J.; Li, C.; Tie, Y.; Zheng, Y.; Cui, Z.; Duan, Y. An Ordinary State-Based Peridynamic Model of Unidirectional Carbon Fiber Reinforced Polymer Material in the Cutting Process. *Polymers* **2023**, *15*, 64. [CrossRef]
15. Uhlmann, E.; Sammler, F.; Richarz, S.; Reucher, G.; Hufschmied, R.; Frank, A.; Stawiszynski, B.; Protz, F. Machining of Carbon and Glass Fibre Reinforced Composites. *Procedia CIRP* **2016**, *46*, 63–66. [CrossRef]
16. Altin Karataş, M.; Gökkaya, H. A Review on Machinability of Carbon Fiber Reinforced Polymer (CFRP) and Glass Fiber Reinforced Polymer (GFRP) Composite Materials. *Def. Technol.* **2018**, *14*, 318–326. [CrossRef]
17. Kroisová, D.; Dvořáčková, Š.; Knap, A.; Knápek, T. Destruction of Carbon and Glass Fibers during Chip Machining of Composite Systems. *Polymers* **2023**, *15*, 2888. [CrossRef]
18. Colligan, K.; Ramulu, M. The Effect of Edge Trimming on Composite Surface Plies. *Manuf. Rev.* **1992**, *5*, 274–283.
19. Chen, Q.; Zhou, J.; Chen, X.; Chen, Y.; Fu, Y. Experimental Study on Delamination during Trimming of CFRP. *Adv. Mater. Res.* **2015**, *1089*, 331–336. [CrossRef]
20. Seeholzer, L.; Kneubühler, F.; Grossenbacher, F.; Wegener, K. Tool Wear and Spring Back Analysis in Orthogonal Machining Unidirectional CFRP with Respect to Tool Geometry and Fibre Orientation. *Int. J. Adv. Manuf. Technol.* **2021**, *115*, 2905–2928. [CrossRef]
21. Hintze, W.; Hartmann, D. Modeling of Delamination During Milling of Unidirectional CFRP. *Procedia CIRP* **2013**, *8*, 444–449. [CrossRef]
22. Henerichs, M.; Voß, R.; Kuster, F.; Wegener, K. Machining of Carbon Fiber Reinforced Plastics: Influence of Tool Geometry and Fiber Orientation on the Machining Forces. *CIRP J. Manuf. Sci. Technol.* **2015**, *9*, 136–145. [CrossRef]
23. Seo, J.; Kim, D.Y.; Kim, D.C.; Park, H.W. Recent Developments and Challenges on Machining of Carbon Fiber Reinforced Polymer Composite Laminates. *Int. J. Precis. Eng. Manuf.* **2021**, *22*, 2027–2044. [CrossRef]
24. Calzada, K.A.; Kapoor, S.G.; DeVor, R.E.; Samuel, J.; Srivastava, A.K. Modeling and Interpretation of Fiber Orientation-Based Failure Mechanisms in Machining of Carbon Fiber-Reinforced Polymer Composites. *J. Manuf. Process.* **2012**, *14*, 141–149. [CrossRef]
25. Hintze, W.; Hartmann, D.; Schütte, C. Occurrence and Propagation of Delamination during the Machining of Carbon Fibre Reinforced Plastics (CFRPs)—An Experimental Study. *Compos. Sci. Technol.* **2011**, *71*, 1719–1726. [CrossRef]
26. Hosokawa, A.; Hirose, N.; Ueda, T.; Furumoto, T. High-Quality Machining of CFRP with High Helix End Mill. *CIRP Ann.* **2014**, *63*, 89–92. [CrossRef]
27. Knápek, T.; Dvořáčková, Š.; Knap, A. Wear Study of Coated Mills during Circumferential Milling of Carbon Fiber-Reinforced Composites and Their Influence on the Sustainable Quality of the Machined Surface. *Coatings* **2022**, *12*, 1379. [CrossRef]
28. Knap, A.; Dvořáčková, Š.; Knápek, T. Study of the Machining Process of GFRP Materials by Milling Technology with Coated Tools. *Coatings* **2022**, *12*, 1354. [CrossRef]

29. ČSN EN ISO 21920-1; Geometrické specifikace produktu (GPS)—Textura povrchu: Profil—Část 1: Indikace textury povrchu. Úřad pro Technickou Normalizaci, Metrologii a Státní Zkušebnictví: Praha, Czech Republic, 2023; Volume 56, p. 014450.
30. Waqar, S.; He, Y.; Abbas, C.; Majeed, A. Optimization of Cutting Tool Geometric Parameters in Milling of Cfrp Laminates. In Proceedings of the 21st International Conference on Composite Materials, Xi'an, China, 20–25 August 2017.
31. Ning, H.; Zheng, H.; Yuan, X. Establishment of Instantaneous Milling Force Prediction Model for Multi-Directional CFRP Laminate. *Adv. Mech. Eng.* **2021**, *13*, 168781402110277. [CrossRef]

Disclaimer/Publisher's Note: The statements, opinions and data contained in all publications are solely those of the individual author(s) and contributor(s) and not of MDPI and/or the editor(s). MDPI and/or the editor(s) disclaim responsibility for any injury to people or property resulting from any ideas, methods, instructions or products referred to in the content.

Article

Manufacturing of Corrosion-Resistant Surface Layers by Coating Non-Alloy Steels with a Polymer-Powder Slurry and Sintering

Grzegorz Matula * and Błażej Tomiczek

Scientific and Didactic Laboratory of Nanotechnology and Material Technologies, Faculty of Mechanical Engineering, Silesian University of Technology, Konarskiego 18a St., 44-100 Gliwice, Poland; blazej.tomiczek@polsl.pl
* Correspondence: grzegorz.matula@polsl.pl

Abstract: This paper describes the combination of surface engineering and powder metallurgy to create a coating with improved corrosion resistance and wear properties. A new method has been developed to manufacture corrosion-resistant surface layers on steel substrate with additional carbide reinforcement by employing a polymer-powder slurry forming and sintering. The proposed technology is an innovative alternative to anti-corrosion coatings applied by galvanic, welding or thermal spraying techniques. Two different stainless-steel powders were used in the research. Austenitic 316 L and 430 L ferritic steel powders were selected for comparison. In addition, to improve resistance to abrasive wear, coatings containing an additional mixture of tetra carbides (WC, TaC, TiC, NbC) were applied. The study investigates the effects of using multicomponent polymeric binders, sintering temperature, and atmosphere in the sintering process, as well as the presence of reinforcing precipitation, microstructure and selected surface layer properties. Various techniques such as SEM, EDS, hardness and tensile tests and corrosion resistance analysis are employed to evaluate the characteristics of the developed materials. It has been proven that residual carbon content and nitrogen atmosphere cause the release of hard precipitations and thus affect the higher mechanical properties of the obtained coatings. The tensile test shows that both steels have higher strength after sintering in a nitrogen-rich atmosphere. Nitrogen contributes over 50% more to the tensile strength than an argon-containing atmosphere.

Keywords: powder metallurgy; pressureless forming; stainless steel; protective coatings; composite

Citation: Matula, G.; Tomiczek, B. Manufacturing of Corrosion-Resistant Surface Layers by Coating Non-Alloy Steels with a Polymer-Powder Slurry and Sintering. *Materials* **2023**, *16*, 5210. https://doi.org/10.3390/ma16155210

Academic Editor: Raul D. S. G. Campilho

Received: 5 July 2023
Revised: 21 July 2023
Accepted: 23 July 2023
Published: 25 July 2023

Copyright: © 2023 by the authors. Licensee MDPI, Basel, Switzerland. This article is an open access article distributed under the terms and conditions of the Creative Commons Attribution (CC BY) license (https://creativecommons.org/licenses/by/4.0/).

1. Introduction

The development of modern engineering materials is dependent on and closely related to the technology of forming and sintering powders [1–3]. The high requirements set by consumers regarding high properties and low costs make it necessary to look for new technological solutions. Steel is still the best material solution for corrosion-resistant elements and relatively high mechanical loads. Unfortunately, alloying additives that determine high corrosion resistance are expensive [4]. Technologies using anti-corrosion protection, based mainly on zinc and paint coatings, are widely used to coat structural steels. Unfortunately, the influence of zinc galvanization on the hardness of covered steels is significant. For example, many types of steel suffer from a considerable decrease in hardness, particularly high-strength steel [5]. In the analyzed zinc-coated steels, the reduction in hardness ranged from about 28% to as much as 55%.

To obtain a material that is resistant to corrosion and, at the same time, has high mechanical properties, stainless steels with a high chromium and nickel content should be used. Unfortunately, a dynamic increase in nickel prices has been caused by its increasing use. This element's average annual price growth rate is 7.29% [6]. The demand for nickel also results from its unique chemical properties that make it useful for various applications, like catalysts in methanation [7] or solid oxide fuel cells [8]. Due to the low-temperature coefficient of resistance, nickel-chromium alloys are used in devices operating at high temperatures [9]. It should therefore be expected that the costs of austenitic steels will increase even more. The price of high-nickel steels is four to six times higher than unalloyed steels and twice as high as high-alloy tool steels [4]. Therefore, searching for new materials and technological solutions is important to reduce the share of costly elements such as nickel in steel.

Duplex steels with a ferritic-austenitic structure are undoubtedly an interesting solution. Due to their optimal variety of mechanical properties and high corrosion resistance, duplex steels have more applications. Obtaining high properties is possible from the balance in dual-phase composition and the steel production method, the parameters of individual processes and the insertion of alloying additives. Only the selection and strict control of each listed aspect make it possible to obtain duplex steel that meets the application requirements. Powder metallurgy is widely used in duplex steel production methods [10–14]. Austenitic-ferritic steel powder can be obtained using conventional base powder mixing, compaction and sintering. A powder with a precise chemical composition can be prepared by atomization [15], or base powders with different chemical compositions can be mixed to obtain the correct ratio. In [10], a mixture of austenitic and martensitic corrosion-resistant steel powders was used, and in [11], a combination of austenitic and ferritic powders was used. Various forming techniques are also used. The metal injection molding technique produced small orthodontic components with complex shapes from duplex steel [12]. Often, during the sintering of duplex steel, the chemical composition is equalized during diffusion at high temperatures, and the share of the ferritic phase concerning the austenitic phase increases, as proved in [16]. Duplex steel can also be produced by additive manufacturing. An example of this is a multilayer steel structure made of austenitic and martensitic stainless-steel wires using wire and arc additive manufacturing equipment based on plasma arc welding [17]. An interesting solution may be using powder metallurgy methods for the manufacturing of corrosion-resistant surface layers by coating non-alloy steels. Previous studies [18] have shown that it is possible to produce a layered material with high mechanical strength using polymer-powder slurry.

The main goal of the undertaken research is the development of layered materials resulting from the combination of surface engineering and powder metallurgy. Materials with a layered structure consisting of a stainless surface layer on steel intended for thermal improvement, with high mechanical properties, were developed using polymer-powder slurry and sintering. In addition, hard carbide particles were introduced to the surface layer's structure to increase the mechanical properties, particularly hardness and resistance to abrasive wear, while maintaining strong corrosion resistance. Particularly noteworthy is the innovative approach to stainless steel, especially 316 L austenitic steel, where it is essential to maintain a low carbon concentration. To prevent intergranular corrosion, it is important to block the precipitation of chromium carbides and the drop in electrochemical potential at the grain boundaries. Because carbon lowers the solidus temperature and initiates the sintering process, which ensures a surface layer's diffusion connection with the non-alloy steel substrate, a local increase in carbon concentration is required. The proposed technology is an innovative alternative to anti-corrosion coatings applied by galvanic, welding or thermal spraying techniques [19,20]. It is worth emphasizing that the developed method of forming layers resistant to corrosion and wear on steel is innovative and is a unique invention of the authors. So far, there have not been any reports in the literature on using similar solutions.

Surface layers based on ferritic 430 L or austenitic 316 L steel developed as part of this research can be used to cover the screws of extruders and injection molding machines for processing plastics. The increasing use of recyclates with a higher viscosity than pure polymers, which may additionally be contaminated with solid particles, causes an increase in the wear of the surfaces of cylinders and screws. Their regeneration by surfacing with alloys with a high proportion of Co and Ni is associated with high costs. Using the presented technology may reduce these costs and maintain comparable tool properties. This technology is expected to be used primarily in producing new components exposed to wear and corrosion. However, using these layers to regenerate or repair worn steel surfaces may also be technologically and economically justified.

2. Materials and Methods

To produce corrosion-resistant surface layers on carbon steels, the powders of austenitic steel 316 L and ferritic steel 430 L marked according to ASTM and manufactured by Sandvik Osprey Ltd. were used. The powders employed had spherical particles typically created by atomizing inert gas [21]. The morphology and particle size distribution are shown in Figure 1. The particle size distribution analysis of the selected powders was performed using the laser particle size analyzer Analysette 22 MicroTec Plus, Fritsch GmbH, and the results are gathered in Table 1.

Table 1. Characteristics of base powders.

Powder	316 L	430 L	Tetra Carbides
Density, g·cm^{-3}	7.94	7.70	11.82
D10, μm	3.78	3.19	0.80
D50, μm	9.88	8.16	2.70
D90, μm	19.99	16.83	9.48
Sw	3.53	3.54	2.74

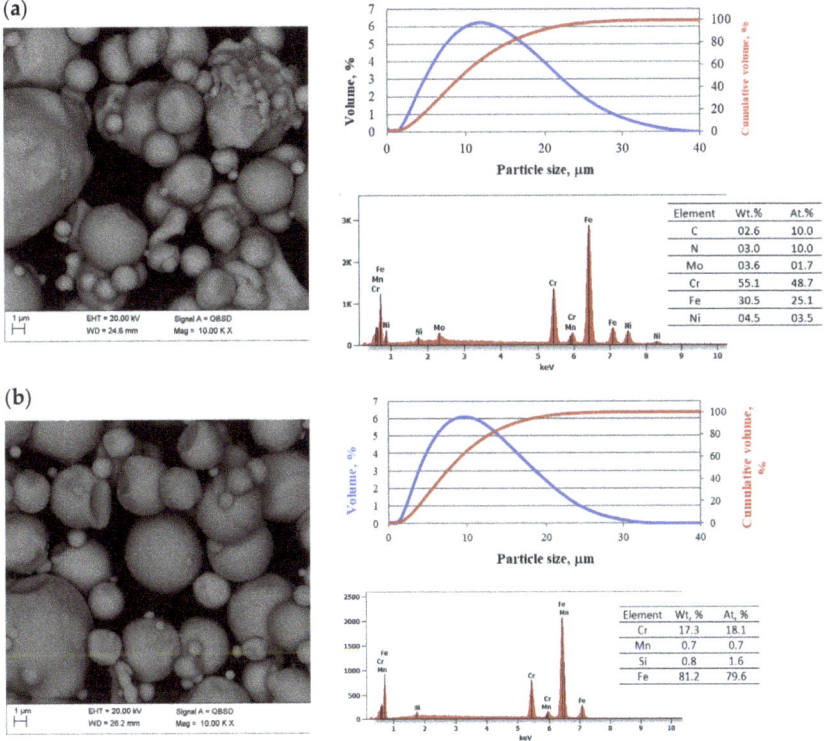

Figure 1. Cont.

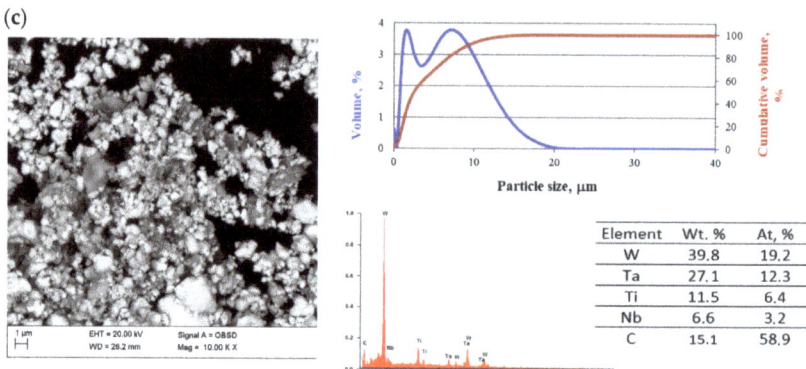

Figure 1. SEM morphology, particle size distribution and EDS chemical composition of (**a**) 316 L steel powder, (**b**) 430 steel powder (**c**) tetra carbides powder. Particle size distribution curve is blue and cumulative curve is red.

The scanning microscope SUPRA 35 by Zeiss (Oberkochen, Germany) was used to determine the powders' morphology and study the sinters' structure. ASTM 4140 steel was used as the substrate. Steels used for the surface layer, especially 316 L steel with an austenitic structure, have low hardness, so to increase it, and in particular to increase resistance to abrasive wear, the steel was reinforced by a mixture of carbides known by the name "Tetra Carbides" and produced by Treibacher Industrie AG, containing 47% WC, 14% TiC, 33% TaC and 6% NbC in volume. The Tetra Carbides powder is referred to as TC. The morphology of these carbides and the particle size distribution are also shown in Figure 1. Their volume fraction concerning steel was 5%. The applied powders of stainless steels, atomized by gas, are generally used for the production of feedstock for powder injection molding. A total of 90% of the 316 L and 430 L steel particles are smaller than approximately 19 and 16 µm. Moreover, the particles are spherical, which improves surface wettability with polymers. A total of 90% of the carbide TC particles used to reinforce the stainless steel were smaller than 9 µm. The size distribution of carbide particles is bimodal, which is caused by the strong aggregation of fine carbide particles, and this can be observed in the structure of the sintered samples. To determine the mechanical properties of the surface layers in the form of steel and carbide steels produced on carbon steels, it was necessary to prepare samples of these materials for which the powder injection molding technology was used. The powders of the used stainless steels are suitable for this technology not only due to their spherical shape and size below 20 µm, but also due to their particle size distribution, which is evidenced by the particle size distribution slope parameter S_w. This parameter is the slope of the log-normal cumulative distribution and can be calculated using Formula (1) [22]. The particle distribution is narrower the higher the value of S_w. A broad particle size distribution (S_w of 2–4) indicates easy-to-mold, low-viscosity material, but a narrow particle size distribution (S_w of 4–7) of powder often results in high feedstock viscosity. This ensures high surface quality and sinter edges and, in particular, low surface roughness. Considering the powder parameters (Table 1) and the values of the S_w coefficient calculated on this basis, it can be concluded that all the powders used can be used in the powder injection molding technology because the particle size distributions are relatively wide.

$$S_w = \frac{2.56}{\log\frac{D_{90}}{D_{10}}} \tag{1}$$

As part of the preliminary research, various powder-forming techniques were used. As a result of these analyses, the non-pressure forming method was selected as the best due to the properties of the finished element. The technology of forming polymer-powder slips on solid steel surfaces allows for a local increase in the share of carbon initiating the powder sintering process, as well as the surface layer and substrate, which guarantees a good connection of the layer with the diffusion substrate. Figure 2 shows a diagram of this process.

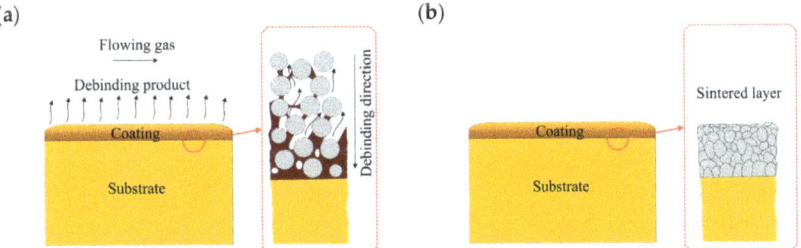

Figure 2. Scheme of (**a**) thermal debinding process of polymer-powder coatings applied on a solid steel substrate and (**b**) manufactured coating with stainless steel sintered layer.

Due to the direction of degradation of the polymer binder from the surface into the layer, the largest share of residual carbon will be found in the area directly above the surface of the substrate. A high proportion of carbon lowers the sintering temperature [23], which reduces the properties of the steel but guarantees good adhesion of the coating to the substrate. An essential issue in this method is the uniform thickness of the layer applied from the polymer-powder slurry. Therefore, it is necessary to introduce automation and control of the coating application process. For example, in the case of components with a circular cross-section, it is possible to dispense the slurry with the simultaneous rotation of the coated bar, which guarantees even distribution of the polymer-powder slurry, as shown in Figure 3.

Figure 3. Scheme of the automation process of applying polymer-powder coatings.

To perform comparative tests of the mechanical properties of steels and carbide steels, polymer-powder slurries based on a binder containing PP and PW were prepared. A Zamak-Mercator MP-30 mixer was used to homogenize the polymer-powder mixtures. The rotational speed of the mixer screws was 20 rpm, the homogenization temperature was 170 °C, and the time was 30 min. The carbides were pre-mixed with the binder by adding stearic acid as a surfactant, which increases the wettability of the carbide powder surface [24]. Using homogenized slurries, samples for testing were produced using Zamak Mercator equipment. A mini-piston injection molding machine with a cylinder capacity of 15 cm^3 from the same company was used for injection molding. The actual injection pressure is much higher, but unfortunately, the device cannot measure it. It is only possible to adjust the pressure of the air supplied to the actuator. The injection conditions depended on the shape of the sample. For the beam intended for bending, the conditions were as follows: cylinder temperature 170 °C, die temperature 40 °C, injection time 5 s, pressure 5 bar. In the case of samples with more complex shapes, such as dog bones intended for tensile testing, the viscosity of the slurry should be lower; hence, the temperature of the cylinder and die were 180 °C and 50 °C, respectively, and the other parameters remained the same. A series of samples, such as tensile paddles and beams for the three-point bending test, were thus prepared.

Regardless of the powder-forming method, the produced samples are characterized by a smooth surface. Figure 4a shows a sample in the form of a dog bone injection molded from 316 L steel. The samples are characterized by high quality and a lack of defects in the form of distortions and external and internal bubbles. Figure 4b shows samples of steel 4140, which were covered with a polymer-powder slip and sintered. Samples were made with holes in the centre to test the ability to cover the surface of the inner holes. The drawing shows a clean steel substrate, slip-coated steel, and the finished sinter.

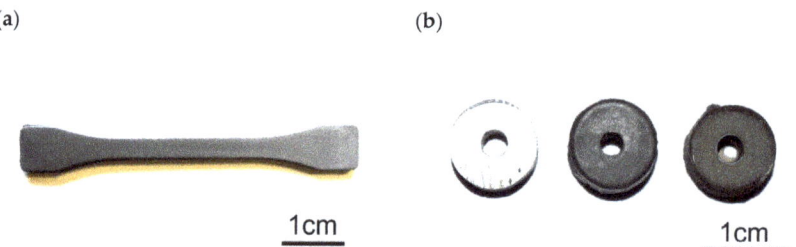

Figure 4. View of (**a**) a 316 L steel sample after powder injection molding and (**b**) pre-coating, post-coating and post-sintering samples with holes.

The injection molded samples were then subjected to binder degradation and sintering. The degradation was carried out in two stages. Initially, solvent degradation in heptane was used for max. 24 h, at a temperature of 25 °C. The first step of the degradation allowed paraffin removal. Then, the samples were placed in a tube furnace, in which thermal degradation of the binder and direct sintering was performed at temperatures between 1150 and 1350 °C, with steps of 50 °C. Solvent degradation generally facilitates thermal degradation, the cycle of which has been selected experimentally. Both thermal degradation and sintering were performed in a Czylok tube furnace in an atmosphere of a flowing gas mixture comprised of N_2-10% H_2 and Ar-10% H_2. The maximum heating rate did not exceed 5 °C/min, and during heating, the thermal degradation temperature was much lower and did not exceed 1 °C/min. Due to the high viscosity of the slurries used for injection and significant technological problems with their low-pressure application on the surface of unalloyed steel, a mixture was prepared in which only paraffin was used as a polymer binder. The proportion of steel to carbide powders was comparable. However, in the case of steel powder and steel-carbide powder, the volume fraction of the paraffin binder was raised by 10% and 15%, respectively, to ensure the low viscosity of the slurry. The use of carbides requires a higher proportion of binders due to their small size, irregular shape and the resulting greater specific surface area that needs to be wetted. Table 2 presents the composition of powders and binder slurries intended for injection molding solid samples and forming surface layers on stainless steel. The coated samples were only subjected to thermal degradation at 200 °C for 1 h and then heated directly to the sintering temperature. The sintering time of injection molded and low-pressure samples was 30 min. Injection molded sinters were subjected to shrinkage and density analysis using the hydrostatic method. The microhardness measurement was carried out in the Vickers Future-Tech FM-700 hardness tester with a load of 100 g. The tensile strength and three-point bending tests were performed using appropriate attachments in the Zwick/Roell Z020 testing machine.

Table 2. Compositions of injected and pressureless formed samples.

Molding Methods	Designation	Powder Volume Fraction, %			Binder Volume Fraction, %			Density of Slurry, g·cm^{-3}
		316 L	430 L	Tetra Carbides	PW	SA	PP	
Powder Injection Molding PIM	PIM316	60	-	-	20	-	20	5.12
	PIM430	-	60	-	20	-	20	4.98
	PIM316TC	54	-	6	19.8	0.4	19.8	5.35
	PIM430TC	-	54	6	19.8	0.4	19.8	5.22
Pressureless forming of powder PLF	PLF316	50	-	-	50	-	-	4.42
	PLF430	50	-	-	50	-	-	4.3
	PLF316TC	40.5	-	4.5	55	-	-	4.27
	PLF430TC	-	40.5	4.5	55	-	-	4.17

Observations of the structure of the produced materials were made in a scanning electron microscope (SEM) ZEISS SUPRA 35, using the detection of secondary electrons and backscattered electrons at an accelerating voltage of 20 kV and a maximum magnification of 50,000×. The corrosion resistance test was conducted on a precise Atlas-Sollich 0531 EU potentiostat according to the PN ISO 17475:2010 standard [25]. In addition to the corrosion tests of sinters, reference samples in commercial steel 316 L and 4140 were also tested. The following parameters were tested: open circuit potential EOCP, corrosion resistance (Ecorr) or breakdown potential (Eb), polarization resistance (Rp), and corrosion current density (icorr). The tribological tests were carried out using equipment for the "ball-on-disc" test, which was performed on a Tribometer CSM. The wear tracks were measured using a Sutronic 25 profilometer from Taylor Hobson and observed on an SEM microscope. A replaceable pin in the form of a small ball with a diameter of 6 mm made from Al_2O_3, loaded with 30 N force, was slid on the flat surface of the sample tested. It must be emphasized here that the ball surface wear was negligibly low.

3. Results

The solvent degradation of injection molded materials allowed the removal of 98% of the paraffin and facilitated thermal degradation at a later stage. The lack of solvent degradation often causes the formation of gas bubbles on the surface of the sinters. Removal of one of the polymer components, paraffin, allows for the partial opening of the pores in the entire volume of injection molded fittings. In the case of low-pressure molded surface layers, it is necessary to use only thermal degradation of the binder. The test results of injection molded materials show that the shrinkage value after sintering increases with the increase in sintering temperature. In addition, the shrinkage depends on the sintering atmosphere used; in particular, it is more significant for samples sintered in the N_2-10% H_2 atmosphere and increases in the case of sinters with additional carbides (Table 3). After sintering at 1250 °C in an Ar-10% H_2 atmosphere, the shrinkage of 316 L and 430 L steels was 7.5 and 10.3%, respectively, and 9.34 and 11.26% in the N_2-10% H_2 atmosphere. Thus, it can be seen that the shrinkage depends on the atmosphere and the type of material. The density of these steels sintered in the Ar-10% H_2 atmosphere is comparable and amounts to 87.5 and 86.8% for steel 316 L and 430 L, respectively. Using an atmosphere of N_2-10% H_2 causes an increase in the density of steel by only about 1.5% in both cases. The results of hardness tests confirm that adding carbides significantly increased the hardness of the tested sinters, but the change in atmosphere did not increase hardness in the case of 316 L steel. In particular, the hardness of 316 L steel after sintering at 1250 °C under Ar-10%H_2 atmosphere was 198 $HV_{0.1}$, and was comparable when sintered in N_2-10%H_2 atmosphere amounting to 196 $HV_{0.1}$. The hardness of this sintered steel was increased to 327 $HV_{0.1}$ by adding carbides at the same temperature and N_2-10%H_2 atmosphere. In the case of 430 L steel, the change in the sintering atmosphere had a much greater effect on hardness. After sintering in an argon-rich atmosphere, the hardness was 172 $HV_{0.1}$ and

412 HV$_{0.1}$ when using a nitrogen-rich atmosphere. Adding carbides to this steel increased the hardness to 302 and 546 HV$_{0.1}$ after sintering in the atmosphere of Ar-10%H$_2$ and N$_2$-10%H$_2$, respectively. Therefore, in the case of 430 L steel, the sintering atmosphere is quite essential. Tensile testing has shown that a nitrogen-rich atmosphere increases the strength of 316 L and 430 L steels. In the case of 316 L steel, the change in atmosphere from Ar-10% H$_2$ to N$_2$-10% H$_2$ during sintering at 1250 °C caused an increase in tensile strength from 410 to 643 MPa and a decrease in elongation from 32 to 6.6% (Figure 5). An atmosphere rich in nitrogen undoubtedly strengthens the structure of this steel. It should be noted that the maximum tensile strength of this steel is 652 MPa and can be achieved by adding carbides and sintering in a nitrogen-rich atmosphere. A similar trend can be observed in the case of 430 L steel. Detailed test results are presented in Table 3.

Table 3. Properties of obtained materials sintered at 1250 °C under different atmospheres.

Material	316 L		316 L/TC		430 L		430 L/TC	
Atmosphere	Ar-10%H$_2$	N$_2$-10%H$_2$	Ar-10%H$_2$	N$_2$-10%H$_2$	Ar-10%H$_2$	N$_2$-10%H$_2$	Ar-10%H$_2$	N$_2$-10%H$_2$
Density	6.92	7.03	6.98	7.027	6.693	6.79	7.68	7.825
Shrinkage	7.54	9.34	9.24	10.05	10.32	11.26	11.12	11.28
Hardness, HV$_{0.1}$	198	196	217	327	172	412	302	546
Tensile strength, MPa	410	643	559	652	432	668	658	679

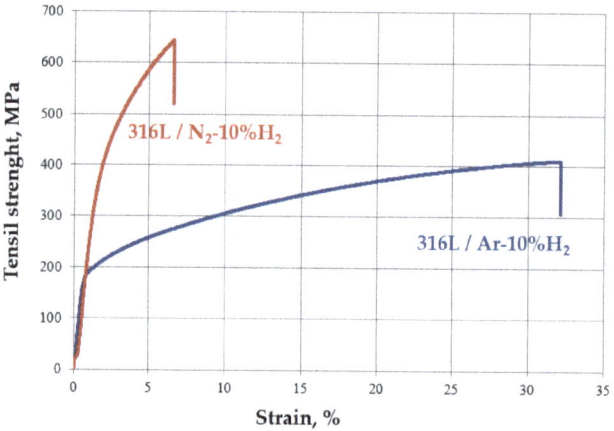

Figure 5. Measured stress–strain curves for 316 L steel sintered under Ar-10%H$_2$ and N$_2$-10%H$_2$.

The scanning microscope tests showed that the increase in the tensile strength of the steel is due to the release of carbonitrides after sintering in the N$_2$-10%H$_2$ atmosphere. In the case of both steels, they are rich in Cr and Fe, which was revealed by scanning microscopy and EDS analysis, as shown in Figures 6 and 7. Determining whether Mo and Ni are also part of these precipitates in 316 L steel is difficult because the phases are less than 1 µm. Similar phases precipitate in 430 L steel sintered in a nitrogen-rich atmosphere. The structure of both steel grades resembles a pearlitic structure due to the presence of fine precipitates of nitrides. Tribology studies (Figure 8) have shown that 316 L stainless steel sintered in an N$_2$-rich atmosphere achieves significantly higher abrasion resistance than steel sintered in an Ar-rich atmosphere. This is undoubtedly the effect of the precipitated nitrides. The width of the trace of abrasion of the sample sintered in a mixture of Ar-10%H$_2$ gases is 1739µm, and the depth is 62.4 µm. For the material sintered in the atmosphere of N$_2$-10%H$_2$, these values are 564 and 11.8 µm, respectively. The depth of the abrasion trace of steel sintered in a nitrogen atmosphere is more than five times lower, which is a surprising result for the authors. Table 4 shows the calculated volume of material removed in the ball-on-disc test. Both materials sintered in a nitrogen-rich atmosphere had lower wear compared to steels sintered in an argon-rich atmosphere.

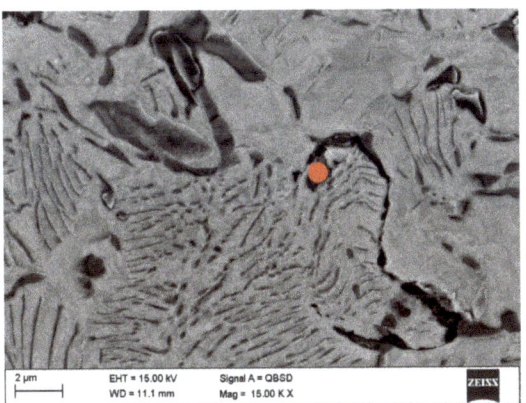 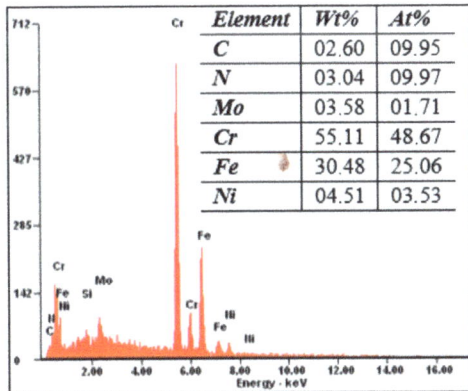

Figure 6. Structure of 316 L steel sintered under N_2-10%H_2 at a temperature of 1250 °C and chemical composition of (marked with red dot) precipitation of investigated materials observed in SEM.

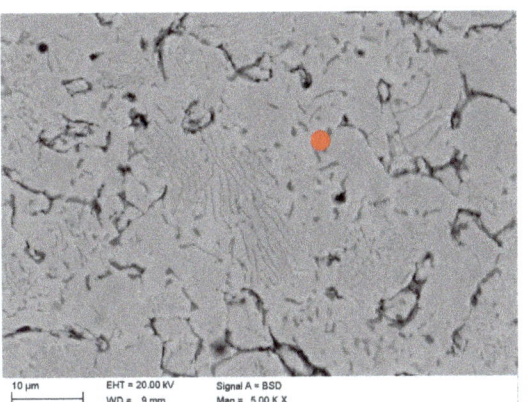 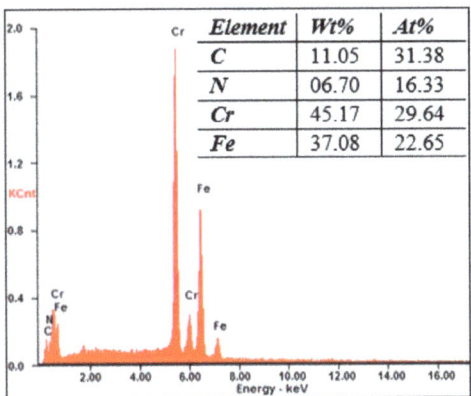

Figure 7. Structure of 430 L steel sintered under N_2-10%H_2 at a temperature of 1250 °C and chemical composition of (marked with red dot) precipitation of investigated materials observed in SEM.

Figure 9 shows carbide steels 316 L/TC and 430 L/TC produced by injection molding powders and sintered at 1250 °C. Comparing the structure of both materials, it can be seen that similar phases are separated in both materials. Grey carbides rich mainly in Cr, Fe and W are marked in both carbon steels as No. 2. Light carbides are rich mainly in W but also in Fe and Cr labelled as No. 3. There are also fine dark precipitates rich in nitrogen, carbon, and oxygen and titanium in these materials. Due to the small particle size, the Cr and Fe content may partly come from the matrix. The chromium concentration in the 430 L/TC carbide matrix is lower than that in the 430 L steel powder, and its mass fraction is 13.9 and 16.5%, respectively. Similarly, in 316 L/TC carbide, chromium concentration in the matrix decreases from 17.3 to 15.2%. This results from the precipitation of carbides rich in this element and the depletion of the matrix, which is a typical effect in stainless steels with a high concentration of carbon. Corrosion tests were also performed on injection molded samples because their surface is flatter and regular. The results of corrosion tests (Table 5) have shown that sintering in a nitrogen-rich atmosphere and adding carbides increases sinters' corrosion resistance. This effect is quite surprising, but it is most likely due to the lower porosity of these materials. In general, residual carbon, a nitrogen-rich atmosphere and other carbides should have the opposite effect, i.e., corrosion resistance should be lower. The residual carbon and nitrogen from the sintering atmosphere give off phases rich in chromium, which are responsible for corrosion resistance. However, it should be noted that both residual carbon and nitrogen cause sinter densification.

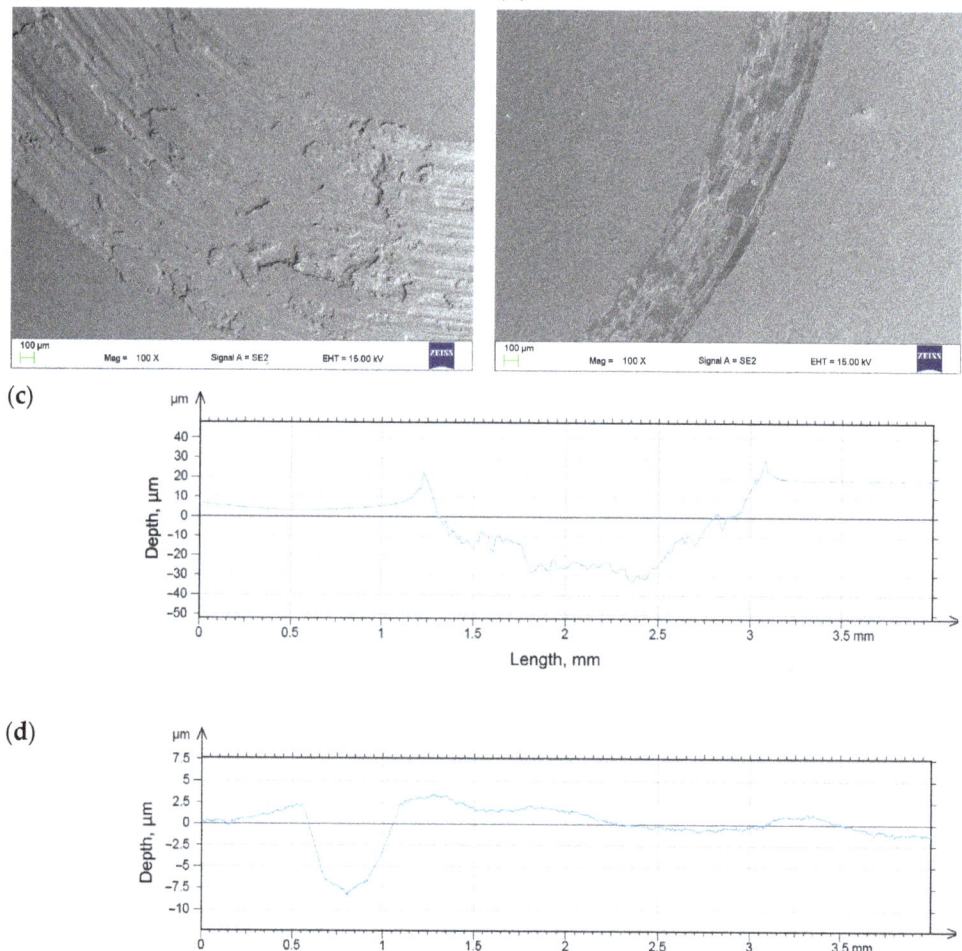

Figure 8. Wear track of the 316 L steel sintered under (**a**) Ar-10%H_2, (**b**) N_2-10%H_2 and its depth profiles (**c**,**d**), respectively.

Table 4. Influence of sintering atmosphere of stainless steel on its wear after ball-on-disc test.

Material	316 L Ar-10%H_2	316 L N_2-10%H_2	430 L Ar-10%H_2	430 L N_2-10%H_2
Volume of wear material, μm^2	1.86	0.106	1.48	0.52

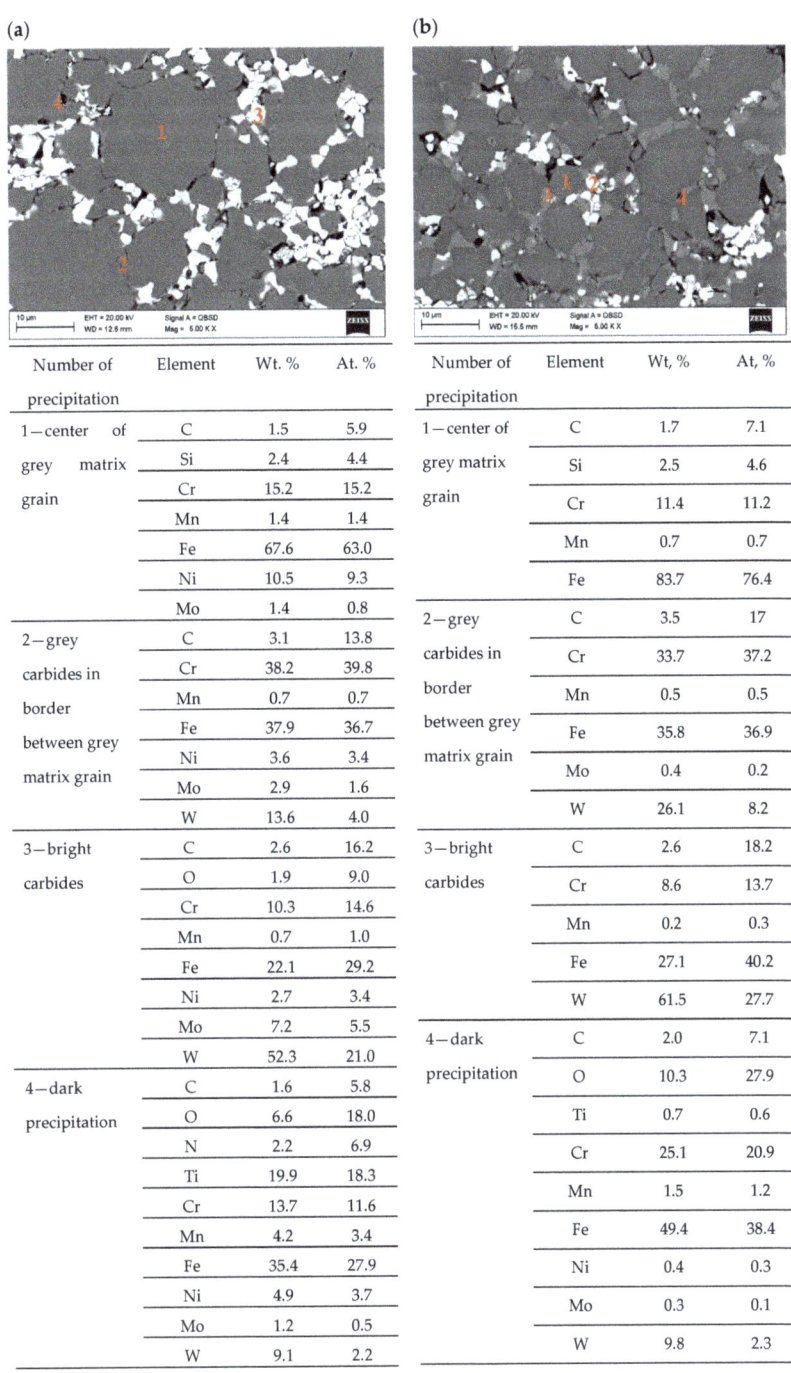

Number of precipitation	Element	Wt. %	At. %	Number of precipitation	Element	Wt, %	At, %
1 — center of grey matrix grain	C	1.5	5.9	1 — center of grey matrix grain	C	1.7	7.1
	Si	2.4	4.4		Si	2.5	4.6
	Cr	15.2	15.2		Cr	11.4	11.2
	Mn	1.4	1.4		Mn	0.7	0.7
	Fe	67.6	63.0		Fe	83.7	76.4
	Ni	10.5	9.3				
	Mo	1.4	0.8	2 — grey carbides in border between grey matrix grain	C	3.5	17
2 — grey carbides in border between grey matrix grain	C	3.1	13.8		Cr	33.7	37.2
	Cr	38.2	39.8		Mn	0.5	0.5
	Mn	0.7	0.7		Fe	35.8	36.9
	Fe	37.9	36.7		Mo	0.4	0.2
	Ni	3.6	3.4		W	26.1	8.2
	Mo	2.9	1.6	3 — bright carbides	C	2.6	18.2
	W	13.6	4.0		Cr	8.6	13.7
3 — bright carbides	C	2.6	16.2		Mn	0.2	0.3
	O	1.9	9.0		Fe	27.1	40.2
	Cr	10.3	14.6		W	61.5	27.7
	Mn	0.7	1.0	4 — dark precipitation	C	2.0	7.1
	Fe	22.1	29.2		O	10.3	27.9
	Ni	2.7	3.4		Ti	0.7	0.6
	Mo	7.2	5.5		Cr	25.1	20.9
	W	52.3	21.0		Mn	1.5	1.2
4 — dark precipitation	C	1.6	5.8		Fe	49.4	38.4
	O	6.6	18.0		Ni	0.4	0.3
	N	2.2	6.9		Mo	0.3	0.1
	Ti	19.9	18.3		W	9.8	2.3
	Cr	13.7	11.6				
	Mn	4.2	3.4				
	Fe	35.4	27.9				
	Ni	4.9	3.7				
	Mo	1.2	0.5				
	W	9.1	2.2				

Figure 9. Structure of carbide steel sintered at 1250 °C and chemical composition of precipitation of investigated materials observed in SEM, (**a**) 316 L/TC/N_2-10%H_2, (**b**) 430 L/TC/N_2-10%H_2.

Table 5. Corrosion test results for investigated materials.

	E_{ocp}, mV	E_{kor}, mV	E_b, mV	J_{kor}, µA/cm^2	R_{pol}, kΩ × cm^2
316 L comparative material	−132	−160	344	0.09	171
316 L/Ar-10%H$_2$	−286	−294	25	0.54	46
316 L/N$_2$-10%H$_2$	−303	−310	−196	18.1	1
316 L/TC/Ar-10%H$_2$	−506	−511	−420	56	0.5
316 L/TC/N$_2$-10%H$_2$	−349	−370	−118	5	4.4
430 L/Ar-10%H$_2$	−469	−434	−157	6.1	3.8
430 L/N$_2$-10%H$_2$	−300	−425	−256	6.6	2.9
430 L/TC/Ar-10%H$_2$	−423	−429	−265	5.2	3.7
430 L/TC/N$_2$-10%H$_2$	−363	−369	−149	3.3	5.8
4140 substrate	−602	−590	−517	25	0.8

The mixture of carbides used has a regular crystalline structure, which is stable at high temperatures. Only WC carbide crystallizes in a hexagonal lattice. It dissolves at high temperatures in the matrix of 316 L or 430 L stainless steel and, unfortunately, forms new carbides often rich in Cr, which was revealed in the tests performed in SEM and using EDS. Adding carbides to 316 L steel sintered in an argon-rich atmosphere makes parameters such as E_{ocp}, E_{kor}, and E_b better than 4140 steel, but the corrosion current density is more than twice as high. The best anti-corrosion properties were found in the sinters produced for 316 L steel at the same sintering atmosphere without adding carbons.

Analyzing the individual parameters, the free potential of the E_{ocp} material and the corrosion resistance of the E_{corr} material is better for 430 L steel with the addition of carbides and a nitrogen-rich atmosphere. At the same time, E_b is worse except for the 430 L/TC/N$_2$-10%H$_2$ material. The tests of injection molded samples and their results confirmed the reasonableness of producing surface layers of these materials on a non-alloy steel substrate. The drawings presented below result from a microscopic examination of selected examples of surface layers. Figures 10 and 11 are particularly noteworthy, proving that a good connection with the diffusive substrate should characterize these layers.

Figure 11 shows an enlargement of the area of the surface layer-substrate boundary with a clear diffusion zone in which there are no carbides but an apparent increase in the concentration of chromium and nickel. It is a layer with a thickness of approx. 15 mm, which separates the substrate from the layer of 316 L steel. The structure of the layer in this figure is rich in carbides, which confirms the thesis that the concentration of residual carbon in this area should be high, which initiates the sintering process of the surface layer with the substrate. Observing the structure of the surface layer on the surface of the sample, it can be seen that it is more porous. Therefore, paraffin degradation in this area is more accessible, and the proportion of residual carbon is lower or absent. Also, the share of carbides in this area is smaller, as evidenced by the Cr concentration distribution presented in the linear distribution of elements. Unfortunately, the porous structure may reduce corrosion resistance. However, it is still higher than the substrate, as confirmed by Figures 12 and 13 which show the surface layer observed under a light microscope before and after etching with Nital. The base structure in 4140 steel is less resistant to Nital.

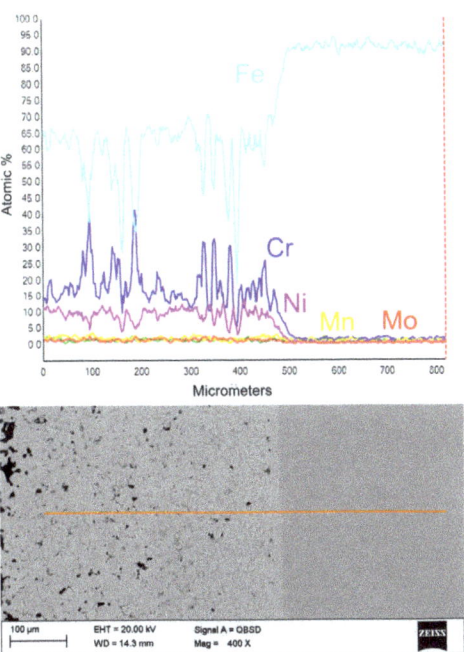

Figure 10. The structure of the surface layer of 316 L steel, manufactured by slurry pressureless forming on the 4140 substrate and sintered at 1250 °C in an atmosphere of N_2-10%H_2 and the distribution of elements in the area marked with a line.

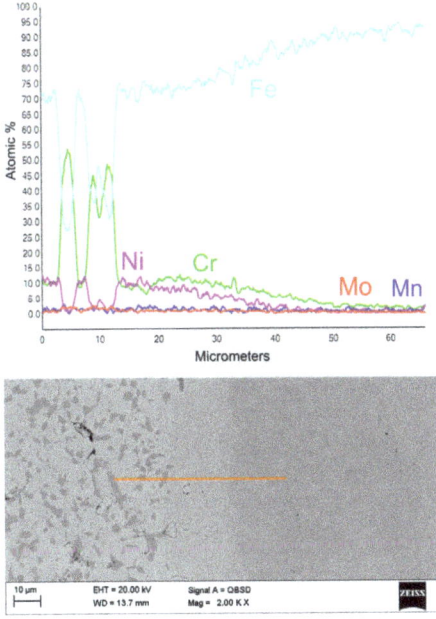

Figure 11. The structure of the surface layer made of 316 L steel, manufactured by slurry pressureless forming on the 4140 substrate and sintered at 1250 °C in an atmosphere of N_2-10%H_2 and the distribution of elements in the area marked with a line—higher magnification of boundary area.

Figure 12. Structure of the surface layer made of 316 L steel, manufactured by slurry pressureless forming on the 4140 substrate and sintered at 1250 °C in N_2-10%H_2 atmosphere—light microscope.

Figure 13. The structure of the material shown in Figure 12 etched with Nital.

Tests of surface layers reinforced with carbides showed that their structure is dense, with few pores that can be observed just below the surface of the layer. Figure 14 shows the 430 L/TC layer sintered at 1250 °C in an N_2-10% H_2 atmosphere. Comparing PIM and PLF, it can be seen that a heterogeneous structure with numerous carbide agglomerates characterizes the produced surface layers enriched with carbides. This results from the manual preparation of the slurry in a mortar. The injection molded material is more homogeneous because although local agglomerates can also be seen, they are not as numerous as in the low-pressure molding method. Undoubtedly, the mixture prepared in the crusher is more homogeneous. Agglomerated carbides are also seen in the bimodal particle size distribution (Figure 1). To continue this research, attention should be paid to homogenizing the structure.

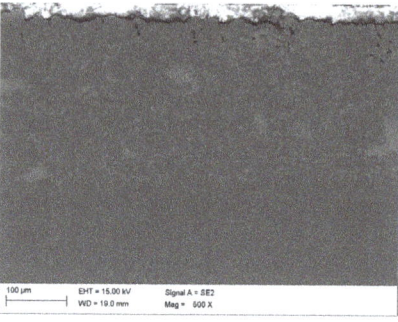

Figure 14. Layer structure of 430 L/TC steel, manufactured by slurry pressureless forming on unalloyed steel and sintered at 1250 °C in Ar-10%H2 atmosphere.

4. Discussion

A newly created technique has been developed that uses polymer-powder slurry forming and sintering to create corrosion-resistant surface layers on steel substrates with additional carbide reinforcement. The materials fabricated this way are characterized by high quality and a lack of defects like distortions and bubbles inside and outside. Injection molded materials' solvent degradation enabled almost all paraffin removal and later promoted heat degradation. A lack of solvent degradation often causes the formation of gas bubbles on the surface of the sinters. This results from accumulating gaseous thermal degradation products formed during pyrolysis [26,27]. The pores in the total volume of injection molded samples can be partially opened by removing one of the polymer components, i.e., paraffin. However, the low-thickness surface layers require only the binder's thermal decomposition [18].

According to geometry measurement results of samples produced through injection molding, the shrinkage value grows with increasing sintering temperature. Even though no quantitative porosity test was performed, the shrinkage results can be considered indirect information about the compaction of the material during sintering. Unfortunately, the decrease in porosity is not entirely dependent on the increase in the tested density. As a result of sintering in an atmosphere rich in nitrogen, this gas diffuses into the sample, and fine precipitations are in the form of nitrides. The precipitation process affects the compaction kinetics and thus affects the density of the material. Of course, the sintered sample density also increases due to the addition of carbide phases. Generally, 316 L austenitic steel is characterized by lower shrinkage than 430 L ferritic steel and correspondingly higher porosity, which is consistent with the results of other authors [28]. According to a comparison of the shrinkage and density of sintered materials in various atmospheres, the gas used has little influence on the degree of densification of the structure of the produced materials. In both instances, using an atmosphere of N_2-10% H_2 only results in a 1.5% increase in steel density.

The results of hardness tests have shown that materials sintered at 1250 °C are characterized by higher hardness than commercial steels [29]. Hardness investigations show that adding carbides greatly increased the hardness of the studied samples, but 316 L steel's hardness was not affected by changes in the gas atmosphere. The difference in sintering atmosphere had a substantially more significant impact on hardness in the case of 430 L steel. Of course, adding carbides increases the hardness to a maximum value of 546 HV0.1 for the 430 L/TC coating sintered in a nitrogen-rich atmosphere.

However, the observations are different regarding the influence of the atmosphere on strength properties. The tensile strength of 316 L steels sintered in the Ar-10% H2 atmosphere was below the strength of commercial steels and had a similar elongation value. According to tensile tests, a nitrogen-rich atmosphere improves the strength of 316 L and 430 L steels. In contrast, the use of N_2-10%H_2 atmosphere caused an increase in strength properties to a value exceeding commercial 316 L steel, i.e., 643 MPa and, unfortunately, a decreasing elongation by nearly five times. This effect corresponds with other authors' investigations [30]. Without a doubt, the nitrogen-rich atmosphere strengthens the structure of stainless steel, which has already been well documented in the literature [31–33]. Unfortunately, all sinters produced are characterized by lower corrosion resistance than commercial steel 316 L, and in the case of 316 L/TC material sintered in an Ar-10% H2 atmosphere, resistance was even worse than the support material. Analyzing the influence of the atmosphere and the share of carbides, a general conclusion can be drawn that the corrosion resistance of these materials depends mainly on porosity.

Furthermore, according to tribology experiments, 316 L stainless steel sintered in an N2-rich atmosphere exhibits significantly better abrasion resistance than steel sintered in an Ar-rich atmosphere. Beyond any uncertainties, this results from the precipitated nitrides, mainly Cr_2N [30]. The wear track of the 316 L sample sintered in an argon-rich atmosphere showed a combination of adhesion, abrasion, and plastic deformation. Small metallic fragments appeared shortly after the test began, unlike the samples sintered in nitrogen gas, where friction produced barely any tiny particles. Visible changes in the friction force suggested that the steel and ball were sticking together and causing adhesive wear. On the sample sintered in Ar+10%H$_2$ gas, severe adhesive and abrasive wear was observed, whereas the specimens sintered in N$_2$+10%H$_2$ revealed only very mild abrasive wear and some plastic deformation as the layer was pressed into the substrate at higher loads. Differences in the behavior of nitrogen-enriched stainless-steel samples are consistent with reports in [34].

Corrosion studies have demonstrated that sintering in an atmosphere rich in nitrogen and adding carbides improved the corrosion resistance of sintered samples. The fact that tests were conducted on the surface of sintered samples that had not been ground or given any other kind of treatment suggests that this impact, however unexpected, is caused by the lower porosity of these materials. The corrosion resistance should generally be reduced in the presence of residual carbon, a nitrogen-rich atmosphere, and other carbides [35]. Chromium-rich phases are produced by the excess carbon and nitrogen from the sintering atmosphere, which prevents corrosion. However, it should be emphasized that sinter densification is brought on by both residual carbon and nitrogen. Nitrogen contributes indirectly to the compaction of the sinter by forming nitrides or carbonitrides, which causes the released non-carbide-forming carbon to initiate the sintering process [36]. Undoubtedly, the lower porosity of sinters increases corrosion resistance, which is confirmed by the results of other authors [37]. The legitimacy of using nitrogen in austenitic stainless steels is indisputable. It can replace expensive nickel. Nitrogen stabilizes the austenitic structure and improves mechanical and anti-corrosion properties, but unfortunately, it reduces plasticity [38–45]. Due to the low solubility of nitrogen in Fe in conventional steels, it is introduced during their melting under high pressure. Some alloy additions also improve the solubility of nitrogen in steel [39]. Sintering these steels in a nitrogen-rich atmosphere is a better solution. Unfortunately, Cr nitrides are released during sintering, which has been confirmed by test results and data in the literature [46]. Similar phases can be seen in Co-Cr-Mo alloys sintered in a nitrogen-rich atmosphere [47]. The latest research results confirm the separation of these phases and, additionally, a decrease in the sinterability of steel, which increases porosity, which in turn decreases mechanical properties and corrosion resistance [32]. Interpretation of some test results is quite difficult. For example, a nitrogen-rich sintering atmosphere for 316 L steel increases tensile strength and wear resistance while maintaining the hardness at the level of steel sintered in an Ar-rich atmosphere. This requires further research of these materials, mainly using higher sintering temperatures, which will increase the density and mechanical properties of the sinters and, most likely, increase corrosion resistance.

The pressureless forming technology enabled the application of polymer-powder slurries containing stainless steels and their mixtures with carbides on the base of unalloyed steels. The structure of the surface layers is similar to the previously produced sinters. A diffusion layer can be observed in the boundary zone with the substrate, which is rich in alloy additions typical of stainless steels, such as Cr and Ni, which diffuse into the substrate made of unalloyed steel. This area is characterized by low porosity. Many carbides can also be observed in this zone, although they were not added to the polymer-powder slurry. This is the effect of an increase in the concentration of carbon, which initiates the sintering process but simultaneously causes the precipitation of Cr-rich carbides. The surface of the produced top layer is characterized by much greater porosity. It can be lowered by adding a mixture of tetra carbides to the slurry, thus increasing the hardness. Irrespective of the type of low-pressure polymer-powder slurry, the sintering surface layers are characterized by an excellent diffusive connection with the substrate and do not show discontinuities in the form of cracks and decohesion. Automating the process of applying the polymer-powder slurry onto elements with a round cross-section or an extended surface is possible. The materials produced this way are also characterized by a continuous surface layer but with a different thickness and structure. Therefore, further research should be conducted to automate the process of forming anti-corrosion or anti-wear surface layers.

5. Conclusions

The aim of the research was to develop layered materials using polymer-powder slurry molding and sintering, which resulted in the creation of steel with a durable, corrosion-resistant surface layer. Several conclusions can be made considering all of the observed results:

1. Sinters such as steels 316 L, 430 L and carbide steels with a matrix of stainless steels, produced by injection molding of powders, are characterized by relatively low porosity, depending mainly on the sintering temperature.
2. The results of hardness tests have shown that materials sintered at 1250 °C are characterized by higher hardness than commercial steels, regardless of the atmosphere used and the addition of carbides.
3. The higher mechanical properties of the obtained stainless-steel coating were influenced by the increase in carbon concentration resulting from residual carbon and nitrogen in the sintering atmosphere, which caused the release of carbonitrides, regardless of the steel grade.
4. Tensile testing shows that 316 L and 430 L steels have higher strength after sintering in nitrogen-rich atmospheres. The atmosphere changes from Ar-10% H_2 to N_2-10% H_2 during sintering improved tensile strength by more than 50% and decreased elongation by almost five times. Nitride precipitations undoubtedly strengthen the stainless steel's structure.

Author Contributions: Conceptualization, G.M.; methodology, G.M.; validation, B.T.; formal analysis, G.M.; investigation, G.M. and B.T.; resources, G.M.; writing—original G.M., writing—review and editing, B.T. and G.M.; supervision, G.M.; project administration, G.M.; funding acquisition, G.M. All authors have read and agreed to the published version of the manuscript.

Funding: Research supported as part of the Excellence Initiative—Research University program implemented at the Silesian University of Technology. Project number 10/110/SDU/10-21-01.

Institutional Review Board Statement: Not applicable.

Informed Consent Statement: Not applicable.

Data Availability Statement: Not applicable.

Conflicts of Interest: The authors declare no conflict of interest.

References

1. German, R.M.; Bose, A. *Injection Molding of Metal and Ceramics*; Metal Powder Industries Federation: Princeton, NJ, USA, 1997.
2. Matula, G.; Jardiel, T.; Levenfeld, B.; Varez, A. Application of powder injection moulding and extrusion process to manufacturing of Ni-YSZ anodes. *J. Achiev. Mater. Manuf. Eng.* **2009**, *36*, 87–94.
3. Petzoldt, F. Metal injection moulding in Europe: Ten facts that you need to know. *Powder Inject. Mould. Int.* **2007**, *1*, 23–28.
4. Available online: https://mepsinternational.com/gb/en/products/world-stainless-steel-prices (accessed on 30 May 2023).
5. Šmak, M.; Kubíček, J.; Kala, J.; Podaný, K.; Vaněrek, J. The Influence of Hot-Dip Galvanizing on the Mechanical Properties of High-Strength Steels. *Materials* **2021**, *14*, 5219. [CrossRef] [PubMed]
6. Available online: https://ycharts.com/indicators/nickel_price (accessed on 30 May 2023).
7. Lach, D.; Polanski, J.; Kapkowski, M. CO_2—A Crisis or Novel Functionalization Opportunity? *Energies* **2022**, *15*, 1617. [CrossRef]
8. Matula, G.; Jardiel, T.; Jimenez, R.; Levenfeld, B.; Jardiel, T.; Varez, A. Microstructure, mechanical and electrical properties of Ni-YSZ anode supported solid oxide fuel cells. *Arch. Mater. Sci. Eng.* **2008**, *32*, 21–24.
9. Ogunbiyi, O.F.; Jamiru, T.; Sadiku, E.R.; Adesina, O.T.; Beneke, L.; Adegbola, T.A. Spark plasma sintering of nickel and nickel based alloys: A Review. *Procedia Manuf.* **2019**, *35*, 1324–1329. [CrossRef]
10. Dobrzanski, L.A.; Brytan, Z.; Grande, M.A.; Rosso, M. Corrosion behavior of vacuum sintered duplex stainless steels. *J. Mater. Process. Technol.* **2007**, *191*, 161–164. [CrossRef]
11. Campos, M.; Bautista, A.; Cáceres, D.; Abenojar, J.; Torralba, J.M. Study of the interfaces between austenite and ferrite grains in P/M duplex stainless steels. *J. Eur. Ceram. Soc.* **2003**, *23*, 2813–2819. [CrossRef]
12. Sotomayor, M.E.; Cervera, A.; Várez, A.; Levenfeld, B. Duplex stainless steel self-ligating Orthodontic Brackets by Micro-powder Injection Moulding. *Int. J. Eng. Res. Sci. (IJOER)* **2016**, *2*, 184–193.
13. Mariappan, R.; Kishore Kumar, P.; Jayavelu, S.; Dharmalingam, G.; Arun Prasad, M.; Stalin, A. Wear properties of PM duplex stainless steels developed from 316L and 430L powders. *Int. J. ChemTech Res.* **2015**, *8*, 109–115.

14. Mariappan, R.; Kumaran, S.; Srinivasa Rao, T. Effect on sintering atmosphere on structure and properties of austeno-ferritic stainless steels. *Mater. Sci. Eng.* **2009**, *517*, 328–333. [CrossRef]
15. Cui, C.; Stern, F.; Ellendt, N.; Uhlenwinkel, V.; Steinbacher, M.; Tenkamp, J.; Walther, F.; Fechte-Heinen, R. Gas Atomization of Duplex Stainless Steel Powder for Laser Powder Bed Fusion. *Materials* **2023**, *16*, 435. [CrossRef] [PubMed]
16. Sotomayor, M.E.; Levenfeld, B.; Várez, A. Powders injection moulding of premixed ferritic and austenitic stainless steels powders. *Mater. Sci. Eng. A* **2011**, *528*, 3480–3488. [CrossRef]
17. Watanabe, I.; Chen, T.-T.; Taniguchi, S.; Kitano, H. Heterogeneous microstructure of duplex multilayer steel structure fabricated by wire and arc additive manufacturing. *Mater. Charact.* **2022**, *191*, 112159. [CrossRef]
18. Matula, G. Application of polymer-powder slurry for fabrication of abrasion resistant coatings on tool materials. *Arch. Mater. Sci. Eng.* **2011**, *48*, 49–55.
19. Singh, M.; Sharma, S.; Muniappan, A.; Pimenov, D.Y.; Wojciechowski, S.; Jha, K.; Dwivedi, S.P.; Li, C.; Królczyk, J.B.; Walczak, D.; et al. In Situ Micro-Observation of Surface Roughness and Fracture Mechanism in Metal Microforming of Thin Copper Sheets with Newly Developed Compact Testing Apparatus. *Materials* **2022**, *15*, 1368. [CrossRef] [PubMed]
20. Thakare, J.G.; Pandey, C.; Mahapatra, M.M.; Mulik, R.S. Thermal Barrier Coatings—A State of the Art Review. *Met. Mater. Int.* **2021**, *27*, 1947–1968. [CrossRef]
21. Koseski, R.P.; Suri, P.; Earhardt, N.B.; German, R.M.; Kwon, Y.S. Microstructural evolution of injection moulded gas- and water-atomized 316L stainless steel powder during sintering. *Mater. Sci. Eng. A* **2005**, *390*, 171–177. [CrossRef]
22. German, R.M. *A-Z of Powder Metallurgy*; Elsevier Science: Amsterdam, The Netherlands, 2006.
23. Wu, Y.; German, R.M.; Blaine, D.; Marx, B.; Schlaefer, C. Effects of residual carbon content on sintering shrinkage, microstructure and mechanical properties of injection molded 17-4 PH stainless steel. *J. Mater. Sci.* **2002**, *37*, 3573–3583. [CrossRef]
24. Matula, G. Carbon effect in the sintered high-speed steels matrix composites—HSSMC. *J. Achiev. Mater. Manuf. Eng.* **2012**, *55*, 90–107.
25. PN-EN ISO 17475:2010; Corrosion of Metal and Alloys—Electro-Chemical Test Methods—Guidelines for Conducting Potentiostatic and Potentiodynamic, Polarization Measurements. ISO: Geneva, Switzerland, 2010; p. 201.
26. Omar, A.; Ibrahim, R.; Sidik, M.I.; Mustapha, M.; Mohamad, M. Rapid debinding of 316L stainless steel injection moulded component. *J. Mater. Process. Technol.* **2003**, *140*, 397–400. [CrossRef]
27. Rafi Raza, M.; Ahmad, F.; Omar, M.A.; German, R.M.; Muhsan, A.S. Defect Analysis of 316LSS during the Powder Injection Moulding Process. *Defect Diffus. Forum* **2012**, *329*, 35–43. [CrossRef]
28. Panda, S.S.; Singh, V.; Upadhyaya, A.; Agrawal, D. Sintering response of austenitic (316L) and ferritic (434L) stainless steel consolidated in conventional and microwave furnaces. *Scr. Mater.* **2006**, *54*, 2179–2183. [CrossRef]
29. Aykac, E.; Turkmen, M. Investigation of the Biocompatibility of Laser Treated 316L Stainless Steel Materials. *Coatings* **2022**, *12*, 1821. [CrossRef]
30. Kurgan, N. Effects of sintering atmosphere on microstructure and mechanical property of sintered powder metallurgy 316L stainless steel. *Mater. Des.* **2013**, *52*, 995–998. [CrossRef]
31. García, C.; Martín, F.; Blanco, Y.; de Tiedra, M.P.; Aparicio, M.L. Corrosion behaviour of duplex stainless steels sintered in nitrogen. *Corros. Sci.* **2009**, *51*, 76–86. [CrossRef]
32. Shen, H.; Zou, J.; Li, Y.; Li, D.; Yu, Y.; Wang, X. Effects of nitrogen on predominant sintering mechanism during the initial stage of high nitrogen nickel-free stainless steel powder. *J. Alloys Compd.* **2023**, *945*, 169230. [CrossRef]
33. Hu, L.; Peng, H.; Baker, I.; Li, L.; Zhang, W.; Ngai, T. Characterization of high-strength high-nitrogen austenitic stainless steel synthesized from nitrided powders by spark plasma sintering. *Mater. Charact.* **2019**, *152*, 76–84. [CrossRef]
34. Liang, W.; Bin, X.; Zhiwei, Y.; Yaqin, S. The wear and corrosion properties of stainless steel nitrided by low-pressure plasma-arc source ion nitriding at low temperatures. *Surf. Coat. Technol.* **2000**, *130*, 304–308. [CrossRef]
35. Meesak, T.; Thedsuwan, C. Corrosion behaviours of stainless steel parts formed by powder metallurgy process. *Mater. Today Proc.* **2018**, *5*, 9560–9568. [CrossRef]
36. Ali, S.; Abdul Rani, A.M.; Ahmad Mufti, R.; Ahmed, S.W.; Baig, Z.; Hastuty, S.; Razak, M.A.A.; Abdu Aliyu, A.A. Optimization of Sintering Parameters of 316L Stainless Steel for In-Situ Nitrogen Absorption and Surface Nitriding Using Response Surface Methodology. *Processes* **2020**, *8*, 297. [CrossRef]
37. Woźniak, A.; Staszuk, M.; Reimann, Ł.; Bialas, O.; Brytan, Z.; Voinarovych, S.; Kyslytsia, O.; Kaliuzhnyi, S.; Basiaga, M. The influence of plasma-sprayed coatings on surface properties and corrosion resistance of 316L stainless steel for possible implant application. *Archiv. Civ. Mech. Eng.* **2021**, *21*, 148. [CrossRef]
38. Simmons, J.W. Overview: High-nitrogen alloying of stainless steels. *Mater. Sci. Eng. A* **1996**, *207*, 159–169. [CrossRef]
39. Reed, R.P. Nitrogen in austenitic stainless steels. *JOM* **1989**, *41*, 16–21. [CrossRef]
40. Maznichevsky, A.; Sprikut, R.V.; Goikhenberg, Y. Investigation of Nitrogen Containing Austenitic Stainless Steel. *Mater. Sci. Forum* **2020**, *989*, 152–159. [CrossRef]
41. Stein, G.; Hucklenbroich, I. Manufacturing and Applications of High Nitrogen Steels. *Mater. Manuf. Process.* **2004**, *19*, 7–17. [CrossRef]
42. Ganesan, V.; Mathew, M.D.; Sankara Rao, K.B. Influence of nitrogen on tensile properties of 316LN SS. *Mater. Sci. Technol.* **2009**, *25*, 614–618. [CrossRef]

43. Bayoumi, F.M.; Ghanem, W.A. Effect of nitrogen on the corrosion behavior of austenitic stainless steel in chloride solutions. *Mater. Lett.* **2005**, *59*, 3311–3314. [CrossRef]
44. Bautista, A.; Velasco, F.; Guzmán, S.; De La Fuente, D.; Cayuela, F.; Morcillo, Y.M. Corrosion behavior of powder metallurgical stainless steels in urban and marine environments. *Rev. Metal.* **2006**, *42*, 175–184. [CrossRef]
45. Matula, G.; Szatkowska, A.; Matus, K.; Tomiczek, B.; Pawlyta, M. Structure and Properties of Co-Cr-Mo Alloy Manufactured by Powder Injection Molding Method. *Materials* **2021**, *14*, 2010. [CrossRef]
46. Singh, M.; Garg, H.K.; Maharana, S.; Yadav, A.; Singh, R.; Maharana, P.; Nguyen, T.V.T.; Yadav, S.; Loganathan, M.K. An Experimental Investigation on the Material Removal Rate and Surface Roughness of a Hybrid Aluminum Metal Matrix Composite (Al6061/SiC/Gr). *Metals* **2021**, *11*, 1449. [CrossRef]
47. Saini, N.; Pandey, C.; Thapliyal, S.; Dwivedi, D.K. Mechanical Properties and Wear Behavior of Zn and MoS_2 Reinforced Surface Composite Al-Si Alloys Using Friction Stir Processing. *Silicon* **2018**, *10*, 1979–1990. [CrossRef]

Disclaimer/Publisher's Note: The statements, opinions and data contained in all publications are solely those of the individual author(s) and contributor(s) and not of MDPI and/or the editor(s). MDPI and/or the editor(s) disclaim responsibility for any injury to people or property resulting from any ideas, methods, instructions or products referred to in the content.

Article

Modeling and Model Verification of the Stress-Strain State of Reinforced Polymer Concrete

Kassym Yelemessov [1], Layla B. Sabirova [2], Nikita V. Martyushev [3,*], Boris V. Malozyomov [4], Gulnara B. Bakhmagambetova [5] and Olga V. Atanova [6]

[1] Institute of Energy and Mechanical Engineering, Satbayev University, Almaty KZ-050000, Kazakhstan; k.yelemessov@satbayev.university
[2] Department of Oil and Gas Production, Satbayev University, Almaty KZ-050000, Kazakhstan; slb2609@mail.ru
[3] Department of Advanced Technologies, Tomsk Polytechnic University, 634050 Tomsk, Russia
[4] Department of Electrotechnical Complexes, Novosibirsk State Technical University, 20, Karl Marks Ave., 630073 Novosibirsk, Russia; borisnovel@mail.ru
[5] Department of Mining, Satbayev University, Almaty KZ-050000, Kazakhstan
[6] Scientific Department, Satbayev University, Almaty KZ-050000, Kazakhstan
* Correspondence: martjushev@tpu.ru

Abstract: This article considers the prospects of the application of building structures made of polymer concrete composites on the basis of strength analysis. The issues of application and structure of polymer-concrete mixtures are considered. Features of the stress-strain state of normal sections of polymer concrete beams are revealed. The dependence between the stresses and relative deformations of rubber polymer concretes and beams containing reinforcement frame and fiber reinforcement has been determined. The main direction of the study was the choice of ways to increase the strength characteristics of concrete with the addition of a polymer base and to increase the reliability of structures in general. The paper presents the results of experimental and mathematical studies of the stress-strain state and strength, as well as deflections of reinforced rubber-polymer beams. The peculiarities of fracture of reinforced rubber-polymer beams along their sections have been revealed according to the results of the experiment. The peculiarities of fracture formation of reinforced rubber-polymer beams have also been revealed. The conducted work has shown that the share of longitudinal reinforcement and the height of the fibrous reinforcement zone are the main factors. These reasons determine the characteristics of the strength of the beams and their resistance to destructive influences. The importance and scientific novelty of the work are the identified features of the stress-strain state of normal sections of rubber-concrete beams, namely, it has been established that the ultimate strength in axial compression and tension, deformations corresponding to the ultimate strength for rubber concrete exceed similar parameters for cement concrete 2.5–6.5 times. In the case of the addition of fiber reinforcement, this increase becomes, respectively, 3.0–7.5 times.

Keywords: geopolymer concrete; rubber polymer concretes; composite materials; concrete structures; strength analysis

1. Introduction

Sustainable economic development of the countries in the world has become one of the main global environmental problems in many countries, and in the last decade, it has been paid much attention. There is an urgent need to develop appropriate strategies for the disposal of various types of waste such as plastic, tires, rubber, and glass. It is estimated that around 1.5 billion rubber tires are disposed of worldwide every year. The materials used in the production of tires are made up of complex mixtures. As a result, the share of discarded recycled tires is relatively low, with over 50% remaining unused.

Rubber industry waste and used tire rubber (UTR) is a non-degradable material that is not properly recycled. UTR also takes a significant amount of time to naturally degrade due to the cross-linked structure of the polymer material and additives such as stabilizers. Burning UTR releases toxic gases. Landfills and waste disposal cause serious environmental pollution of soil, water, and air. It pollutes the soil by killing beneficial bacteria and releasing toxic gases. Waste can be considered a potential resource and valuable material. A logical way to reduce the negative impact of rubber waste and its cost is to use it in construction and industry. Natural aggregates are more valuable than waste. When looking for economically viable deposits to extract non-metallic materials, factors such as transportation costs, the quality of the aggregate suitable for mining, government regulation, and the cost of operating and maintaining vehicles are taken into account. At the same time, transport costs for aggregates are very high. In addition, the energy required to crush the rock into aggregates is proportional to the area of the new surface, and therefore it constitutes a significant part of the energy consumed in the production of aggregates [1]. UTR easily absorb energy and have excellent sound and heat insulation properties, making them suitable for use in a variety of applications screens, and asphalt concrete mixtures [1,2]. Over the past 20 years, researchers have studied the possibility of using UTR as an aggregate in polymer concrete mixes. The use of rubber additives in concrete structures provides concrete products with new properties, including strength, reliability, and resistance to aggressive environments.

Polymer concretes are widely used in the construction industry and other industries. To date, a large amount of experimental data on the study of their structure and properties has been accumulated [1]. At the same time, despite a sufficiently large number of already conducted studies, there are difficulties in selecting optimal compositions for manufacturing the structures. Insufficient elaboration and studying the law of the structure's formation and properties of polymer concrete create complexities in their application [2]. The establishment of such regularities is an extremely difficult task.

Polymer concretes are a type of polymer composite material. The main area of application of such composites is construction. In modern representation, polymer composites are a rather complex hierarchical system formed as a result of physical and chemical interactions between its structural components [3]. The main feature of composites is their ability to form specific structures responsible for the acquisition of non-additive, sometimes unique properties of the composite. Such structures may include fractal, cluster, and lattice structures, the analysis of which is paid more and more attention in modern construction material science [4]. Properties of polymer concrete at the microstructure level are determined by the phenomena occurring during the contact of liquid and solid phases, i.e., they depend on the amount of filler, its dispersity, and physical and chemical activity. There is no universal optimal filler content for composites. Depending on the conditions of application of polymer concretes, this value can take different values. Usually, the optimal content of the filler provides the highest performance indicators of polymer concretes [5]. In this regard, the use of fillers having discontinuous granulometry, i.e., having different geometric fractions, is effective. The study of the influence of aggregates and their role in the structure formation of polymer concretes is an important problem considered in this article.

In this paper, the most promising polystructural theory of construction composites is taken as the basis. It is based on the concept that consists of the representation of construction composites as polystructured. They are composed of many structures, passing one into another according to the "structure-in-structure" principle, in which there are fractions of additional composite materials of several sizes (powder and granules) [6]. There is an organic connection between the structures of different levels and sublevels. The formation of structures of a higher scale level occurs under the influence of structures of a lower level. At the same time, the higher-level structures may determine the substructure formation conditions by the feedback principle. At present, based on the results of numerous practical and theoretical studies, ideas about the optimal structure of construction composites as a matrix medium with dispersed particles evenly distributed in it are being revised [7]. The practical unattainability of such an "ideal" structural situation has been established. On the contrary, during the technological processes of preparation and the curing of construction composites, the structural components of composites tend to combine various kinds of heterogeneities, differently affecting the properties of composites. It is possible to change the structure of construction composites by changing the formulation and technological factors. New structural heterogeneities can be introduced, or existing structural heterogeneities can be changed at various structural levels. The formation of boundary transition layers of the polymer matrix in polymer concretes has a direct impact on their performance properties (flexibility, elastic deformations, cracking resistance). Therefore, the analysis of their formation requires a more detailed consideration. The recommended methods of selecting the particle size distribution of aggregates are difficult to implement and do not guarantee to obtain the smallest intergranular hollowness, since they do not consider the properties of individual fractions of aggregates and their mutual distribution. The properties of polymer concretes depend not only on the quality of the initial components and their mutual arrangement but also on the nature of the interaction between them [8]. To create high-quality polymer concretes, it is necessary to have a strong, chemically and thermally, stable bond between the surface of aggregates and the polymer matrix. The considered modern methods of predicting the properties and calculation of compositions of polymer concretes are almost all based on the selection of the mineral mixture with the lowest hollowness by the method used for cement concrete. Recently, the methods for predicting the properties and calculating compositions of composite materials of reduced polymer capacity, provided that there is a given set of properties based on the models of different types obtained as a result of research, have become widely used. This depends not only on the material properties, but also on the technology of its production and, most importantly, on the purpose and application of polymer concrete structures [9].

To ensure a reliable efficient operation of elements made of new types of concrete, it is necessary to study the stress-strain state arising under the action of forces of various kinds, in particular, a bending moment. In this connection, the study of resistance to the bending moment action of normal sections of beams made of reinforced polymer concrete (ACRP) (beams containing a reinforcement frame and fiber reinforcement located at different heights of the section relative to the bottom edge) is of scientific interest and a practical important research task.

The conducted research analysis of force resistance of polymer concrete and reinforced concrete bendable elements has shown that the application of polymer concretes and structures on their basis is actually because of their inherent high operational characteristics. It is important to study the degree of influence of dispersed reinforcement on the performance characteristics of polymer concrete structures. The main methods existing today for polymer concrete structures' calculation are quite limited in terms of the used materials. Their main scope is structures made of furfural acetone concrete, polyester, and epoxy concrete. This is due to the existing wide experimental base containing these materials and a large number of obtained empirical dependences [10].

In this regard, to improve the strength of polymer concretes and the reliability of structures, as well as to predict the properties of the material and products made of it, the following tasks were set in this article:

- to carry out an analysis of force-resistance studies of polymer concrete and reinforced concrete beams;
- to evaluate the physical conditions of polymer concrete structures based on rubber additives;
- to conduct experimental studies of the stress-strain state and strength and deflections of ACRP beams;
- to reveal the peculiarities of failure of ACRP beams by their cross-sections;
- to reveal the peculiarities of the formation of the failure of ACRP beams;
- to reveal the peculiarities of the development of deflections of ACRP beams.

2. Methods and Materials

There is a sufficiently large production and waste volume of synthetic rubber and rubber products in modern industry [2,3]. Their use as a binder in the production of polymer concrete is therefore of practical interest. They have a low cost in comparison with that of used polymer resins. It is also possible to use liquid rubbers having specified properties. However, in this work, rubber was chosen as a binder. It is important to note that the production of polymer concretes based on rubbers due to the high filling of the mixture can reduce the cost of the composite [11,12].

2.1. Rationale for the Material Choice

Based on previous studies, the optimum ratios of hardening group components containing a binder, aggregate, and filler were determined [13]. It was found that the introduction of fly ash, which is a hard-to-dispose waste, into the composition of rubber polymer concrete (RPC), has a positive effect on its strength and chemical properties. Moreover, in these works, studies were conducted to study the chemical resistance of polymer concrete and it was found that polymer concrete has almost universal chemical resistance to various inherently aggressive environments.

Based on the analysis of literary sources and in the course of experimental studies, the authors obtained a comparative table (Table 1) of polymer concretes based on rubbers, which have the potential to be used for the manufacture of structures [2,3,6,7,12]. The main characteristics of rubber concrete used in the work are presented in Table 1.

Table 1. Main characteristics of used rubber polymer concrete.

Properties	Indicators for Rubber Concrete Based on Rubber Grades	
	Nitrile Butadiene Rubber (NBR)	Cis-Butadiene Low Molecular Weight Rubber (CBLMW-R)
Compressive strength, MPa	60–110	76.9–100.3
Tensile strength, MPa	8–20.0	13–18
Modulus of elasticity, MPa	$(2.0–3.5) \times 10^4$	$(1.5–1.8) \times 10^4$
Compression duration factor	0.77–0.78	0.72–0.76
Poisson's ratio	0.18–0.35	0.2–0.3
Heat resistance, °C	90–100	100–110
Freeze resistance, many cycles of a thawing and freezing process	500	500
Abrasibility, g/cm^2	0.15–0.30	0.25–0.79
Water suction, wt. %	0.05	0.05
Reduction, mm/m	0.17–0.21	-

In practice, the property that determines the scope of polymer concrete is heat resistance at 80–150 °C. At the same time, when the operating temperature of polymer concrete increases, its strength, and modulus of elasticity decrease. Having low heat resistance, polymer concretes nevertheless belong to the class of non-combustible materials, since the content of organic matter in them is low as compared to the proportion of inorganic components.

Since a promising direction in the study of polymer concretes is to reinforce them with dispersed reinforcement, additional reinforcement was used in the fabrication of specimens [14]. As a result, the beam specimens were made of reinforced rubber fiber reinforced concrete (ACPBF). The dispersion reinforcement fiber was made from tire industry waste metal cord. A chaotic orientation of fibers along the volume of the element was used. It allows perceiving and redistributing efforts of different orientations, appearing in the sample, thereby preventing the appearance and development of cracks. It should also be noted that, for a tensile element, the failure of specimens occurs either during fibers rupture or when violating their adhesion to the RPC [15].

2.2. Calculation of Polymer Concrete Structures

The basis of the methodology for calculating polymer concrete structures is set out in [16]. The strength calculation methodology is based on:

1. normal stresses, corresponding in shape to the diagram of mechanical tests for axial compression and tension of polymer concrete;
2. the hypothesis of planar sections [16].
3. Three equations were made by the authors to obtain calculation formulas:
4. moments of forces equilibrium;
5. projections of forces on the neutral plane equilibrium;
6. equations of a ratio of boundary deformations or heights of compressed and stretched zones of the element cross-section. The scheme of forces and the stress diagram in the cross-section normal to the longitudinal axis of the bendable polymer concrete element, when calculating its strength, is shown in Figure 1.

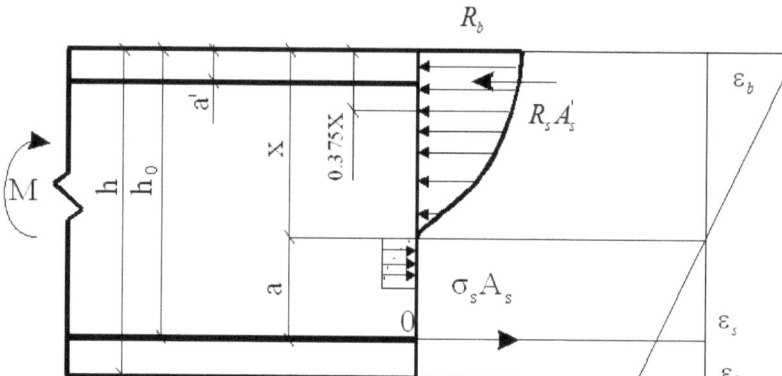

Figure 1. The scheme of forces and the stress diagram in the section normal to the longitudinal axis of the bendable polymer concrete element, when calculating its strength.

First of all, let us write an equation for the moments of forces equilibrium.

When reinforcing the tensile and compressed zones of the bendable polymer concrete element (Figure 1), the calculation formula is as follows:

$$M \leq \sigma_s \cdot A_s (h_0 - 0.375 \cdot x) + R_s \cdot A'_s (0.375 \cdot x - a') \qquad (1)$$

where a' represents the distance between an adjacent compression-deformed fiber and the axis of the compression reinforcement;

a is a tension region height;

x is the height of the compressed zone, determined by the formula:

M is a bending of the beam during tests;

R_s is a design resistance of the longitudinal bar;

A_s is an area of reinforcement in the tension region;

A'_s is an area of reinforcement in the compression region;

ζ is a relative height of the compression region of the cross-section determined by the formula [17]:

$$\zeta = \frac{x}{h_0} = 1.5 \frac{R_s}{R_{compr}} (\mu - \mu'). \qquad (2)$$

where μ is a reinforcement ratio in the tension region;

μ' is a reinforcement ratio in the compression region;

R_{compr} is a design resistance in the compression region.

The calculation procedure [17] for the formation of cracks normal to the longitudinal axis of the element is based on a rectangular stress diagram in the tensile zone and a triangular diagram in the compressed zone at the height of the tensile zone a, equal to:

$$a = h_0 - x. \qquad (3)$$

Let us express x based on Formula (3) and use it in Formula (2). Let us also express $(\mu - \mu')$ using the same formula, transferring the values $1.5\frac{R_s}{R_{compr}}$ to the opposite part of the formula. Then, the difference in reinforcement coefficients in the tensile and compressed zones $(\mu - \mu')$ is determined by the formula:

$$(\mu - \mu') = \frac{0.67\frac{R_{compr}}{R_s}}{1 + \frac{E}{E_s} \cdot \frac{R_s}{R_{compr}}}. \qquad (4)$$

According to [18], the stresses in the longitudinal reinforcement (σ_s) do not reach the yield strength. Let us substitute expression 3 for expression 1 and convert decimals to natural fractions. When converting, the following facts must be considered. To use polymer concrete most completely, it is advisable to provide a clearly defined curved shape of the stress epure in the compression cross-section region, which requires a rather high percentage of reinforcement of about 15% or more. In the case of lower reinforcement percentages, the triangular shape of the epure is to be used. Therefore, the calculation of crack formation is based on a rectangular stress epure in the tension region and a triangular one in the compression region at the height of the tension region. Proceeding from this fact, the bending moment before the formation of cracks is determined by the formula:

$$M \leq \sigma_s \cdot A_s(h_0 - \frac{x}{3}) + R_s \cdot b \frac{3 \cdot h + x}{6}(h - x). \qquad (5)$$

Stresses in the longitudinal reinforcement (σ_s) are found owing to the deformation of the reinforcement (ε_s) and polymer concrete at the reinforcement level of:

$$\sigma_s = \varepsilon_s \cdot E_s. \qquad (6)$$

The reinforcement deformations are equal to:

$$(\varepsilon_s) = \varepsilon_{bt} \frac{h_0 - x}{h - x}, \qquad (7)$$

where ε_{bt} is the deformation of the lower tensile edge of the bendable polymer concrete element. It follows from the hypothesis of flat sections that:

$$\frac{\varepsilon_b}{x} = \frac{\sigma_b}{E_b \cdot x} = \frac{\varepsilon_s}{h_0 - x} = \frac{\varepsilon_{bt}}{h - x}, \qquad (8)$$

where ε_b is the deformation of the upper compressed face of the bendable polymer concrete element;

σ_b is stresses in the upper compressed face of the bendable polymer concrete element. Based on this, we determine the β' as the height of the compressed zone:

$$\beta' = \frac{\varepsilon_s}{\varepsilon_b} = \frac{\varepsilon_s \cdot E_b}{\sigma_b}. \qquad (9)$$

In the first case, the height of the compressed zone is expressed through the deformation of the reinforcement:

$$\beta' = \frac{\varepsilon_s}{\varepsilon_b} = \frac{\varepsilon_s \cdot E_b}{\sigma_b}. \qquad (10)$$

On the other hand, it is possible to make through the deformation of the outermost stretched fiber, that is, through the ultimate tensile strength:

$$\beta = \frac{\varepsilon_{bt}}{\varepsilon_b} = \frac{\varepsilon_{bt} \cdot E_b}{\sigma_b}. \tag{11}$$

By projecting the forces onto the neutral plane, we obtain:

$$\sigma_s \cdot A_s + R_{bt} \cdot b(h-x) = 0.5 \cdot \sigma_b \cdot b \cdot x. \tag{12}$$

The reinforcement factor is determined by the formula:

$$\mu = \frac{A_s}{b \cdot h_0} = \frac{0.5 \cdot x - R_{bt}(h-x)}{\sigma_s \cdot h_0}. \tag{13}$$

The compressive edge stresses (σ_b) are set from the calculation so that they do not go beyond the linear section of the diagram, i.e., "$\sigma - \varepsilon$"$\sigma_b < 0.75R$ [19]. Let us substitute the obtained Formula (6) into expression 5. Let us also consider the fact that the calculation formula for the bending moment, when reinforcing the compression region of the beams before the formation of cracks, will receive an additional summand. The formula for determining the moment will take the following form:

$$M = \varepsilon_s \cdot E_s \cdot A_S \left(h_0 - \frac{x}{3}\right) + R_s \cdot b \frac{3 \cdot h + x}{6}(h-x) + \sigma'_s \cdot A'_s \left(\frac{x}{3} - a'\right), \tag{14}$$

where σ'_s is stress in the compressed reinforcement, determined by the formula:

$$\sigma'_s = \sigma_b \frac{E_s}{E_b} \cdot \frac{x - a'}{x}. \tag{15}$$

2.3. Experimental Design and Research Program

BHP has a high tensile strength [19], as well as a higher ultimate tensile strength, compared to cement concrete. These properties can be effectively used to increase the crack resistance of bendable elements. In addition, if the fiber is added to the bent element in this way, the dispersion increases in the height of the cross-section (partial or full). An element or a two-layer structure having a fibrous reinforcement in the stretching zone is obtained, whose function will increase even more [20]. To achieve this goal, we decided to determine the normal cross-section of the ACPBF bending element, having the performance characteristics of dispersed steel bars over the entire height of the cross-section, and to obtain data on its tensile strength, resistance to cracking, and deformability [21].

The testing of beams made of rubber concrete involved test specimens-beams of all series made having a section size of 6 × 12 cm and a total length of 140 cm. The beams were manufactured according to the Central Asian standards, GOST 948-2016 (ICS 91.080.40), the Interstate Standard of Reinforced-Concrete, and Polymer-Concrete Lintels. The determining parameters of the beams are the following: the length (L) is from 1030 (mm) to 5950 (mm); the width (B) is from 120 (mm) to 250 (mm); the height (H) is from 60 (mm) to 585 (mm). The beams are tested by means of two symmetrically applied forces in the thirds of the span (pure bending test procedure). In this way, the samples of the beam were made of reinforced rubber fiber concrete. The dispersion reinforcing fiber was made of waste metal cord from the tire industry. Various parameters are assigned in the experiment:

- the percentage of the longitudinal gain that has the greatest effect on the resistance of the normal part bending element;
- the height of the dispersion reinforcement zone (measured from the bottom edge) with the same percentage of fiber reinforcement over the volume of the reinforced part ($\mu v = 1\%$, according to composition optimization [22]).

The response function is the strength; they combine the crack resistance of an element of the normal cross-section, as well as their good deformability. The boundary conditions of the area of variation of the experimental parameters are established on the basis of the analysis of the literature [23]. The procedure for testing beams complies with the Central Asian Standards GOST 10180-2012 and ICS 91.100.30, Methods for Strength Determination Using Reference Specimens. The percentage of longitudinal reinforcement gradually increases from 0% (no longitudinal reinforcement) to 8.4% (compression zone failure). The height of the dispersed reinforcement zone is determined based on the testing of AKPBF samples, which are reinforced along the entire length. To determine the physical and mechanical properties of the material used for the manufacture of the test beam, together with the beam, a control sample of a 4 cm × 4 cm × 16 cm prism is made and tested for compression, and a control 8 cm × 4 cm × 40 cm sample is tested for tension.

The test procedure complies with GOST 10180-2012 and ICS 91.100.30, Methods for Strength Determination Using Reference Specifications. In the laboratory, before testing the samples, the temperature is maintained at 20 ± 5 °C and the relative humidity of the air is at least 55%. Under these conditions, the samples were in the dismantle form before the test for at least 24 h. The control samples were tested on the day of testing the beam. The beams were loaded with two equal concentrated loads, applied vertically in the thirds of the span. In the case of this type of load application, the value of the bending moment, arising in the beam, increases from zero on the support to the maximum value under the point of load application. Between the points of load application, the transverse force is zero, and the value of the bending moment is constant and equal to the maximum value (the pure bending zone). A general view of testing the samples of rubber concrete beams is shown in Figure 2.

Figure 2. General view of the bending test of rubber concrete beams.

The beam samples were tested using a laboratory press "INSTRON 600KN" (60 tons), certified and meeting the requirements of GOST 10180-2012 and ICS 91.100.30 "Methods for Strength Determination Using Reference Specifications". The load on the sample was fed at a constant rate until destruction.

2.4. Materials

The following materials were used in the work:
- Cis-polybutadiene low molecular weight CBLMW-R rubber (ISO 6743/4). CBLMW-R rubber has a density of 910 kg/m^3 and a dynamic viscosity of up to 12 Pa × s;

- fine-dispersed filler being ash of a specific surface of 2500...2700 cm^2/g, having the following composition by mass in %: Al O$_{23}$—16.5–22.5; CaO was 5.5–5.5; Fe O$_{23}$ was 13.5–15.5; SiO$_2$ was 47–55; MgO was 2–3; K$_2$ O was 1–2; Na$_2$ O was 1; S O$_{23}$ was 0.4–0.3; others were 6–15;
- vulcanization activator was zinc oxide ZnO (ISO 10262-2016) white powder having a density of 5600–5700 kg/m^3;
- vulcanization gas pedal was tetramethyl thiuram disulfide (Tiuram-D) (ISO 4097 2013) being powder of gray-white color and density of 1300–1400 kg/m^3;
- calcium oxide CaO was a white powder having a density of 2500–2900 kg/m^3;
- Portland cement. It consists of the following ingredients: clinker (calcium silicates); gypsum; plasticizing, hydrophobic, acid-resistant additives; domain fee. The chemical composition of cement: 21.55% silicon oxide and 65.91% calcium oxide. Portland cement is the most durable and high quality, therefore it is widely distributed on the market. It is a cement that is capable of showing an average compressive strength of about 500 kg/m^3 after 28 days of preparation. The mixture can be without additives or with various substances introduced into the composition in a certain proportion. The compressive load is 2500 kg/cm. Frost resistance is more than 100 cycles. The bulk density is 1100–1600 kg/m. True density kg/m^3 3100. The setting speed is from 45 min to 10 h. The average weight of Portland cement is 3 tons/m^3;
- sulfur technical (ISO 3704-76) is a bright yellow powder with a concentration of 2070 kg/m^3, having a melting point of 114 °C;
- metal fibers made from scraps of metal cord by sawing. The fibers obtained in this way are wave-shaped fibers at a ratio of a length to a wire diameter of 1/100;
- sand and crushed granite [24] are selected in accordance with the relevant requirements of GOST 26633-2015. Concrete is heavy and fine-grained. Specifications and ISS 91.100.30. The physical properties of gravel and sand are given in Table 2.
- reinforcing steel bars having a diameter of 8, 10, 12, 14, 16, and 18 mm and a steel wire of a diameter of 5 mm.

Table 2. Physical properties of sand and crushed stone.

Filler	Size of Fractions, mm	Bulk Density, g/cm^3	Raft Density, g/cm^3	Specific Surface Area, cm/g^2	Hollowness, %
Granite rubble	5.00–10.00	1.50	2.67	5.4	41.4
Quartz sand	1.25–2.50	1.61	2.65	33.0	39.1
	0.63–1.25				
	0.32–0.63				

2.5. Mechanical Tests

Determination of the actual physical and mechanical characteristics of steel reinforcement was carried out using the Instron 1500HDX testing system. Measuring the longitudinal parameter with 270 uses a base of a 20-mm strain gauge for deformation. A set of equipment was used to control the stress-strain state (VAR) stage, namely, deformation 2 MG plus design and CATMAN-AP software [25]. The control specimens (3 pieces of twin specimens) were tested for tensile strength before rupture according to the requirements [26].

The component composition of rubber concrete (as a percentage of the element weight) supplemented with dispersed reinforcement, which provides the best strength characteristics and chemical resistance, is shown in Table 3. These compositions were obtained as a result of optimizing the component composition of PBC [27] and are pleasant for the basic ones.

Table 3. Composition of rubber fiber-polymer concrete.

Name	Component Content, wt. %
Quartz sand	24.2
CBLMW-R low molecular weight rubber	8.2
Ashes	7.8
Sulfur technical	4.0
Zinc oxide	1.2
Tiuram-D	0.4
Metal cord fibers (fiber)	2.5
Calcium oxide	0.4
Granite rubble	The rest (51.3)

To manufacture concrete specimens without the filler, the concrete of a composition having a similar modulus of elasticity to that of rubber concretes was chosen. The composition of the cement concrete used for making experimental specimens without the filler (mass per 1 m^3 in tons) [28] was as follows: sand—0.692; water—0.19; cement—0.364; crushed stone—0.121.

2.6. Sample Production Technology

The production of the elements made of fiber-reinforced polymer concrete (FRPP) with the arrangement of fibers along the entire section height (chaotically) was carried out in one stage.

Based on [29], the preparation of the FCPB mixture included the following operations:
- washing the fine and coarse aggregate;
- preparation of curing group components and fibers;
- dosage of ingredients;
- drying of the components;
- mixing of the components.

Dosing of sand and gravel was carried out on the scales VPS-40 M having an accuracy of 5 g. The dosing of other components was carried out on the electronic scales CAS ER JR-15CB having an accuracy of 1 g due to their smaller weight [30].

The components were mixed in a high-speed propeller-type mixer. The polymeric binder was prepared by combining CBLMW-R rubber with hardening group components and the fine filler with fly ash. The mixing time of the binder was 80 s at 1000 rpm. Then fine aggregate and fibers mixed with the coarse aggregate were introduced into the prepared mixture. The resulting polymer concrete mixture was prepared in the same mixer at a speed of 180 rpm for 200 s.

Before laying the FCPB mixture in the form, the surface is smeared with waste oil to facilitate removal later from the formwork. The prepared mixture was placed in the molds and compacted on the vibrating table with a vibration duration of 100 ± 30 s. An indication of sufficient compaction of the polymer concrete mixture is the release of the binder on the surface and the cessation of the intensive formation of air bubbles [31].

After performing the above operations, the mold with the mixture was placed in a dry heating chamber (dryer), where FKPB was cured at 115 ± 5 °C for 12 h (taking into account the heating time of 17 h). Unpacking was performed after the complete polymerization and cooling of the samples.

Fabrication of FKPB elements with the fiber arrangement in the tensile zone (chaotically) was carried out in two similar stages according to the works [32].

The fabrication of FCPB elements took place in one step using metal cord fibers.

The reinforcement of all series of experimental beams was a welded frame, shown in Figure 3.

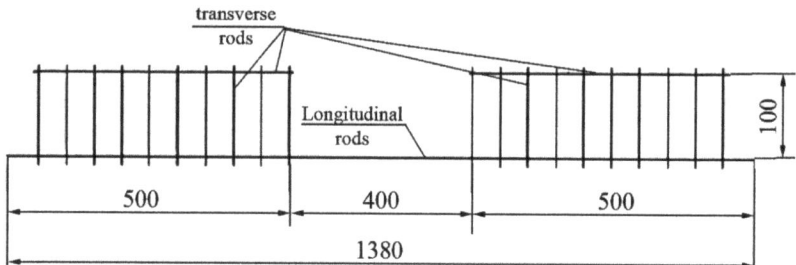

Figure 3. The reinforcement frame of the experimental beams. Dimensions are in mm.

Figure 2 shows longitudinal rods are reinforcement, transverse is the wire (a diameter of 4.5–5.5 mm), and the step of transverse bars is 50 mm in order to prevent fracture in the sloping sections of the experimental beams.

2.7. Test Methods, Basic Instruments and Equipment

The test procedure corresponds to the methods outlined in the regulatory literature [33].

The test of the manufactured beams was carried out using concentrated loads. The load was applied vertically and distributed in the third part of the beams' span. In the case of a loading scheme for beams, the bending moment arising in the beams is 0 on the support. Already under the point of application of the load, the moment increases to a maximum value. Between the points to which the load is applied, the value of the transverse force is zero. The bending moment at the same point has a constant value equal to the maximum value. A diagram of the load distribution on the test specimens is shown in Figure 3.

The beam samples were tested on the laboratory presses of INSTRON800KN (80 tons), certified, and they met the requirements [34]. The load was applied to the sample at a constant rate until it was destroyed (Figure 4).

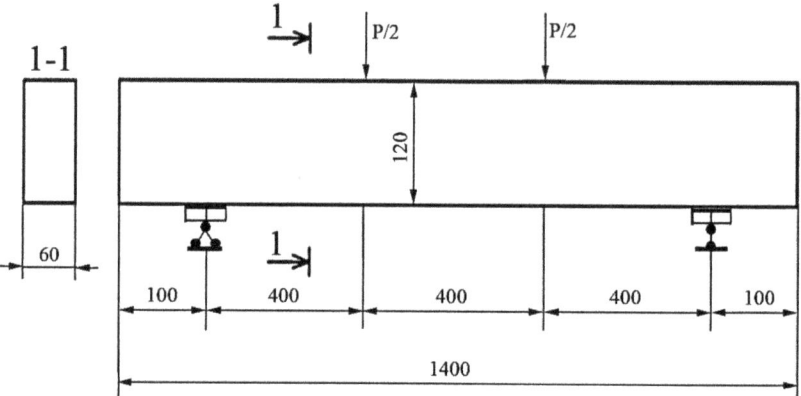

Figure 4. Load diagram and geometric dimensions of beams (unit: mm).

The maximum value of the pressing force was the value of the breaking load during beam testing. It was determined by a force sensor. At this value, the yield strength of the reinforcement was reached. Glued strain gauges made it possible to measure longitudinal deformations. Deformation measurements were carried out in a normal section. The strain gauge base was 2 cm. A linear displacement transducer, a plunger, with an accuracy class of 0.2%, was used to measure the vertical displacements of the beams.

During the given tests, the crack opening width and height were measured. The crack opening width was measured with a micrometer; a duplicate measurement was made with a caliper.

Before testing, the specimens were inspected and measured. Roughness and burrs on the surface of the beam material were removed with an angle grinder. The side surfaces of the beams were ground to facilitate visual observation of the appearance and distribution of cracks due to the dark surface of the material. Before attaching the load cells, the surface of the beam was ground with an angle grinder and degreased with acetone. After that, the load cells were pressed to the surface of the structure with glue.

With each polymer-concrete beam, control samples-prisms (4 cm × 4 cm × 16 cm) and specimens-eights with the size of a working area of 3 cm × 4 cm and a total length of 40 cm were made. To control the deformation-strength characteristics, control samples-cubes (10 cm × 10 cm × 10 cm) and prisms (4 cm × 4 cm × 16 cm) were made with each concrete beam. Three control samples were produced. Studies of prisms and cubes were carried out under central compression and octagonal samples were tested under central tension.

Compression tests of beams, concrete prisms, and concrete cubes were performed according to the requirements [35] on an INSTRON Satec 1200 press (Instron Corporation, New-York, NY, USA) certified and meeting the requirements [36]. Strain gauges were installed on the specimens-prisms on the opposite faces to measure the longitudinal relative deformations arising in the specimen at a constant rate of 60 MPa/min when a short-term compressive load was applied. Before testing, the surface of the specimens was prepared and checked for the absence of defects (cracks, cavities, etc.), and the ends of the specimens were checked to be perpendicular to the longitudinal axis.

Tensile tests of fiber-reinforced polymer concrete (FRPB) specimens were carried out taking into account the requirements of ISO 1920-1:2004 and recommendations on testing methods for polymer concretes [37] on a tensile testing machine INSTRON 5982 (Instron Corporation, New-York, NY, USA), being certified and meeting the requirements [38]. Strain gauges were installed on the eight specimens on the opposite faces to take longitudinal relative deformations. Before testing, the specimens were also carefully prepared and checked for defects. The load was applied uniformly, continuously at a constant rate of 0.15 MPa/s.

3. Discussion

3.1. Results of Experimental Studies of Intensity (Construction of a Material Deformation Diagram "σ-ε")

In order to determine the relationship between stress and relative deformation occurring in FCPB, required for a comprehensive study of the cross-section of the deflection of a curved element, a control sample being a prism and a sample, eight were made from each experimental beam. Test methods of specimens, their geometrical parameters, and schemes of measuring instruments arrangement are given in Section 2.6. On the basis of the compression tests of prism specimens and tension specimens of octahedrons, graphs between the stresses and relative deformations of compression and tension for FKPB were obtained, these graphs are shown in Figures 5 and 6, respectively.

Based on the diagram shown in Figures 5 and 6, equations describing the relationship between stresses and relative strains were derived for the FCPB.

The relationship between compressive stresses and relative strains is shown below.

Based on the analysis of Figures 5 and 6, equations describing the relationship between stresses and relative strains were derived for FKPB products.

The dependencies in the analytical form between compressive stresses and relative deformations of FKPB products are presented in Figure 6. The dependences between the tensile stresses and relative deformations of the FCPB products are shown in Figure 6.

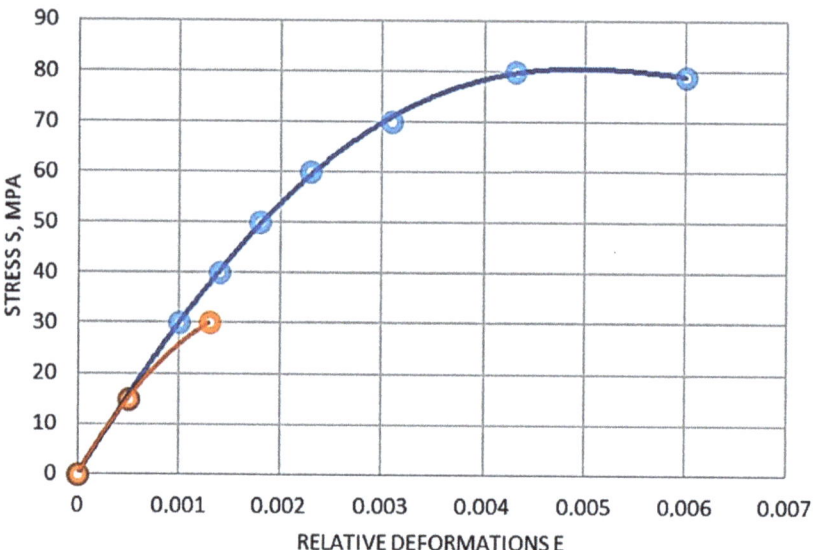

Figure 5. Diagram of the relationship between compressive stresses and relative deformations of FKPB products: red curve—$y = -9 \times 10^6 \, x^2 + 34{,}327 \, x$ ($R^2 = 1$), blue curve—$y = 6 \times 10^{10} \, x^4 - 6 \times 10^8 \, x^3 - 2 \times 10^6 \, x^2 + 32{,}199 \, x$ ($R^2 = 0.9996$).

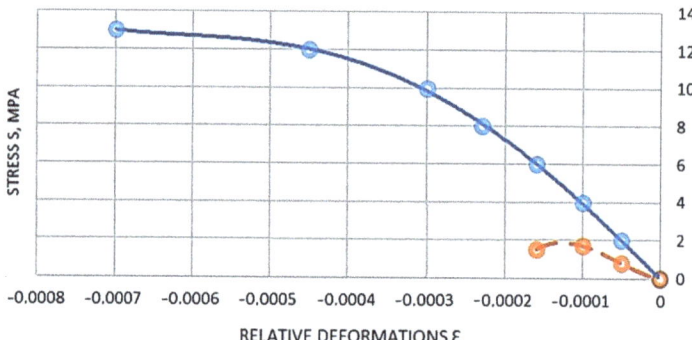

Figure 6. Diagram of the relationship between tensile stresses and relative deformations of FKPB products: red curve—$y = 10^{12} \, x^3 + 2 \times 10^8 \, x^2 - 8314.4 \, x$ ($R^2 = 1$), blue curve—$y = 8 \times 10^{13} \, x^4 + 9 \times 10^3 \, x^3 - 4 \times 10^6 \, x^2 - 40{,}448 \, x$ ($R^2 = 0.9997$).

3.2. Strength of Normal Sections

Experimental studies have shown that the percentage of longitudinal reinforcement and the height of the diffuse reinforcement zone are the main factors affecting the normal cross-sectional strength of the ACPB bending element. The parameters and test procedures of the test sample in the study of strength in normal cross-section are given in Sections 2.5 and 2.6. The destructive bending moment (in case of fracture in the tensile zone) is referred to the moment when the stress in the steel rod reaches the yield point, which also corresponds to a stronger increase in the deformation in the steel rod (Figures 7 and 8). The destructive bending moment (in case of destruction of the compression zone) is the moment when the compression zone collapses, which corresponds to a sharp increase in deformation. In this case, the curve corresponding to the probe deformations has a kink, and the deformations begin to decrease, which indicates the buckling in this area.

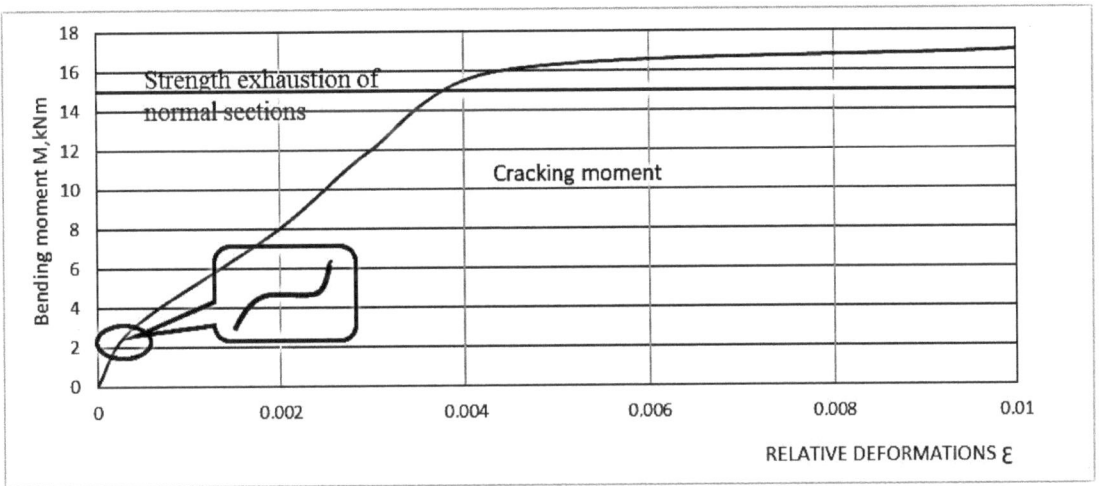

Figure 7. Relative tensile strains in the reinforcement of the FPB beam as a function of bending moments.

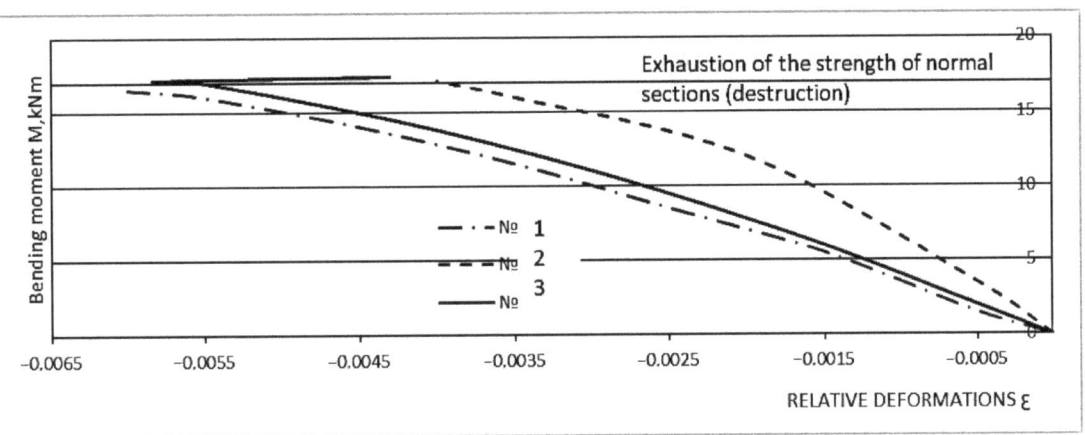

Figure 8. Relative deformations of the compressed zone of FPB as a function of bending moments. Nos. 1–3 is load cell numbers.

As a result of the experimental studies, the values of the breaking bending moments depending on the percentage of longitudinal reinforcement and the height of the fiber reinforcement zone were obtained. These are presented in Appendix A. Based on the data in Table A1 (Appendix A.1), the dependence of the bending moment on each varying factor is plotted graphically (Figures 9 and 10). Figure 10 shows a graph of the dependence of the destructive bending moment occurring in beams with the same rebar content on the height of the fiber reinforcement layer.

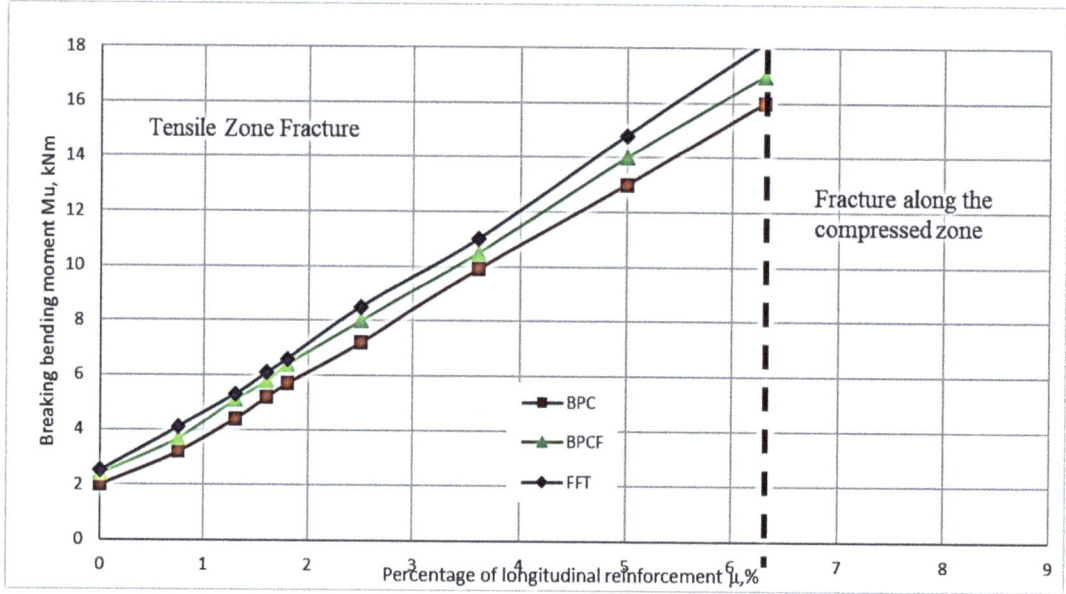

Figure 9. Diagram of the dependence of the value of destructive bending moments on the percentage of longitudinal reinforcement.

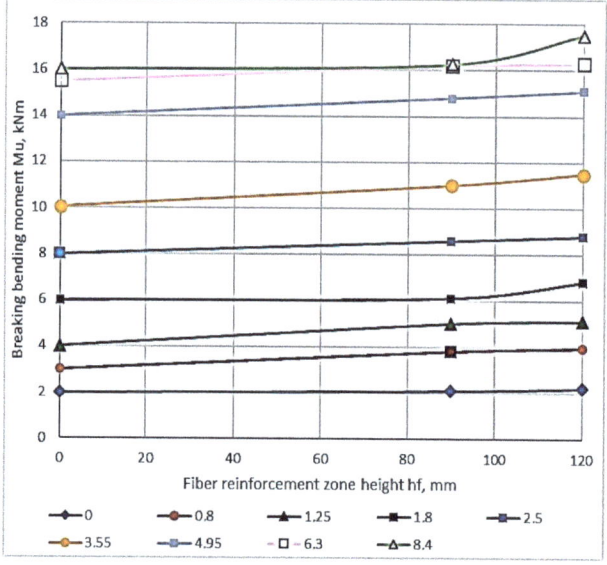

Figure 10. Dependence plot of fracture bending moments on the fiber reinforcement zone height.

The study of the composition in Figures 9 and 10 shows that the proportion of the filling of the longitudinal reinforcement has a significant effect. The reinforcement percentage affects the value of the breaking load more strongly than other factors do. The next parameter in terms of influence is the height of the fiber reinforcement zone. The height's influence on the fiber reinforcement zone is significantly less than the reinforcement percentage influence.

The study of the composition of Figures 9 and 10 confirms that the proportion of the longitudinal reinforcement content has a significant effect. The second varying parameter in the form of the height of the fiber reinforcement zone has an effect on the bearing capacity, but its influence is much smaller than that of the reinforcement longitudinal percentage.

In the range of longitudinal reinforcement percentage values of 0.6–6.5%, the dependence in Figure 9 is linear. An increase in the proportion of the presence of the longitudinal reinforcement in ACPBF beams leads to their destruction in the compression zone. However, in this case, there is no increase in strength characteristics for normalized sections.

The linear dependence between the parameters of the percentage of longitudinal reinforcement and the bending moment is maintained over a sufficiently large segment. This is a segment of the values of the percentage of longitudinal reinforcement equal to 0.8–6.3%. A further increase in the percentage of longitudinal reinforcement leads to a change in the nature of the destruction. The destruction of ACPBF beams occurs already in the compressed zone. The strength of the sections almost does not change in the presence of such destruction. It is important to note that the failure of ACPBF beams having a percentage of longitudinal reinforcement equal to 0.8–6.3% began in the stretched zone, i.e., when the reinforcement reached its yield strength. In the over-reinforced beams ($\mu = 8.4\%$) the failure is brittle, i.e., it occurs in the compressed zone [39].

The failure of ACPBF beams having a percentage of longitudinal reinforcement equal to 0.78–6.3% began in the tensile zone, i.e., when the reinforcement reached its yield strength. In addition, in the over-reinforced beams ($\mu = 8.4\%$) the failure is brittle, i.e., it occurred in the compressed zone. That is, the failure of ACPBF beams is similar to that of ACPBF beams with the addition of fibers. In the tested beams (having a percentage of longitudinal reinforcement of 6.3%), relative strains in the outermost compressed fibers of the rubber and reinforcement reach the limit values almost simultaneously [40].

The failure of ACPBF beams having the zone reinforcement, i.e., possessing 3/4 of the cross-sectional height and longitudinal reinforcement in the range of 0.8–4.95% occurred when the reinforcement reached its yield strength, i.e., in the tensile zone. In the over-reinforced beams ($\mu = 6.4\%$) the failure is brittle in nature, i.e., it occurred in the compressed zone. Consequently, to increase the bearing capacity of ACPBF beams having 3/4 of the cross-sectional height with a percentage of longitudinal reinforcement greater than 4.95%, it is necessary to reinforce the compressed zone. The earlier destruction of the ACPBF elements with zone reinforcement in the compressed zone is conditioned by the fact that the strength of the stretched zone corresponds to the solid ACPBF elements. At the same time, the strength of the compressed zone is lower than that of ACPBF elements, because the ultimate strength of non-dispersion reinforced ACPBF in compression is lower than that of ACPBF with the fiber. This leads to the fact that due to the higher tensile strength of ACPBF, the crack height develops and, consequently, the position of the neutral axis shifts towards the outermost compressed fiber, and the compressed zone is fractured (Figure 11).

Figure 11. Fracture along the compression zone of the CPS beam.

The magnitude of the destructive bending moment during destruction along the compressed zone changes insignificantly. Due to the fact that the deformation of the reinforcing bar does not reach the yield strength compared to the previous series of beams, it, as a result, leads to the curvature of the graph "M_u-μ". It is worth noting that the increase in the difference of bending moment values for CPB and ACPB beams with fiberglass, with µ = 6.3% is due to the fact that in CPB beams the beginning of reinforcement flow almost coincided with the material reaching its yield strength in the compressed zone, but occurred slightly earlier [41].

To compare the load-carrying capacity of ACPB bending elements with longitudinal reinforcement having the load-carrying capacity of reinforced concrete elements, beams containing the B25 class concrete with different reinforcement contents were tested [42].

The results of the pure bending test of reinforced concrete beams are summarized in Table 4.

Table 4. Test results of reinforced concrete bendable elements.

Girder Sample Code	Percentage of Longitudinal Reinforcement µ, %	Destruction Zone	Breaking Bending Moment M_u, kNm
PC 001	0.80	Stretched	2.23
PC 002	1.25	Stretched	3.55
PC 003	1.80	Stretched	4.99
PC 004	3.55	Compressed	5.65

It is worth noting that ACPB beams (having a 3.55% longitudinal reinforcement) failed in the tensile zone. Reinforced concrete beams with the same content as the longitudinal reinforcement failed in the compressed zone [43].

As a result of the experiments, we can say the following:

1. The normal sections of ACPBF bending elements strength exceeds the strength of reinforced concrete elements having µ = 0.8% by 70.56%, µ = 1.25% by 44.8%, µ = 1.8% by 36.8% and by 98.9% (for beams with µ = 3.55%).

2. The normal sections of the durability of the ACPB bending elements is higher than the durability of reinforced concrete elements having µ = 0.8% by 33.7%, µ = 1.25% by 21.5%, µ = 1.8% by 17.2% and by 91.2% (for beams having µ = 3.55%). The strength of normal sections of FRCP bending elements having a zone reinforcement is higher than the strength of reinforced concrete elements having µ = 0.8% by 64.6%, µ = 1.8% by 27.3%, and by 93.4% (for beams possessing µ = 3.55%).

Figure 11 for ACPB beams shows the dependence of the strength of the normal section on the fiber-height reinforcement zone. The dependence of the figure shows that the normal section strength increases along with an increase in the height in the fiber reinforcement zone. It should be considered that the influence of the percentage of longitudinal reinforcement is more significant than the influence of this factor. The fracture along the stretched zone most fully shows the increase in the strength of the normal sections of the CPB. The percent change in longitudinal reinforcement varies with the effect of the fiber reinforcement zone on strength. An increase in the height for the fiber reinforcement zone of the CPB from zero to 120 mm gives a strength increase in the elements by 28% (percentage of longitudinal reinforcement of µ = 0.8%) and 14% (bent elements of µ = 8.4%). In this case, the strength of the normal sections practically does not change.

The bearing capacity of the bending elements with fiber reinforcement in the tensile zone and with fiber reinforcement throughout the cross-section height is higher than that of similar ACPB (without dispersion reinforcement) bending elements [44]. This is due to the fact that the magnitude of reinforcement anchorage in FKPB is higher than that in CPB (i.e., more reliable joint work). This prevents a sharper development of plastic deformations in reinforcement. In addition, disperse reinforcement increases the duration of joint work of reinforcement and polymer concrete of the stretched zone, thereby "postponing" the moment of redistribution of stresses arising in the stretched zone to the reinforcement bar, even in the sections with cracks formed. That is, this happens in the places where there are no cracks, i.e., in the places where the polymer concrete itself is absent, the metal cord fibers continue to resist tension [45].

The bearing capacity of polymer concrete bendable elements is higher than that of similar concrete elements with longitudinal reinforcement. It is related to a number of factors: a part of polymer concrete of the stretched zone above the formed crack participates in the work of normal sections; high adhesion of reinforcement and structural material (CPB, FCPB) prevents sudden development of plastic deformations in reinforcement; higher strength of the compressed zone material. We cannot exclude the fact that the material of the structure in the gap between the cracks takes part in the work of the element as a whole.

It should be noted that CPB is a more plastic material, which is confirmed by the deformation diagrams in Figures 6 and 7 than heavy concrete is. It allows it to work better in tension, and therefore increases the duration of joint work of longitudinal reinforcement bars with polymer concrete before the crack formation, the introduction of steel cord fibers in the mixture further increases these figures, which can also be observed in the deformation diagrams shown in Figures 6 and 7.

3.3. Finite Element Model of Beams Made of Polymer Concrete on a Rubber Binder Implemented in the Ansys Environment

In order to verify the theoretical conclusions and evaluate the results of the experiment of polymer concrete beams, finite element modeling was implemented in the Ansys software package (PC), taking into account the nonlinear properties of materials. Articulated boundary conditions are set on the right support of the beam, prohibiting only vertical movements, but allowing all rotations. Hinged-fixed boundary conditions are set on the left support. The span between the centers of the supports is 1.2 m. The length of the support zone was 60 mm, and the length of the beam overhangs was 70 mm. Two concentrated forces act on the beam, mirrored in the same way as the test scheme (Figure 12).

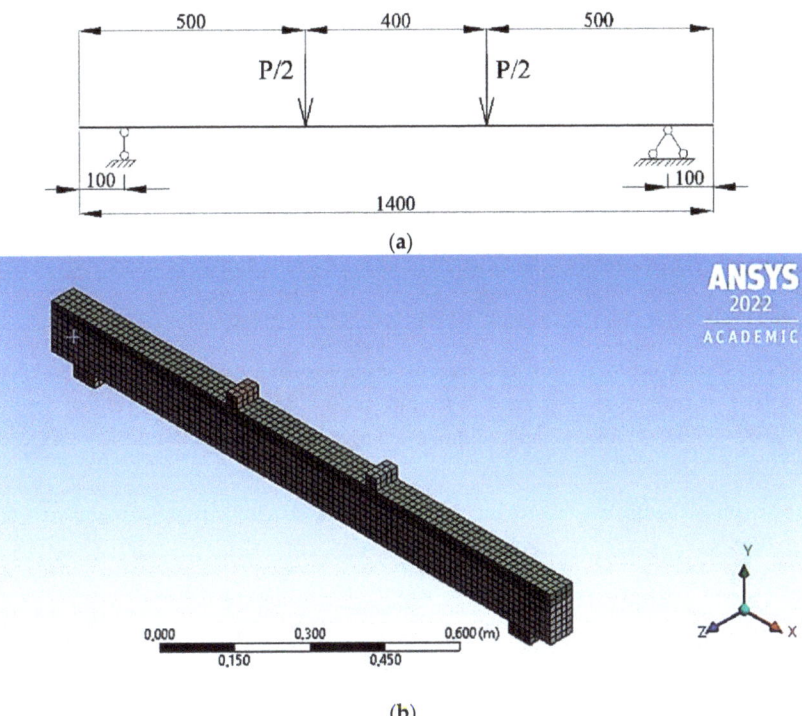

Figure 12. Calculation scheme of the element under study (**a**) and Finite element model of the element under study (**b**).

In the Ansys environment, as an element simulating the concrete body of the sample, an eight-node finite element Solid 65 was chosen, which has the ability to simulate plastic deformations, cracks, and destruction. This finite element implements the Willam-Warnke concrete deformation model. This dependence is an ellipse equation describing the deviatoric section.

When specified in a PC, the model includes the following parameters: stress-strain diagram for compression, modulus of elasticity, ultimate compressive and tensile strength, Poisson's ratio, and shear force transfer coefficient for closed and open cracks.

For modeling longitudinal and transverse reinforcement, the Beam 188 finite element was used, which is a rod, spatial finite element.

Based on the results of the calculation, the values of the destructive bending moments were obtained. The results of numerical studies of armocouton beams in the Ansys environment are graphically presented in Figure 13.

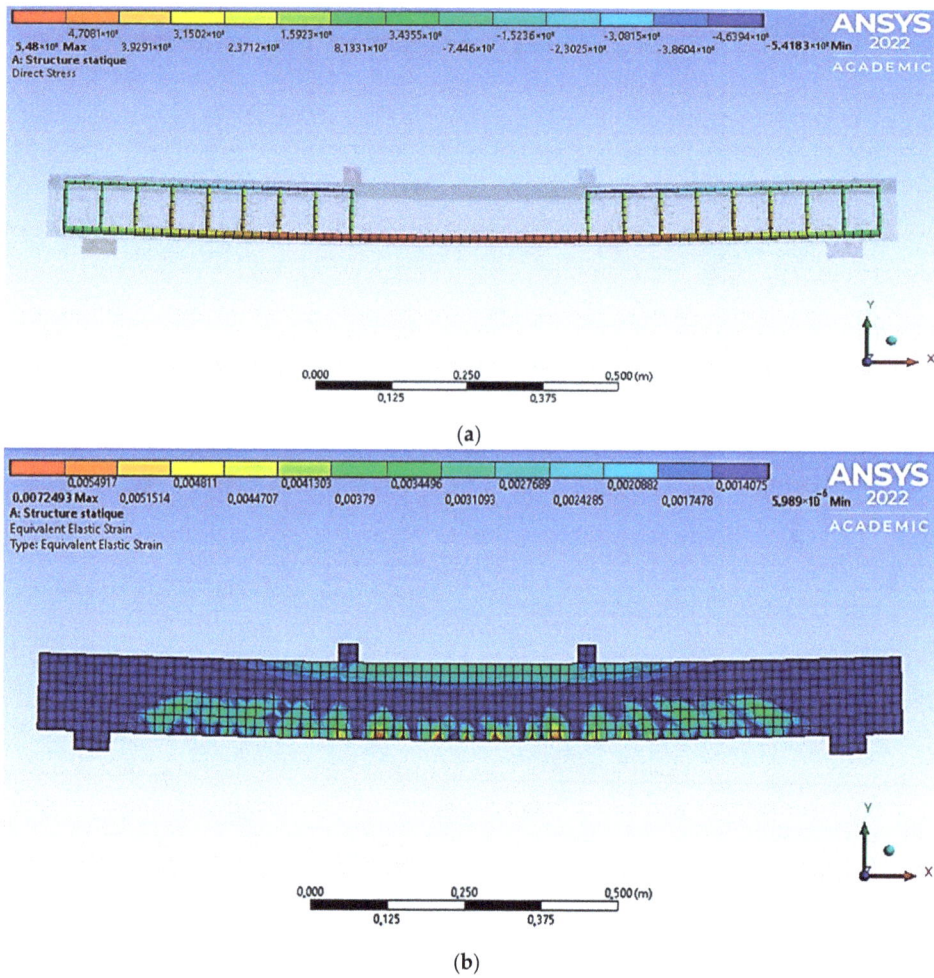

Figure 13. Normal stresses in reinforcement along the abscissa axis in the FE model of a beam (**a**) and inelastic deformations in the simulation model of the ACPBF beam during destruction (reinforcement percentage of 6.3) (**b**).

Simulation of the state of the beam before its destruction shows the general logic of the destruction of bending elements. This logic is consistent with the standard model, which observes the cracks' appearance and their subsequent development.

The study results of concrete strength control (CSC) of bendable elements are summarized in Appendix A.2.

The dependence of the ultimate bending moment on the percentage of longitudinal reinforcement for CPB-reinforced beams is shown in Figure 14; Figure 15 shows CPB-reinforced beams having fibers.

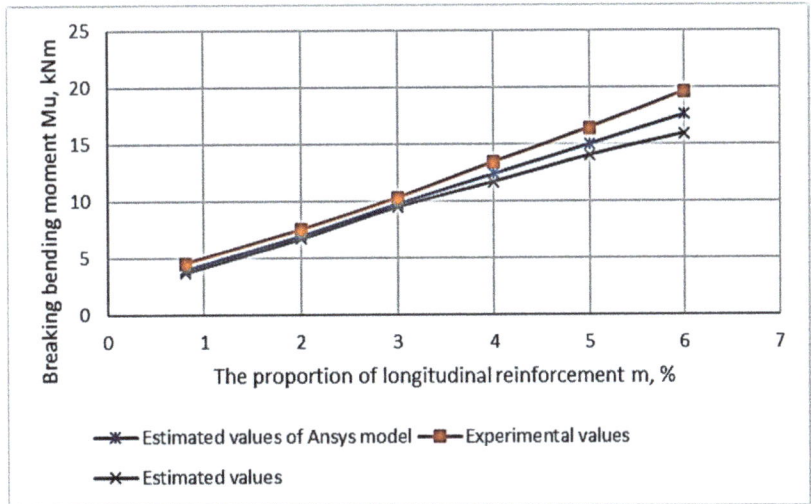

Figure 14. Dependence of the ultimate bending moment on the percentage of reinforcement in ACPB beams.

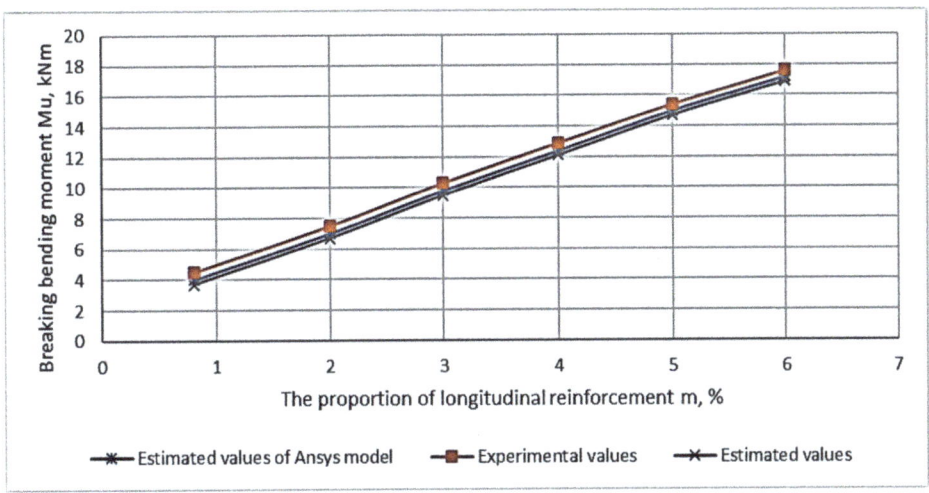

Figure 15. Dependence of the ultimate bending moment on the percentage of reinforcement in ACPBF beams.

The Willam Warnke strength theory developed for composite materials (implemented in this work with the help of Ansys) allows the calculation of ACPB beams. The performed calculations and their comparison with experimental data showed that the maximum discrepancy for the strength values obtained by empirical and calculated methods was 14% (in the series of BPC beams). In addition, a similar deviation of 7.4% for BPC-8 beams was already obtained. The greatest discrepancy between the results was achieved for ACPB BPC beams and amounted to 18.0%.

That in turn will allow the conducted experimental studies to replace the numerical ones, but in conditions of non-compliance with the design requirements or deviation from the norms in the tests, these features should be additionally taken into account in the modeling.

4. Conclusions

1. The features of the stress-strain state of normal sections of ACPB beams are revealed. It has been established that the tensile strength in axial compression and tension, deformations corresponding to the tensile strength for CPB, exceed the similar parameters for the used cement concrete 2.5–6.5 times. According to the results of the work, it was found that the addition of fiber reinforcement increased the tensile strength of the BPS 3.0–7.5 times.

2. The percentage of longitudinal reinforcement and the height of the fiber reinforcement zone are the main factors influencing the strength and crack resistance of normal sections of the ACPB bending elements. The destruction of ACPB and ACPBF bending elements with a percentage of longitudinal reinforcement of less than 6.3% occurs in the tension zone, and with a percentage of longitudinal reinforcement of more than 6.3%, the destruction occurs in the compressed zone.

3. Increasing the height of the fiber reinforcement zone of the CPB from 0 mm to 120 mm, the strength of normal sections of bending elements with a percentage of longitudinal reinforcement $\mu = 0.8\%$ increases by 28%. The strength of normal sections of elements with $\mu = 6.3\%$ increases by 14%, for elements having $\mu = 8.4\%$ the strength of normal sections practically does not increase. It has been established that fiber reinforcement increases the moment of formation of cracks in structures up to 1.5 times, and the bearing capacity of structures as a whole increase up to 1.3 times.

4. It has been experimentally determined that the addition of fiber over the entire height of the element section with the same percentage of longitudinal reinforcement increases the moment of cracking of the ACPBF of the bent elements compared to the elements from the CPB (without dispersed reinforcement) up to 1.5 times. It can be compared to the ACPBF elements with zone reinforcement (with fiber reinforcement at 3/4 of the section height) of up to 1.08 times.

5. Ultimate deflections of ACPB bending elements are less than deflections of ACPBF bending elements with zone reinforcement of up to 18.5%. Deflections of ACPB beams with a percentage of longitudinal reinforcement $\mu \leq 1.8\%$ are similar to reinforced concrete bending elements made of heavy cement concrete.

Author Contributions: Conceptualization, K.Y. and L.B.S.; methodology, G.B.B. and O.V.A.; software, L.B.S.; validation, K.Y. and B.V.M.; formal analysis, N.V.M.; investigation, L.B.S.; resources, G.B.B.; data curation, O.V.A.; writing—original draft preparation, B.V.M.; writing—review and editing, N.V.M.; visualization, K.Y.; All authors have read and agreed to the published version of the manuscript.

Funding: This publication was made as a part of the Sub-project "Production of sodium silicate using innovative energy-saving technology", financed under the "Promotion of Productive Innovation" Project, supported by the World Bank and the Government of the Republic of Kazakhstan. Statements may not reflect the official position of the World Bank and the Government of the Republic of Kazakhstan.

Institutional Review Board Statement: Not applicable.

Informed Consent Statement: Not applicable.

Data Availability Statement: The data presented in this study are available from the corresponding authors upon reasonable request.

Conflicts of Interest: The authors declare no conflict of interest.

Appendix A. Results of Experimental Investigations for Breaking Bending Moments of Beam Samples

Appendix A.1. Results of Experimental Investigations for Breaking Bending Moments of Beam Samples

This appendix updates the results of experiments obtained for the breaking bending moment of beam specimens as a function of the proportion of longitudinal type reinforcement and the height of the fiber reinforcement area, which are presented in Table A1.

Table A1. Values of breaking bending moments obtained from the experiment.

Girder Type	Percentage of Longitudinal of Reinforcement μ, %	Fibre Layer Height Reinforcement h_f, mm	Destruction Zone	Destructive Bending M_u Moment, kNm
FPC 001	0.00	125	Stretched	2.4
FPC 002	0.80	125	Stretched	3.67
FPC 003	1.25	125	Stretched	5.21
FPC 004	1.60	125	Stretched	6.28
FPC 005	1.80	125	Stretched	6.68
FPC 006	2.5	125	Stretched	8.75
FPC 007	3.55	125	Stretched	11.31
FPC 008	4.95	125	Stretched	15.09
FPC 009	6.3	125	Stretched	17.48
FPC 010	8.4	125	Compressed	16.69
FPC 011	0.00	95	Stretched	2.28
FPC 012	0.80	95	Stretched	3.71
FPC 013	1.80	95	Stretched	6.29
FPC 014	3.55	95	Stretched	10.92
FPC 015	4.95	95	Stretched	14.78
FPC 016	6.4	95	Compressed	16.39
PC 001	0.00	0	Stretched	1.88
PC 002	0.80	0	Stretched	3.02
PC 003	1.25	0	Stretched	4.31
PC 004	1.80	0	Stretched	5.92
PC 005	2.5	0	Stretched	7.91
PC 006	3.55	0	Stretched	10.29
PC 007	4.95	0	Stretched	14.11
PC 008	6.3	0	Stretched	15.41
PC 009	8.4	0	Compressed	15.91

Appendix A.2. Experimental Results for Breaking Bending Moments of Beam Samples

This section provides data on the bending moments of beam specimens leading to failure. Based on these data, a beam failure model was formed. Thus, the model shows the appearance and development of more areas of inelastic deformation (cracks). The results of the theoretical experimental and numerical studies of concrete strength control (CBC) of bending members are summarized in Table A2.

Table A2. Results of the studies conducted.

Girder Code	μ, %	Mut, KnM	Mue, KnM	Mum, KnM	ΔMum, %	ΔMut, %
PC 001	0.8	2.51	2.96	2.6	−14.0	−18.0
PC 002	1.25	4.02	4.28	4.2	−1.9	−6.5
PC 003	1.8	5.82	5.8	6.0	3.3	0.3
PC 004	2.5	7.92	7.82	8.0	2.3	1.3
PC 005	3.6	10.86	10.3	11.2	8.0	5.2
PC 006	4.95	14.47	14.0	15.2	7.9	3.2
PC 007	6.3	14.96	15.32	16.8	8.8	−2.4
FPC 001	0.8	4.00	3.78	3.52	−7.4	5.5
FPC 002	1.25	5.52	5.10	5.16	1.2	7.6
FPC 003	1.8	7.39	6.77	6.8	0.4	8.4
FPC 004	2.5	9.85	8.77	8.8	0.3	10.9
FPC 005	3.6	12.63	11.28	12.0	6.0	10.7
FPC 006	4.95	16.64	15.10	15.4	1.9	9.3
FPC 007	6.3	17.49	17.52	17.72	1.1	−0.2

Note: Mut—destructive bending moment according to the results of calculation by the proposed method; Mue—destructive bending moment according to the test results; Mum—destructive bending moment according to the results of calculation in PC "Ansys"; ΔMum—deviation of calculation results according to the proposed method from experimental values; ΔMut—deviation of calculation results in PC "Ansys" from experimental values.

References

1. Jun, Z.; Xiang-Ming, W.; Jian-Min, C.; Kai, Z. Optimization of processing variables in wood-rubber composite panel manufacturing technology. *Bioresour. Technol.* **2008**, *99*, 2384–2391. [CrossRef] [PubMed]
2. Jafari, K.; Toufigh, V. Experimental and analytical evaluation of rubberized polymer concrete. *Constr. Build. Mater.* **2017**, *155*, 495–510. [CrossRef]
3. Siddique, R.; Naik, T.R. Properties of concrete containing scrap-tire rubber-an overview. *Waste Manag.* **2004**, *24*, 563–569. [CrossRef] [PubMed]
4. Adhikari, B.; De, D.; Maiti, S. Reclamation and recycling of waste rubber. *Prog. Polym. Sci.* **2000**, *25*, 909–948. [CrossRef]
5. Fang, Y.; Zhan, M.; Wang, Y. The status of recycling of waste rubber. *Mater. Des.* **2001**, *22*, 123–127. [CrossRef]
6. Issa, C.A.; Salem, G. Utilization of recycled crumb rubber as fine aggregates in concrete mix design. *Constr. Build. Mater.* **2013**, *42*, 48–52. [CrossRef]
7. Hassani, A.; Ganjidoust, H.; Maghanaki, A.A. Use of plastic waste (poly-ethylene terephthalate) in asphalt concrete mixture as aggregate replacement. *Waste Manag. Res.* **2005**, *23*, 322–327. [CrossRef]
8. Isametova, M.E.; Nussipali, R.; Martyushev, N.V.; Malozyomov, B.V.; Efremenkov, E.A.; Isametov, A. Mathe-matical Modeling of the Reliability of Polymer Composite Materials. *Mathematics* **2022**, *10*, 3978. [CrossRef]
9. Busuyi, A.T. Cost evaluation of producing different aggregate sizes in selected quarries in ondo state nigeria. *Int. J. Eng. Adv. Technol. Stud.* **2016**, *4*, 6–19.
10. Tepordei, V.V. *Natural Aggregates-Foundation of America's Future (No. 144-97)*; US Geological Survey: Reston, VA, USA, 1997.
11. Borja, R.; Sanchez, E.; Martin, A.; Jimenez, A.M. Kinetic behavior of waste tyre rubber as microorganism support in an anaerobic digester treating cane molasses distillery slops. *Bioprocess. Eng.* **1996**, *16*, 17–23. [CrossRef]
12. Toutanji, H.A. The use of rubber tire particles in concrete to replace mineral aggregates. *Cem. Concr. Compos.* **1996**, *18*, 135–139. [CrossRef]
13. Collins, K.; Jensen, A.; Mallinson, J.; Roenelle, V.; Smith, I. Environmental impact assessment of scrap tyre artificial reef. *ICES J. Mar. Sci.* **2002**, *59*, 243–249. [CrossRef]
14. Eldin, N.N.; Senouci, A.B. Rubber-tire particles as concrete aggregate. *J. Mater. Civil Eng.* **1993**, *5*, 478–496. [CrossRef]
15. Khatib, Z.K.; Bayomy, F.M. Rubberized portland cement concrete. *J. Mater. Civil Eng.* **1999**, *11*, 206–213. [CrossRef]
16. Gong, Y.; Yang, J.; He, X.; Lyu, X.; Liu, H. Structural Design Calculation of Basalt Fiber Polymer-Modified RPC Beams Subjected to Four-Point Bending. *Polymers* **2021**, *13*, 3261. [CrossRef]
17. Topcu, I.B. The properties of rubberized concretes. *Cem. Concr. Res* **1995**, *25*, 304–310. [CrossRef]
18. Topcu, I.B.; Avcular, N. Collision behaviors of rubberized concrete. *Cem. Concr. Res.* **1997**, *27*, 1893–1898. [CrossRef]
19. Zheng, L.; Huo, X.; Yuan, Y. Experimental investigation on dynamic properties of rubberized concrete. *Constr. Mater.* **2008**, *22*, 939–947. [CrossRef]
20. Khaloo, A.R.; Dehestani, M.; Rahmatabadi, P. Mechanical properties of concrete containing a high volume of tire-rubber particles. *Waste Manag.* **2008**, *28*, 2472–2482. [CrossRef] [PubMed]
21. Mohammed, B.S.; Azmi, N.J. Strength reduction factors for structural rubbercrete. *Front. Struct. Civil Eng.* **2014**, *8*, 270–281.

22. Bravo, M.; de Brito, J. Concrete made with used tyre aggregate: Durability- related performance. *J. Clean. Prod.* **2012**, *25*, 42–50. [CrossRef]
23. Al-Tayeb, M.M.; Bakar, B.A.; Ismail, H.; Akil, H.M. Effect of partial replacement of sand by recycled fine crumb rubber on the performance of hybrid rubberized- normal concrete under impact load: Experiment and simulation. *J. Clean. Prod.* **2013**, *59*, 284–289. [CrossRef]
24. Bedi, R.; Chandra, R. Reviewing some properties of polymer concrete. *Indian Concr. J.* **2014**, *88*, 47–68.
25. Barbuta, M.; Diaconu, D.; Serbanoiu, A.A.; Burlacu, A.; Timu, A.; Gradinaru, C.M. Effects of tire wastes on the mechanical properties of concrete. *Proc. Eng.* **2017**, *181*, 346–350. [CrossRef]
26. Bedi, R.; Chandra, R.; Singh, S.P. Mechanical properties of polymer concrete. *J. Compos.* **2013**, *2013*, 948745. [CrossRef]
27. El-Hawary, M.M.; Abdul-Jaleel, A. Durability assessment of epoxy modified concrete. *Constr. Build. Mater.* **2010**, *24*, 1523–1528.
28. Bai, W.; Zhang, J.; Yan, P.; Wang, X. Study on vibration alleviating properties of glass fiber reinforced polymer concrete through orthogonal tests. *Mater. Des.* **2009**, *30*, 1417–1421. [CrossRef]
29. Reis, J.M.L.; Ferreira, A.J.M. The effects of atmospheric exposure on the fracture properties of polymer concrete. *Build. Environ.* **2006**, *41*, 262–267. [CrossRef]
30. Bruni, C.; Forcellese, A.; Gabrielli, F.; Simoncini, M. Hard turning of an alloy steel on a machine tool with a polymer concrete bed. *J. Mater. Proc. Technol.* **2008**, *202*, 493–499. [CrossRef]
31. Orak, S. Investigation of vibration damping on polymer concrete with polyester resin. *Cem. Concr. Res.* **2000**, *30*, 171–174. [CrossRef]
32. Gupta, K.; Mani, P.; Krishnamoorthy, S. Interfacial adhesion in polyester resin concrete. *Int. J. Adhes. Adhes.* **1983**, *3*, 149–154. [CrossRef]
33. Rebeiz, K.S. Precast use of polymer concrete using unsaturated polyester resin based on recycled PET waste. *Const. Build. Mater.* **1996**, *10*, 215–220. [CrossRef]
34. Bulut, H.A.; Sahin, R. A Study on mechanical properties of polymer concrete containing electronic plastic waste. *Compos. Struct.* **2017**, *178*, 50–62. [CrossRef]
35. Kim, H.S.; Park, K.Y.; Lee, D.G. A study on the epoxy resin concrete for the ultraprecision machine tool bed. *J. Mater. Proc. Technol.* **1995**, *48*, 649–655. [CrossRef]
36. Martyushev, N.V.; Malozyomov, B.V.; Khalikov, I.H.; Kukartsev, V.A.; Kukartsev, V.V.; Tynchenko, V.S.; Tynchenko, Y.A.; Qi, M. Review of Methods for Improving the Energy Efficiency of Electrified Ground Transport by Optimizing Battery Consumption. *Energies* **2023**, *16*, 729. [CrossRef]
37. Chen, D.H.; Lin, H.H.; Sun, R. Field performance evaluations of partial-depth repairs. *Constr. Build. Mater.* **2011**, *25*, 1369–1378. [CrossRef]
38. Chen, D.H.; Zhou, W.; Kun, L. Fiber reinforced polymer patching binder for concrete pavement rehabilitation and repair. *Constr. Build. Mater.* **2013**, *48*, 325–332. [CrossRef]
39. Cortes, F.; Castillo, G. Comparison between the dynamical properties of polymer concrete and grey cast iron for machine tool applications. *Mater. Des.* **2007**, *28*, 1461–1466. [CrossRef]
40. San-Jose, J.T.; Vegas, I.; Ferreira, A. Reinforced polymer concrete: Physical properties of the matrix and static/dynamic bond behavior. *Cem. Concr. Compos.* **2005**, *27*, 934–944. [CrossRef]
41. Fowler, D.W. Future trends in polymer concrete. *Spec. Publ.* **1989**, *116*, 129–144.
42. Rebeiz, K.S.; Serhal, S.P.; Fowler, D.W. Structural behavior of polymer concrete beams using recycled plastic. *J. Mater. Civil Eng.* **1994**, *6*, 150–165. [CrossRef]
43. Kumar, R. A review on epoxy and polyester based polymer concrete and exploration of Polyfurfuryl Alcohol as polymer concrete. *J. Polym.* **2016**, *13*, 1–13. [CrossRef]
44. Toufigh, V.; Shirkhorshidi, S.M.; Hosseinali, M. Experimental investigation and constitutive modeling of polymer concrete and sand interface. *Int. J. Geomech.* **2016**, *17*, 04016043. [CrossRef]
45. Martyushev, N.V.; Malozyomov, B.V.; Sorokova, S.N.; Efremenkov, E.A.; Qi, M. Mathematical Modeling of the State of the Battery of Cargo Electric Vehicles. *Mathematics* **2023**, *11*, 536. [CrossRef]

Disclaimer/Publisher's Note: The statements, opinions and data contained in all publications are solely those of the individual author(s) and contributor(s) and not of MDPI and/or the editor(s). MDPI and/or the editor(s) disclaim responsibility for any injury to people or property resulting from any ideas, methods, instructions or products referred to in the content.

Article

Mechanics of Pure Bending and Eccentric Buckling in High-Strain Composite Structures

Jimesh D. Bhagatji *, Oleksandr G. Kravchenko * and Sharanabaseweshwara Asundi

Department of Mechanical and Aerospace Engineering, Old Dominion University, Norfolk, VA 23529, USA; sasundi@odu.edu
* Correspondence: jbhag001@odu.edu (J.D.B.); okravche@odu.edu (O.G.K.)

Abstract: To maximize the capabilities of nano- and micro-class satellites, which are limited by their size, weight, and power, advancements in deployable mechanisms with a high deployable surface area to packaging volume ratio are necessary. Without progress in understanding the mechanics of high-strain materials and structures, the development of compact deployable mechanisms for this class of satellites would be difficult. This paper presents fabrication, experimental testing, and progressive failure modeling to study the deformation of an ultra-thin composite beam. The research study examines the deformation modes of a post-deployed boom under repetitive pure bending loads using a four-point bending setup and bending collapse failure under eccentric buckling. The material and fabrication challenges for ultra-thin, high-stiffness (UTHS) composite boom are discussed in detail. The continuum damage mechanics (CDM) model for the beam is calibrated using experimental coupon testing and was used for a finite element explicit analysis of the boom. It is shown that UTHS can sustain a bending radius of 14 mm without significant fiber and matrix damage. The finite element model accurately predicts the localized transverse fiber damage under eccentric buckling and buckling stiffness of 15.6 N/mm. The results of the bending simulation were found to closely match the experimental results, indicating that the simulation accurately shows deformation stages and predicts damage to the material. The findings of this research provide a better understanding of the structure characteristics with the progressive damage model of the UTHS boom, which can be used for designing a complex deployable payload for nano-micro-class satellites.

Keywords: composite boom; bending characterization; buckling; continuum damage mechanics

Citation: Bhagatji, J.D.; Kravchenko, O.G.; Asundi, S. Mechanics of Pure Bending and Eccentric Buckling in High-Strain Composite Structures. *Materials* **2024**, *17*, 796. https://doi.org/10.3390/ma17040796

Academic Editor: Raul D. S. G. Campilho

Received: 28 December 2023
Revised: 30 January 2024
Accepted: 2 February 2024
Published: 7 February 2024

Copyright: © 2024 by the authors. Licensee MDPI, Basel, Switzerland. This article is an open access article distributed under the terms and conditions of the Creative Commons Attribution (CC BY) license (https://creativecommons.org/licenses/by/4.0/).

1. Introduction

Composites were used as a building block of the traditional satellite that debuted in the early 1970s. It was used on the Apollo capsule as an Avcoat ablative heat shield, fiberglass honeycomb structure [1], etc. Since then, advanced composites have been the choice of various space programs such as reusable launch vehicles, observation satellites, and the International Space Station (ISS) [2]. Fast-forward to today where advanced composites still play a significant role in the advancement of space programs. Emerging small satellite technology also cannot escape the use of composites. Composite materials are used in various avenues of small satellite technology like solar structure panels [3], high gain antenna [3], momentum/reaction wheel [4], deorbiting [5], etc. These composite structures are typically used in rigid structural elements, while novel nano- and micro-class satellites, which are constrained by size, weight, and power, require a new class of high-strain composite structures. The small satellite technology demands a high post-deployment surface area to packaging volume as well as compact deployable mechanisms, which require further advancement in the high-strain composite structures.

The early development of the boom started with simpler deployable structures with a tape spring [6] cross-section made of Carbon Fiber-Reinforced Polymer (CFRP). Composites enabled adaptive performance, a coefficient of thermal expansion, and reduced mass.

Furthermore, various cross-section booms were attempted in order to improve the structure performance of the boom, considering increasing the (stored) packaging volume. Tubular Extendible Member (STEM) booms [7] and tape springs [6] are characterized under a single-shell storable boom and Triangular Rollable and Collapsible (TRAC) booms [8], lenticular shearless booms, and Collapsible Tubular Mast (CTM) boom [9] are characterized as double-shell storable booms. The selection of the boom is highly subjective to the loading conditions for the payload. A parametric analysis [10] was conducted to identify the optimal boom geometries that maximize stiffness for various cross-sections of boom. Results indicated that the CTM provides a structural advantage with a maximum second moment of inertia in all three axes when compared among the various double-shell booms.

NASA has been investigating the performance of CTM for the Advanced Composite Solar Sail System (ACS3) [11] technology demonstration mission. To develop high-strain composite structures with improved packaging efficiency and deployed structural performance, accurate prediction of strain-stress states and failure modes of flexural members composed of thin composite laminates is necessary. To explore this potential benefit, various test setups [12–14] have been conducted on flat coupons of woven-ply CFRP materials subjected to pure moments. In typical flat coupon bending, the high-strain composite material exhibits fiber tensile stiffening and compression softening, with a net effect that leads to a gradual decrease in bending stiffness as strain increases. The advantage of using this method was its simplified experimental setup; however, under high deformation structure bending of UTHS/CTM composite, the material experiences a bi-directional strain that was not accurately captured by the current test setup.

In contrast to flat coupon bending, an experimentally intensive technique [15] can be used to formulate a stress-based failure criterion in terms of failure parameters. This approach considers a repeating unit cell of a symmetric two-ply plain weave laminate and the stress resultants from a homogenized plate model. Five sets of tests were conducted to estimate the failure parameters, and five additional combined loading test configurations were tested for validation. This approach has the advantage of finding a failure locus for a two-ply plain weave laminate in terms of force and moment resultants, making a six-dimensional loading space with an experimental intensive approach.

It is essential to find the bending characteristics of the deployable composite boom to create an effective design. A simple and precise approach needs to be taken to bridge this gap. Few studies, such as [16], observed contradictory modeling results from the experimental results due to an inappropriate methodology to determine the flexural modulus of the material for boom flattening lengths in the range of 40–100 mm. For higher flattening lengths of 250–500 mm with a material thickness range of 0.2–0.4, the material remains in the elastic region under large deformation structure bending. It was important to understand the relation between the size factor of UTHS/CTM composite boom and stress in the material. Another study [17] conducted a flattening test for the deployable composite boom of a flattening length (approx.) 280 mm, where nonlinearity was observed mostly due to geometry rather than to material nonlinearity itself, which can include plasticity and localized damage. Gaining a precise knowledge of strain behavior and non-linear effects under large deformation loads will allow for the design of the deployable structures more effectively, preventing potential failure.

This paper presents analysis, design, and fabrication approach of UTHS/CTM structure for small satellite applications and validates its performance. The boom structure is a basic building block for in-space assembly and Lunar exploration missions. They offer structural support and stability for post-deployment payload assembly in low packaging volume for in-space assembly of telescope or antenna [18]. Similarly, in lunar exploration, booms are used as deployable towers for critical instruments/equipment for power, communication, and weather research [19]. It is important to understand structural failure behavior post-deployment and determine safety factors for such deployable payloads. Hence, it is essential to develop predictive modeling tools that can consider the mechanical behavior under bending and buckling of thin booms in the post-deployed configuration.

The first section discusses the composite material and design of the lenticular cross-section boom (CTM), as well as elaborates on the fabrication process and challenges that were addressed to develop a consistent deployable structure. The structural level testing of the boom was investigated for the large deformation behavior of the composite boom using four-point bending tests and coupled bending-buckling behavior through an eccentric compression test. The large deformation behavior was used to evaluate damage progression during pure bending and to determine the critical bending radius. This analysis was crucial for understanding the folding deformation and the associated damage that occurs during rolling. Moreover, complex deployable mechanisms [3] often subject UTHS booms to intricate loading conditions, leading to localized bending failure. Therefore, it becomes crucial to comprehend the deformation behavior under short spans to simulate localized pure bending.

Additionally, another mechanical test was conducted where the coupled bending and compression behavior was assessed using an eccentric buckling setup. This test allowed for determining global structural failure during post-deployment loading conditions, where the deployed UTHS boom underwent compressive loading with pure bending deformation. In the second section, an experimental approach for calibrating finite element analysis (FEA) Continuum Damage Mechanics (CDM) model is discussed, which used 0° and 45° coupon tensile testing of the weave. The validation of FEA CDM of the boom was performed using both four-point bending and eccentric buckling by comparing with the experimental results. Modeling results of four-point bending tests and eccentric buckling of ultra-thin composites agreed closely with the mechanical test data. FE analysis revealed that during the snapping of the boom, no significant damage was induced in the material, allowing for a reliable UTHS boom deployment. In the case of eccentric buckling, FEA accurately captured the global buckling with transverse fiber damage. The proposed material and structural testing, along with modeling methodology, can be adopted to the UTHS boom for the various deployable payloads used in small-class satellites and in space structures.

2. Material and Fabrication for Mechanical Test

2.1. Materials

To create high-deformation structures, it was necessary to select a material that was both lightweight and able to withstand large non-linear deformation of UTHS boom. Carbon fiber was chosen as the material in this case due to its excellent strength-to-weight ratio. However, in contrast to glass fiber, carbon fiber provides better strength but at a lower strain, which opposes the high-strain material requirement.

Therefore, thin laminates of only one or two plies were of particular interest in our study to meet the material requirement. To maximize the laminate surface strain and axial modulus (E_1) of the laminate, a single-ply twill-weave was selected. Twill weave at a 45° weave orientation provides high shear strain and better formability during manufacturing, especially on curved surfaces. To simplify the handling of the material during manufacturing, high-quality carbon fiber prepreg 3 K (200 gsm), 2 × 2 twill weave was procured from Fibreglast, Brookville, OH, USA.

Coupon testing was carried out for both single-ply weave and two-ply weave before deciding to use the single-ply weave, as presented later in the paper. Uniaxial tension tests of a 0° weave laminate and 45° weave laminate were prepared from weave laminate panels. These tensile specimens were cut into dogbone specimens of 20 mm width and 200 mm long (including tabs) using a ProtoMAX waterjet system by Omax, Kent, WA, USA. These specimens were tested in an MTS Alliance RF/300 machine with a load cell of 300 kN capacity by MTS System, Eden Prairie, MN, USA. Testing was conducted with a crosshead displacement of 1 mm/min. The test was recorded using a GOM 3D Digital Image Correlation (DIC) system to accurately capture the strain distribution during testing.

2.2. Composite Boom Fabrication

The cross-section of the specimen can be parametrized into six independent parameters, as shown in Figure 1a. Here, in order to reduce the inconsistency of in-plane strain along the cross-section during flattening and wrapping:

1. α_1 and α_2 were selected as same value ($\alpha_1 = \alpha_2 = \alpha$).
2. r_1 and r_2 were selected as same value ($r_1 = r_2 = r$).

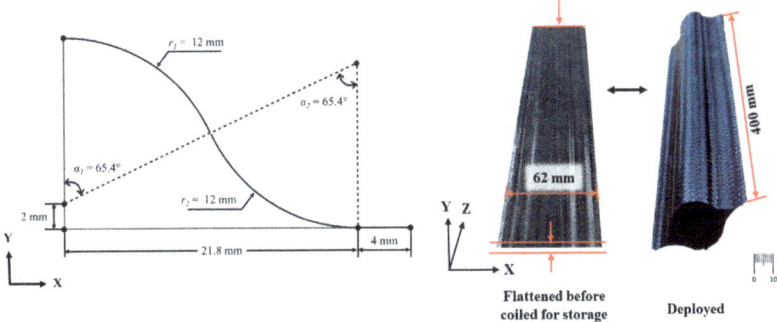

(a) Quarter of boom cross-section   (b) Flattened and deployed boom

Figure 1. Design parameters of the UTHS/CTM deployable structure.

The design of the specimen has been simplified to three parameters—α, r, and t. These parameters determine the structural properties of the composite boom. The selection of these three parameters was primarily determined by the bending radius of the composite structure (boom), flattening height, and second moment of area, which were obtained from the mission requirements. The design of the optimal boom must consider the stability of its structure under post-deployment loading conditions and its ability to fit within the volume constraints specified by the mission requirements. In this study, Old Dominion University's 3U CubeSat constraints were considered [20]. The final selected design parameters are shown in Figure 1a. With these parameters, the structure has a second moment of inertia of 2.78×10^4 mm^4, with a flattened height of 62 mm, shown in Figure 1b.

This lenticular cross-section was achieved by joining a flat end (web) of two halves (omega-shaped) made of carbon fiber prepreg twill weave. In the literature [9,11], a rigid aluminum or rubber internal mold was employed for the fabrication of the boom, alongside two split molds. However, this approach requires additional tooling to support the internal surface of the boom. Hence, we decided to examine the co-cured boom structure using internal vacuum bagging with two split molds. Multipurpose 6061 Aluminum order from McMaster-Carr, Elmhurst, Illinois, USA was used to fabricate 400 mm long mold as shown in Figure 2a. Low tolerance was provided in the mold assembly for the alignment of two halves when closed. One of the key requirements for the deployable structures is the ability to withstand the high bending strain of the boom structure, which does not produce large non-linear deformation in the composite material, similar to Von Karman non-linear beam model [21] and Carrera Unified Formulation (CUF) [22]. A single layer was selected for this study, as it can provide the necessary bending strain and stiffness demanded by the structure. A 45° angle woven orientation was selected to improve the structure's bending flexibility, as well as to provide additional benefits in terms of torsional stiffness and in-plane shear stiffness.

  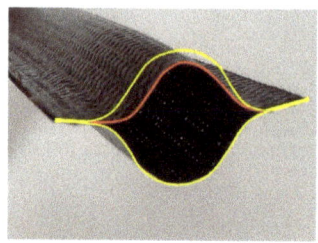

(**a**) Fabrication mold (**b**) Local matrix squeezed out (**c**) Imperfection in top ply

Figure 2. Fabrication of composite boom.

A single cure cycle is recommended by the manufacturer at 155 °C (310 °F) at a ramp at 5 °C/min with a hold stage for 120 min and subsequent cool down. During the early iterations of the process, a simplified manufacturing process was first conducted without vacuum bagging. The layup prepreg sheets were laid down on the two-half mold and compressed against the mold using the hand roller to obtain the desired lenticular shape. Then, the mold was carefully closed and placed in an autoclave for a single-stage cure cycle at 80 Psi pressure. However, this process resulted in a significant deviation from the desired geometry due to the absence of internal support for the top woven ply. As the temperature increased, the viscosity of the epoxy resin reduced, resulting in the peeling of the unsupported ply from the mold surface. This caused the deformed lenticular shape as shown in Figure 2c where the upper ply was in a deformed shape (in red) in comparison to the bottom ply. This issue was resolved by providing internal support to the top ply during the curing process by using internal support provided by the vacuum bag film, which was used as a sleeve and was inserted into the hollow space of the lenticular shape. This process provided the desired lenticular shape but was not able to achieve uniform thickness near the bond line of two halves of the boom. Upon close observation of the sample, an accumulation of epoxy resin was found near the web as shown in Figure 2b. As the two ends of the lenticular shape were joined with the help of mold pressure, the matrix tended to squeeze out [23,24] at elevated temperatures. This accumulated epoxy resin near the web resulted in a local increase in thickness, which upon bending resulted in cracks. To solve the issue of non-uniform thickness, a co-curing cycle was adopted. The initial cure cycle allowed for the matrix to partially cure by reaching the gelation point. Therefore, during the co-cure stage, when the mold was closed, matrix squeezing was prevented. As a result, a uniform thickness joint was achieved between the two halves of the boom. In summary, the proposed two-stage co-curing process with an inner vacuum bag offered an alternative fabrication method with simplified tooling by eliminating the need for an internal mold. This approach provided uniform pressure on the inner surface of the lenticular shape and prevented matrix squeezing near the web, which was observed when two-piece mold was used (Figure 2b).

2.3. Mechanical Testing

The damage in the material is mainly by bi-directional strains during localized boom bending. It was crucial to quantify the damage to the composite boom caused by large deformation bending, which can cause the matrix and the fiber damage due to the complex local state of stress. To evaluate the large deformation capability of the boom structure, four-point bending testing was conducted on the boom with a 160 mm span. A load–displacement plot was chosen for analyzing bending performance to avoid the need for calculating varying moments of inertia or centroid shift. This approach also simplifies the comparative analysis of experimental data with reaction force versus displacement plots in FEA simulations.

To understand the potential of local damage in the material during the rolling of the boom, two loading cycles were performed during four-point bending. Upon bending for two successive cycles, any damage that was caused by the first bending will be revealed in the load vs. deformation plot as reduced bending stiffness and potential reduction in the critical load.

The fixed bottom roller span was 124 mm of four-point bending setup (as shown in Figure 3a). To experimentally test the effect of varying critical radius of curvature upon local bending, the UTHS boom specimens were evaluated under with varying top roller span lengths: 20, 30, and 50 mm using a 10 kN MTS test machine by MTS System, Eden Prairie, MN, USA. (as shown in Figure 3b). The different spans were considered to understand any damage with varying bending curvature would in the boom. A preload was applied to the boom to avoid boom to roll over and capture the large deformation of the boom without slipping and provided a constant moment at the center of the boom. To demonstrate localized high-strain deformation of the boom, a crosshead was displaced to induce a strain of 2% in the structure when subjected to bending.

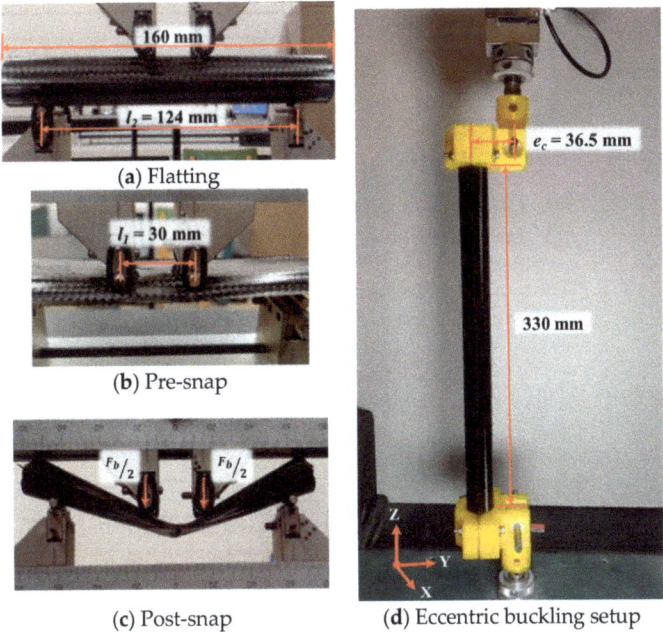

Figure 3. (**a**–**c**) Four-point bending test setup and deformation modes. (**d**) Custom 3D printed end attachment for eccentric buckling.

The UTHS boom was also tested under coupled buckling (F_c) and bending ($M_c = F_c \times e_c$) conditions to simulate post-deployment global bending scenarios through an eccentric buckling test. A custom 3D printed end attachment was designed to feature fixed end condition at the UTHS boom and bearing-supported load-cell adapter, eccentrically positioned at e_c = 36.5 mm from the center of the boom. A displacement-controlled eccentric load was applied until failure at a rate of 1 mm/min on a 330 mm long boom.

3. Modeling of Ultra-Thin Composite Beam

To develop and test a design strategy for ultra-thin composite booms, CDM material model was validated and used for non-linear analysis of the UTHS boom bending simulation. The FEA modeling was validated using experimental results on the coupon scale and using composite boom geometry. The CDM material model of composite boom allowed to capture the non-linear material behavior during large deformation experienced in four-point bending.

3.1. Material Model Calibration

A progressive damage analysis (PDA) was performed using a finite element model discussed in the following section to determine potential failure modes in composite material during the boom bending. The CDM model discussed in [25] was used for fabric-reinforced composites with a non-linear response to matrix shear, assuming orthogonal fiber directions and using orthotropic damaged elasticity for in-plane stress-strain relations shown in Equation (1). This analysis used CDM to model damage in warp and weft directions, as well as matrix damage. Within this framework, the damage variables d_1 and d_2 were utilized to represent fiber damage in the warp and weft directions, respectively. These variables effectively capture the extent of damage resulting from both tensile and compressive loading. The analysis also incorporated matrix shear plasticity by introducing the matrix damage variable d_{12}. This variable accounts for the extent of damage in the matrix material due to shear forces.

$$\begin{bmatrix} \frac{1}{(1-d_1)E_1} & \frac{-v_{12}}{E_1} & 0 \\ \frac{-v_{21}}{E_2} & \frac{1}{(1-d_2)E_2} & 0 \\ 0 & 0 & \frac{1}{(1-d_{12})2G_{12}} \end{bmatrix} \begin{bmatrix} \sigma_{11} \\ \sigma_{22} \\ \sigma_{12} \end{bmatrix} = \begin{bmatrix} \varepsilon_{11} \\ \varepsilon_{22} \\ \varepsilon_{12}^{el} \end{bmatrix} \quad (1)$$

The CDM material model was accessed by creating material with suffix ABQ_PLY_FABRIC, an embedded user subroutine (VUMAT) in Abaqus/Explicit 2021 was used to model woven ply. The stress component of FEM elements is transferred onto the fiber failure criteria, which updates the damage threshold to determine the fiber damage activation function. The damage threshold satisfies the Kuhn–Tucker complementary conditions, ensuring its monotonically increasing behavior [25]. By utilizing the damage threshold and fracture energy, the fiber damage variable is determined. Figure 4 presents the test results obtained from the uniaxial tension test of coupons used to calibrate the CDM model. A difference in modulus and strength was observed between single-ply and double-ply laminates for both 0° weave laminate and 45° weave laminate.

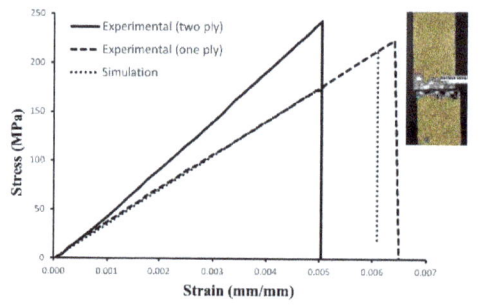

(**a**) 0° weave coupon testing

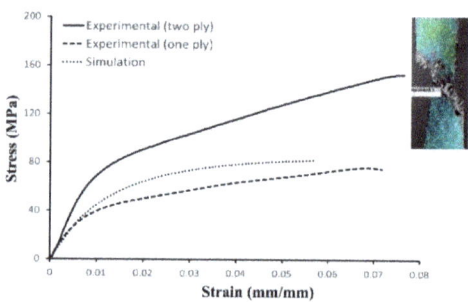

(**b**) 45° weave coupon testing

Figure 4. Experimental testing results of uniaxial tension test.

The double-ply laminate shows minimal to no pinholes (voids), leading to increased modulus and strength. On the contrary, the single-ply laminate had detected pinholes, affecting its properties. A micromechanical model [26] with pinholes demonstrates a similar increase in effective mechanical properties as the number of ply increases. However, for this study, a single layer was chosen based on high-deformation structure requirements, hence, despite the presence of pinholes, single-ply laminate's equivalent properties were used to calibrate the material model.

The elastic modulus (E_1) and strength (X_1) of the CDM material model were calibrated using a uniaxial tension test of a 0° weave laminate, resulting in values of 36.9 GPa and 240 MPa, respectively, as shown in Figure 4a. These effective properties are relatively low compared to the properties of carbon fiber warp/weft tows, and can be attributed to the twill weave architecture. Notably, the effective mechanical properties obtained from the micromechanical model [26] align well with the properties determined through tensile experiment.

Shear modulus (G_{12}) was calculated using Equation (2), resulting in values of 1.56 GPa, where E_x is elastic modulus and v_{xy} is Poisson's ratio, determined from the stress-strain curve of 45° weave coupon (shown in Figure 4b). Shear damage threshold (S) and initial effective shear yield stress (σ_y^0) are calculated using Equation (3), resulting in values of 15 MPa and 25 MPa, where σ_e is the elastic limit and σ_y is the yield strength of the 45° weave coupon.

$$G_{12} = \frac{Ex}{2(1+v_{xy})} \quad (2)$$

$$S = \frac{\sigma_e}{2}, \; \sigma_y^0 = \frac{\sigma_y}{2} \quad (3)$$

The shear hardening behavior of the material is characterized by two parameters: the coefficient of plastic hardening (C) and the power term of plastic hardening (p). These parameters were calibrated using uniaxial tension tests on 45° weave coupons to curve fit data with the shear hardening function [25]. The fracture toughness properties needed for the CDM material model in FEA were established based on Refs. [27,28], and are presented in Table 1.

Table 1. CFRP twill weave fabric mechanical properties.

Symbol	Material Constants (Units)	Magnitude
	Elastic properties	
E_1	Warp Young's modulus (GPa)	36.90
E_2	Weft traction Young's modulus (GPa)	32.60
V_{12}	Poisson coefficient	0.053
G_{12}	Shear modulus (GPa)	1.560
	Strength properties (damage initiation coefficients)	
X_1	Warp strength (MPa)	240
X_2	Weft strength (MPa)	234
S	In-plane shear damage threshold (MPa)	15
	Fracture toughness (damage evolution coefficients)	
Gf^{1+}_f	Energy rate per unit area warp tension (mJ/mm^2)	20
Gf^{1-}_f	Energy rate per unit area warp compression (mJ/mm^2)	40
Gf^{2+}_f	Energy rate per unit area weft tension (mJ/mm^2)	10
Gf^{2-}_f	Energy rate per unit area weft compression (mJ/mm^2)	20
	Shear damage and hardening parameters (shear plasticity coefficients)	
α_{12}	Parameter for in-plane shear damage	0.316
α^{max}_{12}	Maximum in-plane shear damage	1
σ_y^0	Initial effective shear yield stress (MPa)	25
C	Coefficient in the hardening equation	500
P	Power term in the hardening equation	0.42

To validate the calibrated material model, a virtual coupon was simulated using FEA. Figure 4 also presents the simulation results for the 0° and 45° weave coupons, demonstrating fiber and shear failure, respectively, in agreement with the results of the tensile test. Figure 4a shows that the simulation data for the 0° weave coupon match the experimental data, while the 45° weave coupon exhibits a slight offset in the shear plastic zone despite a good match in initial modulus. Given that the applied strain for bending is limited to 2%, this calibration remains satisfactory for further analysis.

3.2. Boom Finite Element Simulation

This section presents a bending simulation of a composite boom under high deformation, using both linear elastic model and the CDM material model. The linear elastic analysis was used to determine the (reaction) force–displacement plot in the composite boom, while the PDA was used to evaluate the damage caused by the matrix and fiber components of the boom.

3.2.1. Boom Bending Simulation

Firstly, to capture quasi-static linear elastic bending, the FEM model was developed in Abaqus/Implicit for the four-point bending setup as described in Section 2.3, using linear–elastic material properties (shown in Table 1). The model consisted of 18,560 linear quadrilateral shell elements (S4R), representing ply thickness with 1 mm mesh size. The choice of mesh size was determined through a mesh convergence approach. Since the boom was modeled using shell elements, a two-ply laminate was assigned to the flat region and a single-ply laminate to the remaining region. Surface–surface contact was activated for interactions between the rigid roller and the boom, and self-contact was assigned within the area of the lenticular section. The bottom set of rollers was fixed, while the top set provided displacement to apply 2% strain on the UTHS boom.

The modeling of PDA was developed to predict the damage initiation and evolution behavior of the UTHS boom under bending. An explicit modeling approach was chosen to capture the snapping behavior during bending. The calibrated CDM material model was implemented in the 3D (solid) model of the UTHS boom using Abaqus/Explicit. The entire model was defined with a single-ply weave orientation, with thickness being incorporated into the 3D solid model. Cohesive elements were not employed at the flat interface of plies, as no signs of delamination were observed based on experimental testing. Therefore, it was assumed that the inter-adhesion of the ply exhibited linear elastic behavior. The model consisted of 24,780 continuum shell (SC8R) elements, with a mesh size of 1 mm. For a quasi-static bending using dynamic explicit analysis, time, mass scaling parameters, and mesh size were selected to keep the kinetic energy a smaller fraction (5%) of the external work. A displacement rate was assigned to the top pair of rollers to apply 2% strain to the UTHS boom within the given step time. The bottom set of rollers were fixed, similar to the implicit analysis. As shown in Figure 5, major deformation occurred in the region between the top two rollers during high deformation bending. A damage process zone measuring 20 mm in length and 62 mm in width (comprising 2480 elements and 19,842 nodes) was selected to analyze the distribution of damage of the UTHS boom under bending.

3.2.2. Boom Buckling Simulation

The same Abaqus/Explicit model used bending was employed for the simulation of boom eccentric buckling with SC8R elements and 1 mm mesh size. The rationale for choosing this modelling approach was to account for high-order geometric non-linearity, which involves capturing the true experimental scale lenticular cross-section (as depicted in Figure 1a) and boom length (as shown in Figure 3d) without any simplification. We believe by adopting this approach, the explicit solver effectively considers the P-delta effect [29,30] in eccentric buckling. Additionally, the approach addresses structural non-linearity using of a dynamic explicit solver to accurately determine changes in cross-section and structural deformation. Material non-linearity using CDM material model was also incorporated to FEM simulation accuracy. To replicate the end boundary condition of eccentric buckling, a flat rigid shell element was fixed to the boom ends, simulating a fixed boundary with zero slope. The upper flat shell was assigned rotating degrees of freedom along the X-axis (as depicted in Figure 3d) and a downward displacement of 8 mm along the Z-axis. Meanwhile, the lower flat shell had fixed displacement and rotation, excluding rotation along the X-axis.

For quasi-static bending, mass scaling was employed, and the step time was adjusted to ensure kinetic energy constituted a smaller fraction (5%) of the external work.

4. Analysis of Pure Bending Deformation Modes and Eccentric Buckling Collapse in Composite Boom

The linear–elastic model used in implicit simulation was able to accurately capture all deformation features observed during the test (shown in Figure 5). The load–displacement plot of the implicit analysis shows various deformation modes of the UTHS boom during pure bending. The bending of the UTHS boom began with pre-flattening stage, where the structure undergoes bending with indentation. The phenomenon of bending with indentation was also observed in the three-point bending of a thin-walled rectangular beam [31], as the ratio of top roller distance to beam height ranged between 3 and 7. Compared to the flattening stage, the pre-flattening stage exhibited a stiffer behavior (as shown by the slope (K_1 = 2.96 N/mm, K_2 = 1.64 N/mm) in the load–displacement plot depicted in Figure 5). During flattening, the lenticular cross-section of the boom was narrowed down until it reached a point of pre-snap, at which the load–displacement curve exhibited a plateau. The boom behaved linearly, with an increase in load until the pre-snap phase, at which point the boom achieved its maximum load. At the snap, a rapid decrease in load was observed, and the sample transformed into a folded shape, where the top and bottom surfaces of the boom between the top rollers were pressed together. As the applied displacement increased in the post-snap phase, the folded boom region expanded outwards from the top rollers into the span direction. During this mode of deformation, the load displacement showed plateau-like behavior with the initial stiffening (up until 28 mm) followed by minor softening.

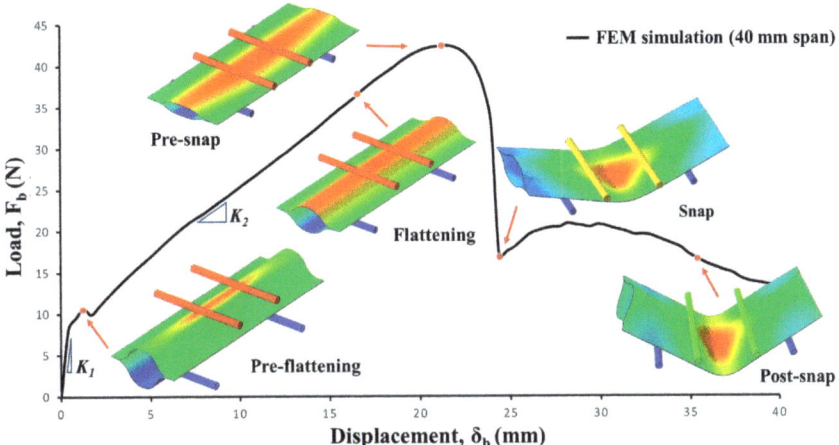

Figure 5. Load–displacement plot of a large deformation structure under pure bending (colors represent vertical displacement).

Figure 6 also presents a comparison between the experimental and implicit simulation results in a load vs. deformation plot. During the experimentation, an 11 N preload was applied to ensure proper alignment of the sample and prevent the boom from rolling before the displacement of the top roller commenced. Consequently, the initial slope (K_1) in the load–displacement curve observed in the FEM simulation, attributed to bending with indentation, was not accounted for in the experimental load–displacement plot. To rectify this, the preload was subtracted from the load–displacement curve obtained through the FEM simulation.

The slope of the flattening stage of the simulation (K_2 = 1.76 N/mm) was observed to be consistent with the experimental results (K_2 = 1.64 N/mm); however, FEA overestimated the pre-snap load for the 20 mm span sample and underestimated for the 40 mm span sample. This overestimation/underestimation of pre-snap load can be attributed to the imperfections of the geometry, as snapping is a stability phenomenon that was not accurately capturing it in an implicit FEM simulation. When comparing the load–displacement behavior of FEM simulations for different l_1 (shown in Figure 6), minimal increases in the peak pre-snap load were observed for an increase in top roller span: 30.84 N for 20 mm span and 32.54 N for 40 mm span. Additionally, an increase in the top roller span resulted in a delay of the snap-stage location. Analyzing the effect of increases in the roller span for experimental results revealed similar trends in the peak pre-snap load but at significantly higher factors: 28 N for 20 mm and 38.41 N for 40 mm. As shown in Figure 6, it is evident from the experimental load–displacement plot that the pre-snap load increases with an increase in roller span, l_1. Moreover, a clear delay in the snapping stage was observed for increases in the span.

The experimental test involved two consecutive loadings for the four-point bending setup, and the load displacement of the second loading cycle is also shown in Figure 6. Upon analyzing the second loading cycle with the first loading cycle, minimal degradation of the UTHS boom was observed: the load–displacement slope of remained unaffected, while peak snapping load only reduced by 4.16 N for l_1 = 40 mm. This reduction in the second loading cycle was caused by high deformation bending in the previous cycle. The post snap plateau showed an overall increase with the top span length l_1. To gain a more comprehensive understanding of the qualitative and quantitative nature of the damage modes, PDA was used.

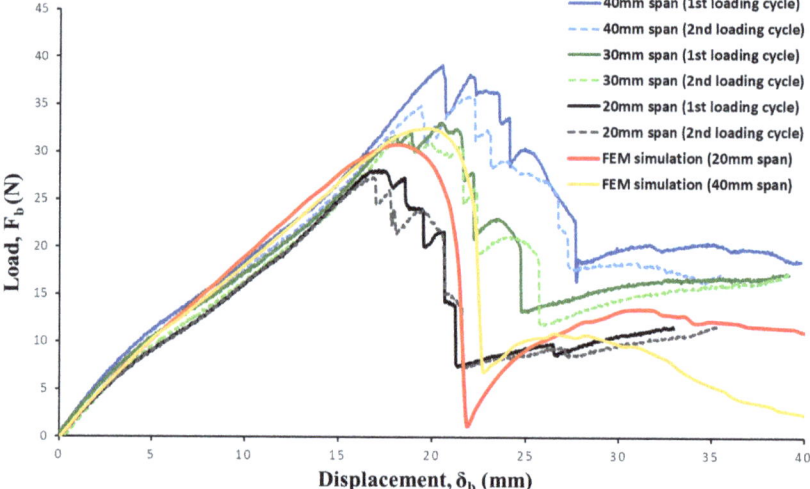

Figure 6. Load–displacement plots for experimental and FEM simulated results for varying effective span lengths.

The damage analysis of different span lengths was performed on a UTHS boom using explicit analysis. Quantitative analysis of fiber damage and matrix damage at the damage process zone for the top and bottom ply was performed. The FEM simulation results show that the boom damage process zone undergoes high-strain deformation under snapping, leading to damage initiation and propagation (Figure 7a). The spread of fiber and matrix damage in the damage process zone is presented in Figure 7c,d. The deformation profile of the edge of the lenticular boom is presented in Figure 7b. The deformation profile is shown from pre-snap at roller displacement, δ_b = 18 mm to post-snap δ_b = 42 mm. The post-snap deformed shape was used to calculate the maximum bending radius (r_c) = 13.86 mm at δ_b = 42 mm. The damage was initiated at the snapping-stage and progressed during the post-snap deformation. The concentration of fiber damage distribution (shown in Figures 7c and 8) in the top-ply was observed at the center, while minor damage was found in the bottom ply. The difference in damage distribution between the top and bottom ply was due to the top ply being under compression and the bottom ply being under tensile loads. The top and bottom plies only connected through the contact region; therefore, any crack initiated in the top ply due to compression did not propagate into the bottom ply, as the crack was arrested at the free surface of the top ply. Figure 7c displays a visual distribution of fiber damage, while Figure 8 shows the mean fiber damage, $\bar{d}_f$, across the boom width and corresponding standard deviation for the top and bottom ply. Similarly, more matrix damage was observed in the top ply and less damage developed in the bottom ply during snapping: Figure 7d shows a visual matrix distribution of the top and bottom ply.

The analyses of damage for various span lengths were conducted by determining the relative damage frequency and cumulative relative frequency of fiber and matrix damage variables in the damage process zone. The results are presented in Figures 8 and 9, which display the damage variable for different span lengths of the UTHS boom. In Figure 8, it was observed that the mean and standard deviation (damage variation) of fiber damage distributions were higher for the 20 mm span when compared to the 40 mm span. A higher damage was observed because the damages were more concentrated over for a shorter span.

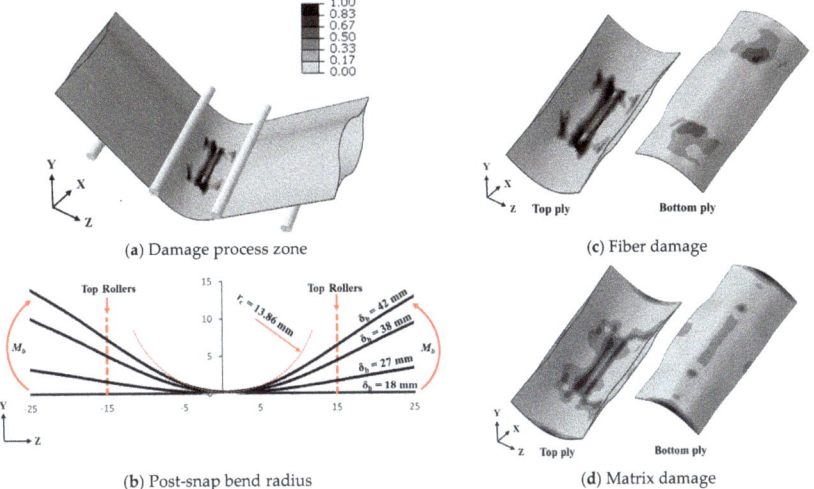

Figure 7. Damage analysis for 30 mm span length.

The cumulative relative frequency plots of the matrix and fiber damage in top and bottom plies is shown in Figure 9. Booms with the l_1 = 40 mm span in comparison to l_1 = 20 mm span (Figure 9a) indicated a slightly higher number of elements with fiber damage: 10% vs. 8%. Specifically, there were more damaged elements with low fiber damage variable (between 0.1–0.3). This result explains the marginal reduction in the peak load of the boom after the first loading cycle (Figure 6) for l_1 = 40 mm span, which can be attributed to the fact that for shorter spans, the damage was highly concentrated and less dispersed. Overall results indicate that the damage in the boom was not significantly changed for span lengths ranging from 20 mm to 40 mm for both fiber and matrix.

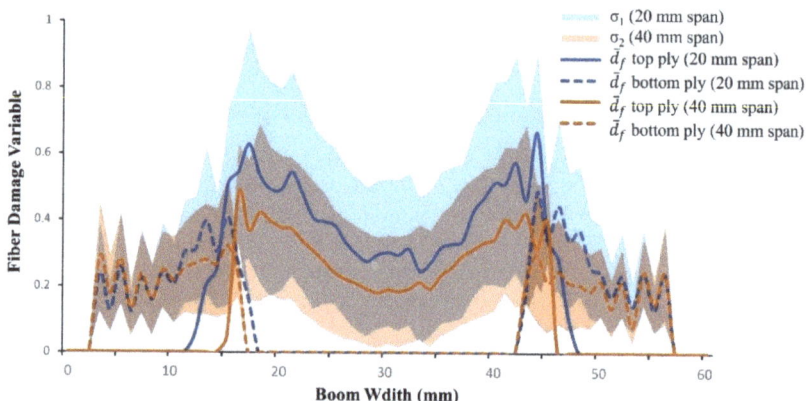

Figure 8. Fiber damage distribution across the boom width in DPZ.

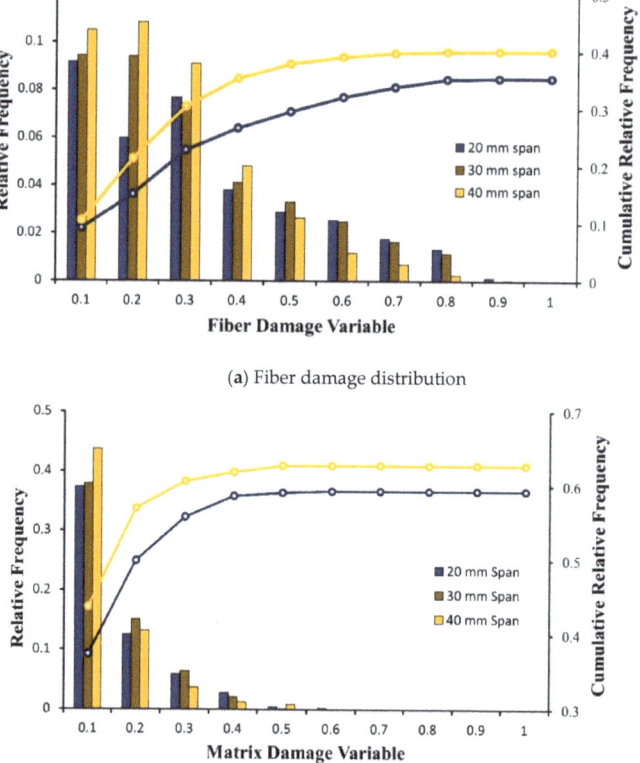

(a) Fiber damage distribution

(b) Matrix damage distribution

Figure 9. FEA results of fiber and matrix damage comparative analysis for different span lengths.

The PDA model was further utilized to analyze the damage in the UTHS boom under eccentric buckling load. The eccentric buckling performance was assessed based on force (F_c)–displacement (δ_c) plot. As depicted in Figure 10, the experimental results revealed an initial buckling stiffness of 18.9 N/mm, which momentarily increased and progressively decreased to 11.9 N/mm until failure. The FEM simulation accurately captured this behavior, initiating with a stiffness of 18.8 N/mm and steadily declining to 10.7 N/mm until failure. In terms of strength, the FEM simulation slightly outperformed the experimental results, at 118 N compared to 109 N.

A bending collapse with non-linear behavior (as depicted in Figure 11a) was observed, indicating a transverse crack due to fiber damage. Upon visual inspection, fiber failure was observed in the post-mortem local dimple region; thus, the overall structure underwent global buckling. The PDA model aligned with the visual inspection, revealing fiber failure in the two center regions of the boom, as illustrated in Figure 11b. The major fiber failure occurred on the compressive side of the boom, indicating similar behavior to four-point bending results, which also showed the majority of damage on the compressive side of the flattened boom (Figure 7).

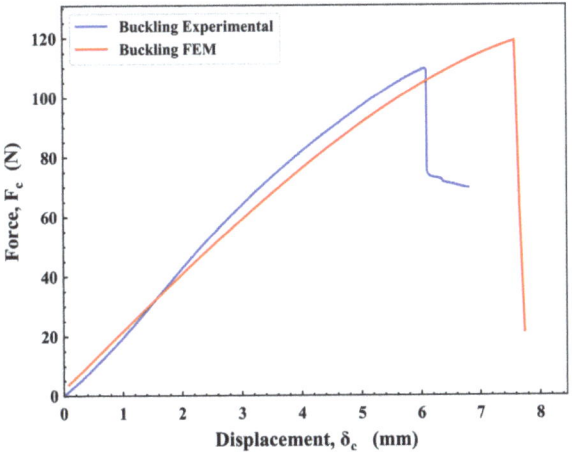

Figure 10. Force–displacement plot from the eccentric buckling test.

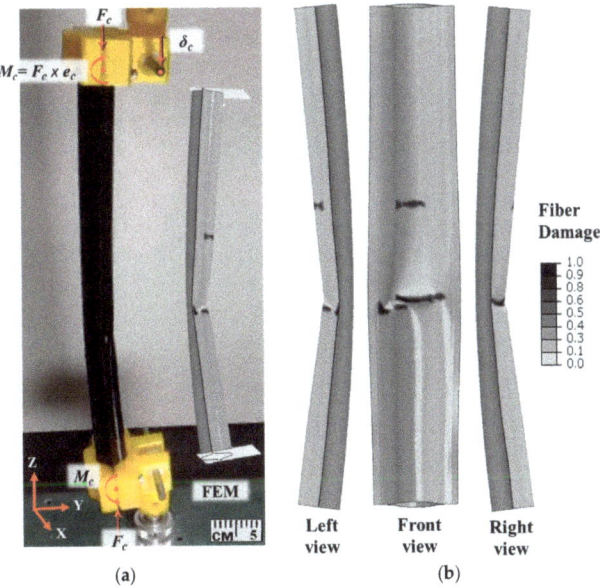

Figure 11. (a) Bending collapse comparison: experiment vs. finite element simulation, (b) fiber damage distribution in three views under eccentric buckling.

5. Conclusions

In conclusion, the fabrication process, structure characterization, and PDM of the UTHS composite boom with a lenticular cross-section was presented for small satellite payloads. The study employs a scalable lenticular boom fabrication process to achieve uniform thickness and presents the challenges encountered during the development of this process. The structure characterization was performed to investigate:

I. Large deformation analysis during localized pure bending to determined critical bending radius, crucial for understanding folding deformation and damage during rolling.
II. The eccentric buckling test assessed coupled bending and compression behavior, revealing global structural bending under post-deployment loading conditions.

The four-point bending test revealed four stages—flattening, pre-snap, snap, and post-snap—which were effectively captured by the FEA's modelling approach. The two consecutive bending cycles with different top roller spans revealed sub-critical damage, as validated by the PDA model for a critical bending radius (r_c) of 13.86 mm. The PDA revealed differences in damage distribution between the top and bottom plies due to compression and tension. The FEA model was also able to capture the bending collapse under eccentric buckling load with clear resemblances of the transverse crack fiber damage in the boom. Upon correlating these mechanical tests, the bending capability under pure bending was found to be 0.78 Nm, while under coupled bending and buckling load (M_c), it measured 3.98 Nm. The validated PFA model was able to predict the non-linear global buckling of the boom under eccentric loading, while capturing the complex failure behavior. These mechanical tests facilitated the characterization of the UTHS boom to determine load-bearing capacity for conditions both pre-deployment and post-deployment.

The study's results have practical implications for the design and manufacturing of ultra-thin composite booms for small satellite applications. The post-snap bend radius during the four-point bending test provides a valuable metric for determining the wrapping radius that can be achieved without significant mechanical degradation, which can be used in achieving a high payload packaging efficiency of composite booms in small satellite systems. The coupled bending-compressive deformation showed the global buckling, which resulted in transverse fiber cracking on the compressive side of the boom. Therefore, the modeling of boom PDM can serve the purpose of structural dynamic analysis for complex deployments and post deployment, aiding in the assessment of fiber and matrix damage. The progressive damage modeling is crucial for designing booms for intricate loading conditions applied in the different new-age deployment of payloads for small satellites and space structures.

Author Contributions: Conceptualization, J.D.B. and O.G.K.; Validation, J.D.B.; Formal analysis, J.D.B. and O.G.K.; Investigation, S.A.; Writing—original draft, J.D.B.; Writing—review & editing, O.G.K. and S.A.; Visualization, J.D.B.; Supervision, S.A.; Project administration, O.G.K. and S.A. All authors have read and agreed to the published version of the manuscript.

Funding: This research was funded by the Virginia Space Grant Consortium (VSGC) and the Virginia Institute of Spaceflight and Autonomy (VISA).

Data Availability Statement: Data are contained within the article.

Conflicts of Interest: The authors declare no conflict of interest.

References

1. Tenney, D.R.; Davis, J.G., Jr.; Pipes, R.B.; Johnston, N. Nasa composite materials development: Lessons learned and future challenges. In Proceedings of the NATO RTO AVT-164 Workshop on Support of Composite Systems, Bonn, Germany, 19–22 October 2009.
2. Tenney, D.R.; Davis, J.G., Jr.; Johnston, N.J.; Pipes, R.B.; McGuire, J.F. *Structural Framework for Flight I: NASA's Role in Development of Advanced Composite Materials for Aircraft and Space Structures*; Technical Report; NASA: Washington, DC, USA, 2019.

3. Wang, B.; Zhu, J.; Zhong, S.; Liang, W.; Guan, C. Space Deployable Mechanics: A Review of Structures and Smart Driving. *Mater. Des.* **2024**, *237*, 112557. [CrossRef]
4. Bhagatji, J.; Asundi, S.; Agrawal, V.K. Design, Simulation and Testing of a Reaction Wheel System for Pico/Nano-Class CubeSat Systems. In Proceedings of the 2018 AIAA SPACE and Astronautics Forum and Exposition, Orlando, FL, USA, 17–19 September 2018.
5. Asundi, S.; Bhagatji, J.; Tailor, P. Genesis of a multi-function drag measurement system to facilitate atmosphere modeling and space debris mitigation. In Proceedings of the 2017 AIAA SPACE and Astronautics Forum and Exposition, Orlando, FL, USA, 12–14 September 2017.
6. Yee, J.; Soykasap, O.; Pellegrino, S. Carbon fibre reinforced plastic tape springs. In Proceedings of the 45th AIAA/ASME/ASCE/AHS/ASC Structures, Structural Dynamics & Materials Conference, Palm Springs, CA, USA, 19–22 April 2004; p. 1819.
7. Wu, C. Dynamic Analysis of Extended Bistable Reeled Fibre-Reinforced Composite Booms for Space Applications. Ph.D. Thesis, University of Surrey, Guildford, UK, 2017.
8. Roybal, F.; Banik, J.; Murphey, T. Development of an elastically deployable boom for tensioned planar structures. In Proceedings of the 48th AIAA/ASME/ASCE/AHS/ASC Structures, Structural Dynamics, and Materials Conference, Honolulu, HI, USA, 23–26 April 2007; p. 1838.
9. Herbeck, L.; Leipold, M.; Sickinger, C.; Eiden, M.; Unckenbold, W. Development and test of deployable ultra-lightweight cfrp-booms for a solar sail. In *Spacecraft Structures, Materials and Mechanical Testing*; European Space Agency: Paris, France, 2001; Volume 468, p. 107.
10. Lee, A.; Fernandez, J.M. Mechanics of bistable two-shelled composite booms. In Proceedings of the 2018 AIAA Spacecraft Structures Conference, Kissimmee, FL, USA, 8–12 January 2018; p. 0938.
11. Wilkie, W.K. Overview of the nasa advanced composite solar sail system (acs3) technology demonstration project. In Proceedings of the AIAA Scitech 2021 Forum, Virtual, 11–15, 19–21 January 2021; p. 1260.
12. Murphey, T.W.; Peterson, M.E.; Grigoriev, M.M. Large strain four-point bending of thin unidirectional composites. *J. Spacecr. Rocket.* **2015**, *52*, 882. [CrossRef]
13. Fernandez, J.M.; Murphey, T.W. A simple test method for large deformation bending of thin high strain composite flexures. In Proceedings of the 2018 AIAA Spacecraft Structures Conference, Kissimmee, FL, USA, 8–12 January 2018; p. 0942.
14. Sharma, A.H.; Rose, T.; Seamone, A.; Murphey, T.W.; Jimenez, F.L. Analysis of the column bending test for large curvature bending of high strain composites. In Proceedings of the AIAA Scitech 2019 Forum, San Diego, CA, USA, 7–11 January 2019; p. 1746.
15. Mallikarachchi, H.; Pellegrino, S. Failure criterion for two-ply plain-weave CFRP laminates. *J. Compos. Mater.* **2013**, *47*, 1357. [CrossRef]
16. West, S.T.; White, C.; Celestino, C.; Philpott, S.; Pankow, M. Design and testing of deployable carbon fiber booms for cubesat non-gossamer applications. In Proceedings of the 56th AIAA/ASCE/AHS/ASC Structures, Structural Dynamics, and Materials Conference, Kissimmee, FL, USA, 5–9 January 2015; p. 0206.
17. Hu, Y.; Chen, W.; Gao, J.; Hu, J.; Fang, G.; Peng, F. A study of flattening process of deployable composite thin-walled lenticular tubes under compression and tension. *Compos. Struct.* **2017**, *168*, 164. [CrossRef]
18. Belvin, W.K. In-space structural assembly: Applications and technology. In Proceedings of the 3rd AIAA Spacecraft Structures Conference, San Diego, CA, USA, 4–8 January 2016.
19. Lordos, G.C.; Amy, C.; Browder, B.; Dawson, C.; do Vale Pereira, P.; Dolan, S.I.; Hank, T.; Hinterman, E.D.; Martell, B.; et al. Autonomously Deployable Tower Infrastructure for Exploration and Communication in Lunar Permanently Shadowed Regions. In Proceedings of the ASCEND 2020, Virtual, 16–18 November 2020.
20. Bhagatji, J.; Kravchenko, O.; Asundi, S. Large Deformation Bending of Ultralight Deployable Structure for Nano-Micro-Class Satellites. In Proceedings of the American Institute of Aeronautics and Astronautics SCITECH Forum, National Harbor, MD, USA, 23–27 January 2023.
21. Pai, P.F.; Nayfeh, A.H. A nonlinear composite beam theory. *Nonlinear Dyn.* **1992**, *3*, 273. [CrossRef]
22. Pagani, A.; Carrera, E. Large-deflection and post-buckling analyses of laminated composite beams by Carrera Unified Formulation. *Compos. Struct.* **2017**, *170*, 40. [CrossRef]
23. Jamora, V.C.; Rauch, V.; Kravchenko, S.G.; Kravchenko, O.G. Effect of Resin Bleed Out on Compaction Behavior of the Fiber Tow Gap Region during Automated Fiber Placement Manufacturing. *Polymers* **2023**, *16*, 31. [CrossRef] [PubMed]
24. Ravangard, A.; Jamora, V.; Bhagatji, J.; Kravchenko, O.G. Origin and Significance of Non-Uniform Morphology in AFP Composites. In Proceedings of the 38th Technical Conference of the American Society for Composites, Woburn, MA, USA, 18–20 September 2023; Destech Publications, Inc.: Lancaster, PA, USA, 2023.
25. Johnson, A.F. Modelling fabric reinforced composites under impact loads. *Compos. Part A Appl. Sci. Manuf.* **2001**, *32*, 1197. [CrossRef]
26. Bacarreza, O.; Abe, D.; Aliabadi, M.H.; Kopula Ragavan, N. Micromechanical Modeling of Advanced Composites. *J. Multiscale Model.* **2012**, *4*, 1250005. [CrossRef]
27. Lombarkia, R.; Gakwaya, A.; Nandlall, D.; Dano, M.L.; Lévesque, J.; Vachon-Joannette, P. Experimental investigation and finite-element modeling of the crushing response of hat shape open section composite. *Int. J. Crashworthiness* **2022**, *27*, 772. [CrossRef]

28. Larson, R.; Bergan, A.; Leone, F.; Kravchenko, O.G. Influence of Stochastic Adhesive Porosity and Material Variability on Failure Behavior of Adhesively Bonded Composite Sandwich Joints. *Compos. Struct.* **2023**, *306*, 116608. [CrossRef]
29. Thombare, C.N.; Sangle, K.K.; Mohitkar, V.M. Nonlinear Buckling Analysis of 2-D Cold-Formed Steel Simple Cross-Aisle Storage Rack Frames. *J. Build. Eng.* **2016**, *7*, 12–22. [CrossRef]
30. Thai, H.T.; Uy, B.; Khan, M. A Modified Stress-Strain Model Accounting for the Local Buckling of Thin-Walled Stub Columns under Axial Compression. *J. Constr. Steel Res.* **2015**, *111*, 57–69. [CrossRef]
31. Huang, Z.; Zhang, X. Three-point bending collapse of thin-walled rectangular beams. *Int. J. Mech. Sci.* **2018**, *144*, 461–479. [CrossRef]

Disclaimer/Publisher's Note: The statements, opinions and data contained in all publications are solely those of the individual author(s) and contributor(s) and not of MDPI and/or the editor(s). MDPI and/or the editor(s) disclaim responsibility for any injury to people or property resulting from any ideas, methods, instructions or products referred to in the content.

Article

Unified Failure Criterion Based on Stress and Stress Gradient Conditions

Young W. Kwon *, Emma K. Markoff and Stanley DeFisher

Department of Mechanical & Aerospace Engineering, Naval Postgraduate School, Monterey, CA 93943, USA
* Correspondence: ywkwon@nps.edu

Abstract: Specimens made of various materials with different geometric features were investigated to predict the failure loads using the recently proposed criterion comprised of both stress and stress gradient conditions. The notch types were cracks and holes, and the materials were brittle, ductile, isotropic, orthotropic, or fibrous composites. The predicted failure stresses or loads were compared to experimental results, and both experimental and theoretically predicted results agreed well for all the different cases. This suggests that the stress and stress-gradient-based failure criterion is both versatile and accurate in predicting the failure of various materials and geometric features.

Keywords: failure criterion; notch; crack; brittle; ductile; composite

1. Introduction

Structural members are designed to avoid unexpected failure during their service life. To achieve this, these members are tested experimentally or analyzed using proper modeling and simulation techniques. Extensive physical testing is time-consuming and costly, so computational modeling and simulation are frequently used to replace or minimize unnecessary testing. In order to have confidence in the accuracy of the results, however, computational modeling should be reliable. To this end, many failure theories have been developed to predict failure loads based on material type, including isotropic and anisotropic ones, subjected to a variety of loading conditions, including static and cyclic scenarios [1].

Important data in designing load-carrying structural members include the maximum load that they can carry without failure. Hence, failure criteria are necessary to predict the maximum failure load, but the load also depends on the geometry and material of the structure. If the structures have notches such as holes or cracks, their load-carrying capacity is significantly limited.

In the past, different failure theories were used to predict failure loads of structural members depending on the state of notches they contained. For example, many different failure criteria were proposed for structural members without any notches and subjected to combined loading (i.e., multiaxial loading) [1–8]. Those criteria were to apply the failure strength obtained from uniaxial testing to the prediction of failure under combined loading. However, if a structural member has a notch, those failure criteria are not reliable.

A structural member with a crack has stress singularity at the crack tip if the material behaves linearly elastically. Thus, fracture mechanics was also developed for structural members with cracks [9–16]. On the other hand, if structural members have holes, an entirely different set of failure criteria was used because fracture mechanics is suitable for holes. The critical distance failure criteria, for example, is often applied for structural members containing holes [17–25]. Some used the stress at the critical distance, while others used the average value up to the critical distance from the notch tip to predict failure at the notch tip.

The cohesive zone model was also developed to predict failure loads better [26–33]. The model considers a localized zone around the potential failure location, which is called

Citation: Kwon, Y.W.; Markoff, E.K.; DeFisher, S. Unified Failure Criterion Based on Stress and Stress Gradient Conditions. *Materials* **2024**, *17*, 569. https://doi.org/10.3390/ma17030569

Academic Editor: Raul D. S. G. Campilho

Received: 27 December 2023
Revised: 23 January 2024
Accepted: 24 January 2024
Published: 25 January 2024

Copyright: © 2024 by the authors. Licensee MDPI, Basel, Switzerland. This article is an open access article distributed under the terms and conditions of the Creative Commons Attribution (CC BY) license (https://creativecommons.org/licenses/by/4.0/).

the cohesive zone. A traction–separation relationship is applied to the cohesive zone and is used to predict failure. For the cohesive zone model to be accepted as a failure theory, the same traction–separation relationship must be applicable to structural members independent of the notch shape, like a crack or a hole, as well as its size.

Recently, a unified failure criterion was proposed by the authors' team [34–37]. The unified failure criterion can be applied to structural members regardless of the existence of notches, as well as their shapes and sizes. This failure criterion was validated against different experimental data on brittle specimens with cutouts, including holes and slits. The objective of this study is to validate the unified failure criterion further to determine whether the theory can apply to various cases, which include ductile or brittle materials; isotropic materials, 3D-printed orthotropic materials, or laminated fibrous composites; and holes, long slits, or cracks. The next section describes the new unified failure criterion, and it is followed by various sections that discuss the failures of different cases with subsequent conclusions.

2. Failure Criterion Based on Stress and Stress Gradient

The recently proposed failure criterion uses both stress and stress gradient to determine failure. Both stress and stress gradient conditions must be satisfied for failure to occur [34–37]. First, the stress condition must be checked. This condition states that the effective stress at any material point should not be less than the failure strength of the material, which is stated below:

$$\sigma_e \geq \sigma_f \tag{1}$$

where σ_e and σ_f are the effective stress and failure strength, respectively. The effective stress is different depending on the material behavior. The maximum normal stress is usually used for the effective stress for an isotropic brittle or quasi-brittle material. On the other hand, the maximum shear or the octahedral shear stress is selected as the effective stress for an isotropic ductile material. If a material is anisotropic, multiple effective stresses are considered depending on the directions of the material properties and the loading.

Once the stress condition is satisfied, the stress gradient condition must also be checked. This condition is expressed as

$$\sigma_e \geq \left(2EY\left(\frac{d\sigma_e}{ds}\right)\right)^{1/3} \tag{2}$$

where E is the modulus; Y is a material failure value, which is discussed below; and s is the failure path. For a brittle material, the failure path is normal to the maximum principal axis at the failure location. When both stress and stress gradient conditions are satisfied at any point in the sample in question, failure is deemed to have occurred. In other words, even if the effective stress is greater than the failure strength, failure does not occur unless the stress gradient condition is also satisfied.

To investigate the material failure value Y, which is different from σ_f, we consider a crack under the first mode of fracture. The stress field very near a crack is expressed as

$$\sigma_e = \frac{K}{\sqrt{s}} \tag{3}$$

in which s is along the crack orientation measured from the crack tip, which is perpendicular to the loading direction. Equation (3) and its derivative are substituted into Equation (2), which, at the onset of failure, results in the following:

$$Y = \frac{K^2}{E} \tag{4}$$

Thus, the material failure value Y is equivalent to the critical energy release rate in fracture mechanics.

3. Ductile Aluminum Alloy with Notches

3.1. Description of Specimens

All test specimens were 140 mm long, 24 mm wide, and 1 mm thick. The grip-to-grip distance was 100 mm, meaning that each specimen was captured over 20 mm on each end. Any notch was introduced at the center of every specimen, as sketched in Figure 1. The size of the circular hole varied from 1 to 18 mm in diameter incrementally. All the holes were located at the center of the specimens. There were six specimens for every size of the hole. All the specimens were tested under tensile loading using INSTRON 5982 (Norwood, MA, USA). In addition, six dog-bone shapes of specimens were also tested, without any notch, to determine the stress–strain curves of the 5000 series aluminum alloy. Strain gauges were attached to the dog-bone shape of specimens to measure both longitudinal and transverse strains. Figure 2 shows the stress–strain curve of the aluminum alloy. The graph shows the very ductile nature of the material with a low tangential modulus for strain hardening.

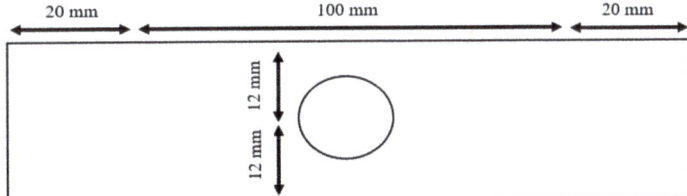

Figure 1. Aluminum specimen with a center hole.

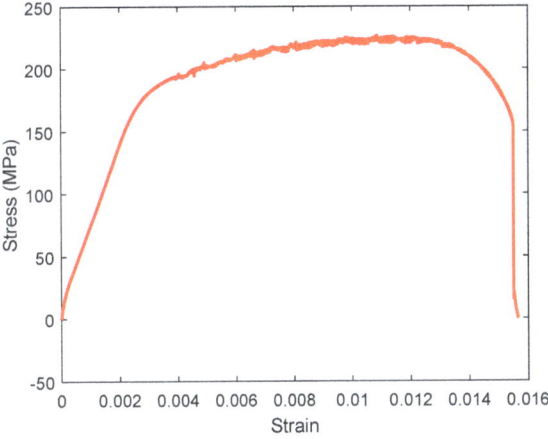

Figure 2. Stress–strain curve of 5000 series aluminum alloy.

The next set of specimens had the same dimensions as before, but a slit was introduced at the center instead of a circular hole, as sketched in Figure 3. The slit length was either 4 mm or 6 mm. The 6 mm slit had four different orientations with respect to the width direction of the specimens. The orientation angles were 0°, 15°, 30° and 45°, respectively. Each type of specimen with a slit was called Lx/y°, where x is the crack length in mm and y is the orientation angle in degrees. Six of every specimen type were prepared for testing.

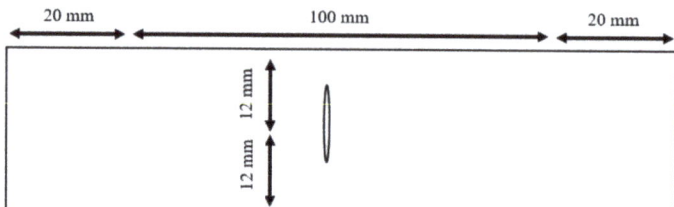

Figure 3. Aluminum specimen with a center slit.

3.2. Results

Tensile tests of the dog-bone-shaped specimens, as well as the specimens with variable diameters of center holes, were conducted to determine their failure loads. Then, the applied failure stresses were computed from the failure loads divided by the cross-sections at the grips, which do not consider the hole. The applied failure stresses are plotted in Figure 4 for different hole sizes. The plot shows that the applied failure stresses decrease almost linearly as a function of the hole diameter, and the standard deviations of the notched specimens were very small as compared to that of the dog-bone-shaped specimens. In addition, the specimens with a hole of a 1 mm diameter had applied failure stresses almost the same as that of the dog-bone specimens, even though the dog-bone-shaped specimens had a larger standard deviation than the perforated specimens. The dog-bone-shaped specimen data were included in the figure if the failure occurred at the mid-section of the specimens. Otherwise, the data were excluded. Figure 4 confirms that a very small hole, as compared to the specimen width, does not have a noticeable effect on the load-carrying capacity, and the load-carrying ability decreases as the hole size further increases.

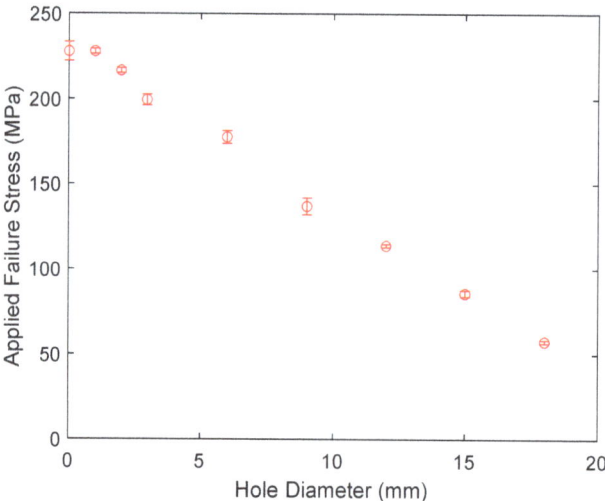

Figure 4. Applied failure stress of aluminum specimens with different hole sizes.

The data in Figure 4 were re-calculated such that the applied failure stress was replaced by the nominal failure stress at the minimum cross-section across each hole. The nominal failure stress is the average stress across the minimum cross-section at the onset of failure. Table 1 shows the results. It is reasonable to state that the nominal failure stress was almost constant, independent of the hole size. This suggests that for this geometry, the stress failure condition dominates the failure of the aluminum test specimen. Therefore, the stress gradient condition was easily satisfied because the failure value from the stress gradient condition is much smaller than that from the stress condition. In order to demonstrate this, the stress gradient was computed at the edge of the hole of different sizes using finite element analysis to investigate the stress gradient condition.

Table 1. Nominal failure stress of aluminum specimens with different sizes of holes.

Hole Size (mm)	0	1	2	3	6	9	12	15	18
Average Failure Stress (MPa)	228	237	236	228	236	219	227	228	230
Standard Deviation (MPa)	5.5	1.6	1.3	3.2	3.7	4.8	0.94	1.8	0.96

Specimens of 100 mm × 24 mm with a center hole were modeled for the plane stress condition using four-node quadrilateral elements in Ansys [38]. The elastic–plastic analysis was conducted using the stress–strain curve obtained from the dog-bone-shaped specimen. Because the tangential modulus during the plastic deformation is quite small compared to the elastic modulus, the stress gradient at the edge of the hole becomes smaller, along with more plastic deformation around the edge of the hole. As a result, the stress gradient with the ductile aluminum alloy becomes much smaller than that of any elastic analysis of brittle materials. That is, the stress gradient of the former was at least an order of magnitude less than that of the latter. Thus, the stress gradient condition was already satisfied at the edge of the holes. This indicates that the failure of the ductile aluminum specimens with a low tangential modulus of plastic deformation was also governed by the stress condition rather than the stress-gradient condition, even though they contain a circular hole.

The aluminum specimens containing slits were also plotted in Figure 5 for the applied failure stress. This stress was calculated by dividing the failure load divided by the cross-sectional area without considering the slit. The results show an increase in the failure stress along with the slit angle. Then, the nominal failure stresses were computed across the minimum cross-section of the specimen. This minimum cross-section is defined in Figure 6 for slits with nonzero orientation angles. The nominal failure stresses were also almost constant for different slit lengths and orientations, as seen in Figure 7. This also indicates that the failure of the specimens with slits was predicted exclusively by the stress failure condition instead of the stress gradient failure condition, as explained for the aluminum specimens with circular holes. In other words, the stress gradients were so small that the failure stress resulting from the stress gradient condition was smaller than the failure stress from the stress condition. Thus, the failure stress from the stress condition is the failure strength of the specimen.

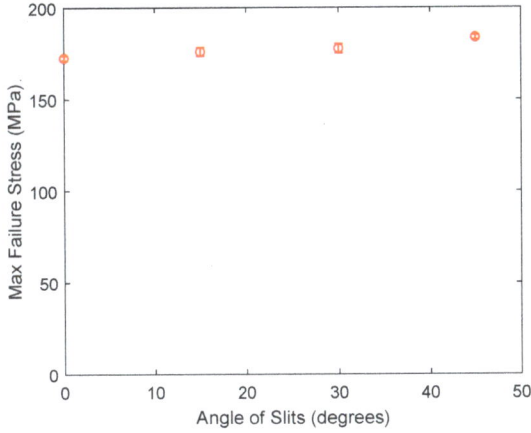

Figure 5. Plot of applied failure stress vs. the slit angle of slit size of 6 mm.

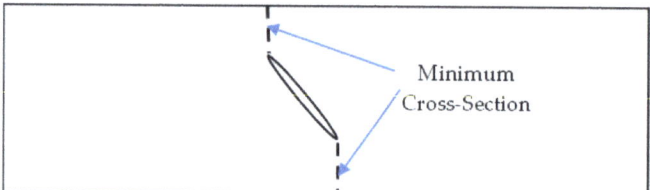

Figure 6. Minimum cross-section for a slit with an orientational angle.

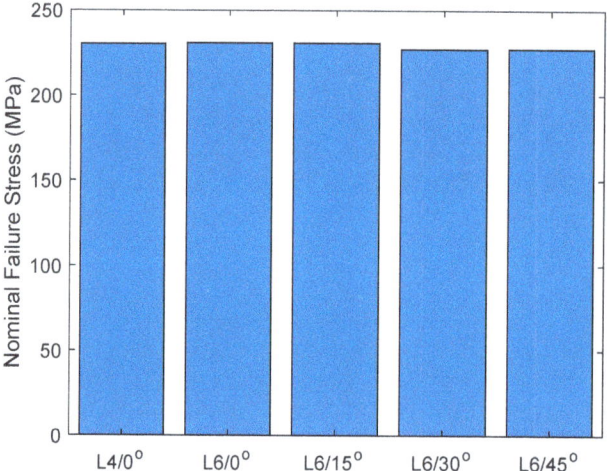

Figure 7. Nominal failure stress of aluminum specimens with different slit lengths and angles.

4. Hardened Cement Pastes with Cracks

4.1. Description of Specimens

The next set of specimens was hardened cement paste with cracks, which were studied experimentally in Ref. [39]. All the specimens are 3-dimensional blocks of L mm × H mm × W mm, where L is the length, H is the height, and W is the width of the specimen. Each specimen has a crack of the depth 'a' in the middle of the length and at the bottom side. The crack is through the width of every specimen, which was tested under the three-point bending setup. Figure 8 shows half of the hardened cement paste model because of symmetry.

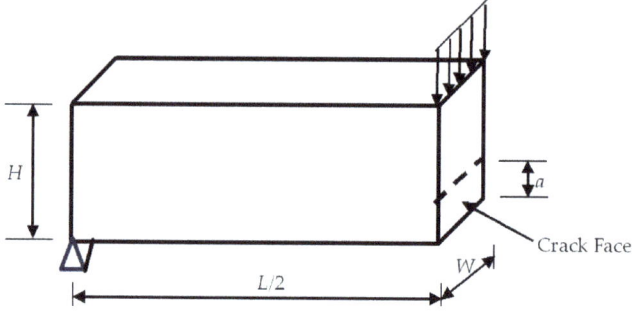

Figure 8. Half of the hardened cement paste model.

All the specimens made of the hardened cement paste had the width $W = 100$ mm, and the length to the depth ratio remained as $L/H = 4$ while the depth was varied with different ratios of the crack to specimen depth a/H. The hardened cement paste behaved in a brittle manner with an elastic modulus of 20.8 GPa.

4.2. Modeling

First, half of the specimen, as sketched in Figure 8, was modeled using 3D solid elements. The mesh was uniform, with the element length around 0.3 mm. The supporting and symmetric boundary conditions were applied to the model, as well as the applied load across the width of the specimen. Figure 9 shows the 3D finite element mesh of the model with boundary and loading conditions. The right face had a symmetric boundary condition except for the crack face at the bottom side.

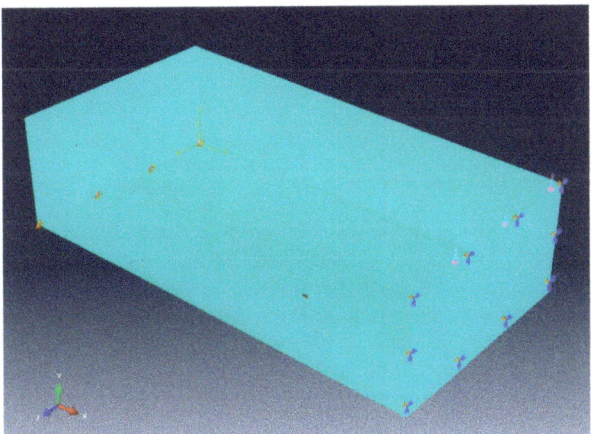

Figure 9. Three-dimensional three-point bending model of hardened cement paste.

After linear elastic analysis, the stress profiles were examined along the crack line across the width of the specimen. The results showed that the stress variation from the edge of the crack to the vertical direction was very close across the specimen width. This suggests that 2D analysis would be acceptable to save computational time. Therefore, 2D analyses of three-point bending were conducted using four-node quadrilateral elements to predict the failure loads of the specimens made of the hardened cement paste. After some mesh sensitivity study, the final mesh was around 10,000 elements.

4.3. Results

The failure loads were predicted for the hardened cement paste specimens with initial cracks from the 2D FEA, as discussed in the previous section. Because of the crack with stress singularity, the stress gradient condition is used for the prediction of the failure loads. In other words, the stress condition is already satisfied at the crack tip.

Since the failure value Y, which is related to the critical energy release rate, is not known for the given material, one of the test results in Ref. [39] was used to extract the value. Then, the same failure value Y was used for the remainder of the specimens to predict the failure loads. The stress intensity factor K was obtained from 2D finite element analyses. Because conventional FEA, in general, does not provide accurate stresses very near the crack tip, a curve fit was conducted using Equation (3) to determine the stress intensity factor K for a given applied load. Figure 10 shows an example of the curve fit to the FEA solution. Then, the failure loads are determined using Equation (4).

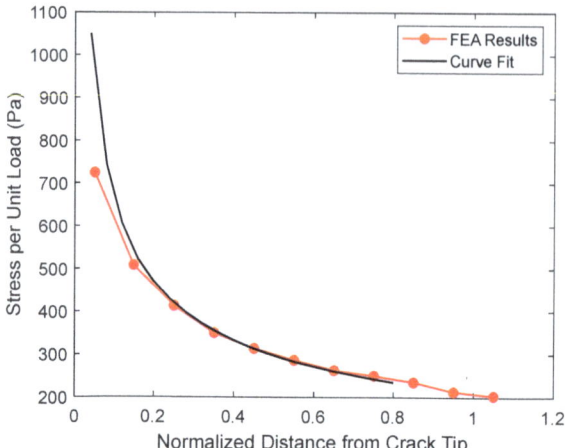

Figure 10. Curve fit of the FEA solution to determine the stress intensity factor.

Figure 11 shows the comparison between the predicted failure loads and the experimentally measured values. Because the specimen with a crack-to-depth ratio of 0.1 and the specimen height of 100 mm was used to determine the failure value Y, both theoretical and experimental values agreed exactly for the specimen in this case. All other specimens show close agreement between the two results except for one specimen with a crack ratio of 0.3 with a specimen height of 100 mm. Because the authors do not have additional information on that test specimen, no further study could be conducted to understand the difference between the two results. Overall, the results confirmed the applicability of the present failure criterion to predict the failure of cracked specimens.

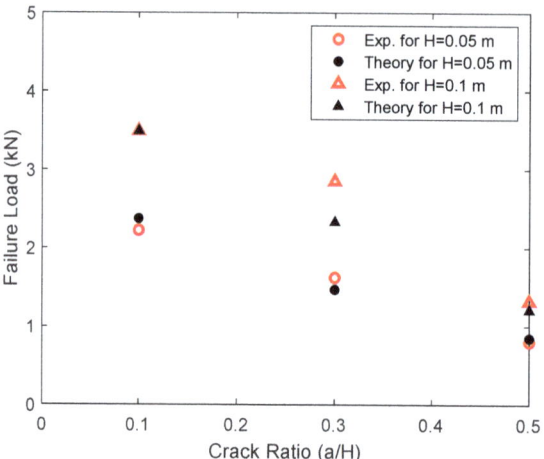

Figure 11. Comparison of failure loads of hardened cement paste specimens with cracks.

5. Three-Dimensional Printed PLA with Holes

5.1. Description of Specimens and Experiments

PLA specimens were printed by a 3D printer using the fused filament fabrication technique. The printing conditions influence the material properties of the 3D-printed PLA specimens. For example, the printing temperatures affected the strength of the printed specimens. In other words, the strength in the printing direction was much greater than that in its transverse direction, such that the 3D printed specimens behaved like an orthotropic material, such as a unidirectional fibrous composite.

The printing conditions for the present PLA specimens are provided in Table 2. First, rectangular shapes of specimens were printed in 0°, 90°, and ±45° with a length of 140 mm, a width of 24 mm, and a thickness of 2 mm. The printing angle is with respect to the length direction of the specimens, which is also the loading direction. This means that the 0° specimens were printed along the loading direction, and those specimens were stronger than other specimens. For uniaxial testing, tabs of 20 mm × 24 mm were attached to both sides of every specimen on both ends. That is, each specimen had four tabs so as not to fail at the grip section during the tests. This made the gauge length of every specimen 100 mm.

Table 2. Printing conditions for PLA specimens.

Print temperature	185 °C
Bed temperature	55 °C
Print speed	45 mm/s
Line thickness	0.2 mm
Line width	0.35 mm

The stiffness and strength were determined from the tensile tests of the rectangular shapes of specimens with tabs. Figure 12 shows the typical stress–strain curves of the PLA specimens printed in different orientations relative to the loading direction. The test results indicate a larger difference in strength than in stiffness. Those graphs provided the material properties of the PLA specimen. To determine Poisson's ratio, strain gauges were also attached to the specimens in both longitudinal and transverse directions. The details are given in Ref. [37].

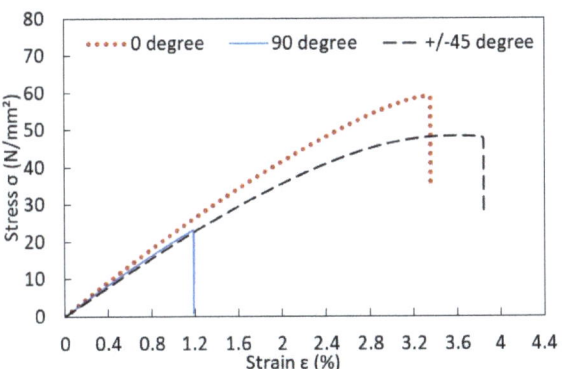

Figure 12. Stress-strain curves of rectangular PLA specimens.

In the previous study [37], rectangular specimens with the same geometry were tested with different sizes of holes and different printing angles relative to the loading direction. Then, the experimental results were compared to the predicted failure loads. The agreement was very good. In this paper, much larger sizes of dog-bone-shaped specimens were printed using PLA. Those specimens were tested to investigate the effect of the hole size relative to the specimen width on the failure load. Figure 13 shows the dog-bone shape of a PLA specimen.

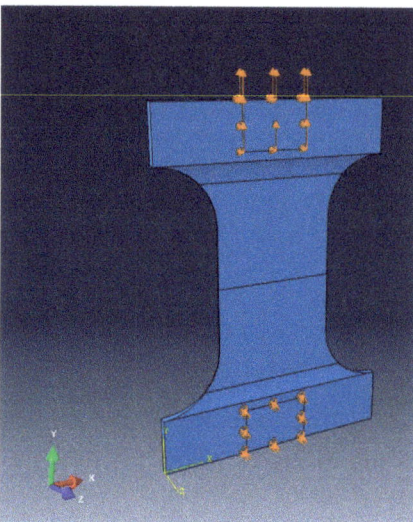

Figure 13. Dog-bone shape PLA specimens.

All the dog-bone shapes of PLA specimens were printed along the loading direction, i.e., at a 0° angle. All specimens had a test section of 80 mm wide and 1 mm thick. The top and bottom portions of the specimen were 3 mm thick, while the thickness gradually decreased to 1 mm in the test section of the specimen. This is to prevent failure around the grip sections of the specimens because the testing equipment could hold only a small portion of the end sections, as sketched in Figure 13. In other words, there was no grip available for the testing, which was wide enough to hold the whole width of the specimen. The hole was drilled at the center of every specimen, and the hole size was 3 mm, 4 mm, 5 mm, or 6 mm, respectively. At the minimum, three specimens were tested for the same size of the hole.

All the dog-bone shapes of PLA specimens were subjected to tensile loading until failure. The maximum forces were obtained as the failure loads from which the applied stresses at failure were computed. Furthermore, a high-speed video was used to capture the locations of the initial failures of the specimens. The video was set to 50,000 frames per second. The observed failure locations were later compared to the predicted failure location.

5.2. Results

The dog-bone specimens were modeled using 2D quadrilateral elements only for a quarter of their geometry because of double symmetries. To emulate the physical test condition, uniform displacements were applied to the FEA model at the grip section of each specimen, as shown in Figure 13. The applied displacement was increased gradually until both stress and stress-gradient failure conditions were satisfied at any material point that would be the location of the initial failure. The analyses were conducted for specimens with different hole sizes.

Figures 14–16 show both the analytical predictions of failure locations for the 3, 4, and 6 mm holes, respectively. Each figure has three graphs. The first graph is the failure strength from the stress failure condition, the second graph is the failure strength from the stress-gradient condition, and the third graph is the induced effective stress from the applied loading. All the graphs were normalized with respect to the failure strength from the stress condition. The failure strength from the stress condition is constant, independent of the location in the specimen. However, failure strength from the stress-gradient condition, as well as the induced equivalent stress, varies across the specimen from the hole.

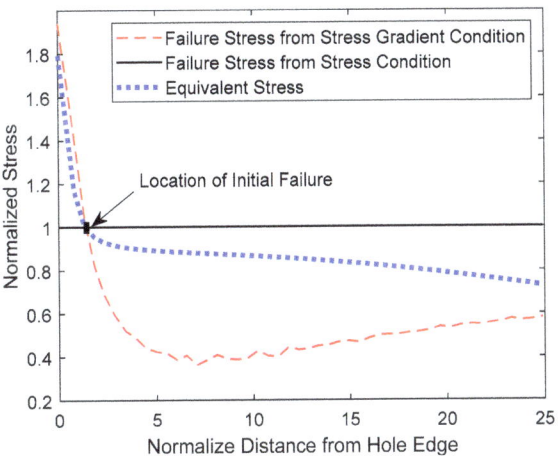

Figure 14. Analytical failure prediction of a PLA dog bone with a 3 mm diameter center hole.

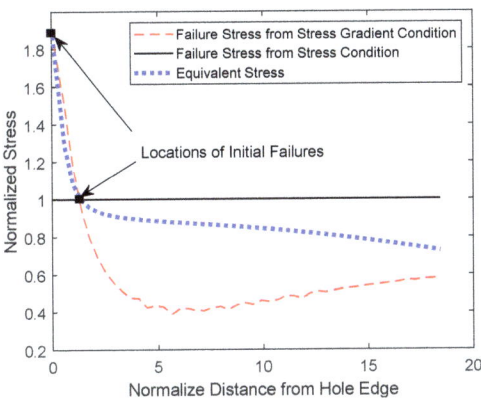

Figure 15. Analytical failure prediction of a PLA dog bone with a 4 mm diameter center hole.

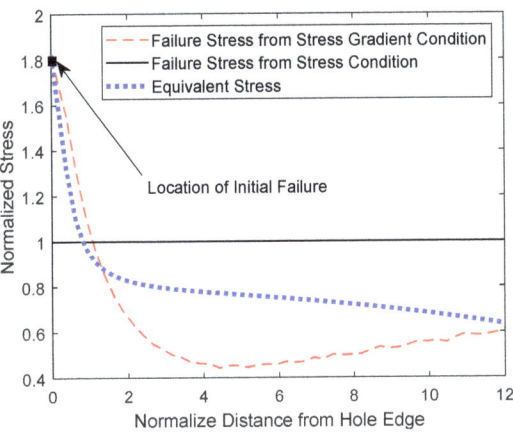

Figure 16. Analytical failure prediction of a PLA dog bone with a 6mm diameter center hole.

The FEA analysis gave the failure strength computed from the stress-gradient condition at every node or element starting from the edge of the hole. The stress-gradient-based failure strength varies at different locations because the stress gradients change from point to point. Those varying failure stresses from the stress-gradient condition are plotted in the figures along the minimum cross-section of the specimen from the edge of each hole.

The equivalent stress is the maximum normal stress along the same minimal sections of the specimens. The equivalent stress increases as the applied displacement to the specimen models increases. For the failure to initiate, the equivalent stress at one location must be equal to or greater than both failure stresses from the stress and stress-gradient failure conditions. Figures 14–16 show the plots of the equivalent stresses that just meet both failure conditions at the initial failure locations for different sizes of the center holes from 3 mm to 6 mm.

Figure 14 is for the 3 mm hole, which shows that the equivalent stress meets both failure strengths from the stress and stress-gradient conditions, respectively, at a distance away from the edge of the hole. Before that failure location, the stress-gradient condition is not satisfied, and after the location, the stress condition is not satisfied. Thus, the initial failure does not occur at the edge of the 3 mm hole but rather some distance away from it. Hence, the stress concentration at the hole edge did not influence the failure, and the failure load was not affected by such a small hole in the specimen.

When the hole size was increased to 4 mm, Figure 15 shows there are two potential failure locations: one at the edge of the hole and the other at a distance away from it. That is, both stress and stress-gradient conditions could be satisfied at both locations almost simultaneously. This suggests that the failure would occur at one site, immediately followed by the other site. On the other hand, failure occurs at the edge of the hole when the hole size grows to 6 mm, as shown in Figure 16.

Experiments were conducted using PLA specimens with different hole sizes. As the tensile load was applied to each specimen until failure, a high-speed video was used to capture the moments of initial failures. Figure 17 shows the video clip of the failure progression of the specimen with a 3 mm hole just after the initial failure. The experiment agrees with the theory showing the failure initiation away from the edge of the 3 mm hole.

Figure 17. Progression of initial failure of the specimen with a 3 mm hole.

For the 4 mm hole size, some specimens showed initial failure starting from the edge of the hole, and others showed it at a distance away from the edge. Figure 18 shows the first failure at the edge of the hole, which was followed immediately at a distance away from the 4 mm diameter hole. On the other hand, the specimen containing a 6 mm central hole showed failure initiation from the edge of the hole, as seen in Figure 19. Because of uncontrollable asymmetry, the initial failure occurred on one side of the hole, and then failure also followed on the other side of the hole. Thus, the experimental results confirmed the theoretical predictions based on both stress and stress-gradient failure criteria.

Figure 18. Progression of initial failure of the specimen with a 4 mm hole.

Figure 19. Initial failure of the specimen with a 6 mm hole.

The applied failure stresses of the PLA specimens at the onset of failure were also compared between the theory and the experimental results. Figure 20 shows this comparison. Both results agreed well with each other.

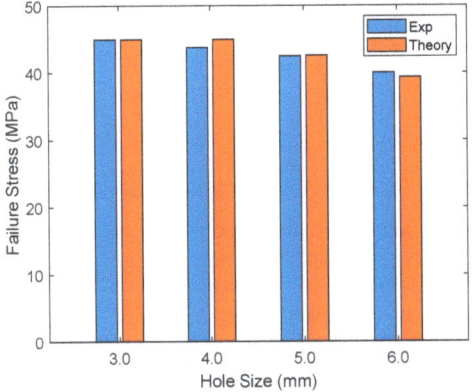

Figure 20. Comparison of theoretical and experimental failure stresses of PLA specimens with different hole sizes.

6. Laminated Glass Fiber Composites with Holes

6.1. Description of Specimens

Test specimens were cut out of a quasi-isotopically laminated glass fiber composite (GFC) plate, which has the following layer angles: $0°/45°/−45°/90°/90°/−45°/45°/0°$. Here, $0°$ denotes that fibers are orientated along the loading direction. The overall dimensions of the GFC specimens were identical to the previously tested aluminum alloy specimens. The GFC specimens had 3 mm, 6 mm, and 9 mm diameter holes at their centers. Three specimens were prepared for the same hole size, and all the GFC specimens were tested under tensile loading until failure.

6.2. Multiscale Failure Modeling

Failure of laminated composite structures was modeled using a multiscale approach, which links the microscale and the macroscale of the composite materials and structures. The microscale indicates the fiber and matrix materials, while the macroscale is the smeared or homogenized composite material. The multiscale approach is sketched in Figure 21, and it consists of two processes bridging the microscale and macroscale. The first process is to transfer the information at the microscale to that at the macroscale. This process, called the upscaling or stiffness process, computes the effective composite material properties from the material properties of the fiber and matrix and their volume fractions. The second process occurs in the opposite direction and transfers information from the macroscale down to the microscale. This is called the downscaling process or the strength process. This second process determines the stresses and strains in the fiber and matrix materials from those at the composite material level. The main reason for this is to apply failure criteria at the fiber and matrix material level. The three different failure modes at the microscale are fiber failure, matrix failure, and fiber/matrix interface failure; for example, interlaminar delamination is described as matrix failure and/or fiber/matrix interface failure.

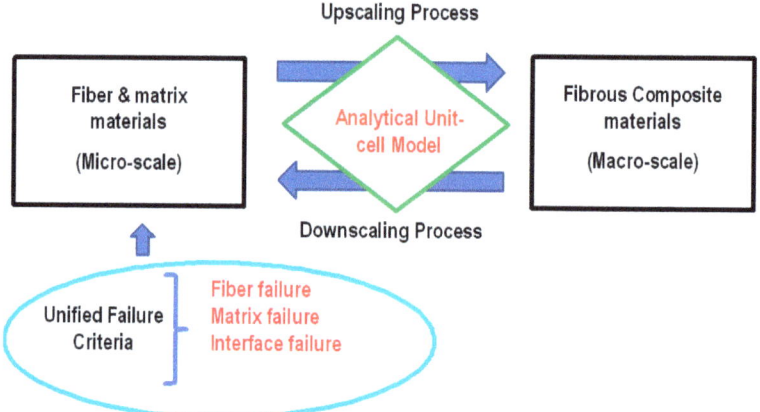

Figure 21. Multiscale approach.

The overall analysis occurs in the following manner.

1. First, the composite material properties are computed from virgin fiber and matrix materials using the upscaling process.
2. The composite material properties are used for the analysis of the given composite structure with an applied loading. Because the structural analysis is complex, FEA is mostly used for the structural analysis, which provides the stresses and strains in the composite structure.
3. Then, the composite level stresses and strains are decomposed into the stresses and strains at the fiber and matrix materials using the downscaling process.
4. The unified failure criteria are applied to the stresses and strains of the fiber and matrix materials.
5. If there is a failure, then the corresponding material properties are degraded based on the specific failure, and the degraded material properties are used for the next upscaling process.
6. The analysis cycle repeats as failure progresses locally or the applied load increases.

Because both upscaling and downscaling processes are used iteratively, the computational cost of the multiscale analysis may be quite high. To overcome this, both upscaling and downscaling processes use analytical solutions without any additional numerical model. The derivations of the analytical solutions are based on a unit cell model as described in Refs. [40,41]. The unit cell consists of subcells. Some of the subcells represent the embedded fibers, and the remaining subcells represent the surrounding matrix material. Stresses and strains were assumed constant with every subcell for mathematical simplicity. Then, stress equilibrium and deformation compatibility are applied to the subcells to derive the equations necessary for the upscaling and downscaling processes. The details of the analytical derivations for the up and downscaling processes are omitted here. In summary, the upscaling process has the following analytical expression:

$$E^c_{ijkl} = f\left(E^f_{ijkl}, E^m_{ijkl}, \nu^f, \nu^m\right) \tag{5}$$

where E_{ijkl} is the material property tensor; ν is the volume fraction; and superscripts 'c', 'f', and 'm' denote the homogenized composite, fiber, and matrix materials, respectively. This equation computes t homogenized composite material properties directly from the fiber and matrix material properties.

The analytical expression used for the downscaling process is expressed as below:

$$\varepsilon^f_{ij} = g_1\left(\varepsilon^c_{ij}\right) \text{ and } \varepsilon^m_{ij} = g_2\left(\varepsilon^c_{ij}\right) \tag{6}$$

in which ε_{ij} is the strain tensor, and the same superscripts as before were used. Once the strains at the fiber and matrix materials are determined from Equation (6), the stresses at the fiber and matrix materials are computed as below:

$$\sigma^{f \text{ or } m}_{ij} = E^{f \text{ or } m}_{ijkl} \varepsilon^{f \text{ or } m}_{ij} \tag{7}$$

where σ_{ij} is the stress tensor.

The failure criteria at the microscale level are given for the fiber breakage/buckling, matrix cracking, and fiber/matrix interface debonding. Fiber failure is the major catastrophic failure of fibrous composites because fibers are the major load-carrying elements. The effective stress for the fiber failure is expressed as

$$\sigma^f_e = \sqrt{\left(\sigma^f_{11}\right)^2 + \left(\frac{E^f_{11}}{G^f_{12}}\right)^2\left[\left(\sigma^f_{12}\right)^2 + \left(\sigma^f_{13}\right)^2\right]} \tag{8}$$

This effective stress is applied to the failure criterion based on the stress and stress gradient as given in Equations (1) and (2). In Equation (8), '1' is the fiber orientation, and '2' is the transverse direction normal to the fiber orientation. In addition, E and G are elastic and shear moduli. All the components in Equation (8) are for the fiber material.

The matrix material used in this study is an isotropic brittle material, so as a result, the maximum normal stress in the matrix material is used for the effective stress. Finally, the effective stress to check the fiber/matrix interface failure is given below:

$$\sigma^{int}_e = \sqrt{\left(\frac{\sigma^m_{12} + \sqrt{\nu^f}\left(\sigma^m_{22} - \sigma^m_{11}\right)}{\tau^{int}_{fail}}\right)^2 + \left\langle\frac{\sigma^m_{22}}{\sigma^{int}_{fail}}\right\rangle^2} \tag{9}$$

where τ^{int}_{fail} and σ^{int}_{fail} are the tangential and normal failure strength of the interface. In addition, <...> in Equation (9) is the Macaulay function. Hence, this function is used to indicate that only the tensile but not compressive normal stress at the fiber/matrix interface contributes to the interface failure.

6.3. Results

The GFC specimens behaved like quasi-brittle material. In order to view the failure surfaces closely, GFC specimens were sputter-coated with approximately 15 nm of Pt/Pd. Because GFC is non-conductive, it must be coated with a conductive metal to achieve high-quality images using scanning electron microscopy. Additionally, sputter coating the GFC samples prevents the GFC from absorbing the energy, which could result in deforming the samples.

Figure 22 shows that the GFC samples, regardless of hole size, had fiber fractures in the 0° layers at the edge of the hole with minimal cross-section. The fiber fracture initiated in the perpendicular direction to the applied loading, and then it propagated at approximately 45°. On the other hand, the ±45° layers showed no fiber fracture but indicated that fiber pull-out had occurred, as shown in Figure 23. The failure of the ±45° layers occurred after the failure of the 0° layer. Thus, the applied failure stresses of the GFC specimens were obtained at the onset of the initial fiber fracture of the 0° layers.

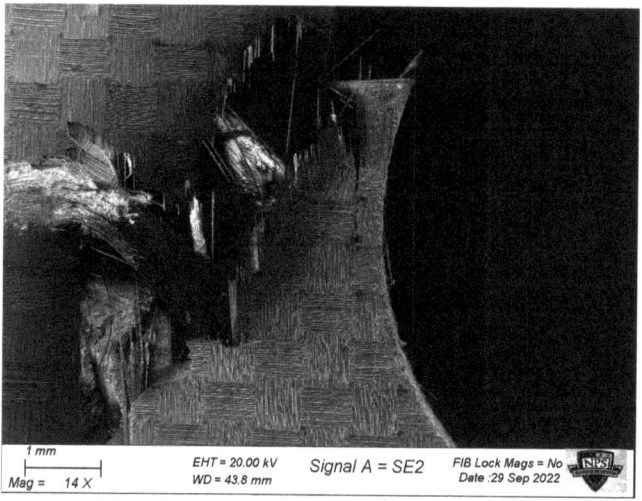

Figure 22. Left side edge of fracture of the GFC specimen with a hole.

Figure 23. Fiber pull-out failure of ±45° fibers of the GFC specimen with a hole.

The multiscale analysis, as described in the previous section, was conducted for the GFC specimens with holes. As the applied load increased incrementally, failure criteria at the fiber and matrix material levels were checked, respectively, using the effective stress and the stress gradient condition. Then, when the 0° layers initiated the fiber fracture at the edge of each hole, it was the onset of the main failure. The applied failure stress was then determined at that applied load.

Figure 24 shows the comparison between the experimental and predicted failure stresses, which were computed from the applied load divided by the specimen cross-section at the grip locations, i.e., the section without holes. Because the fiber failure value Y for the stress-gradient condition was obtained from the 6 mm hole specimens, the theoretical prediction is exactly on top of the mean experimental failure stress. Using the same failure value, failure stresses were predicted for the specimens with a 3 mm or 9 mm hole. As shown in Figure 24, the theoretically predicted failure stresses agreed well with the experimental stresses, which also suggests that the unified failure criterion in association with the multiscale approach is useful for predicting the failure of the quasi-brittle laminated composites.

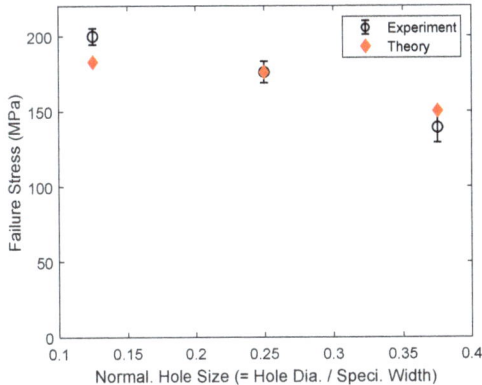

Figure 24. Comparison of experimental and theoretical failure stresses of GRP specimens with center holes.

7. Summary and Conclusions

A new unified failure criterion was proposed to predict the failure loads of structural members. The criterion uses both stress and stress gradient to determine failure. For failure to occur, both stress and stress gradient conditions must be satisfied simultaneously. Various scenarios were examined to validate the new failure criterion. The cases included brittle or quasi-brittle materials and a ductile aluminum alloy. The former materials were isotropic hardened cement pastes, orthotropic PLA, and laminated glass fiber composites. Different geometric features were also tested, including a crack, a long slit, and a circular hole of different sizes. The new failure criterion predicted the failure loads satisfactorily for all the cases examined in this study as compared to their corresponding experimental results. The failure criterion showed that the failure load was ultimately determined by either the stress condition or stress-gradient condition depending on the material or geometric conditions of the tested specimens, even though both conditions were satisfied simultaneously for all the specimens.

Author Contributions: Y.W.K. was responsible for conceiving this study, providing guidance, and writing the manuscript. E.K.M. and S.D. prepared the specimens, conducted tests, performed FEA modeling, and analyzed the results. All authors have read and agreed to the published version of the manuscript.

Funding: This research was funded by the Office of Naval Research for Young Kwon. The funding document numbers are N0001423WX01033.

Institutional Review Board Statement: Not applicable.

Informed Consent Statement: Not applicable.

Data Availability Statement: All the data were provided in this paper in the form of figures and tables.

Acknowledgments: The authors acknowledge the technical support by Chanman Park.

Conflicts of Interest: Authors do not have any conflicts of interest associated with this research.

References

1. Christensen, R.M. *The Theory of Materials Failure*; Oxford University Press: Oxford, UK, 2016.
2. Podgórski, J. General Failure Criterion for Isotropic Media. *J. Eng. Mech.* **1985**, *111*, 188–201. [CrossRef]
3. Soden, P.; Hinton, M.; Kaddour, A. A comparison of the predictive capabilities of current failure theories for composite laminates. *Compos. Sci. Technol.* **1998**, *58*, 1225–1254. [CrossRef]
4. Kaddour, A.; Hinton, M.; Smith, P.; Li, S. The background to the third world-wide failure exercise. *J. Compos. Mater.* **2013**, *47*, 2417–2426. [CrossRef]
5. Tsai, S.W.; Wu, E.M. A General Theory of Strength for Anisotropic Materials. *J. Compos. Mater.* **1971**, *5*, 58–80. [CrossRef]
6. Hashin, Z.; Rotem, A. A Fatigue Failure Criterion for Fiber Reinforced Materials. *J. Compos. Mater.* **1973**, *7*, 448–464. [CrossRef]
7. Sun, C.; Quinn, B.; Tao, J.; Oplinger, D. *Comparative Evaluation of Failure Analysis Methods for Composite Laminates*; DOT/FAA/AR-95/109; School of Aeronautics and Astronautics, Purdue University: West Lafayette, IN, USA, 1996.
8. Daniel, I.M. Failure of composite materials. *Strain* **2007**, *43*, 4–12. [CrossRef]
9. Irwin, G.R. Analysis of stresses and strains near the end of a crack traversing a Plate. *J. Appl. Mech.* **1957**, *79*, 361–364. [CrossRef]
10. Irwin, G.R.; Kies, J.A.; Smith, H.L. Fracture strengths relative to onset and arrest of crack propagation. *Proc. ASTM* **1958**, *58*, 640–660.
11. Griffith, A.A., VI. The phenomena of rupture and flow in solids. *Philos. Trans. R. Soc. London Ser. A Contain. Pap. A Math. Or Phys. Character.* **1921**, *221*, 163–198.
12. Griffith, A.A. The theory of rupture. In Proceedings of the 1st International Congress of Applied Mechanics, Delft, The Netherlands, 22–26 April 1924; pp. 55–63.
13. Anderson, T.L. *Fracture Mechanics: Fundamentals and Applications*, 4th ed.; CRC Press: Boca Raton, FL, USA, 2017.
14. Han, Q.; Wang, Y.; Yin, Y.; Wang, D. Determination of stress intensity factor for mode I fatigue crack based on finite element analysis. *Eng. Fract. Mech.* **2015**, *138*, 118–126. [CrossRef]
15. Sih, G.C. Some basic problems in fracture mechanics and new concepts. *Eng. Fract. Mech.* **1973**, *5*, 365–377. [CrossRef]
16. Atkinson, C.; Craster, R.V. Theoretical aspects of fracture mechanics. *Prog. Aerosp. Sci.* **1995**, *31*, 1–83. [CrossRef]
17. Neuber, H. *Theory of Notch Stresses*; Springer: Berlin, Germany, 1958.
18. Peterson, R.E. Notch Sensitivity. In *Metal Fatigue*; Sines, G., Waisman, J.L., Eds.; McGraw Hill: New York, NY, USA, 1959; pp. 293–306.
19. Whitney, J.M.; Nuismer, R.J. Stress fracture criteria for laminated composites containing stress concentrations. *J. Compos. Mater.* **1974**, *8*, 253–265. [CrossRef]
20. Taylor, D. Geometrical effects in fatigue: A unifying theoretical model. *Int. J. Fatigue* **1999**, *21*, 413–420. [CrossRef]
21. Sapora, A.; Torabi, A.R.; Etesam, S.; Cornetti, P. Finite fracture mechanics crack initiation from a circular hole. *Fatigue Fract. Eng. Mater. Struct.* **2018**, *41*, 1627–1636. [CrossRef]
22. Braun, M.; Müller, A.M.; Milaković, A.-S.; Fricke, W.; Ehlers, S. Requirements for stress gradient-based fatigue assessment of notched structures according to theory of critical distance. *Fatigue Fract. Eng. Mater. Struct.* **2020**, *43*, 1541–1554. [CrossRef]
23. Taylor, D. *The Theory of Critical Distances: A New Perspective in Fracture Mechanics*; Elsevier: London, UK, 2007.
24. Camanho, P.P.; Ercin, G.H.; Catalanotti, G.; Mahdi, S.; Linde, P. A finite fracture mechanics model for the prediction of the open-hole strength of composite laminates. *Compos. Part A* **2012**, *43*, 1219–1225. [CrossRef]
25. Taylor, D.; Cornetti, P.; Pugno, N. The fracture mechanics of finite crack extension. *Eng. Fract. Mech.* **2004**, *72*, 1021–1038. [CrossRef]
26. Dugdale, D.S. Yielding of steel sheets containing slits. *J. Mech. Phys. Solids* **1960**, *8*, 100–104. [CrossRef]
27. Barenblatt, G.I. The mathematical theory of equilibrium cracks in brittle fracture. *Adv. Appl. Mech.* **1962**, *7*, 55–129.
28. Shang, F.; Yan, Y.; Yang, J. Recent advances in cohesive zone modelling of fracture. *Int. J. Aeronaut. Aerosp. Eng.* **2019**, *1*, 19–26. [CrossRef]
29. Elices, M.; Guinea, G.; Gomez, J.; Planas, J. The cohesive zone model: Advantages, limitations and challenges. *Eng. Fract. Mech.* **2002**, *69*, 137–163. [CrossRef]
30. Park, K.; Paulino, G.H. Cohesive zone models: A critical review of traction- separation relationships across fracture surfaces. *Appl. Mech. Rev.* **2011**, *64*, 060802. [CrossRef]
31. Park, K.; Paulino, G.H.; Roesler, J.R. A unified potential-based cohesive model of mixed-mode fracture. *J. Mech. Phys. Solids* **2009**, *57*, 891–908. [CrossRef]
32. Schwalbe, K.-H.; Scheider, I.; Cornec, A. *Guidelines for Applying Cohesive Models to the Damage Behaviour of Engineering Materials and Structures*; Springer Science and Business Media LLC: Dordrecht, The Netherlands, 2012.
33. Needleman, A. Some Issues in Cohesive Surface Modeling. *Procedia IUTAM* **2014**, *10*, 221–246. [CrossRef]
34. Kwon, Y.W. Revisiting Failure of Brittle Materials. *J. Press. Vessel. Technol.* **2021**, *143*, 064503. [CrossRef]

35. Kwon, Y.W.; Diaz-Colon, C.; Defisher, S. Failure Criteria for Brittle Notched Specimens. *J. Press. Vessel. Technol.* **2022**, *144*, 051506. [CrossRef]
36. Kwon, Y.W. Failure Prediction of Notched Composites Using Multiscale Approach. *Polymers* **2022**, *14*, 2481. [CrossRef]
37. Schmeier, G.E.C.; Tröger, C.; Kwon, Y.W.; Sachau, D. Predicting Failure of Additively Manufactured Specimens with Holes. *Materials* **2023**, *16*, 2293. [CrossRef]
38. Ansys. *R1, User Manual*; Ansys Inc.: Canonsburg, PA, USA, 2021.
39. Karihaloo, B.L.; Abdalla, H.M.; Xiao, Q.Z. Size effect in concrete beams. *Eng. Fract. Mech.* **2003**, *70*, 979–993. [CrossRef]
40. Kwon, Y.W.; Park, M.S. Versatile micromechanics model for multiscale analysis of composite structures. *Appl. Compos. Mater.* **2013**, *20*, 673–692. [CrossRef]
41. Kwon, Y.W.; Darcy, J. Failure criteria for fibrous composites based on multiscale modeling. *Multiscale Multidiscip. Model. Exp. Des.* **2018**, *1*, 3–17. [CrossRef]

Disclaimer/Publisher's Note: The statements, opinions and data contained in all publications are solely those of the individual author(s) and contributor(s) and not of MDPI and/or the editor(s). MDPI and/or the editor(s) disclaim responsibility for any injury to people or property resulting from any ideas, methods, instructions or products referred to in the content.

Article

Numerical Simulations of the Low-Velocity Impact Response of Semicylindrical Woven Composite Shells

Luis M. Ferreira [1,2], Carlos A. C. P. Coelho [3] and Paulo N. B. Reis [4,*]

1. Grupo de Elasticidad y Resistencia de Materiales, Escuela Técnica Superior de Ingeniería, Universidad de Sevilla, Camino Descubrimientos, 41092 Sevilla, Spain; lmarques@us.es
2. Escuela Politécnica Superior, Universidad de Sevilla, C/Virgen de África, 7, 41011 Sevilla, Spain
3. Unidade Departamental de Engenharias, Escola Superior de Tecnologia de Abrantes, Instituto Politécnico de Tomar, Rua 17 de Agosto de 1808, 2200-370 Abrantes, Portugal; cccampos@ipt.pt
4. University of Coimbra, CEMMPRE, ARISE, Department of Mechanical Engineering, 3030-194 Coimbra, Portugal
* Correspondence: paulo.reis@dem.uc.pt

Abstract: This paper presents an efficient and reliable approach to study the low-velocity impact response of woven composite shells using 3D finite element models that account for the physical intralaminar and interlaminar progressive damage. The authors' previous work on the experimental assessment of the effect of thickness on the impact response of semicylindrical composite laminated shells served as the basis for this paper. Therefore, the finite element models were put to the test in comparison to the experimental findings. A good agreement was obtained between the numerical predictions and experimental data for the load and energy histories as well as for the maximum impact load, maximum displacement, and contact time. The use of the mass-scaling technique was successfully implemented, reducing considerably the computing cost of the solutions. The maximum load, maximum displacement, and contact time are negligibly affected by the choice of finite element mesh discretization. However, it has an impact on the initiation and progression of interlaminar damage. Therefore, to accurately compute delamination, its correct definition is of upmost importance. The validation of these finite element models opens the possibility for further numerical studies on of woven composite shells and enables shortening the time and expenses associated with the experimental testing.

Keywords: low-velocity impact; finite element method (FEM); woven-fabric composites

1. Introduction

Due to its distinctive combination of high strength, low weight, and exceptional fatigue resistance, composite materials have grown in popularity. Nevertheless, low-velocity impacts that may happen during handling, transit, maintenance, and service might harm them. These collisions may result in localized structural damage, which may diminish the composite material's strength, stiffness, and durability. To ensure the dependability and safety of composite structures, it is crucial to comprehend the behavior of composite materials under low-velocity impacts.

In a low-velocity impact event, composite materials undergo various stages of damage. Firstly, when the impactor contacts with the material, a sudden increase in stress and strains is experienced in a very localized region. Secondly, microcracks start to develop in the material matrix, which then propagate through the material and spread to the adjacent laminas and neighboring interface regions. At this stage, the damage progression causes the separation of the adjacent laminas of the composite material, which is known as delamination. This damage mechanism, alongside with matrix cracking, fiber failure and perforation, can significantly reduce the strength and stiffness of the material. The extent and severity of damage depends on the properties of the composite material, the

impactor's properties, and the impact energy. The development of reliable models capable of predicting these damage stages is of highest importance for the analysis and design of composite structures that are resistant to low-velocity impacts.

In this context, the behavior of composite structures has been extensively studied using finite element (FE) models in a variety of industries, including the marine, automotive, and aerospace sectors [1–5]. The impact response of composite flat plates has been extensively studied using numerical models to examine the impacts of various parameters such as material characteristics, impact energy, and geometry. Yet, there are relatively few numerical studies that analyze the impact dynamics on cylindrical shells, particularly when it comes to composites made of woven fabric. The most relevant numerical studies that were conducted on cylindrical shells include the work developed by Kim et al. [6] in which they observed that the contact force increases with the curvature of shell-shaped composite laminates, while the deflection and contact time decrease. It was also observed that the impactor's velocity has a greater influence on the contact force than the impactor's mass, which is similar to the impact response of composite plates [6–8]. This is justified by the fact that the kinetic energy of the impactor increases linearly with the increase in mass and quadratically with velocity. The contact force and deflection histories for composite laminated cylindrical shells with convex and concave shapes were analyzed by Choi [9]. The author found the same contact force and central deflection histories for both shapes. Kistler and Wass [10] performed a numerical study on unidirectional (UD) laminated cylindrical shells and identified scaling relationships between impact energy, momentum, mass, and velocity, while Zhao and Cho [11] investigated the impact-induced damage initiation and propagation of UD composite shells and found that the damage propagates differently from composite flat plates. Another study was performed by Kumar et al. [12] to study the impact response and impact-induced damages of cylindrical UD composite laminate shells using a 3D finite element formulation. Recently, Khalili et al. [13] used FE analysis to investigate the impact response of UD composite laminate plates and shells structures under low-velocity impact loads and optimize the procedure for future work. Albayrak et al. [14] conducted an experimental and numerical investigation to study the geometrical effect on low-velocity impact behavior for curved composites with a rubber interlayer. They found that the curved surface geometry affects the absorbed energy and that increasing the width of the laminate while keeping the height constant results in higher impact energy absorption.

Overall, these studies provide valuable insights into the behavior of composite laminate structures under low-velocity impacts and can inform about the design and optimization of such structures for various applications. Nevertheless, none of these numerical studies was dedicated to the development of FE models capable of analyzing the low-velocity impact response of semicylindrical woven fabric composite shells. Therefore, the main goal of this paper is to develop reliable and efficient FE models capable of predicting the impact behavior of these materials. For this purpose, constitutive modes that consider the intralaminar and interlaminar progressive damage were implemented in the explicit finite element approach using ABAQUS/Explicit [15]. The validation of the FE models was carried out using the authors' previous work on the experimental assessment of the effect of thickness on the multi-impact response of semicylindrical composite laminated shells [16]. To facilitate the understanding, the nomenclature used in this paper is listed in the Nomenclature.

2. Material and Experimental Procedure

Composite semicylindrical shells were produced using a matrix based on an AROPOL FS 1962 polyester resin and a MEKP-50 hardener (both supplied by SF Composites, Mauguio, France). A bi-directional E-glass woven fabric (taffeta with 210 g/m^2) was used as reinforcement, and the composite was produced by hand lay-up with 9 woven fabric layers (corresponding to 1.6 mm of final thickness). In order to ensure a constant fiber volume fraction and uniform thickness, as well as to eliminate any air bubbles, the laminates were placed inside a vacuum bag immediately after impregnation. The manufacturing process culminated with curing at 40 °C for 24 h. More details about the materials and manufacturing process can be found in [16–18]. Figure 1 shows the specimens' dimensions and the schematic view of the test conditions.

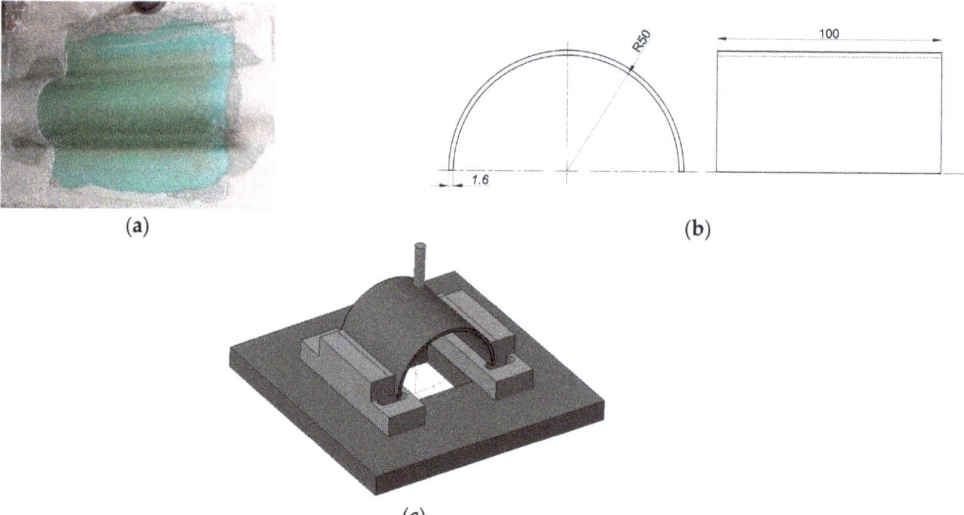

Figure 1. (**a**) Manufacturing process; (**b**) Geometry and dimensions of the specimens (in mm); (**c**) Schematic view of the test conditions.

Finally, low-velocity impact tests were performed on a drop weight testing machine IMATEK-IM10 (Old Knebworth, UK), which is described in detail in [19]. These tests were carried out according to ASTM D7136 standard, at room temperature, and using an impactor diameter of 10 mm with a mass of 2.826 kg. The energy of 5 J was used to promote visible damage but without full perforation. More details about these tests can be found in [16].

3. Damage Models

Two constitutive models were used in this study to simulate the material damage caused by low-velocity impact loads in semicylindrical composite laminate shells: (i) a continuum damage model (CDM) at the lamina level to account for intralaminar damage; and (ii) a surface-based cohesive damage model (S-BCM) at the lamina interface to account for interlaminar damage.

The built-in constitutive model for fabric-reinforced composites available in Abaqus/Explicit [15], developed by Johnson [20] and based on Ladeveze and Ledantec work [21], was used to evaluate the complex damage progression at the intralaminar level. When the user-defined material is named with the string "ABQ PLY FABRIC", this model can be used as a built-in VUMAT user subroutine [22].

The maximum stress criterion determines the damage initiation, and the fracture energies serve as the basis for the damage evolution model, which regulates the decline in stiffness. In this way, the following damage activation functions, F_α and F_{12}, are used to compute the elastic domain at any given time,

$$F_\alpha = \frac{\tilde{\sigma}_\alpha}{X_\alpha} - r_\alpha \leq 0 \text{ with } \alpha = 1\pm, 2\pm \quad (1)$$

$$F_{12} = \frac{\tilde{\sigma}_{12}}{S_{12}} - r_{12} \leq 0 \quad (2)$$

where $\tilde{\sigma}_\alpha$ and $\tilde{\sigma}_{12}$ are the effective normal and shear stresses, respectively, X_α is the tensile/compressive strength, S_{12} is the shear strength, and r_α and r_{12} are the corresponding damage thresholds, which are initially set to 1. Once damage is predicted, the elastic–stress–strain relations are given by,

$$\varepsilon = \begin{bmatrix} \frac{1}{(1-d_1)E_1} & -\frac{v_{12}}{E_1} & 0 \\ -\frac{v_{12}}{E_2} & \frac{1}{(1-d_2)E_2} & 0 \\ 0 & 0 & \frac{1}{(1-d_{12})2G_{12}} \end{bmatrix} \sigma \quad (3)$$

where d_1 and d_2 are the damage variables associated with fiber fracture along directions 1 and 2, respectively, and d_{12} is the damage variable associated with the matrix microcracking due to shear deformation. These variables are determined with Equations (4) and (5),

$$d_\alpha = 1 - \frac{1}{r_\alpha} e^{-A_\alpha(r_\alpha - 1)} \text{ with } A_\alpha = \frac{2g_0^\alpha L_e}{G_f^\alpha - g_0^{\alpha,1,2} L_e} \text{ and } g_0^\alpha = \frac{X_\alpha^2}{2E_\alpha} \quad (4)$$

$$d_{12} = min[\alpha_{12} \ln(r_{12}), d_{12}^{max}] \quad (5)$$

where r_α and r_{12} stand for the damage thresholds for axial and shear loads, respectively, G_f^α stands for the fracture energies, g_0^α to the elastic energy density, L_e stands for the characteristic length of the element, and α_{12} stands for the shear damage parameter.

The non-linear behavior of the matrix, due to microcracking, dominates the shear damage response at the intralaminar level and includes both stiffness reduction and plasticity. The latter is defined with the following yield and hardening functions; see Equations (6) and (7), respectively.

$$F_{pl} = |\tilde{\sigma}_{12}| - \tilde{\sigma}_0\left(\bar{\varepsilon}^{pl}\right) \leq 0 \quad (6)$$

$$\tilde{\sigma}_0\left(\bar{\varepsilon}^{pl}\right) = \tilde{\sigma}_{y0} + C\left(\bar{\varepsilon}^{pl}\right)^p \quad (7)$$

where $\tilde{\sigma}_{y0}$ and $\tilde{\sigma}_0$ are the initial effective shear stress and shear yield stress, $\bar{\varepsilon}^{pl}$ is the plastic strain due to shear deformation, and C and the superscript p correspond to the coefficient and power term in the hardening function, respectively.

Using the two-step homogenization methodology described by Liu et al. [23], the stiffness properties of the woven fabric composite laminas were estimated from the constituents' properties of the tested specimens in order to validate the numerical model based on the experimental evidence presented in [16]. The values of the remaining intralaminar material properties were taken from the literature. In this way, the fracture toughness values, the damage evolution parameters, α_{12} and d_{12}^{max}, and the shear plasticity parameters, $\tilde{\sigma}_{y0}$, C and p, were taken from impact studies on E-glass laminates [24–26], whereas the strength properties were taken from impact studies on woven E-glass/polyester composite laminates [27–29]. The intralaminar material parameters needed to specify the material model in the VUMAT subroutine are shown in Table 1. It is noteworthy to mention that a preliminary parametric study was performed using a coarser FE mesh to assess how a reasonable variation of the shear plasticity parameters could impact the results. It was found that they have a negligible effect on the numerical solutions. Regarding the strength properties, the data found in the literature vary substantially, depending on the manufacturing process, type e-glass fiber, volume fraction, etc. Taking this fact into consideration, the values employed are based on averaged values and again, a preliminary parametric study was performed with a coarser FE mesh to assess which values best fit the experimental evidence. Finally, to account for the interlaminar damage, the bond between the laminas of the composite laminate was modeled using cohesive surfaces. This approach is primarily intended for negligible small interface thicknesses and offers very similar capabilities to cohesive elements. The cohesive behavior is defined as a surface interaction property, and identically to the cohesive elements, it is governed by a traction–separation constitutive model. The properties employed in the surfaced-based cohesive model shown in Table 2 were extracted from the literature [30–38].

Table 1. Intralaminar properties.

Property	Symbol	Units	Value
Density	ρ	kg/m^3	1900
Stiffness properties	$E_{1+} = E_{1-}$	GPa	21.9
	$E_{2+} = E_{2-}$	GPa	21.9
	E_3	GPa	8.6
	G_{12}	GPa	3.4
	G_{13}	GPa	2.4
	ν_{12}	-	0.14
Strength properties	$X_{1+} = X_{2+}$	MPa	250
	$X_{1-} = X_{2-}$	MPa	200
	S	MPa	40
Fracture toughness	G_f^α	N/mm	4500
Shear plasticity	d_{12}^{max}	-	1
	$\tilde{\sigma}_{y0}$	MPa	25
	C	-	800
	p	-	0.552

Table 2. Interlaminar properties.

Property	Symbol	Units	Value
Stiffness properties	$k_n = k_s = k_t$	N/mm^3	10^6
Strength properties	τ_n^0	MPa	15
	$\tau_s^0 = \tau_t^0$	MPa	30
Fracture toughness	G_{Ic}	N/mm	0.3
	$G_{IIc} = G_{IIIc}$	N/mm	0.6
	η	-	1.45

The cohesive stiffness in this study is set at 10^6 N/mm^3, as suggested by Camanho et al. [30]. In addition, it is considered that its value is the same for all directions, that is, $k_n = k_s = k_t$, as used in [31–33] with satisfactory results. It is noteworthy to mention that considering high values for the cohesive stiffness potentially results in convergence problems. On the other hand, the use of low values may affect the global stiffness and thus compromise the validation of the FE model [32]. A value of $\eta = 1.45$ was considered for the interaction parameter in the definition of the cohesive model [34–38]. Identically to the intralaminar properties, there is a wide range in the data for the interlaminar strength parameters and fracture toughness. Given this information, the values used are averages, and preliminary parametric analysis using a coarser FE mesh was completed to determine which values the best-matched experimental data.

4. Finite Element Model

The FE model was created using the ABAQUS/Explicit FE code [15] taking into account the dimensions of the nine-layer laminate specimens evaluated in [16]. The tested specimens had a semicircular internal radius of 50 mm, a length of 100 mm and an average thickness of 1.6 mm (Figure 2).

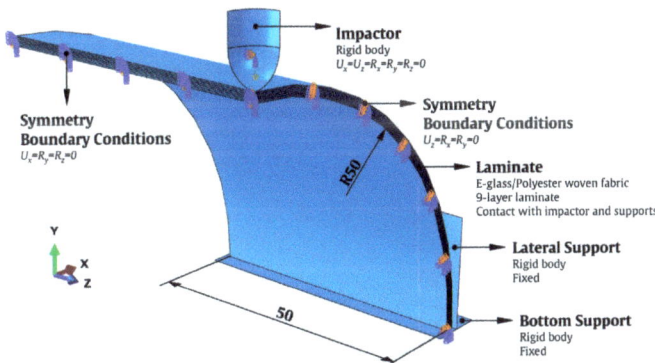

Figure 2. FE model with geometric parameters and boundary conditions.

To replicate the experiments, two fixed rigid body supports (a lateral and a bottom support) were included in the FE model. Only one-fourth of the semicylindrical composite laminate was generated, taking use of the geometric symmetry of the model to reduce the computing cost of the numerical simulations. The yz-plane face $(U_x = R_y = R_z = 0)$ and one of the xy-plane faces $(U_z = R_x = R_y = 0)$ were therefore added to the symmetry boundary conditions. The impactor was modeled with a lumped mass fixed on a reference point at its center of mass equivalent to the experiments and with a hemispherical head with a diameter of 10 mm. Only the displacements in the y-direction were permitted $(U_x = U_z = 0)$, and all the rotations of the impactor were constrained $(R_{x,y,z} = 0)$.

Each lamina was discretized with SC8R continuum shell elements (eight-node hexahedron) with reduced integration and stiffness hourglass formulation. The orientations of the materials along the semicircular cross-section were taken into account when defining the local coordinate system of the laminas. R3D4 discrete rigid elements were used to model the impactor. The lamina was modeled with cohesive surfaces; thus, no element specification was required.

The surface-to-surface contacts between the composite laminate, the metal impactor, and the metal supports were simulated using the penalty enforcement contact method from Abaqus/Explicit [15]. In the interface of the composite laminas, which experience friction after being entirely delaminated, this contact formulation was also defined. The friction coefficient values, μ, used in this work for fully damaged interfaces and metal–composite contacts were taken from [39,40]. A value of $\mu = 0.3$ was specified for the contact between the metal hemispherical head of the impactor and the upper surface of the composite laminate, and a value of $\mu = 0.7$ was specified for the contact between the metal surfaces of the supports and the composite laminate surfaces. For the interfaces, a value of $\mu = 0.5$ was considered.

5. Numerical Results

Several numerical simulations were performed to determine the influence of the FE mesh discretization and mass scaling on the efficiency and reliability of the FE model. To be able to analyze how these parameters affect the numerical predictions, the most relevant load and energy histories are presented, as well as the maximum force, displacement, and contact time.

The use of cohesive surfaces implies that its elements' characteristic length matches the characteristic length of the continuum shell elements defined for the laminas. Given that the interlaminar damage surface-based cohesive damage model implementation yields mesh-dependent results, the FE mesh discretization of the laminas needs to be defined based on the characteristic length of the cohesive surface elements. In other words, the FE mesh size of the interface defines the size of the FE mesh employed in the whole model.

To find a balance between the computational cost of the solution and the accurate computation of the fracture toughness in the interlaminar damage model, which results in delamination, a parametric study was conducted to optimize the characteristic length of the elements of the FE mesh. Based on the work of Hilleborg et al. [41], Turon et al. [31] purposed using Equations (8) and (9) to calculate the characteristic length of the element in the direction of the crack propagation for fracture modes I, II, and III in orthotropic composite materials,

$$l_{e,I} = \frac{ME_3 G_{Ic}}{N_e \left(\tau_n^0\right)^2} \quad (8)$$

$$l_{e,II} = l_{e,III} = \frac{MG_{13} G_{IIc}}{N_e \left(\tau_s^0\right)^2} \quad (9)$$

where M is a parameter that depends on the cohesive zone model, and N_e is the number of elements in the cohesive zone. The lowest value derived from the equations is $l_{e,II} = 0.31$ mm, with the assumptions that $M = 1$, as suggested in [23,33,35], $N_e = 5$ to properly establish the cohesive zone [42–44], and the baseline properties of the laminas and of the cohesive zone. Consequently, the baseline FE mesh was generated with $l_e = 0.3$ mm, which corresponds to an aspect ratio of 1.6. Notice that this value is in good agreement with that used by Lopes et al. in [35].

This baseline FE model contains approximately half a million elements and one million nodes. Therefore, to shorten the computing time for the solutions, a semi-automatic mass scaling was uniformly applied to the entire model with a target time increment of 1×10^{-7}. Notice that mass scaling artificially increases the mass of the structure to reduce the frequency of the dynamic response and allows the time step of the simulation to be increased. In this study, a mass increase of 1.8% was obtained for the baseline FE model, resulting in a reduction in the computational cost of the solutions of about 51%. To assess the impact of mass scaling on the load and energy history curves, the results obtained with the baseline FE model are shown in Figure 3 (force–time), Figure 4 (energy–time) and Figure 5 (force–displacement).

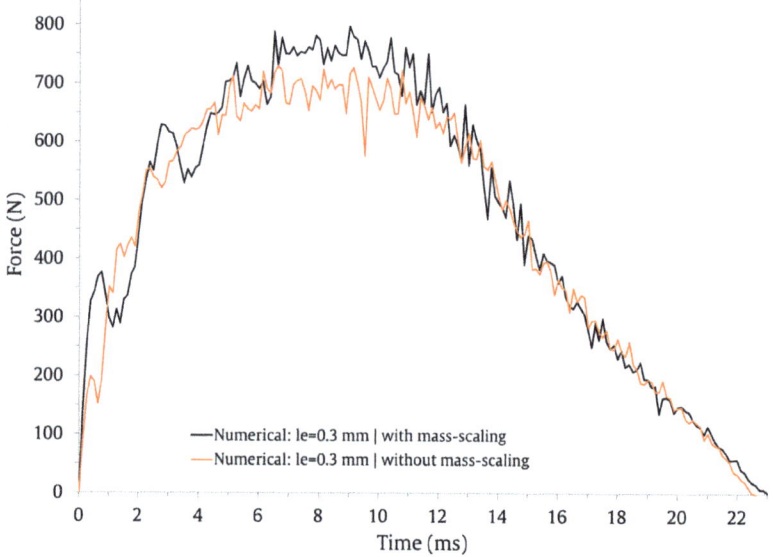

Figure 3. Effect of mass scaling on the force–time impact curves.

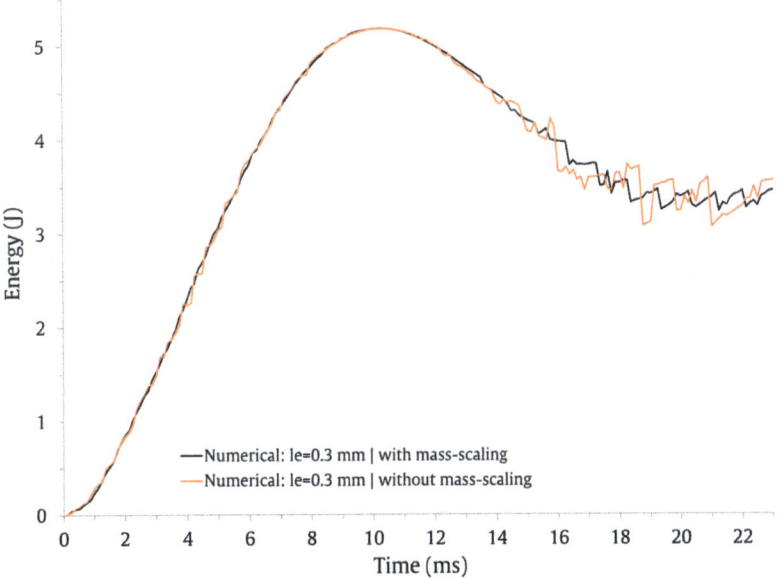

Figure 4. Effect of mass scaling on the energy–time impact curves.

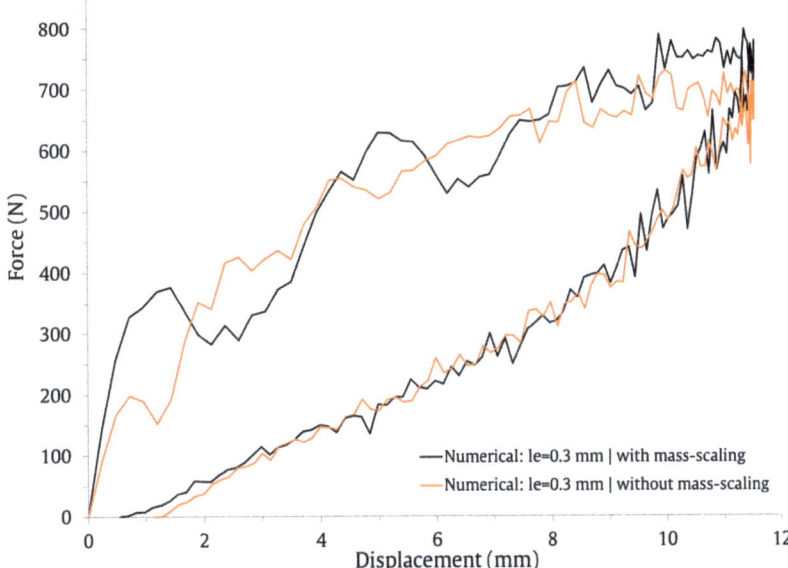

Figure 5. Effect of mass scaling on the force–displacement impact curves.

It is possible to observe that these curves include oscillations brought on by the elastic wave and produced by the models' vibrations [45,46]. The numerical predictions for the numerical maximum load, maximum displacement, and contact time are compared in Table 3.

Table 3. Effect of mass scaling on the numerical predictions of maximum load, maximum displacement, and contact time.

	Max. Load (N)	Dif. [1] (%)	Max. Displacement (mm)	Dif. [1] (%)	Contact Time (ms)	Dif. [1] (%)
With mass scaling	797	-	11.5	-	23	-
Without mass scaling	730	8.8	11.5	0	22.6	1.8

[1] Dif. = Difference.

It can be observed that the impact response of the curves is similar with and without mass scaling. Its application has a negligible effect on the contact time and no impact on the maximum displacement. Only for the maximum force do the numerical predictions differ by 8.8%; its value is higher when mass scaling is employed. Overall, the results indicate that the defined semi-automatic mass-scaling parameters have an acceptable impact on the numerical predictions and significantly lower the computational cost of the solutions.

To determine if a coarser FE mesh discretization would produce a good trade-off between the accuracy of the solution and computing cost, a parametric study was performed using increasingly coarser FE meshes, ranging from $l_e = 0.3$ mm to $l_e = 2$ mm. The FE models employed to analyze the effect of the FE mesh discretization are shown in Figure 6.

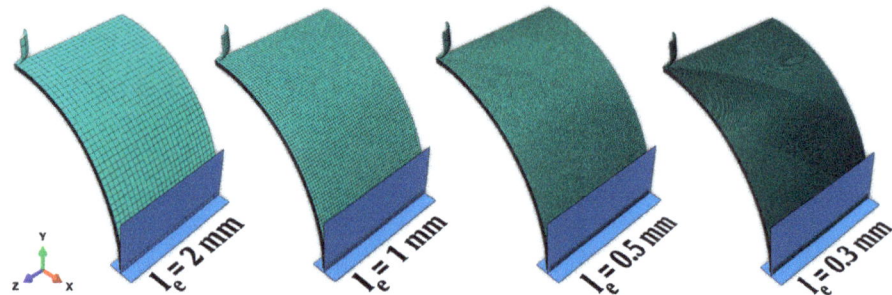

Figure 6. FE models used to study the effect of the FE mesh discretization.

The numerically predicted force–time and energy–time results are shown in Figures 7 and 8, respectively, and the force–displacement results are shown in Figure 9. The results show that the use of coarser FE meshes ($l_e = 1$ mm and $l_e = 2$ mm) induces higher oscillations on the force–time and force–displacement curves. Nonetheless, the maximum force, maximum displacement, and contact time values are barely affected. This can be observed in Table 4, where the values and percentage difference between $l_e = 0.3$ mm and the remaining element lengths are presented.

Therefore, it is clear that the FE mesh discretization, within the studied range, has a negligible impact on the load history maximum values. However, it is important to assess its effect on the interlaminar damage predictions. For this purpose, the output identifier CSQUADSCRT was used to measure the damage initiation in the cohesive surfaces. This variable indicates if the quadratic contact stress damage initiation criterion presented in Equation (9) has been satisfied. When its value reaches 1, damage in the cohesive surface is predicted to initiate. The scalar stiffness degradation for cohesive surfaces, output identifier CSDMG, was used to measure delamination after damage initiation. When it reaches the value of 1, the interface can be considered as fully delaminated (complete debonding).

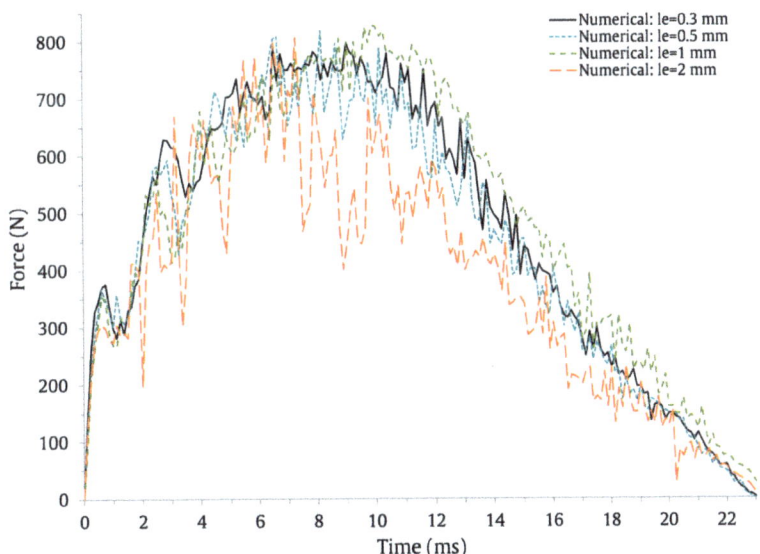

Figure 7. Effect of the FE mesh discretization on the low-velocity impact response of the force–time impact curves.

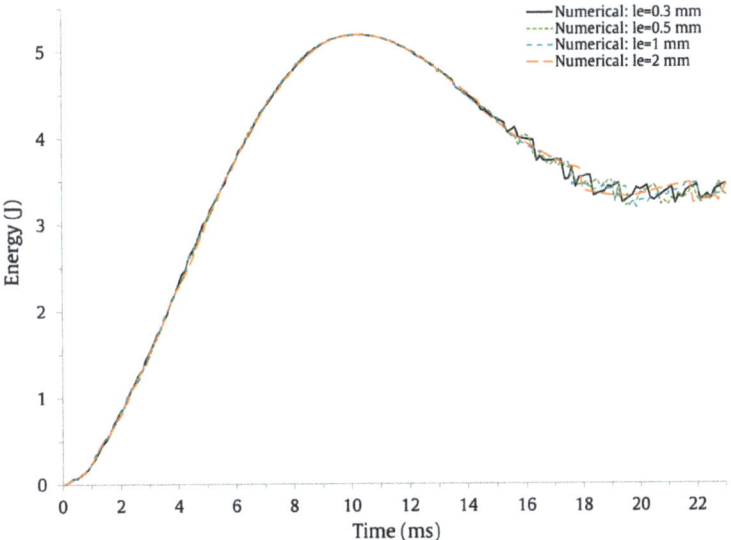

Figure 8. Effect of the FE mesh discretization on the low-velocity impact response of the energy–time impact curves.

The effect of the FE mesh discretization and mass scaling on the delamination initiation and progression is shown in Figure 10. The results are expressed in terms of the percentage of nodes of the 3D FE mesh for which the CSQUADSCRT is equal to 1 and for those where the CSDMG is higher than 0.6. It can be appreciated that the percentage of nodes where interlaminar damage initiation is predicted decreases for finer FE mesh discretization. If mass scaling is employed, this behavior will be especially obvious. The results indicate different behavior for the fraction of delaminated nodes with and without mass scaling. With mass scaling, the percentage of delaminated nodes slightly increases with the FE mesh refinement but reduces without it. However, for the baseline FE mesh discretization, that is, $l_e = 0.3$ mm, the percentage of delaminated nodes is comparable: 2.8% and 3.22% with mass scaling and without mass scaling, respectively.

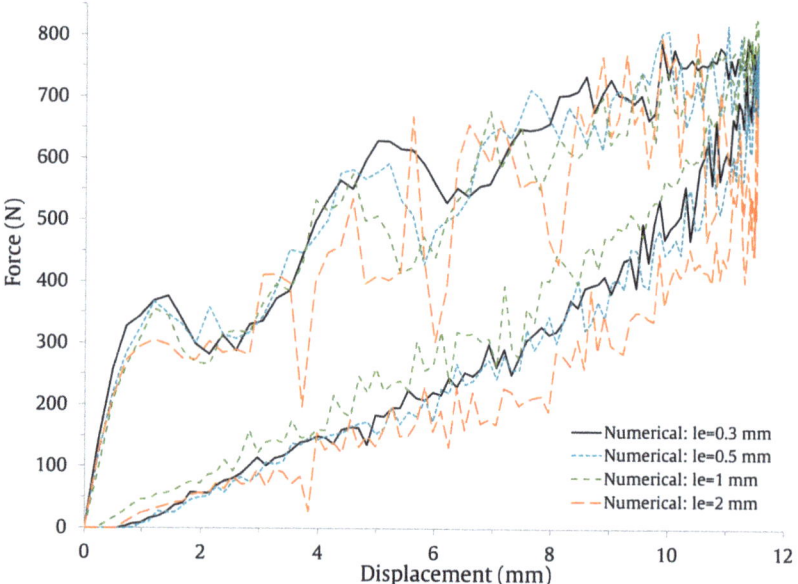

Figure 9. Effect of the FE mesh size on the low-velocity impact response of the force–displacement impact curves.

Table 4. Effect of FE mesh discretization on the numerical predictions of maximum load, displacement, and contact time.

Mesh Size (mm)	Max. Load (N)	Difference (%)	Max. Displacement (mm)	Difference (%)	Contact Time (ms)	Difference (%)
0.3	797	-	11.5	-	23	-
0.5	818	2.6	11.5	0	23	0
1	828	3.8	11.5	0	23.5	2.2
2	806	1.1	11.5	0	23.1	0.4

Data suggest that the choice of the FE mesh size does not have a significant impact on the maximum load, maximum displacement, or contact time. However, it affects the interlaminar damage initiation and progression. Consequently, it is recommended to apply the equations suggested by Turon et al. [31] to compute the FE mesh size in order to assure accurate numerical predictions of delamination. The results indicate that the defined semi-automatic mass scaling parameters have no significant impact on the numerical predictions. In addition, taking into consideration the fact that using mass scaling reduces the computational cost of the solutions by around 51%, its application is recommended.

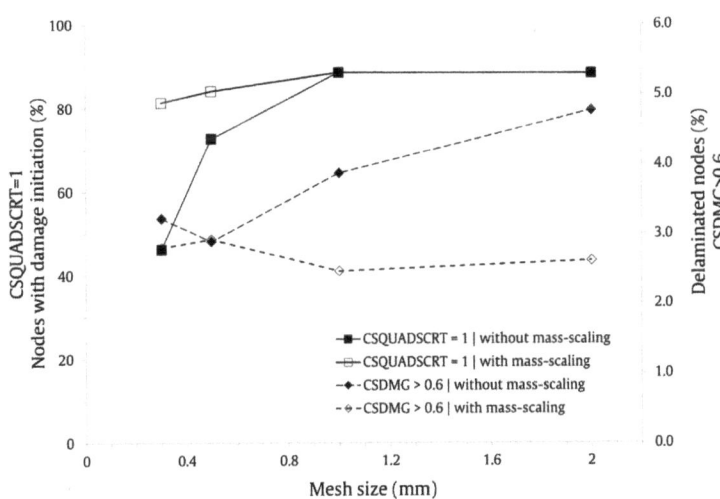

Figure 10. Effect of FE mesh size and mass scaling on the interlaminar damage.

6. Numerical–Experimental Correlation

The numerical results obtained with the nine-layer composite laminates FE model with mass scaling and with the different FE mesh sizes are compared with the experimental data presented in [16]. Notice that a higher post-processing by the drop-tower is responsible for the experimental curves' higher smoothness. The numerical predictions and experimental results are summarized in Table 5 (maximum force, maximum displacement, contact time). It can be observed that the maximum load is slightly overestimated by all the numerical models, presenting errors ranging from 5% to 8.6%, among which the FE model with $l_e = 0.3$ mm was the closest to the experimental averaged value. The maximum displacement is not affected by the FE mesh. Its value is correctly predicted with an error of 2.7% for all FE models. This is due to the fact that the displacement is controlled by the same experimental velocity–time curve that was incorporated to the FE models. The contact time is negligibly affected by the FE mesh size. The numerical predictions are slightly overestimated, presenting an error ranging from 10.4% to 12.3%. Although all FE models present comparable values, the one that best fits the experimental evidence is obtained with $l_e = 0.3$ mm. Moreover, taking also into consideration the results presented in Figure 10, in which it was observed that the FE mesh size considerably affects the interlaminar damage initiation and progression, the use of $l_e = 0.3$ mm, calculated using the equations of Turon et al. [31], is recommended.

Table 5. Numerical and experimental comparison.

Mesh Size (mm)	Maximum Load (N)			Maximum Displacement (mm)			Contact Time (ms)		
	Num.	Exp.	Error (%)	Num.	Exp.	Error (%)	Num.	Exp.	Error (%)
0.3	797	757	5.0	11.5	11.2	2.7	23	20.6	10.4
0.5	818		7.5	11.5		2.7	23		10.4
1	828		8.6	11.5		2.7	23.5		12.3
2	806		6.1	11.5		2.7	23.1		10.8

The numerical and experimental results with $l_e = 0.3$ mm are compared graphically in Figure 11 (force–time), Figure 12 (energy–time) and Figure 13 (force–displacement). The numerical and experimental curves, during the loading and unloading stages, show a very satisfactory agreement.

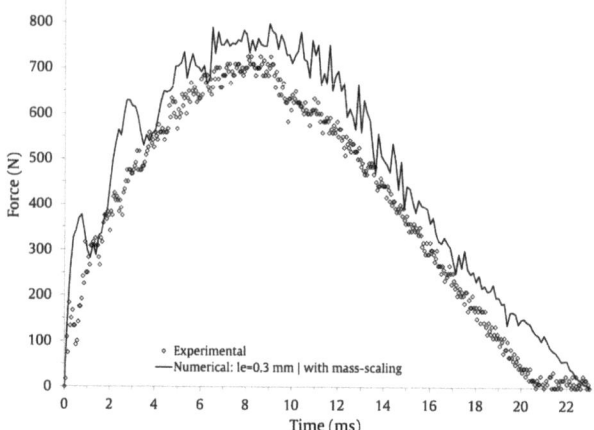

Figure 11. Numerical and experimental force-time results.

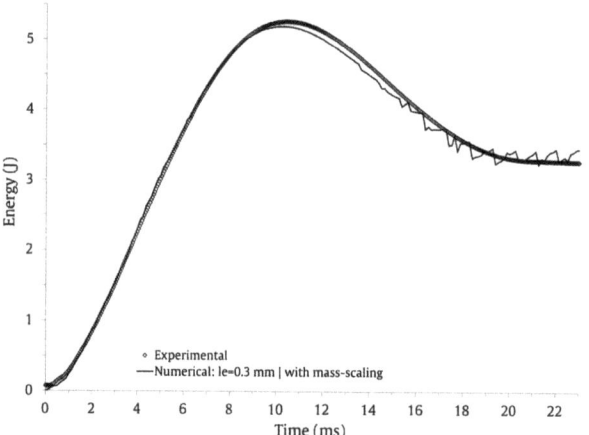

Figure 12. Numerical and experimental energy-time results.

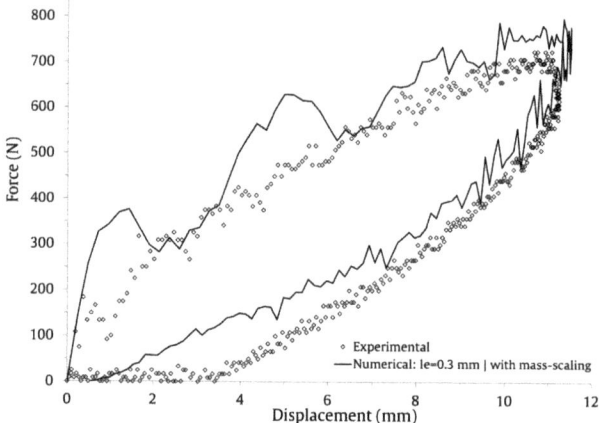

Figure 13. Numerical and experimental force-displacement results.

7. Conclusions

Finite element models were generated to study the low-velocity impact response of semicylindrical woven composite laminate shells using ABAQUS/Explicit. These FE models represent a nine-layer laminate with a thickness of 1.6 mm. A continuum damage model at the lamina level to account for intralaminar damage and a surface-based cohesive damage model at the laminas' interface to account for interlaminar damage were used as the two constitutive models to simulate the material damage brought on by low-velocity impact load. The premise for this study was the authors' previous work on the experimental evaluation of the effect of thickness on the impact response of semicylindrical composite laminated shells. As a result, the FE models were tested against the experimental findings for validation purposes. In this way, the stiffness properties of the woven fabric composite laminas were estimated from the constituents' properties of the tested specimens using a homogenization process, while the reaming properties were obtained in the literature. The developed FE models require a fine FE mesh discretization to properly define the interlaminar damage behavior, making them computational expensive. Therefore, numerical simulations were carried out to find an acceptable trade-off between the accuracy of the predictions and the computational cost of the solutions. An efficient approach is employed taking advantage of the model symmetries, continuum shell elements, which have lower computing costs than solid elements, and cohesive surfaces, which do not require element definition. Moreover, the FE mesh discretization and mass-scaling technique were also examined to ascertain how they affect the effectiveness and reliability of the FE model.

The results obtained indicate that the low-velocity impact response of semicylindrical woven composite laminate shells can be reliably predicted using the aforementioned FE models. Furthermore, the mass-scaling technique is successfully used to increase the stable time increment without compromising the accuracy of the dynamic response. Its implementation reduced the computing cost of the solutions by about 51%. The maximum load, maximum displacement, and contact time do not appear to be significantly affected by the choice of FE mesh size. Yet, it has an impact on the initiation and development of interlaminar damage. In order to ensure accurate numerical predictions of delamination, it is advised to use the formula proposed by the literature to compute the characteristic length of the elements of the FE mesh.

The load and energy histories were put to the test in comparison to the experimental findings from low-velocity impacts on specimens with identical elastic properties. A satisfactory correlation between the numerical outcomes and the experimental data was observed. The results show that the maximum load and contact time are slightly overestimated by the FE model, while the maximum displacement is correctly predicted. Nevertheless, all the numerical predictions are comparable with the experimental data.

The numerical simulations complement the previous experimental work, and the developed FE models can be used for further studies, for example, to analyze the response to multi-impacts, the effect of boundary condition or how the geometric parameters of the shells, such as thickness, length, or curvature, affect the impact response of woven fabric composite laminate shells.

Author Contributions: Conceptualization, L.M.F. and P.N.B.R.; methodology, L.M.F., P.N.B.R. and C.A.C.P.C.; software, L.M.F.; validation, L.M.F., P.N.B.R. and C.A.C.P.C.; formal analysis, L.M.F. and P.N.B.R.; investigation, L.M.F. and P.N.B.R.; L.M.F., P.N.B.R. and C.A.C.P.C.; writing—original draft preparation, L.M.F. and P.N.B.R.; writing—review and editing, L.M.F. and P.N.B.R. All authors have read and agreed to the published version of the manuscript.

Funding: This research received no external funding.

Institutional Review Board Statement: Not applicable.

Informed Consent Statement: Not applicable.

Data Availability Statement: Not applicable.

Acknowledgments: This research was sponsored by national funds through FCT—Fundação para a Ciência e a Tecnologia, under the project UIDB/00285/2020 and LA/P/0112/2020.

Conflicts of Interest: The authors declare no conflict of interest.

Nomenclature

Symbol	Description	Symbol	Description
ρ	Density	$\tilde{\sigma}_0$	Shear hardening function
E_1	Young's modulus along fiber direction 1	r_α	Axial damage thresholds
E_2	Young's modulus along fiber direction 2	r_{12}	Shear damage threshold
E_2	Though-thickness Young's modulus	X_α	Tensile/compressive strength along the fiber directions
G_{12}	In-plane shear modulus	d_1	Tensile/compressive damage variable along direction 1
G_{13}	Out of plane shear modulus	d_2	Tensile/compressive damage variable along direction 2
v_{12}	In-plane Poisson's ratio	d_{12}	Shear damage variable
X_{1+}	Tensile strength along direction 1	g_0^α	Elastic energy density
X_{1-}	Compressive strength along direction 1	L_e	Characteristic length of the element
X_{2+}	Tensile strength along direction 2	$\bar{\varepsilon}^{pl}$	Plastic strain due to shear deformation
X_{2-}	Compressive strength along direction 2	k_n	Elastic normal interlaminar stiffness
S_{12}	In-plane shear strength	k_s	Elastic shear interlaminar stiffness
G_f^α	Intralaminar fracture toughness along direction 1 and 2	k_t	Elastic tangential interlaminar stiffness
α_{12}	Parameter in the equation of shear damage	τ_n^0	Maximum normal contact stress
d_{12}^{max}	Maximum shear damage	τ_s^0	Maximum 1st shear contact stress
C	Coefficient in hardening equation	τ_t^0	Maximum 2nd shear contact stress
lp	Power term in hardening equation	G_{Ic}	Interlaminar normal fracture toughness
F_α	Axial damage activation function	G_{IIc}	Interlaminar 1st shear fracture toughness
F_{12}	Shear damage activation function	G_{IIIc}	Interlaminar 2nd shear fracture toughness
F_{pl}	Plasticity activation function	η	Benzeggagh–Kenane exponent
$\tilde{\sigma}_\alpha$	Effective tensile/compressive stress	μ	Friction coefficient
$\tilde{\sigma}_{y0}$	Initial effective shear yield stress	M	Parameter defined for the cohesive zone model
$\tilde{\sigma}_{12}$	Effective shear stress	N_e	Number of elements in the cohesive zone

References

1. Ferreira, L.M.; Graciani, E.; París, F. Predicting Failure Load of a Non-Crimp Fabric Composite by Means of a 3D Finite Element Model Including Progressive Damage. *Compos. Struct.* **2019**, *225*, 111115. [CrossRef]
2. Ferreira, L.; Graciani, E.; París, F. Progressive Damage Study of NCF Composites under Compressive Loading. In Proceedings of the 16th European Conference on Composite Materials, Seville, Spain, 22–26 June 2014; Volume 14, pp. 22–26.
3. Ferreira, L.M.; Graciani, E.; París, F. Three Dimensional Finite Element Study of the Behaviour and Failure Mechanism of Non-Crimp Fabric Composites under in-Plane Compression. *Compos. Struct.* **2016**, *149*, 106–113. [CrossRef]
4. Ferreira, L.M.; Graciani, E.; París, F. Modelling the Waviness of the Fibres in Non-Crimp Fabric Composites Using 3D Finite Element Models with Straight Tows. *Compos. Struct.* **2014**, *107*, 79–87. [CrossRef]
5. Ferreira, L.; Coelho, C. Modelling Progressive Damage in NCF Composites Using the Continuum Damage Mechanics Method. In Proceedings of the 2022 Advances in Science and Engineering Technology International Conferences (ASET), Dubai, United Arab Emirates, 21–24 February 2022; IEEE: New York, NY, USA; pp. 1–4.
6. Kim, S.J.; Goo, N.S.; Kim, T.W. The Effect of Curvature on the Dynamic Response and Impact-Induced Damage in Composite Laminates. *Compos. Sci. Technol.* **1997**, *57*, 763–773. [CrossRef]
7. Her, S.-C.; Liang, Y.-C. The Finite Element Analysis of Composite Laminates and Shell Structures Subjected to Low Velocity Impact. *Compos. Struct.* **2004**, *66*, 277–285. [CrossRef]
8. Krishnamurthy, K.; Mahajan, P.; Mittal, R. Impact Response and Damage in Laminated Composite Cylindrical Shells. *Compos. Struct.* **2003**, *59*, 15–36. [CrossRef]
9. Choi, I.H. Finite Element Analysis of Low-Velocity Impact Response of Convex and Concave Composite Laminated Shells. *Compos. Struct.* **2018**, *186*, 210–220. [CrossRef]
10. Kistler, L.S.; Waas, A.M. Impact Response of Cylindrically Curved Laminates Including a Large Deformation Scaling Study. *Int. J. Impact Eng.* **1998**, *21*, 61–75. [CrossRef]
11. Zhao, G.; Cho, C. On Impact Damage of Composite Shells by a Low-Velocity Projectile. *J. Compos. Mater.* **2004**, *38*, 1231–1254. [CrossRef]
12. Kumar, S.; Nageswara Rao, B.; Pradhan, B. Effect of Impactor Parameters and Laminate Characteristics on Impact Response and Damage in Curved Composite Laminates. *J. Reinf. Plast. Compos.* **2007**, *26*, 1273–1290. [CrossRef]

13. Khalili, S.M.R.; Soroush, M.; Davar, A.; Rahmani, O. Finite Element Modeling of Low-Velocity Impact on Laminated Composite Plates and Cylindrical Shells. *Compos. Struct.* **2011**, *93*, 1363–1375. [CrossRef]
14. Albayrak, M.; Kaman, M.O.; Bozkurt, I. Experimental and Numerical Investigation of the Geometrical Effect on Low Velocity Impact Behavior for Curved Composites with a Rubber Interlayer. *Appl. Compos. Mater.* **2023**, *30*, 507–538. [CrossRef]
15. *Abaqus*, version 2021; Dassault Systemes Simulia Corporation: Providence, RI, USA, 2021.
16. Reis, P.N.B.; Sousa, P.; Ferreira, L.M.; Coelho, C.A.C.P. Multi-Impact Response of Semicylindrical Composite Laminated Shells with Different Thicknesses. *Compos. Struct.* **2023**, *310*, 116771. [CrossRef]
17. Silva, M.P.; Santos, P.; Parente, J.M.; Valvez, S.; Reis, P.N.B.; Piedade, A.P. Effect of Post-Cure on the Static and Viscoelastic Properties of a Polyester Resin. *Polymers* **2020**, *12*, 1927. [CrossRef] [PubMed]
18. Ferreira, L.; Coelho, C.; Reis, P. Impact Response of Semi-Cylindrical Composite Laminate Shells Under Repeated Low-Velocity Impacts. In Proceedings of the 2022 Advances in Science and Engineering Technology International Conferences (ASET), Dubai, United Arab Emirates, 21–24 February 2022; pp. 1–5.
19. Amaro, A.M.; Reis, P.N.B.; Magalhães, A.G.; de Moura, M.F.S.F. The Influence of the Boundary Conditions on Low-Velocity Impact Composite Damage. *Strain* **2011**, *47*, e220–e226. [CrossRef]
20. Johnson, A.F.; Pickett, A.K.; Rozycki, P. Computational Methods for Predicting Impact Damage in Composite Structures. *Compos. Sci. Technol.* **2001**, *61*, 2183–2192. [CrossRef]
21. Ladeveze, P.; LeDantec, E. Damage Modelling of the Elementary Ply for Laminated Composites. *Compos. Sci. Technol.* **1992**, *43*, 257–267. [CrossRef]
22. User, A.A. *VUMAT for Fabric Reinforced Composites, Dassault Systèmes*; Dassault Systemes Simulia Corporation: Providence, RI, USA, 2008.
23. Liu, X.; Rouf, K.; Peng, B.; Yu, W. Two-Step Homogenization of Textile Composites Using Mechanics of Structure Genome. *Compos. Struct.* **2017**, *171*, 252–262. [CrossRef]
24. Sridharan, S.; Pankow, M. Performance Evaluation of Two Progressive Damage Models for Composite Laminates under Various Speed Impact Loading. *Int. J. Impact Eng.* **2020**, *143*, 103615. [CrossRef]
25. Bodepati, V.; Mogulanna, K.; Rao, G.S.; Vemuri, M. Numerical Simulation and Experimental Validation of E-Glass/Epoxy Composite Material under Ballistic Impact of 9 Mm Soft Projectile. *Plast. Impact Mech.* **2017**, *173*, 740–746. [CrossRef]
26. Esnaola, A.; Elguezabal, B.; Aurrekoetxea, J.; Gallego, I.; Ulacia, I. Optimization of the Semi-Hexagonal Geometry of a Composite Crush Structure by Finite Element Analysis. *Compos. Part B Eng.* **2016**, *93*, 56–66. [CrossRef]
27. Kostopoulos, V.; Markopoulos, Y.P.; Giannopoulos, G.; Vlachos, D.E. Finite Element Analysis of Impact Damage Response of Composite Motorcycle Safety Helmets. *Compos. Part B Eng.* **2002**, *33*, 99–107. [CrossRef]
28. Hoo Fatt, M.S.; Lin, C. Perforation of Clamped, Woven E-Glass/Polyester Panels. *Compos. Part B Eng.* **2004**, *35*, 359–378. [CrossRef]
29. Alonso, L.; Martínez-Hergueta, F.; Garcia-Gonzalez, D.; Navarro, C.; García-Castillo, S.K.; Teixeira-Dias, F. A Finite Element Approach to Model High-Velocity Impact on Thin Woven GFRP Plates. *Int. J. Impact Eng.* **2020**, *142*, 103593. [CrossRef]
30. Camanho, P.P.; Davila, C.G.; de Moura, M.F. Numerical Simulation of Mixed-Mode Progressive Delamination in Composite Materials. *J. Compos. Mater.* **2003**, *37*, 1415–1438. [CrossRef]
31. Turon, A.; Dávila, C.G.; Camanho, P.P.; Costa, J. An Engineering Solution for Mesh Size Effects in the Simulation of Delamination Using Cohesive Zone Models. *Eng. Fract. Mech.* **2007**, *74*, 1665–1682. [CrossRef]
32. Song, K.; Dávila, C.G.; Rose, C.A. Guidelines and Parameter Selection for the Simulation of Progressive Delamination. In Proceedings of the 2008 ABAQUS User's Conference, Newport, RI, USA, 19–22 May 2008; Volume 41, pp. 43–44.
33. Nguyen, K.-H.; Ju, H.-W.; Truong, V.-H.; Kweon, J.-H. Delamination Analysis of Multi-Angle Composite Curved Beams Using an out-of-Autoclave Material. *Compos. Struct.* **2018**, *183*, 320–330. [CrossRef]
34. Zhou, J.; Wen, P.; Wang, S. Numerical Investigation on the Repeated Low-Velocity Impact Behavior of Composite Laminates. *Compos. Part B Eng.* **2020**, *185*, 107771. [CrossRef]
35. Lopes, C.S.; Camanho, P.P.; Gürdal, Z.; Maimí, P.; González, E.V. Low-Velocity Impact Damage on Dispersed Stacking Sequence Laminates. Part II: Numerical Simulations. *Compos. Sci. Technol.* **2009**, *69*, 937–947. [CrossRef]
36. Baluch, A.H.; Falcó, O.; Jiménez, J.L.; Tijs, B.H.A.H.; Lopes, C.S. An Efficient Numerical Approach to the Prediction of Laminate Tolerance to Barely Visible Impact Damage. *Compos. Struct.* **2019**, *225*, 111017. [CrossRef]
37. Falcó, O.; Lopes, C.S.; Sommer, D.E.; Thomson, D.; Ávila, R.L.; Tijs, B.H.A.H. Experimental Analysis and Simulation of Low-Velocity Impact Damage of Composite Laminates. *Compos. Struct.* **2022**, *287*, 115278. [CrossRef]
38. Wang, F.; Wang, B.; Kong, F.; Ouyang, J.; Ma, T.; Chen, Y. Assessment of Degraded Stiffness Matrices for Composite Laminates under Low-Velocity Impact Based on Modified Characteristic Length Model. *Compos. Struct.* **2021**, *272*, 114145. [CrossRef]
39. Schön, J. Coefficient of Friction of Composite Delamination Surfaces. *Wear* **2000**, *237*, 77–89. [CrossRef]
40. Bresciani, L.M.; Manes, A.; Ruggiero, A.; Iannitti, G.; Giglio, M. Experimental Tests and Numerical Modelling of Ballistic Impacts against Kevlar 29 Plain-Woven Fabrics with an Epoxy Matrix: Macro-Homogeneous and Meso-Heterogeneous Approaches. *Compos. Part B Eng.* **2016**, *88*, 114–130. [CrossRef]
41. Hillerborg, A.; Modéer, M.; Petersson, P.-E. Analysis of Crack Formation and Crack Growth in Concrete by Means of Fracture Mechanics and Finite Elements. *Cem. Concr. Res.* **1976**, *6*, 773–781. [CrossRef]

42. Camanho, P.P.; Dávila, C.G. *Mixed-Mode Decohesion Finite Elements for the Simulation of Delamination in Composite Materials*; NASA Langley Research Center: Hampton, VA, USA, 2002.
43. Carpinteri, A.; Cornetti, P.; Barpi, F.; Valente, S. Cohesive Crack Model Description of Ductile to Brittle Size-Scale Transition: Dimensional Analysis vs. Renormalization Group Theory. *Eng. Fract. Mech.* **2003**, *70*, 1809–1839. [CrossRef]
44. Falk, M.L.; Needleman, A.; Rice, J.R. A Critical Evaluation of Cohesive Zone Models of Dynamic Fractur. *J. Phys. IV* **2001**, *11*, Pr5-43–Pr5-50. [CrossRef]
45. Schoeppner, G.A.; Abrate, S. Delamination Threshold Loads for Low Velocity Impact on Composite Laminates. *Compos. Part Appl. Sci. Manuf.* **2000**, *31*, 903–915. [CrossRef]
46. Belingardi, G.; Vadori, R. Low Velocity Impact Tests of Laminate Glass-Fiber-Epoxy Matrix Composite Material Plates. *Int. J. Impact Eng.* **2002**, *27*, 213–229. [CrossRef]

Disclaimer/Publisher's Note: The statements, opinions and data contained in all publications are solely those of the individual author(s) and contributor(s) and not of MDPI and/or the editor(s). MDPI and/or the editor(s) disclaim responsibility for any injury to people or property resulting from any ideas, methods, instructions or products referred to in the content.

Article

Multi-Objective Optimization of Adhesive Joint Strength and Elastic Modulus of Adhesive Epoxy with Active Learning

Paripat Kraisornkachit [1,2], Masanobu Naito [1,2,*], Chao Kang [3] and Chiaki Sato [3]

1 Data-Driven Polymer Design Group, Research Center for Macromolecules and Biomaterials, National Institute for Materials Science (NIMS), Ibaraki 305-0047, Japan; kraisornkachit.paripat@nims.go.jp
2 Program in Materials Science and Engineering, Graduate School of Pure and Applied Sciences, University of Tsukuba, Ibaraki 305-8577, Japan
3 Institute of Innovative Research (IIR), Tokyo Institute of Technology, Kanagawa 226-8503, Japan; kang.c.ab@m.titech.ac.jp (C.K.); csato@pi.titech.ac.jp (C.S.)
* Correspondence: naito.masanobu@nims.go.jp

Citation: Kraisornkachit, P.; Naito, M.; Kang, C.; Sato, C. Multi-Objective Optimization of Adhesive Joint Strength and Elastic Modulus of Adhesive Epoxy with Active Learning. *Materials* **2024**, *17*, 2866. https://doi.org/10.3390/ma17122866

Academic Editor: Mariana Cristea

Received: 19 April 2024
Revised: 28 May 2024
Accepted: 5 June 2024
Published: 12 June 2024

Copyright: © 2024 by the authors. Licensee MDPI, Basel, Switzerland. This article is an open access article distributed under the terms and conditions of the Creative Commons Attribution (CC BY) license (https://creativecommons.org/licenses/by/4.0/).

Abstract: Studying multiple properties of a material concurrently is essential for obtaining a comprehensive understanding of its behavior and performance. However, this approach presents certain challenges. For instance, simultaneous examination of various properties often necessitates extensive experimental resources, thereby increasing the overall cost and time required for research. Furthermore, the pursuit of desirable properties for one application may conflict with those needed for another, leading to trade-off scenarios. In this study, we focused on investigating adhesive joint strength and elastic modulus, both crucial properties directly impacting adhesive behavior. To determine elastic modulus, we employed a non-destructive indentation method for converting hardness measurements. Additionally, we introduced a specimen apparatus preparation method to ensure the fabrication of smooth surfaces and homogeneous polymeric specimens, free from voids and bubbles. Our experiments utilized a commercially available bisphenol A-based epoxy resin in combination with a Poly(propylene glycol) curing agent. We generated an initial dataset comprising experimental results from 32 conditions, which served as input for training a machine learning model. Subsequently, we used this model to predict outcomes for a total of 256 conditions. To address the high deviation in prediction results, we implemented active learning approaches, achieving a 50% reduction in deviation while maintaining model accuracy. Through our analysis, we observed a trade-off boundary (Pareto frontier line) between adhesive joint strength and elastic modulus. Leveraging Bayesian optimization, we successfully identified experimental conditions that surpassed this boundary, yielding an adhesive joint strength of 25.2 MPa and an elastic modulus of 182.5 MPa.

Keywords: epoxy; adhesive; elastic modulus; machine learning; active learning; experimental testing; multi-objective optimization; sandwich-structured material

1. Introduction

In the realm of materials science and engineering, the study of adhesive materials plays a pivotal role in advancing technologies across diverse industries [1], for example, automotive [2], aerospace [3] and construction materials [4]. The adhesive joint strength, representing the force required to break or deform a bonded interface, and the elastic modulus, signifying a material's resistance to deformation under applied stress, are two fundamental properties that have important influence on the performance and reliability of adhesive systems. The adhesive joint strength serves as a crucial metric in assessing the integrity of bonded structures. A robust and durable bond is often synonymous with high adhesive joint strength, ensuring the stability of assemblies in various environments and under different loading conditions [5]. However, this strength is not isolated from the elastic modulus of the adhesive material. The elastic modulus, or stiffness, defines how the material responds to external forces [6], influencing the distribution of stresses within the adhesive joint. As such, the interplay between adhesive joint strength and elastic modulus becomes a critical factor in determining the overall performance and longevity of bonded structures [7]. Understanding the trade-offs and synergies between adhesive joint strength and elastic modulus is imperative for optimizing material selection based on specific application requirements. The relationship between adhesive joint strength and elastic modulus is often complex. Several studies have reported that high-elastic-modulus materials have a higher adhesive joint strength [8,9]. However, lower-modulus adhesives can provide the ability to absorb external forces, and this ability is an important factor in adhesive materials used in bonding parts that are easily broken or damaged [7]. In architectural applications like structural glazing systems, low elastic adhesives are used to bond glass panels to the structural framework. These low modulus adhesives can handle high stress gradients at glass interfaces [10].

Achieving an optimal balance between these properties is a multifaceted challenge, as conventional optimization approaches often focus on a singular property, potentially neglecting the complex interdependencies that exist. Multi-objective optimization presents a paradigm shift by simultaneously addressing the enhancement of many properties. Recently, high productivity of heat-resistant epoxy matrix systems was successfully achieved using multi-objective optimization along with machine learning. This accomplishment has never been achieved, even though the conventional trial-and-error experiment has been attempted up to three hundred times [11]. Furthermore, because the required properties frequently conflict with one another in nature, multi-objective optimization can be a candidate method from the material design point of view. The maximum molecular weight and the maximum of number average degree of branching of polymers were achieved in the polymerization process with the assistance of a multi-objective optimization approach. This can reduce unnecessary costs and time consumption in the experimental stage [12]. Moreover, the multi-objective optimization approach was applied in optimizing material removal rate and taper during electrochemical discharge machining of the silicon carbide-reinforced epoxy composites, and it was reported that there was a 15% improvement in taper reduction, together with the material removal rate decreasing by 0.91%. In addition, utilizing multi-objective optimization is able to address the trade-off problem [13]. This approach not only extends the edge of discovery but also accelerates the study process for humankind.

The adhesive joint strength has been measured previously by using single-lap shear joints [14]. On the other hand, indentation hardness measurements were used to evaluate the elastic modulus of the same adhesives, owing to the significant convenience in sample preparation and measurements compared to the widely used tensile test. The relationship between hardness and elastic modulus has been thoroughly studied, and several methods for calculating elastic modulus from hardness measurements have been reported previously [15]. Nevertheless, before measuring the hardness of polymeric specimens, it is important to prepare appropriate samples to ensure accurate and reliable results. The polymeric specimens have to be prepared in standardized shapes and sizes suitable for the specific hardness testing method being used. Additionally, the surface of the specimens must be prepared to ensure flatness and smoothness, as irregularities or rough surfaces can directly affect hardness measurements. Moreover, voids and bubbles lead to inaccurate results; thus, homogeneity of the polymer material is one of the important factors.

The adhesive joint strength and elastic modulus are key mechanical properties that directly influence the adhesion performance of adhesive materials. However, simultaneously studying multiple properties can be time-consuming and costly. In this work, we propose to investigate the adhesive joint strength and elastic modulus of epoxy adhesives using a multi-objective optimization approach. A metal mold is designed and introduced to prepare the polymeric specimens according to the French standard NFT 76-142 [16] to eliminate porosity. The elastic modulus of the polymeric specimens is calculated according to Shore hardness which is measured by two types of durometers referring to ASTM-D2240 [17]. The experimental conditions are designed and conducted to obtain a small initial dataset of the first 32 conditions. The machine learning model is trained and validated for optimizing accuracy. Then, prediction of the extended 256 conditions is carried out together with the active learning method. Active learning is a strategy that can be employed to improve model accuracy and reduce deviations in predicted results. Active learning is a machine learning approach that involves selecting the most informative data points for labeling or further training, with the goal of enhancing the model performance while minimizing the amount of labeled data needed. After that, the trade-off boundary is obtained after the improved prediction of the 256 conditions. Finally, Bayesian optimization is employed to identify experimental conditions with predicted results that can overcome the trade-off boundary. The results are confirmed by checking the actual experiments. Adhesive epoxies with desired adhesive joint strength and elastic modulus properties can be fabricated with the provided conditions from the proposed multi-objective optimization approaches.

2. Materials and Methods

2.1. Experiments

2.1.1. Materials

In this study, a commercial epoxy resin, Diglycidyl ether of bisphenol A-based epoxy resin (DGEBA) from Mitsubishi Chemical Corporation, Tokyo, Japan, with 4 different molecular weights, and a commercial diamine curing agent (Jeffamine™), Poly(propylene glycol) bis(2-aminopropyl ether) from Sigma-Aldrich, Tokyo, Japan, with 4 different molecular weights, were used. The molecular weights of DGEBA and Jeffamine™ are M_{wE} = {370, 1650, 2900 and 3800} g/mol and M_{wC} = {230, 400, 2000 and 4000} g/mol, respectively. The appearance of DGEBA with M_{wE} of 230 g/mol is in liquid phase; in contrast, it is in solid phase for M_{wE} of 1650, 2900 and 3800 g/mol. Additionally, the appearance of Jeffamine™ with M_{wC} of 230 and 400 g/mol is viscous liquid; on the other hand, it is a highly viscous liquid for 2000 and 4000 g/mol. All chemicals were used as received without further purification or pretreatment. Generally, the stoichiometric mixing ratio of epoxy resin is two moles of epoxy resin per one mole of curing agent. The chemical structures and reactions of DGEBA and Jeffamine™ are illustrated in Figure 1.

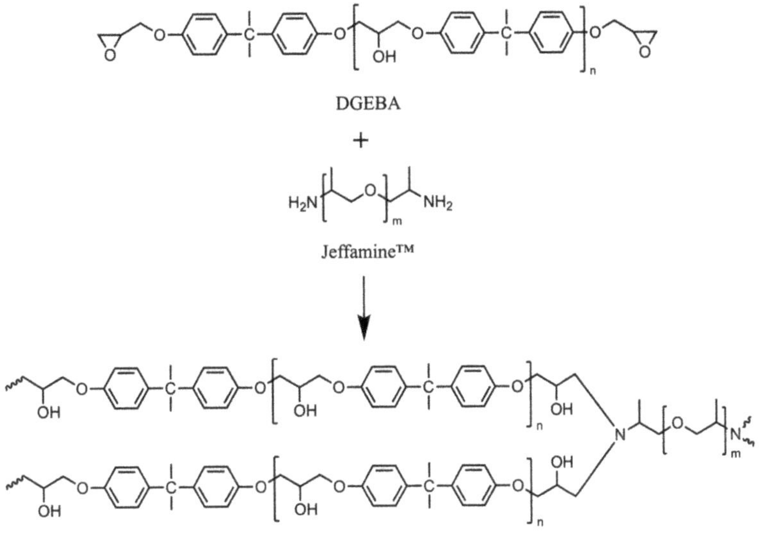

Figure 1. Chemical structures of DGEBA, Jeffamine™ and cured epoxy.

2.1.2. Preparation of Specimens

A metal mold was designed and fabricated specifically for this study according to the French standard NFT76-142 [16]; the illustration is shown in Figure S1 (Supplementary Materials). In principle, the objective of using this mold is not only to produce cured adhesive specimens with a flat and smooth surface according to the standard but also to remove the voids and bubbles entrapped inside the specimens as much as possible. This metal mold consists of a metal lid and a metal base, and the adhesive specimen is fitted into a 6 mm thick silicon rubber frame (obtained from Axel 4-1371-08) at the center of the mold. This silicon rubber is bordered by a metal box which contains 4 small gaps on the topside of each edge; these gaps allow the adhesive mixture to overflow during high-pressure curing conditions. An overflow of adhesive mixture is potentially able to remove gas bubbles entrapped in the specimens [18]. This mold created a square-shaped specimen with a dimension of $40 \times 40 \times 6$ mm^3 according to the standard test method for rubber property ASTM-D2240 [17]; the illustration is shown in Figure S1. Teflon tape (obtained from Axel 3-5579-09) was pasted onto the metal lid and metal base on the contact side of the adhesive specimen in order to reduce the difficulty in removing the cured adhesive specimens from the mold.

DGEBA and Jeffamine™ were separately preheated in the oven at 190 °C for 30 min; this process is to ensure that solid epoxy is melted into liquid prior to the mixing step. After that, it was rapidly mixed together in the disposal bottle glass for a couple seconds until the mixture achieved a homogenous phase, and then, the mixture was poured into a metal mold. In total, four specific amine-to-epoxide ratios were investigated in this study: $r = \{0.75, 1.00, 1.25 \text{ and } 1.50\}$. For example, $r = 1$ represents the stoichiometric mixing ratio of the epoxy with the curing agent: $r = 0.75$ represents 25% less amine curing agent than that used in the stoichiometric mixing ratio; on the other hand, $r = 1.25$ represents 25% excess amine curing agent over epoxy resin. After pouring the mixture into the mold, a metal lid was placed on top, and the assembled mold was put into hot-press machine with applying force of 2 MPa to the mold. The curing time for every condition was set to 60 min. In addition, the effect of curing temperature was investigated. The mixture was cured at different specific curing temperatures of $T_C = \{90, 130, 170 \text{ and } 210\}$ °C. The mold was then kept under applying force until it cooled down to room temperature prior to taking out the specimens for further measurement. The total variable parameters in this study followed our previous work [14] and are summarized in Table 1. For the conditions of liquid DGEBA, the weight of the cured epoxies was set to be sufficient for the metal mold at 16 g, whereas the weight of the cured epoxies with the conditions of solid DGEBA was set at 24 g.

Table 1. Summary of variable parameters for experiments [14].

Epoxy Resin		Curing Agent		Amine to Epoxide Ratio (r)	T_C (°C)
M_{wE} (g/mol)	Appearance	M_{wC} (g/mol)	Appearance		
370	Liquid	230	Viscous	0.75	90
1650	Solid	400	Viscous	1.00	130
2900	Solid	2000	Highly viscous	1.25	170
3800	Solid	4000	Highly viscous	1.50	230

The single-lap shear specimens were prepared and tested for adhesive joint strength using the same procedure as described in our previous work. The epoxy resin and curing agent were preheated and then hand-mixed for a few seconds until achieving a homogeneous phase. Subsequently, the mixture was poured and spread onto an aluminum substrate (A6061P-T6, dimensions: $100 \times 25 \times 2$ mm) over an area of 25×12.5 mm. The adhesive thickness was controlled by adding 0.1 parts per hundred resin of spherical glass beads (Fujiseisakujo, Tokyo, Japan). A second aluminum substrate was then placed on top of the adhesive overlapping the first substrate, creating a sandwich specimen. The specimens were clamped and subjected to the specific curing temperature in an oven for

60 min. Adhesive joint strength testing was conducted using a 10-kN AG-X plus series universal tensile testing machine (Shimadzu, Kyoto, Japan). To ensure data reliability, at least two specimens were fabricated for each measurement. The results are reported as the average value along with the standard deviation [14].

2.1.3. Modulus Testing Technique

Shore hardness measurements of the specimens were performed using durometer hardness testing tools. After pressing the indentation on top of the specimen surface, the indenter of the durometer pierced the specimens, and then, the hardness values were observed by the gauge of the durometer. There are 2 types of durometer hardness testing tools: Shore A and Shore D, which were performed in this study. Shore A is appropriate for the soft materials that are in the range between 20A and 80A; on the other hand, Shore D is suitable for the hard materials that are in the range of 80A to 85D. After measuring hardness values, it was converted into S by the following Equation (1). Then, the elastic modulus was calculated by the relationship between hardness and elastic modulus via Equation (2) as follows [15]:

$$S = \begin{cases} \text{Shore A} & 20A < S < 80A \\ \text{Shore D} + 50 & 80A < S < 85D \end{cases} \quad (1)$$

$$\log E_0 = 0.0235S - 0.6403 \quad (2)$$

The specimens were placed on the rigid surface prior to measuring the hardness at five points on the topside of the specimens in an effort to minimize variation. Every measurement point was at least 6.0 mm away from the other measuring points and at least 12.0 mm away from any edge of the specimens as illustrated in Figure S1b [17]. Areas of the specimens with trapped gas bubbles must be avoided when measuring the hardness. The elastic modulus of each point was independently calculated, and then, the average value with standard deviation was reported for each specimen.

2.2. Multi-Objective Optimization Approaches

2.2.1. Dataset Preparation

The total number of possible conditions in this study was 256, which consisted of 4 M_{wE} times 4 M_{wC} times 4 amine-to-epoxide ratios times 4 curing temperatures (Table 1). However, the initial dataset for machine learning model training was 32 conditions according to our previous work [14]. According to our previous study, these 32 conditions of the initial dataset were selected by applying experimental techniques using four-by-four Graeco–Latin square design in order to uniformly distribute the experimental conditions [19,20]. There were two properties studied for each condition, which consisted of adhesive joint strength and elastic modulus. Datasets of adhesive joint strength were obtained from our previous work; on the other hand, datasets of elastic modulus were obtained by conducting experiments with the same conditions.

2.2.2. Cross-Validation Technique

Machine learning modeling was carried out using the Python programming language. Several Python packages were applied in this work from the scikit-learn library, for example, data splitting, cross-validation, and machine learning model training and testing. An initial dataset of 32 conditions, each consisting of four variable parameters with two properties, was split into a training set and a testing set in order to train the machine learning model and evaluate its accuracy. In order to uniformly utilize all 32 conditions in the initial dataset as a training set and testing set, the K-fold cross-validation technique was introduced to split these initial datasets. K-fold cross-validation is a technique used for evaluating the performance of a model by dividing the dataset into multiple smaller folds. This method helps to reduce overfitting and provides a more accurate estimate of a model's performance on unseen data. The initial dataset was randomly split into K equal-sized folds. In this study, the initial dataset was divided into K = 32 folds. Then, the machine learning model was trained on the data from the remaining K − 1 = 31 folds and evaluated for accuracy of models on the data from the remaining fold. This process was repeated K times, with each fold being used as the validation set only once. After that, the accuracy score of the machine learning models for each K times was calculated and then averaged to obtain the accuracy score of that model. The model with the highest accuracy score was chosen for further use. K-fold cross-validation is able to reduce the possibility of overfitting by training and evaluating the model on different folds of the data. This technique provides a more reliable estimate of the model's performance on unseen data, as it reduces the impact of any single-fold peculiarities on the overall performance score [21].

2.2.3. Modeling and Prediction

Seven machine learning algorithms were investigated in order to find the best model in this study. The Ridge regression algorithm is generally used to deal with a problem called multicollinearity. This problem occurs when the predictor variables—variables that are used to predict the outcome variable—in a regression model are highly correlated with each other. As a result, this can be difficult to estimate the effects of each predictor variable on the outcome variable accurately. The Ridge regression solves this problem by adding a small number to the diagonal of the matrix of predictor variables. By adding this number, the Ridge regression shrinks the estimates of the coefficients towards zero, which reduces their impact on the outcome variable. Thus, the Ridge regression algorithm is a technique that helps to improve the accuracy and reliability of estimates in regression models when there is multicollinearity among predictor variables [22]. Likewise, the Lasso regression algorithm is commonly used to prevent overfitting in linear regression models. However, the main difference between the Lasso and Ridge regression is the method used to prevent overfitting. Unlike the Ridge algorithm, the Lasso shrinks some of the coefficients to exactly 0, effectively performing variable selection and making the model simpler [23]. The Elastic net algorithm combines the strengths of both the Ridge and Lasso methods, while the Ridge regression does not perform variable selection and the Lasso can struggle with highly correlated predictors, to provide an improved approach for regularization and variable selection [24]. A new K-nearest neighbor (k-NN) algorithm utilizes the neighborhood points in the training sample. For each new data point, the algorithm finds the K nearest neighbors based on Euclidean distance. Then, the output variable for a new data point is predicted as the average of the output variables of its K nearest neighbors. This k-NN regression algorithm is a simple and intuitive algorithm that can work well for small datasets [25]. In cases of complex nonlinear relationships between the input and output variables, the Decision Tree Regression (DTR) algorithm can handle these cases efficiently. The DTR algorithm creates a tree-like structure where each internal node represents a decision based on a feature and threshold value, and each leaf node represents a prediction for the output variable. Then, the DTR algorithm selects the feature and threshold value that best split the data into two subsets that are as homogeneous as possible with respect to the output variable [26]. In order to improve the accuracy of the model, the Random Forest algorithm was introduced as an extension of the DTR algorithm. The principle of the Random Forest is to construct multiple decision trees at the training step and combine their predictions to make a final output variable prediction. Each tree is built using a random subset of the features and data points, which helps to reduce the correlation between the trees and improve the diversity of the forest. By combining the predictions of multiple trees, the Random Forest can capture more of the underlying patterns in the data and make more accurate predictions [27]. Lastly, the Gradient boosting algorithm was one of the candidate algorithms in this study. This algorithm works by iteratively adding decision trees to the model, with each tree attempting to correct the errors of the previous tree. The algorithm uses gradient descent optimization to minimize the loss function, which measures the difference between the predicted values and the actual values [28].

All seven machine learning algorithms were trained independently by using the initial dataset with the K-fold cross-validation technique. The accuracy of each algorithm was evaluated by the calculation of three tools: the coefficient of determination (R^2 score), the mean absolute error (MAE) and the root-mean-square error (RMSE). The algorithms with a R^2 score close to 1 show higher accuracy. On the other hand, the algorithms with a lower value of MAE and RMSE show higher accuracy. The calculation of these three tools utilized the prediction results from the testing dataset and the measured results from the experiment. Moreover, the predictions of both adhesive joint strength and elastic modulus were carried out simultaneously in this step.

The model with the highest accuracy was selected to perform further prediction of the total 256 conditions. Furthermore, the K-fold cross-validation technique was also applied in this step in order to average the prediction results of each fold in each condition. Hence, the relationship between adhesive joint strength and elastic modulus for each condition was observed. In addition, the standard deviation of each prediction result was obtained.

Three predicted results with high deviations from three regions: low adhesive joint strength with low elastic modulus, low adhesive joint strength with high elastic modulus and high adhesive joint strength with high elastic modulus. These were selected to perform an active learning approach. The experimental conditions at these three selected points were conducted for the experiment. Then, the additional dataset ($m_i = 3$) of the measured adhesive joint strength and elastic modulus was added to the initial dataset ($n_i = 32$) to make a new dataset ($n = n_i + m_i = 32 + 3 = 35$) for the second active learning cycle, and so on. The overall process started with machine learning model training once again in order to improve the model accuracy as well as the standard deviation of the predicted results; this process is called the active learning approach. This active learning approach was repeated until the average values of the prediction errors (the standard deviation of predicted results) were comparable to the experimental errors; then, the active learning loop was terminated.

After termination of the active learning loop, machine learning model accuracy and the standard deviation of the predicted results were collected. In addition, the correlation between adhesive joint strength and elastic modulus properties was observed by plotting the predicted 256 conditions from the last active learning cycle. Moreover, the trade-off line of these two properties, represented by the Pareto frontier line, was drawn by connecting the boundary points. These data were kept for investigation in the next step.

2.2.4. Bayesian Optimization

In this study, Bayesian optimization was performed using PHYSBO [29], a Python library for Bayesian optimization, in order to search for extended conditions outside the Pareto frontier line. In this step, the variable parameters of amine-to-epoxide ratios together with curing temperatures were studied as a continuous value. The amine-to-epoxide ratios ranged from 0.75 to 1.50 with 0.01 increment steps. On the other hand, the curing temperatures ranged from 90 °C to 210 °C with an increment step of 1 °C. Nevertheless, the M_{wE} and M_{wC} parameters were maintained as discrete values because of the limitations in the supply of these commercially available materials. The summary of variable parameters for Bayesian optimization is shown in Table 2. All of measured data was utilized as an input dataset (n) in the PHYSBO with the adjusted settings for a multi-objective optimization case with two objective numbers. Additionally, the Thompson sampling method [30] was selected to search for conditions beyond the Pareto frontier line from the active learning stage. In the end, the conditions proposed by PHYSBO were conducted the experiments to confirm the results. The pipeline of the overall workflow is illustrated in Figure 2.

Table 2. Summary of variable parameters for Bayesian optimization.

Parameters	Values				Step	Type
M_{wE} (g/mol)	370	1650	2900	3800	-	Discrete
M_{wC} (g/mol)	230	400	2000	4000	-	Discrete
r	0.75–1.50				0.01	Continuous
T_C (°C)	90–120				1	Continuous

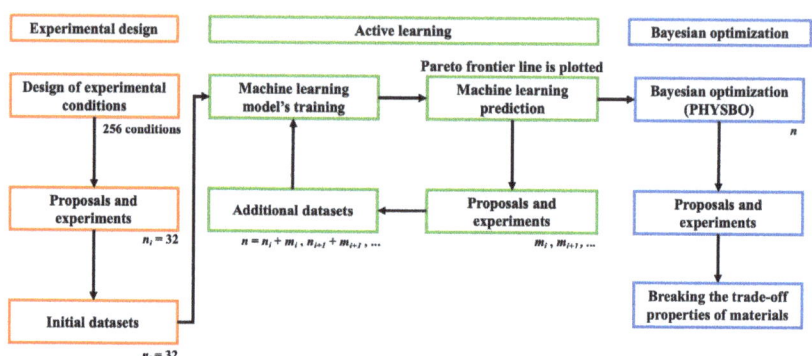

Figure 2. Illustration of the overall workflow, consisting of three stages: (i) experimental design, (ii) active learning cycle and (iii) Bayesian optimization, where n refers to the total number of datasets, i refers to the first cycle and m refers to the number of additional datasets.

3. Results

3.1. Experiments

The polymeric specimens were fabricated successfully, ensuring smooth surfaces and uniformity. Eye inspection revealed the absence of voids and bubbles within the specimens. The example of specimens is illustrated in the Figure S2. The specimens were then measured for the hardness. The elastic modulus was calculated after measuring the hardness of each specimen. The highest elastic modulus was observed at 363.9 MPa, whereas the lowest elastic modulus was observed at 0 MPa because of the incompletely cured specimen. All of the experimental results with variable parameters are listed in Table S1 (Supplementary Materials). The distribution of elastic modulus was plotted as the percentage of the total 32 specimens within specific ranges of elastic modulus values to examine the distribution of the dataset. For example, 22% indicates that 7 out of 32 specimens have elastic modulus values in the range of 0 to 46 MPa. This distribution demonstrated a well-spread dataset, as illustrated in Figure 3a. Together with the adhesive joint strength results from previous work [14], these two properties were observed as the characteristics in Figure 3b.

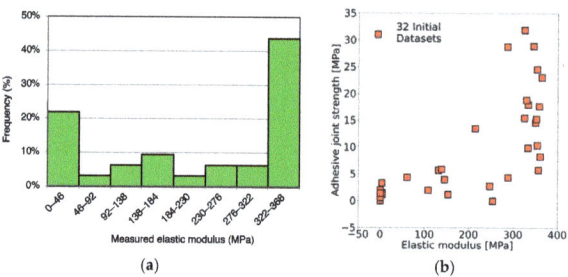

Figure 3. Experimental results of adhesive joint strength (MPa) and elastic modulus (MPa); (a) distribution chart of elastic modulus from 32 conditions of the initial dataset, (b) characteristics of adhesive joint strength (MPa) and elastic modulus (MPa) properties on 32 conditions.

3.2. Multi-Objective Optimization

3.2.1. Machine Learning Model Selection and Model Training

The accuracy of seven machine learning algorithms was evaluated by applying the 32 initial datasets as a training and testing set along with the K-fold cross-validation technique. Three evaluation tools, R^2 score, MAE and RMSE, were used to check for accuracy, and the results are reported in Table 3. The Ridge, Lasso and Elastic net algorithms manifested a similar level of accuracy. As a result, the fundamental structure of these three algorithms is a linear regression model; thus, they assumed a linear relationship between the input features and the target variables. Even though the regularization techniques of these three models are different, the accuracies are close to each other. Therefore, the initial dataset showed a non-linear relationship. The k-NN algorithm provided a slightly lower accuracy compared to the linear regression models because the k-NN model makes predictions based on the similarity of data points in the input space without assuming a specific functional form for the underlying relationship. Therefore, these four models were not selected for the reason that they are inappropriate algorithms for the dataset and provide low accuracy. The DTR model reported the most obvious lowest accuracy among the others; for this reason, it was discarded. The Random Forest and Gradient boosting models showed high accuracy on both adhesive joint strength and elastic modulus properties; however, the Gradient boosting model could achieve higher accuracy in adhesive joint strength properties. Although the Random Forest and Gradient boosting algorithm are both ensemble learning techniques, their approaches to building and combining individual models are different. Random Forest creates diverse and independent trees in parallel, while Gradient boosting builds trees sequentially, focusing on correcting errors made by the ensemble. Considering its better performance, the Gradient boosting model was nominated for further prediction.

Table 3. Comparison of the accuracy of seven machine learning models represented by R^2 score, MAE and RMSE.

		Ridge	Lasso	Elastic Net	k-NN	DTR	Random Forest	Gradient Boosting
Adhesive joint strength	R^2 score	0.42	0.43	0.42	0.40	0.18	0.51	0.60
	MAE	5.7	5.6	5.7	5.4	5.7	4.7	4.3
	RMSE	7.2	7.1	7.2	7.3	8.5	6.6	6.0
Elastic modulus	R^2 score	0.57	0.58	0.58	0.54	0.76	0.82	0.82
	MAE	70.5	69.3	70.8	70.8	43.9	41.9	40.0
	RMSE	91.7	91.3	91.5	95.3	68.5	60.2	59.5

3.2.2. Machine Learning Prediction and Proposals for Experiments

The Gradient boosting model was trained by using an initial dataset of 32 conditions as well as applying the K-fold cross-validation technique. Firstly, the averaged prediction on both adhesive joint strength and elastic modulus of a testing dataset from each fold was reported by comparing it with measured results from the experiment as plotted in Figure S3. The diagonal dash line refers to the same value between the prediction and experimental results. Secondly, the averaged prediction results of adhesive joint strength and elastic modulus for the 256 possible conditions, consisting of four values of the four variable parameters as shown in Table 1, were conducted and reported together with the initial dataset as shown in Figure 4(a1–c1). Lastly, the standard deviation of predicted adhesive joint strength and elastic modulus at each prediction point was averaged to be reported as a representative of deviation value, and it is shown in the color scale in Figure 4(a2–c2).

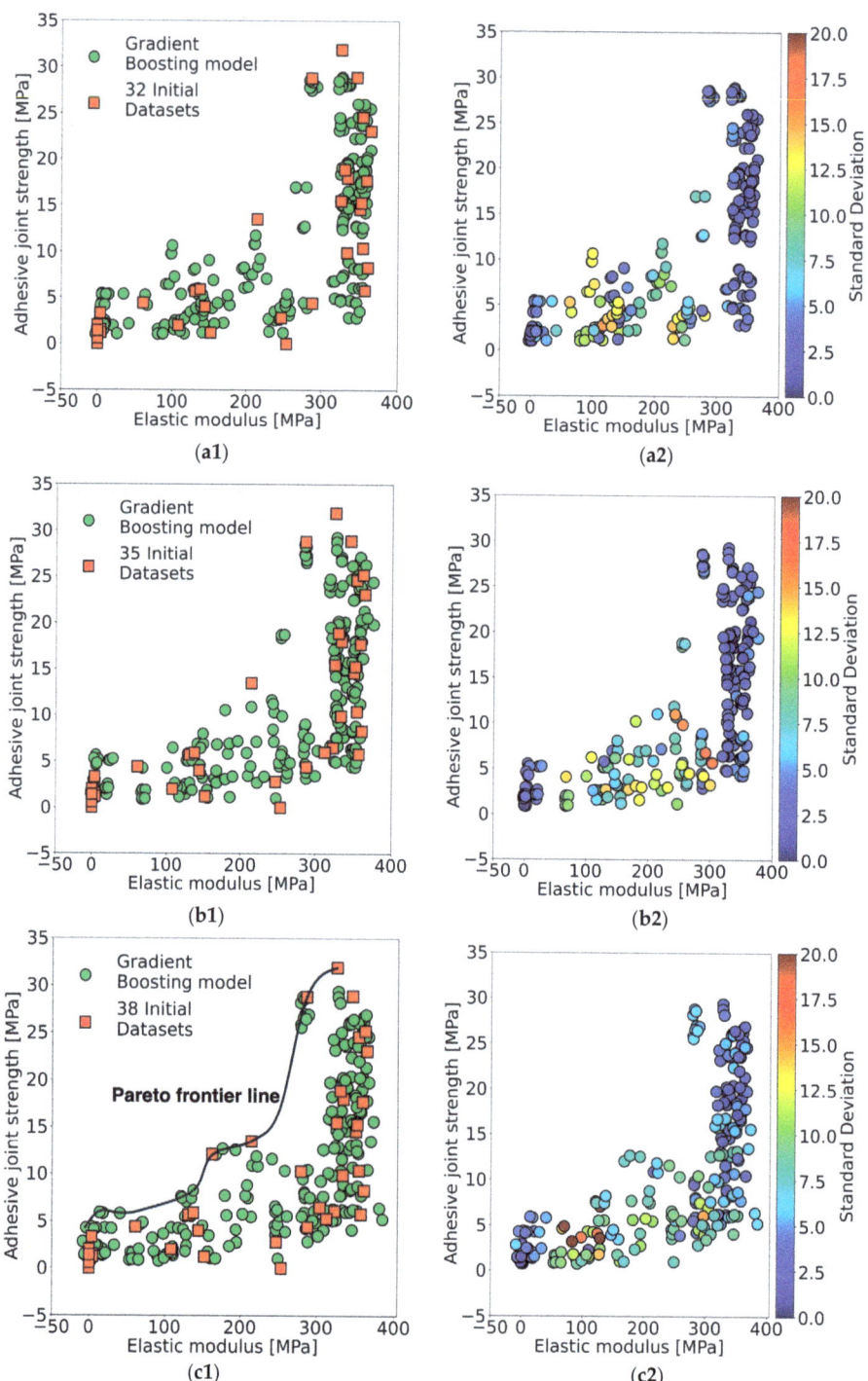

Figure 4. Prediction of the adhesive joint strength (MPa) and elastic modulus (MPa) on 256 conditions compared with initial datasets after (**a1**) active learning cycle 1; $n_i = 32$, (**b1**) cycle 2; $n = 35$, and (**c1**) cycle 3; $n = 38$, as well as the averaged standard deviation of each predicted result (**a2**) cycle 1, (**b2**) cycle 2 and (**c2**) cycle 3.

In the first cycle (Figure S3a), the prediction of the testing dataset showed a high deviation at the high adhesive joint strength of above 10 MPa; on the other hand, a high deviation could be observed at the middle range of elastic modulus from 100 to 350 MPa. In addition, the prediction of a total of 256 possible conditions on both adhesive joint strength and elastic modulus was successfully carried out as shown in Figure 4(a1). The prediction results showed good distribution along with the initial dataset (n_i = 32); however, prediction could not be observed in the area of high adhesive joint strength with low elastic modulus (top-left zone). The missing data indicate the possibility of a trade-off characteristic between these two properties. In the Figure 4(a2), the standard deviation of prediction results from the first active learning cycle showed the highest deviation at 15.5 MPa. Moreover, regarding the percentage of high-deviated prediction results, higher than half of the highest deviation, was observed at 9%. After that, the high-deviation conditions from three regions—low–low, low–high and high–high adhesive joint strength and elastic modulus, respectively—were selected for further experiments as listed in Table 4 for condition numbers 33, 34 and 35. The experimental results from the proposed conditions indicated almost similar results to the predictions except for the elastic modulus of condition number 35. This is because high deviation leads to low accuracy in the prediction.

Table 4. Prediction of adhesive joint strength (MPa) and elastic modulus (MPa) of the proposed conditions and the experimental results.

From Cycle	No.	Variable Parameters				Predicted Adhesive Joint Strength (MPa)	Predicted Elastic Modulus (MPa)	Measured Adhesive Joint Strength (MPa)	Measured Elastic Modulus (MPa)
		M_{wE} (g/mol)	M_{wC} (g/mol)	r	T_C (°C)				
1	33	370	230	1.00	130	28.1 ± 2.5	335.8 ± 8.5	25.2 ± 2.3	361.5 ± 5.2
	34	1650	400	1.25	90	6.8 ± 0.7	314.4 ± 8.0	6.1 ± 1.8	322.7 ± 8.8
	35	2900	2000	1.50	170	6.4 ± 0.8	95.3 ± 24.4	5.3 ± 2.7	312.3 ± 14.9
2	36	370	400	0.75	130	18.4 ± 1.3	253.7 ± 18.0	12.1 ± 1.8	162.8 ± 11.8
	37	2900	2000	1.50	210	7.5 ± 0.8	291.9 ± 36.0	10.3 ± 2.9	279.7 ± 20.1
	38	3800	2000	1.50	90	6.4 ± 1.2	108.3 ± 25.0	6.5 ± 2.1	303.8 ± 18.3

The measured results of both adhesive joint strength and elastic modulus from the experiment according to conditions 33, 34 and 35 were added to the initial dataset for the second cycle of active learning. These new initial datasets (n = 32 + 3 = 35) were utilized to train the machine learning model of Gradient boosting once again. Accompanying the K-fold cross-validation technique, the predictions of a testing set were compared to the measured testing set itself, and they were plotted as shown in Figure S3b. The predicted adhesive joint strength still performed with a high deviation at the high predicted values which were the same as those of the first cycle. In contrast, the model accuracy could be improved by observing a better R^2 score, MAE and RMSE after applying this active learning approach. On the other hand, the elastic modulus properties not only showed high-deviation prediction results in the middle range but also obtained lower accuracy on the R^2 score, MAE and RMSE. In Figure 4(b1), the prediction of 256 conditions was reported together with a new initial dataset (n = 35) in this second cycle. The absence of the predicted results could still be discovered in the area of low elastic modulus with high adhesive joint strength. However, the predictions were very well distributed along with the initial dataset. In the deviation point of view, the averaged standard deviation between adhesive joint strength and elastic modulus at 18.1 MPa was observed to have the highest values as shown in Figure 4(b2). Moreover, the predicted results at the area of high elastic modulus performed with an outstandingly low deviation according to the prediction of the testing dataset from Figure S3b. Nevertheless, comparing the percentage of high-deviation prediction results between the first cycle and the second cycle, it could be found that there was a 2% improvement in decreasing the high-deviation prediction results as shown in

Figure 5. After that, a new set of three conditions with high deviation, as listed in Table 4 for the conditions 36, 37 and 38, were chosen again in order to conduct an experiment for a further active learning cycle. Additionally, the measured results from the experiment are reported in Table 4. The experimental results indicated an improvement in the deviation compared to the predictions. Subsequently, these three results were appended to the initial dataset for the third active learning cycle.

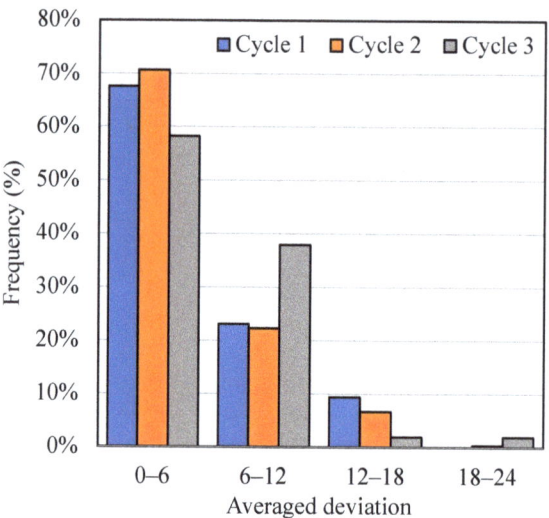

Figure 5. Distribution of averaged deviation from active learning cycle 1, cycle 2 and cycle 3.

The total 38 conditions ($n = 35 + 3 = 38$) were utilized as an initial dataset in this third active learning loop. The process was repeated by training the Gradient boosting model along with applying the K-fold cross-validation technique, then predicting the testing set and evaluating the model accuracy. The third cycle prediction of the testing set is reported in Figure S3c together with the model accuracies of both adhesive joint strength and elastic modulus. The model accuracy of the adhesive joint strength slightly decreased from the second cycle; however, it could exhibit better performance compared to the first cycle. On the other hand, the accuracy of the elastic modulus kept decreasing from the first cycle. In spite of that, it is obvious that the prediction of a testing set deviated from the diagonal dash line much less compared to that of the first and second cycles. Then, the prediction of 256 conditions was carried out and reported together with an initial dataset of 38 conditions as shown in Figure 4(c1). The same tendency as seen in the previous cycle was observed: the predictions could not be found in the area of high adhesive joint strength with low elastic modulus. Therefore, this could imply a trade-off between these two properties after three cycles of the active learning loop. Additionally, the Pareto frontier line was drawn to represent a trade-off boundary connecting all of the results at the top-left edge. In Figure 4(c2), the maximum deviation of these two properties is 21.0 MPa, and the extremely high-deviation results show less than 10 conditions. Moreover, the deviation of the overall predicted results improved because of the clearly seen shift in color scale from yellow in the first cycle to light blue in this cycle. Furthermore, a significant enhancement in prediction accuracy was observed, with a 50% reduction in high-deviation outcomes from the first to the third cycle, as illustrated in Figure 5. The active learning cycle terminated once the average standard deviation of predictions closely matched that of experimental results for both adhesive joint strength and elastic modulus. Thus, the errors in the predictions were acceptable according to the errors from the experiments. In conclusion, the active learning approach was able to balance machine learning model accuracy and the deviation of each single predicted result. Consequently, the Pareto frontier line could be obtained from the predictions of 256 conditions (Figure 4(c1)), and it was highly reliable after achieving three cycles of the active learning loop.

The influence of each variable parameter on the prediction of both properties is illustrated in Figure S4. It was observed that the molecular weight of the epoxy resin (M_{wE}) exhibited minimal impact on both properties as indicated by its ability to perform consistently across a wide range of predicted adhesive joint strength and elastic modulus (Figure S4a). However, the epoxy resin with molecular weight of 370 g/mol offers significant advantages in processability due to its liquid phase state. In contrast, the molecular weight of the curing agent (M_{wC}) demonstrated a significant impact on the predicted properties. Lower-viscosity curing agents with M_{wC} values of 230 and 400 g/mol resulted in a prediction of high adhesive joint strength and elastic modulus. Conversely, higher-viscosity curing agents with M_{wC} values of 2000 and 4000 g/mol only achieved a maximum adhesive joint strength of 11.8 MPa. Regarding elastic modulus, predictions with a modulus below 300 MPa were challenging to obtain using curing agents with an M_{wC} of 230 g/mol, suggesting the necessity for higher molecular weight curing agents for low modulus adhesives. Moreover, the control of predicted adhesive joint strength and elastic modulus properties is more readily achieved through the molecular weight of the curing agent (Figure S4b). The impact of the amine-to-epoxide ratio is shown in Figure S4c, indicating its uniform distribution across the range of predicted properties. Higher ratios notably enhanced adhesive joint strength, particularly surpassing 16 MPa, while elastic modulus values between 100 and 200 MPa were predominantly attained with a ratio of 0.75. Additionally, it was observed that curing temperatures at 90 °C resulted in predicted adhesive joint strengths below 12 MPa. Conversely, curing temperatures exceeding 90 °C facilitated a more favorable distribution for both predicted adhesive joint strength and elastic modulus properties, as illustrated in Figure S4d.

3.2.3. Bayesian Optimization

According to the trade-off behavior proposed in the previous section (Figure 4(c1)), adhesive epoxy materials exhibiting high adhesive joint strength but low elastic modulus have yet to be developed through the machine learning approach. However, such a formulation, which shows high adhesive joint strength and low elastic modulus, could offer considerable advantages, particularly in applications requiring efficient force absorption in adhesive materials. To address this, a new initial dataset comprising 38 conditions was utilized for further investigation, employing Bayesian optimization to identify conditions conducive to achieving an adhesive epoxy with the desired properties. In this stage, PHYSBO which is a Python library for Bayesian optimization studies focusing on the physics, chemistry and materials science fields was performed to optimize this multi-objective problem. Four variable parameters with discrete and continuous types, as summarized in Table 2, were studied. The proposed conditions from Bayesian optimization were separately reported in four figures, as shown in Figures S5–S8. The predicted adhesive joint strength and elastic modulus from Bayesian optimization were plotted individually by varying the continuous parameters of amine-to-epoxide ratio and curing temperature; however, the predictions by varying discrete parameters of M_{wE} and M_{wC} were plotted one by one in separate charts. The predicted adhesive joint strength and elastic modulus were mainly changed by varying M_{wE} and M_{wC}. Particularly, changing the M_{wC} along its four values could obtain a variety of predicted adhesive joint strength and elastic modulus results. This was confirmed by the influence of M_{wC} on prediction of both properties after the final active learning cycle (Figure S4). On the other hand, the influence of amine-to-epoxide ratio and curing temperature exhibited a notable influence on the predicted results, particularly noticeable at lower values of M_{wE} and M_{wC}. Afterward, the conditions proposed by Bayesian optimization, where the predicted results exceeded the trade-off boundary, were selected to overcome the Pareto frontier line. Considering the discussion from the previous section regarding the influence of each variable parameter, three conditions with predicted results beyond the Pareto frontier line were selected as listed in Table 5. Additionally, the experimental results corresponding to these conditions were reported in Table 5 and plotted alongside the Pareto frontier line in Figure 6. Notably, conditions 39 and 40 exhibited adhesive joint strengths of 10.2 MPa and 25.2 MPa, respectively, along with elastic modulus of 74.2 MPa and 182.5 MPa, positioning them further from the trade-off limit. Furthermore, a specimen with desirable properties as of high adhesive joint strength together with low elastic modulus was successfully fabricated as polymeric specimen number 40. A comparison of the appearance between specimen number 40 and the specimen with high adhesive joint strength and elastic modulus (number 26) was conducted to confirm the difference, as reported in Figure S9. Significantly, specimen number 40 displayed superior ductile characteristics compared to specimen number 26, despite both exhibiting the same level of adhesive joint strength values. Additionally, elasticity behavior was demonstrated, as illustrated in Figure S10. Specimen number 26, with a higher elastic modulus, retained its shape when subjected to a 100 g load, whereas specimen number 40, with a lower elastic modulus, exhibited deformation under the same load. Therefore, not only was the trade-off between adhesive joint strength and elastic modulus optimized, but also, a specimen with the desired properties was successfully fabricated by adhering to the suggested conditions derived from the Bayesian optimization approach.

Table 5. Proposed conditions from the Bayesian optimization step, and the experimental results according to the conditions.

No.	Variable Parameters				Predicted Adhesive Joint Strength (MPa)	Predicted Elastic Modulus (MPa)	Measured Adhesive Joint Strength (MPa)	Measured Elastic Modulus (MPa)
	M_{wE} (g/mol)	M_{wC} (g/mol)	r	T_C (°C)				
39	370	2000	1.46	129	11.5	21.5	10.2 ± 0.6	74.2 ± 3.3
40	370	400	0.75	168	17.4	177.5	25.2 ± 1.3	182.5 ± 7.9
41	370	230	1.13	160	32.3	346.8	30.3 ± 0.9	319.5 ± 17.4

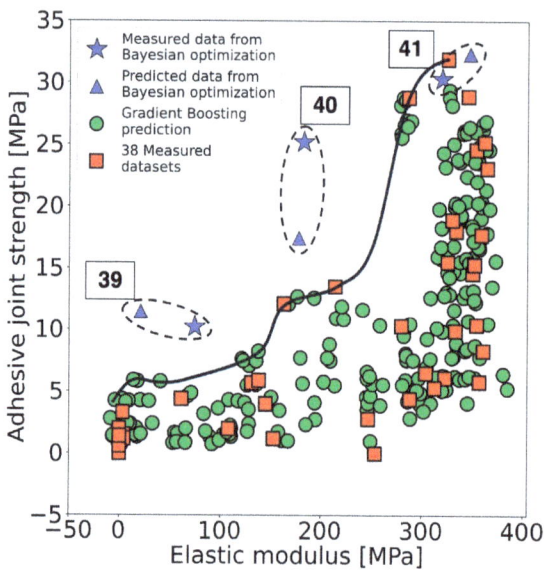

Figure 6. The experimental results and proposed conditions for adhesive joint strength (MPa) and elastic modulus (MPa) from multi-objective optimization using PHYSBO along with the predictions from active learning and a Pareto frontier line.

4. Conclusions

Multi-objective optimization was employed to analyze the adhesive joint strength and elastic modulus properties of commercially available epoxy resin (DGEBA) with an amine-terminated curing agent (Jeffaime™). Four variable parameters, consisting of molecular weight of DGEBA (M_{wE}), molecular weight of Jeffamine™ (M_{wC}), amine-to-epoxide ratio (r) and curing temperature (T_C), were investigated. In this study, the elastic modulus was measured using a specially designed metal mold apparatus. The polymeric specimens were fabricated with smooth surfaces and homogeneity. The elastic modulus values were derived from hardness measurements of each specimen. Subsequently, these values, along with the adhesive joint strength values from a previous study, constituted the initial dataset of 32 conditions to train the machine learning model. Among the seven machine learning algorithms evaluated, the Gradient Boosting model exhibited the highest accuracy and was selected for further analysis. The initial dataset was then utilized in an active learning approach to train the Gradient Boosting model and make predictions, resulting in a 50% reduction in deviation of predicted results and improved balance in predictive accuracy across adhesive joint strength and elastic modulus properties after the third cycle of the active learning approach. Consequently, the Pareto frontier line showing the trade-off boundary between these two properties was able to be reliably presented. The missing prediction area at high adhesive joint strength with low elastic modulus was observed as the trade-off area. The influence of each variable parameter on both adhesive joint strength and elastic modulus properties was examined. The results indicated that an epoxy resin with M_{wE} of 370 g/mol offered optimal processability because of its liquid phase state, the M_{wC} of the Jeffamine™ curing agent played a crucial role in controlling properties to achieve specimens with high adhesive joint strength and low elastic modulus, an amine-to-epoxide ratio (r) of 0.75 was suitable for fabricating adhesive epoxy with lower elastic modulus, and a curing temperature (T_C) above 90 °C was necessary to maintain adhesive ability.

Bayesian optimization was employed to address the trade-off challenges. Utilizing a dataset of n = 38 (initial 32 datasets plus an additional 6), PHYSBO was employed to optimize the adhesive joint strength and elastic modulus properties of the adhesive epoxy system. Molecular weights of DGEBA (M_{wE}) and Jeffamine™ (M_{wC}) were treated as discrete variables, while amine-to-epoxide ratio (r) and curing temperature (T_C) were explored as continuous variables. The PHYSBO approach effectively predicted and suggested properties for this adhesive epoxy system across a vast array of conditions, up to 147,136 in total. Upon experimentation, the optimized conditions were validated, successfully transcending the trade-off boundary (Pareto frontier line) with adhesive joint strength reaching 25.2 MPa and elastic modulus at 182.5 MPa, respectively. The fabricated specimen exhibited superior elastic behavior. This multi-objective optimization strategy provides valuable insights into the curing conditions for the studied adhesive epoxy system, elucidating the impact of each variable parameter on mechanical properties. Moreover, Bayesian optimization demonstrates its capability to efficiently suggest conditions for desired properties, accelerating the overall process, reducing costs, and time consumption, while also enabling the breakthrough of trade-off constraints in the adhesive epoxy system.

Supplementary Materials: The following supporting information can be downloaded at: https://www.mdpi.com/article/10.3390/ma17122866/s1, Figure S1: An illustration of (a) a metal mold according to French standard NFT76-142, (b) dimension of the specimens and the measuring points according to ASTM-D2240; Figure S2: An example of the specimen's appearance with conditions 26 and 31; Figure S3: The averaged prediction of adhesive joint strength and elastic modulus from 32-fold cross-validation compared to its measured results from experiments as well as the accuracies of each active learning cycle; (a) 1st cycle, (b) 2nd cycle and (c) 3rd cycle; Figure S4: The influence of each variable parameter on the prediction of 256 conditions from the 3rd active learning cycle: (a) molecular weight of epoxy resin, (b) molecular weight of curing agent, (c) amine to epoxide ratio and (d) curing temperature; Figure S5: The proposed conditions from Bayesian optimization for adhesive joint strength (MPa) and elastic modulus (MPa) by varying three variable parameters—molecular weight of Jeffamine™ (g/mol), amine-to-epoxide ratio and curing temperature (°C)—with the molecular weight of DGEBA at 370 g/mol;

Figure S6: The proposed conditions from Bayesian optimization for adhesive joint strength (MPa) and elastic modulus (MPa) by varying three variable parameters—molecular weight of Jeffamine™ (g/mol), amine-to-epoxide ratio and curing temperature (°C)—with the molecular weight of DGEBA at 1650 g/mol; Figure S7: The proposed conditions from Bayesian optimization for adhesive joint strength (MPa) and elastic modulus (MPa) by varying three variable parameters—molecular weight of Jeffamine™ (g/mol), amine-to-epoxide ratio and curing temperature (°C)—with the molecular weight of DGEBA at 2900 g/mol; Figure S8: The proposed conditions from Bayesian optimization for adhesive joint strength (MPa) and elastic modulus (MPa) by varying three variable parameters—molecular weight of Jeffamine™ (g/mol), amine-to-epoxide ratio and curing temperature (°C)—with the molecular weight of DGEBA at 3800 g/mol; Figure S9: The appearance of specimens after tearing by tensile force of the specimen with condition number 26 (high elastic modulus) shows less ductile behavior compared to the specimen with condition number 40 (less elastic modulus); Figure S10: Demonstration of the elasticity behavior of the specimens: (a) high elastic modulus specimen no. 26 before adding load, (b) after adding 100 g load, (c) low elastic modulus specimen no. 40 before adding load and (d) after adding 100 g load; Figure S11: A picture of actual metal mold consisting of a metal lid, Teflon tape, metal box, silicon rubber frame and adhesive epoxy specimen; Table S1: Experimental results of adhesive joint strength (MPa) and elastic modulus (MPa) of each condition consist of four variable parameters: molecular weight of DGEBA M_{wE} (g/mol), molecular weight of Jeffaime™ curing agent M_{wC} (g/mol), amine-to-epoxide ratio; r, and curing temperature T_C (°C). Initial dataset size n_i = 32 samples. Refs [14,17] are cited in the supplementary materials.

Author Contributions: Conceptualization, P.K. and M.N.; data curation, P.K. and C.K.; formal analysis, P.K.; investigation, P.K. and C.K.; resources, M.N. and C.S.; software, P.K.; supervision, M.N. and C.S.; writing—original draft preparation, P.K. and C.K.; writing—review and editing, P.K. and M.N.; visualization, P.K.; funding acquisition, M.N. All authors have read and agreed to the published version of the manuscript.

Funding: This work was supported by the Core Research for Evolutional Science and Technology (CREST) program 'Revolution material development by fusion of strong experiments with theory/data science' of the Japan Science and Technology Agency (JST), Japan, under Grant JPMJCR19J3, KAKENHI Grant-in-Aid for Scientific Research (B): 23H02031 and MEXT Program: Data Creation and Utilization-Type Material Research and Development Project Grant Number JPMXP1122714694.

Institutional Review Board Statement: Not applicable.

Informed Consent Statement: Not applicable.

Data Availability Statement: Data are contained within the article and Supplementary Materials.

Acknowledgments: We thank Sirawit Pruksawan, a former member from our group, for the initial dataset of adhesive joint strength.

Conflicts of Interest: The authors declare no conflicts of interest.

References

1. Wei, Y.; Jin, X.; Luo, Q.; Li, Q.; Sun, G. Adhesively Bonded Joints—A Review on Design, Manufacturing, Experiments, Modeling and Challenges. *Compos. B Eng.* **2024**, *276*, 111225. [CrossRef]
2. Cole, G.S.; Sherman, A.M. Light Weight Materials for Automotive Applications. *Mater. Charact.* **1995**, *35*, 3–9. [CrossRef]
3. Campbell, F.C. *Manufacturing Technology for Aerospace Structural Materials*; Elsevier: Amsterdam, The Netherlands, 2006; ISBN 9781856174954.
4. Van Straalen, I.J.J.; Van Tooren, M.J.L. Building and Construction—Steel and Aluminium. In *Adhesive Bonding*; Elsevier: Amsterdam, The Netherlands, 2005; pp. 305–327.
5. Xu, W.; Wei, Y. Strength and Interface Failure Mechanism of Adhesive Joints. *Int. J. Adhes. Adhes.* **2012**, *34*, 80–92. [CrossRef]
6. Jones David, R.H.; Ashby Michael, F. *Engineering Materials 1*, 5th ed.; Elsevier: Amsterdam, The Netherlands, 2019; ISBN 9780081020517.
7. Gordon, T.L.; Fakley, M.E. The Influence of Elastic Modulus on Adhesion to Thermoplastics and Thermoset Materials. *Int. J. Adhes. Adhes.* **2003**, *23*, 95–100. [CrossRef]
8. Loureiro, A.L.; da Silva, L.F.M.; Sato, C.; Figueiredo, M.A.V. Comparison of the Mechanical Behaviour Between Stiff and Flexible Adhesive Joints for the Automotive Industry. *J. Adhes.* **2010**, *86*, 765–787. [CrossRef]
9. Kanani, A.Y.; Liu, Y.; Hughes, D.J.; Ye, J.; Hou, X. Fracture Mechanisms of Hybrid Adhesive Bonded Joints: Effects of the Stiffness of Constituents. *Int. J. Adhes. Adhes.* **2020**, *102*, 102649. [CrossRef]

10. Rocha, J.; Sena-Cruz, J.; Pereira, E. Influence of Adhesive Stiffness on the Post-Cracking Behaviour of CFRP-Reinforced Structural Glass Beams. *Compos. B Eng.* **2022**, *247*, 110293. [CrossRef]
11. Taniguchi, S.; Uemura, K.; Tamaki, S.; Nomura, K.; Koyanagi, K.; Kuchii, S. Multi-Objective Optimization of the Epoxy Matrix System Using Machine Learning. *Results Mater.* **2023**, *17*, 100376. [CrossRef]
12. Mogilicharla, A.; Chugh, T.; Majumdar, S.; Mitra, K. Multi-Objective Optimization of Bulk Vinyl Acetate Polymerization with Branching. *Mater. Manuf. Process.* **2014**, *29*, 210–217. [CrossRef]
13. Antil, P. Modelling and Multi-Objective Optimization during ECDM of Silicon Carbide Reinforced Epoxy Composites. *Silicon* **2020**, *12*, 275–288. [CrossRef]
14. Pruksawan, S.; Lambard, G.; Samitsu, S.; Sodeyama, K.; Naito, M. Prediction and Optimization of Epoxy Adhesive Strength from a Small Dataset through Active Learning. *Sci. Technol. Adv. Mater.* **2019**, *20*, 1010–1021. [CrossRef] [PubMed]
15. Qi, H.J.; Joyce, K.; Boyce, M.C. Durometer Hardness and the Stress-Strain Behavior of Elastomeric Materials. *Rubber Chem. Technol.* **2003**, *76*, 419–435. [CrossRef]
16. NFT 76-142; Méthode de Préparation de Plaques D'adhésifs Structuraux Pour la Réalisation D'éprouvettes D'essai de Caractérisation. Afnor EDITIONS: La Plaine Saint-Denis, France, 1988.
17. ASTM-D2240; Standard Test Method for Rubber Property-Durometer Hardness. American Society for Testing and Materials (ASTM): West Conshohocken, PN, USA, 2021.
18. da Silva, L.F.M.; Dillard, D.A.; Blackman, B.; Adams, R.D. (Eds.) *Testing Adhesive Joints*; Wiley: Hoboken, NJ, USA, 2012; ISBN 9783527329045.
19. Cooper, B.E. *Statistics for Experimentalists*; Elsevier: Amsterdam, The Netherlands, 1969; ISBN 9780080126005.
20. Keppel, G. *Design and Analysis: A Researcher's Handbook*, 1st ed.; Prentice-Hall, Inc: Englewood Cliffs, NJ, USA, 1991; ISBN 0-13-200775-4. (In Hardcover)
21. Refaeilzadeh, P.; Tang, L.; Liu, H. Cross-Validation. In *Encyclopedia of Database Systems*; Springer: Boston, MA, USA, 2009; pp. 532–538.
22. Hoerl, A.E.; Kennard, R.W. Ridge Regression: Biased Estimation for Nonorthogonal Problems. *Technometrics* **2000**, *42*, 80. [CrossRef]
23. Tibshirani, R. Regression Shrinkage and Selection Via the Lasso. *J. R. Stat. Soc. Ser. B Methodol.* **1996**, *58*, 267–288. [CrossRef]
24. Zou, H.; Hastie, T. Regularization and Variable Selection Via the Elastic Net. *J. R. Stat. Soc. Ser. B Stat. Methodol.* **2005**, *67*, 301–320. [CrossRef]
25. Hastie, T.; Tibshirani, R.; Friedman, J. *The Elements of Statistical Learning*; Springer Series in Statistics; Springer: New York, NY, USA, 2009; ISBN 978-0-387-84857-0.
26. Breiman, L.; Friedman, J.H.; Olshen, R.A.; Stone, C.J. *Classification And Regression Trees*; Routledge: New York, NY, USA, 2017; ISBN 9781315139470.
27. Breiman, L. *Random Forests*; Springer: Berlin/Heidelberg, Germany, 2001; Volume 45.
28. Friedman, J.H. Greedy Function Approximation: A Gradient Boosting Machine. *Ann. Stat.* **2001**, *29*, 1189–1232. [CrossRef]
29. Motoyama, Y.; Tamura, R.; Yoshimi, K.; Terayama, K.; Ueno, T.; Tsuda, K. Bayesian Optimization Package: PHYSBO. *Comput. Phys. Commun.* **2022**, *278*, 108405. [CrossRef]
30. Verleysen, M.; Université catholique de Louvain; Katholieke Universiteit Leuven; European Symposium on Artificial Neural Networks, Computational Intelligence and Machine Learning; ESANN. *Proceedings/23rd European Symposium on Artificial Neural Networks, Computational Intelligence and Machine Learning, ESANN 2015, Bruges, Belgium, 22–24 April 2015*; Ciaco: Ottignies-Louvain-la-Neuve, Belgium, 2015; ISBN 978-287587014-8.

Disclaimer/Publisher's Note: The statements, opinions and data contained in all publications are solely those of the individual author(s) and contributor(s) and not of MDPI and/or the editor(s). MDPI and/or the editor(s) disclaim responsibility for any injury to people or property resulting from any ideas, methods, instructions or products referred to in the content.

Article

Evaluation of True Bonding Strength for Adhesive Bonded Carbon Fiber-Reinforced Plastics

Maruri Takamura [1], Minori Isozaki [1], Shinichi Takeda [2], Yutaka Oya [1] and Jun Koyanagi [1,*]

[1] Department of Materials Science and Technology, Tokyo University of Science, Tokyo 125-8585, Japan; 8217056@alumni.tus.ac.jp (M.T.); 8219012@alumni.tus.ac.jp (M.I.); oya@rs.tus.ac.jp (Y.O.)
[2] Japan Aerospace Exploration Agency, Tokyo 181-0015, Japan; takeda.shinichi@jaxa.jp
* Correspondence: koyanagi@rs.tus.ac.jp

Abstract: Carbon fiber-reinforced thermoplastics (CFRTPs) have attracted attention in aerospace because of their superior specific strength and stiffness. It can be assembled by adhesive bonding; however, the existing evaluation of bonding strength is inadequate. For example, in a single-lap shear test, the weld zone fails in a combined stress state because of the bending moment. Therefore, the strength obtained experimentally is only the apparent strength. The true bonding strength was obtained via numerical analysis by outputting the local stress state at the initiation point of failure. In this study, the apparent and true bonding strengths were compared with respect to three types of strength evaluation tests to comprehensively evaluate bonding strength. Consequently, the single-lap shear test underestimates the apparent bonding strength by less than 14% of the true bonding strength. This indicates that care should be taken when determining the adhesion properties for use in numerical analyses based on experimental results. We also discussed the thickness dependence of the adhesive on the stress state and found that the developed shear test by compression reduced the discrepancy between apparent and true strength compared with the single-lap shear test and reduced the thickness dependence compared with the flatwise tensile test.

Keywords: numerical simulation; bond strength test; interfacial strength; CFRP ultrasonic welding

1. Introduction

Carbon fiber-reinforced plastics (CFRPs) have attracted attention in the automotive and aerospace industries because of their superior specific strength and stiffness [1–3]. For example, its light weight and high strength improve the fuel efficiency of automobiles and aircraft, thereby reducing their environmental impact. Among CFRPs, those whose resin portion is made of thermoplastic resin are called carbon fiber-reinforced thermoplastics (CFRTPs). CFRTP has advantages over CFRP, which is made of thermoplastic resin, including easier molding and recycling. In addition, CFRTPs can be assembled by welded joints because they melt when heated. In contrast to fastening with bolts, welding eliminates the need for bolt drilling, resulting in weight reduction and improving mechanical properties by reducing stress concentrations [4–6].

Once the weld joining of CFRTPs is completed, the strength of these joints needs to be verified to evaluate the strength of the joints and design the structure using numerical simulations. For example, the single-lap shear (SLS) and flatwise tensile (FW) tests are known strength evaluation methods for bonded materials [7,8]. SLS evaluates the strength of a joint by applying shear stress to bonded surfaces, whereas FW evaluates the strength of a joint by applying tensile stress to bonded surfaces. Both are often used in bond strength tests. However, they have the following problems. The stress state at the bonded interface depends on the specimen size [9–14]. The bond fails under the combined stress state [15–19]. The fracture stress obtained by dividing the experimental reaction force at failure by the bonded area is the average apparent strength, and if this apparent strength is used as the strength of the interface in the numerical analysis as it is, the experiment cannot be reproduced correctly [20,21]. The size of the specimens must be large to comply with test standards such as ISO and ASTM, and the joints are prone to combined stresses and expensive to test.

To address these problems, we must use numerical analysis to investigate the stress state at the initiation point of failure in strength evaluation tests and perform a comprehensive evaluation to identify the apparent strength obtained experimentally and to what extent it reproduces the true strength. The stresses applied to the interface during the strength evaluation testing are distributed isotropically, and it is difficult to apply pure shear or normal stress to bonded CFRTP [22–29]. However, the nominal stress, which is obtained by dividing the breaking load by the bonded area, assumes that the stress distribution is uniform. Therefore, the obtained fracture stress is a weaker estimate than the strength of the interface for pure shear or normal stress. The apparent strength is referred to as the stress at failure, which is the experimental reaction force at failure divided by the weld area, and the localized failure stress at the initiation point of failure is referred to as the true strength. In a previous study, DeVries et al. discussed the stress state in SLS from a material-mechanics perspective [30]. Villegas et al. demonstrated that in SLS, the bond strength at the edge of the bonded surface directly affects the strength of the entire interface [31]. Redmann et al. proposed a block shear test method to reduce these effects [32].

The following is the background of our research: We have been conducting a combined experimental and analytical study on the ultrasonic welding of CFRTP and investigated the effects of temperature increase and adhesive shape during welding [33]. Ultrasonic welding is a method of welding materials by bringing a metal transducer into contact with overlapping adherends and propagating ultrasonic vibrations of around 20 kHz while applying pressure. The ultrasonic vibration is a longitudinal vibration perpendicular to the welding surface, and the frictional heat generated by the vibration between materials and molecules raises the temperature and melts the thermoplastic resin. Welding can be performed only by applying ultrasonic vibration for a short period of time of 5 s or less under ambient temperature and pressure, making it a highly efficient and low-energy joining method. Generally, when ultrasonic welding is used, a thermoplastic resin called an energy director is placed between the adherends as an adhesive to serve as the starting point for melting, thereby enabling stable welding. In ultrasonic welding, rapid temperature changes occur in a short period of time, so various parameters, such as ultrasonic frequency, welding time, and welding pressure, are complicated and affect the welding conditions [34–39]. Therefore, it is important to select these parameters appropriately. Hence, specimens after ultrasonic welding need to be carefully evaluated. In addressing this issue, we noticed that the existing strength evaluation test methods did not adequately evaluate the adhesive strength. Thus, we embarked on a comprehensive evaluation of bond strength using numerical simulation aiming to predict the local true joint strength from the experimentally obtained apparent strength of ultrasonically welded CFRTP.

In this study, by comparing the true joint interface strength with the apparent strength, we suggest the dangers of existing strength evaluation tests and provide guidelines for a comprehensive interface strength evaluation method. We studied the relationship between the apparent strength in bond strength tests of welded specimens that failed at the adhesive–adherend bonding interface, and the true strength at the fracture initiation point was investigated using a numerical simulation of the stress state at the fracture initiation point. Based on these results, a fracture envelope was drawn to comprehensively evaluate the bond strength. In addition, we propose a novel shear test by compression that reduces the discrepancy between the apparent strength and true strength and is less dependent on the thickness of the adhesive. Through comparison of the apparent strength in this shear test by compression with that in the SLS and FW tests, respectively, the percentage of underestimation of apparent strength, the stress state at the actual fracture initiation point, and the effect of adhesive thickness on apparent strength were discussed.

2. Methods

2.1. Determination of Interfacial Strength between Adhesive and Adherend

In order to investigate the stress state at the bond interface at the onset of failure in various bond strength tests by numerical simulation, the value of the interfacial strength between the adhesive and the adherend was first determined. This interfacial strength was determined on an experimental basis, as reasonable values were used.

2.1.1. Specimen Welding and Shear Test by Compression

We developed a shear test using compression (SC) to conduct bond strength tests using small specimens. SC is a test method in which a bonded specimen is fractured by applying displacement in the shear direction to the bonded surface. This is similar to SLS; however, because SC applies a compressive load, whereas SLS applies a tensile load, the adherend does not require a certain length to hold the tensile jig in place. The use of smaller specimens reduces the bending moments and lowers the cost of the experiment. A schematic of the jig used for the SC, as shown in Figure 1, fixes one side of the adherend and compresses the other test piece from above.

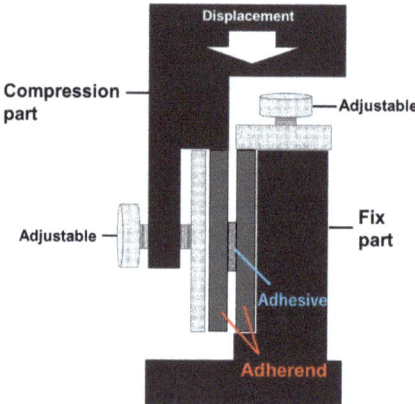

Figure 1. Illustration of the fixture developed for the shear test by compression. One side of the specimen is fixed with a fixed part, and the other side is compressed from the top of the specimen with a compression part to shear failure at the adhesive interface.

In the welding experiment, quasi-isotropic polyether ether ketone reinforced with carbon fiber (CF/PEEK) (IMS/PEEK, Teijin, Osaka, Japan) with a thickness of 2 mm and a length of 30 × 30 cm per side was used as the adherend, and three layers of PEEK mesh (#25, Clever Co., Ltd., Toyohashi, Japan) were sandwiched between the adherends as an adhesive. The specimens were heated by applying 19.5 kHz ultrasonic waves for 3 s under a pressure of 0.1 MPa using an ultrasonic welding machine (JP80s, SEIDENSHA ELECTRONICS Co., LTD, Tokyo, Japan). A circular horn with a diameter of 10 mm was used as a transducer. After the ultrasonic vibration was unloaded, adhesion was performed by maintaining a holding load of 0.1 MPa for 5 s. Then, to confirm that the welding was completed, the condition of the adhesive in the welded area was observed by obtaining CT scan imaging of the welded specimen using a nondestructive inspection device (inspeXio MX-225CTS, SHIMADZU, Kyoto, Japan). A jig for SC was set up on a universal testing machine (Model 8802, INSTRON, Kawasaki, Japan) to evaluate the bond strengths of the welded specimens. SC was performed to obtain the reaction forces at failure, and the fracture surfaces of the specimens were observed using a Digital Microscope (VHX-6000, KEYENCE, Osaka, Japan).

2.1.2. Numerical Analysis to Fit Cohesive Properties to Experimental Results

In numerical simulations, a parabolic criterion can be used as the failure criterion for the interface, as expressed in Equation (1) proposed by Ogihara et al. [40], where Y_n is the pure normal strength of the interface, Y_s is the pure shear strength of the interface, and t_n and t_s are the normal and shear stress at the interface, respectively. Here, normal stress indicates the vertical stress to the interface. When the value on the left-hand side reached one, the interface began to fail. Furthermore, Koyanagi et al. proved that the interfacial strength can be expressed by Equation (2) as the ratio of shear strength to normal strength [41].

$$\frac{t_n}{Y_n} + \left\{\frac{t_s}{Y_s}\right\}^2 = 1 \quad (1)$$

$$\sqrt{2} Y_n = Y_s \quad (2)$$

Therefore, appropriate, cohesive properties can be obtained by numerically simulating the SC and fitting the interfacial strength at which the reaction force is equal to the experimental value. SC, like other bond strength tests, fails at the interface in a combined stress state, but by reproducing the experiment with a numerical model, the pure shear strength at the interface and normal strength can be predicted. The stress at debonding in the material properties of this cohesive element represents the true interfacial strength. We used Abaqus2020, a numerical analysis software, to simulate the model with the same dimensions as in the experiment, with cohesive elements introduced at the adhesive–adherend interface. Here the cohesive element is a sufficiently stiff offset element with a thickness of 0 mm. It represents the separation behavior of the interface depending on the stress state. The interface strength can be determined by specifying Y_n, the pure vertical strength, and Y_s, the pure shear strength. The properties of the materials are listed in Table 1. However, the cohesive property in this model is a fitting parameter. Here, for the material property anisotropy of CF/PEEK, E1 and E2 are assigned to the in-plane direction of the material and E3 to the material thickness direction. The area of the welded part was obtained through CT scanning of the welded specimen using a nondestructive inspection device (inspeXio SMX-225CTS, SHIMADZU, Kyoto, Japan) before the test. The cohesive properties were fitted such that the reaction force at the edge of the model at the onset of failure was equal to the experimental value according to the shear strength–normal strength ratio in Equation (2). In this analytical model, we focused only on the fracture initiation point and hypothesized that interface separation would be introduced, but no material damage would occur.

Table 1. Material properties of CF/PEEK adherend, PEEK adhesive, and interface bonding cohesive.

	Young modulus (MPa)						Poisson's ratio		
	E1	E2	E3	G12	G13	G23	Nu12	Nu13	Nu23
CF/PEEK [41]	56,800	56,800	8210	43,600	3000	3000	0.25	0.35	0.35
PEEK	Young modulus (MPa)							Poisson's ratio		
	3000							0.37		
Cohesive	Interfacial strength (MPa)								
	Y_n						Y_s		
	48.0126						67.9		

2.2. Numerical Analysis to Derive True and Apparent Bond Strength

Numerical simulations of CS, SLS, and FW were performed to investigate changes in the relationship between true interfacial strength and apparent strength across different test methods, with the adhesive thickness being consistent in each test. In addition, we investigated changes in the apparent strength in each test when the adhesive thickness varied.

2.2.1. Comparison of Underestimation of Apparent Strength in Three Different Bond Strength Tests

Numerical simulations were performed for the three types of bond strength tests using the cohesive properties determined in the previous section, as shown in Figure 2. We modeled three types of tests: the ISO standard (ISO 4587:2003) [42] SLS, SC, and FW, which is a test with a different stress state at failure as a reference. The models were constructed using the cohesive elements introduced at the adhesive–adherend interface. As shown in Figure 3, the adhesive and adherend thicknesses were 0.2 mm and 2.0 mm, respectively, for all models. As for the mesh size dependence, the mesh-dependent effect due to stress concentration was reduced by filleting the edge of the welded area. Displacement in the shear direction to the adhesive surface was applied to the edge of the adherend in SLS and SC, and displacement in the normal direction in FW. Then, the stress state at the point of failure initiation and the apparent strength, obtained by dividing the reaction force at the start of fracture at the specimen end under displacement by the initial adhesive area, were calculated. A parabolic criterion was adopted as the destruction criterion. In these models, moreover, we focused only on the cases where the interface was the initiation point of fracture; thus, material fracture was not introduced. The properties of the materials are listed in Table 1.

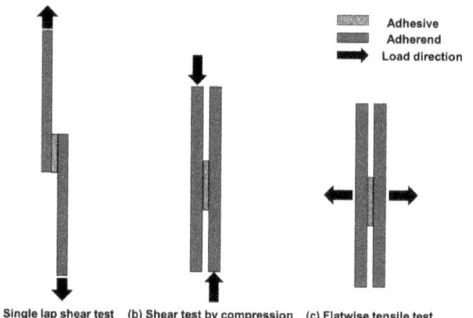

Figure 2. Schematic of the three types of bond strength tests discussed in this study. (**a**) Single-lap shear test and (**b**) shear test by compression apply load in shear direction to the interface, and (**c**) flatwise tensile test applies load vertically to the interface.

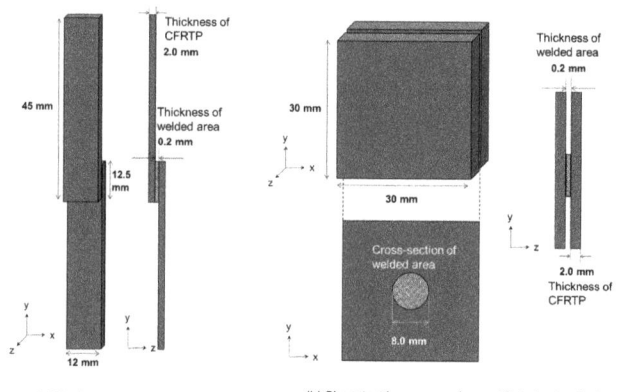

Figure 3. Dimensions and geometry of the simulation model. (**a**) Single-lap shear test was performed according to ISO standards (ISO 4587:2003). (**b**) Shear tests by compression and flatwise tensile tests gave different displacements for the same model.

2.2.2. Effect of Adhesive Thickness on Stress State at Adhesive Interface in Three Different Bond Strength Tests

To discuss the dependence of the stress state in the bond strength test on the adhesive thickness, numerical simulations were used to compare thinner and thicker adhesive thicknesses, which were set at 0.2 mm in the last section. The apparent strengths of the three models (SLS, CS, and FW) were investigated when the adhesive thicknesses were changed to 0.1, 0.2, and 0.3 mm, respectively. Other dimensions were the same as the models in the previous chapter, as shown in Figure 3. The parabolic criterion adopted as the failure criterion was also the same as that adopted for the previous model, so the true interface strength is the same for all models. The properties of the materials are listed in Table 1.

3. Results and Discussion

3.1. Determination of Interfacial Strength between Adhesive and Adherend

The CT scan of the welded specimen showed that the resin melted evenly in a circular pattern, with a welded area of approximately 50 mm^2, and that the welding was complete. The reaction force at failure of the SC was obtained as 1.34 kN. Thus, the apparent strength was 26.8 MPa. The images of the fracture surface of the specimen taken with a digital microscope are shown in Figure 4, which shows that the adhesive and adherend showed little damage and that the specimen failed at the interface between the adhesive and the adherend. Therefore, it can be assumed that material failure of the adhesive does not occur when welded under the present welding conditions, and the fracture envelope can be drawn by focusing only on the interface between the adhesive and adherend. Based on the apparent strength obtained from the experimental results, the interface strength was fitted using a numerical simulation, and the cohesive properties listed in Table 1 were obtained. Y_n and Y_s were 48.0 and 67.9, respectively. The parabolic criterion was determined as the fracture envelope of Figure 5 according to this result and Equations (1) and (2).

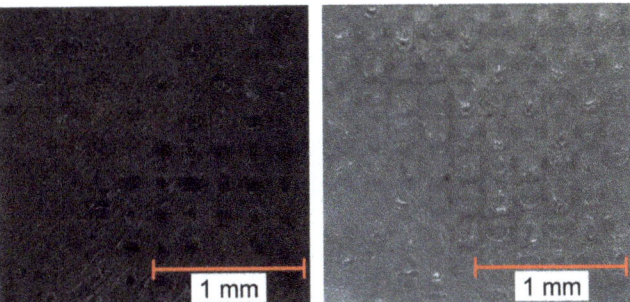

(a) Almost undamaged adherend   (b) Adhesive with little damage

Figure 4. Digital microscope image of fracture surface after shear test using compression. There was little damage to (**a**) the adherend or (**b**) the adhesive, indicating that failure occurred at the interface between the adherend and the adhesive.

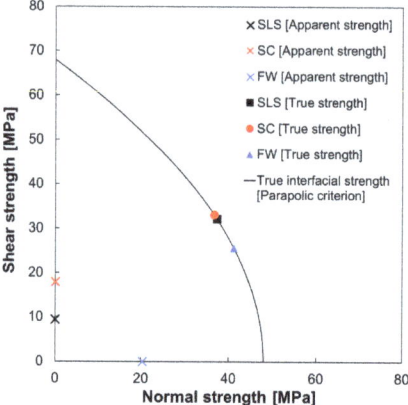

Figure 5. Local stress state at the fracture initiation point, which is the true interface strength. The average apparent strength is obtained through numerical simulations of SLS (single-lap shear test), SC (shear test using compression), and FW (flatwise tensile test), and the CFRP-polymer bonding interface strength (parabolic criterion) are plotted.

3.2. Numerical Analysis to Derive True and Apparent Bond Strength

3.2.1. Comparison of Underestimation of Apparent Strength in Three Different Bond Strength Tests

The results of the numerical simulation outputting S33, S13, and S23 for the CS of the cohesive elements are shown in Figure 6. In this model, S33 represents the vertical stress against the adhesive–adherend interface, while S13 and S23 represent the shear stress. This indicates that not only shear stress but also tensile and compressive stresses are applied to the adhesive interface, indicating that tensile stresses contribute to the fracture at the fracture initiation point marked by the arrows in Figure 6. Numerical simulations of SLS, CS, and FW were performed, and the stress state at the failure initiation point is shown in Figure 5, with the shear stress on the vertical axis and the normal stress on the horizontal axis. The apparent strengths are also shown in Figure 5. The true interfacial strength was expressed using a parabolic criterion. Because the model is three-dimensional, only S33 is the normal stress for the cohesive elements, whereas transverse shear stress S13 and longitudinal shear stress S23 are the output of the shear stress. In this case, $\sqrt{S13^2 + S23^2}$ was treated as the shear stress because the shear stress about the interface is expressed along one axis by coordinate transformation. The stress state at the fracture initiation point in the three tests lies in the line of the parabolic criterion used in this study, and it is clear from these plots in Figure 5 that the fracture is due to combined stresses. Focusing on the percentage of the combined stress state at the fracture initiation point for each bond strength test, we found that the t_n/t_s percentage ratio for the SLS was approximately 54/46%. As SLS and CS are intended to fail only under shear stress, their respective apparent strengths $\left(Y_n^{App}, Y_s^{App}\right)$ are plotted at (0, 9.39) and (0, 17.90) on the shear strength axis in Figure 5. A comparison of the true interface strength to the apparent strength shows that SLS underestimates by 13.8%, and SC underestimates by 26.4%. This underestimation occurs because the SLS and SC are subjected to a bending moment in addition to shear stress against the bonded surface, resulting in normal stress [43]. SC tends to have higher shear stresses at the interface than SLS because of the shorter longitudinal direction of the specimen in SLS, which suppresses rotation. The t_n/t_s percentage ratio for FW was approximately 62/38%. As FW is intended for tensile failure, the apparent strength $\left(Y_n^{App}, Y_s^{App}\right)$ was plotted at (20.4, 0) on the normal strength axis. The apparent strength underestimates the true interfacial strength by 42.4%, owing to the shear stress caused by the difference in Poisson's ratio between the adhesive and the adherend. The apparent strength obtained using FW is closer to the local interface strength than the other two types of shear tests but still shows that the apparent strength does not account for the combined stress state.

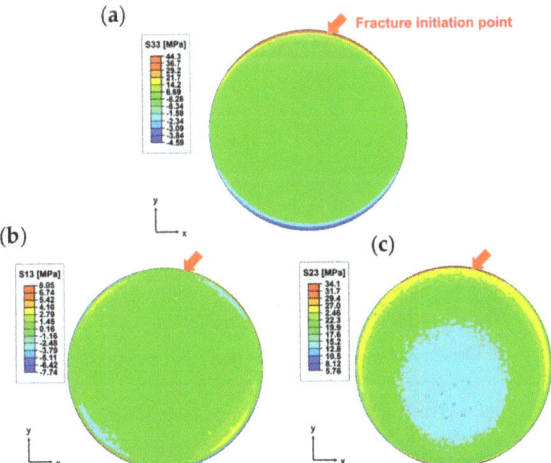

Figure 6. Numerical simulation results of cohesive elements in shear test by compression. (**a**) vertical stress S33, (**b**) transverse shear stress S13, and (**c**) longitudinal shear stress S23 are the output, and the break initiation point is indicated using arrows. There is a distribution of stress over the entire welded surface, indicating that tensile stress, in addition to shear stress, contributes to fracture.

3.2.2. Effect of Adhesive Thickness on Stress State at Adhesive Interface in Three Different Bond Strength Tests

The local stress state and apparent strength change at the fracture initiation point for the three types of bond strength tests in relation to the thickness of the adhesive are listed in Table 2. Figure 7 shows a graph comparing the simulation results of the three different tests with Y_s^{App}/Y_s on the vertical axis and the adhesive thickness on the horizontal axis. The closer the value on the vertical axis is to 1, the smaller the degree of underestimation of apparent strength. Because the reaction force in FW is in the tensile direction toward the adhesive interface, the value on the vertical axis is the ratio of the apparent shear strength to the true shear strength, as predicted by Equation (2). The normal apparent strength of CS and SLS in Table 2 was also determined in the same way outlined in Equation (2).

Table 2. Local stress state and apparent strength variation at fracture initiation point with adhesive thickness in three bond strength tests: SLS (single-lap shear test), SC (shear test using compression), and FW (flatwise tensile test).

Types of Bond Strength Tests	Thickness of Adhesive [mm]	t_n [MPa]	t_s [MPa]	Y_n^{App} [MPa]	Y_s^{App} [MPa]
SLS	0.1	36.05	33.90	6.12	8.65
	0.2	37.28	32.11	6.64	9.39
	0.3	38.55	30.14	6.37	9.01
CS	0.1	35.25	35.01	12.08	17.08
	0.2	36.69	32.98	12.66	17.90
	0.3	38.48	30.25	11.85	16.76
FW	0.1	44.23	19.06	6.02	8.52
	0.2	41.18	25.62	20.36	28.80
	0.3	40.20	27.39	34.22	48.39

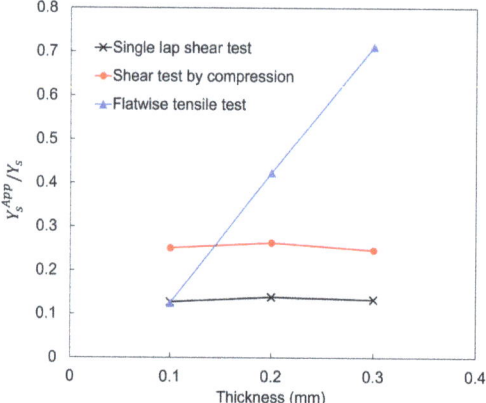

Figure 7. Dependence on weld thickness of the ratio of the apparent pure shear strength to the pure shear strength at the fracture initiation point for three different adhesive thicknesses (0.1, 0.2, and 0.3 mm). The thickness dependence is small in the single-lap shear test and the Shear test by compression, but the thickness of the adhesive (thickness-to-welded area ratio) has a significant effect on the apparent strength in the flatwise tensile test.

Focusing on the variation in the Y_s^{App}/Y_s ratio due to the adhesive thickness, the standard deviations for SLS, SC, and FW were 4.4×10^{-3}, 7.1×10^{-3}, and 2.4×10^{-1}, respectively. SLS and CS are less thickness-dependent but, on average, underestimate the apparent strength to approximately 13.3% and 25.4% of the true strength, respectively. SC is inferior to FW but reduces the discrepancy between the apparent and true strengths more than SLS. FW underestimates the apparent strength by an average of 42.1% of the true bond strength, and the apparent strength reproduces the true bond strength by approximately 71.3% at an adhesive thickness of 0.3 mm. When the adhesive is thin, the rate of shear stress is particularly high and should be used with caution.

4. Conclusions

This study proposed a comprehensive method for evaluating the strength of the adhesive–adherend interface of ultrasonically welded specimens by drawing a fracture envelope based on a numerical simulation of the stress state at the fracture initiation point. We have clarified the following points:

1. SLS underestimates the apparent strength to less than 14% of the true strength.
2. FW provides an apparent strength of approximately 42% of the true adhesive strength and has a lower degree of underestimation than the other two shear tests; however, care should be taken because the accuracy of the apparent strength evaluation depends on the adhesive thickness and can vary significantly.
3. A new test method, the shear test using compression (SC), is developed and proves comparable to SLS in terms of adhesive thickness dependence. Although the apparent strength is still underestimated, it is improved to approximately 26% of the true interfacial strength, allowing the test to be performed on smaller specimens.
4. The SC we proposed has the advantages of being able to conduct experiments with small specimens, which makes it easy to apply shear stress to the interface, and of being able to conduct experiments at low cost. However, the amount of data is limited because the fixture is currently in the development stage, and a combination of numerical analyses is still necessary.
5. This paper clarifies the inadequacy of conventional methods for evaluating interface strength in bond strength tests and provides a method for obtaining true interface strength and guidelines for predicting true interface strength from experimental

results. This enables accurate numerical simulation of interfacial strength and reliable structural design.

Future considerations include drawing a fracture envelope that accounts for the fracture of the adhesive itself, in addition to the adhesive strength of the adhesive–adherend interface. Under different welding conditions, failure can initiate from material failure rather than from the bonding interface. This approach allows us to determine whether the interface or adhesive will fracture first under specific stress conditions.

Author Contributions: Conceptualization, J.K.; methodology, J.K. and S.T.; software, M.T.; validation, J.K. and M.T.; formal analysis, M.T.; investigation, M.I.; resources, M.I.; data curation, M.T.; writing—original draft preparation, M.T.; writing—review and editing, M.T and Y.O.; visualization, M.T. and M.I.; supervision, J.K. and S.T.; project administration, J.K.; funding acquisition, Y.O. All authors have read and agreed to the published version of the manuscript.

Funding: Parts of this study were financially supported by the Japan Keirin Autorace Foundation (JKA).

Institutional Review Board Statement: Not applicable.

Informed Consent Statement: Not applicable.

Data Availability Statement: Data are contained within the article.

Conflicts of Interest: The authors declare no conflict of interest.

References

1. Hashish, M.; Kent, W. Trimming of CFRP aircraft components. In Proceedings of the 2013 WJTA-IMCA Conference and Expo, Houston, TX, USA, 9–11 September 2013.
2. Soutis, C. Fibre reinforced composites in aircraft construction. *Prog. Aerosp. Sci.* **2005**, *41*, 143–151. [CrossRef]
3. Sakuma, K.; Yokoo, Y.; Seto, M. Study on Drilling of Reinforced Plastics (GFRP and CFRP): Relation between Tool Material and Wear Behavior. *Bull. JSME* **1984**, *27*, 1237–1244. [CrossRef]
4. Darwish, S.M.H.; Ghanya, A. Critical assessment of weld-bonded technologies. *J. Mater. Process. Technol.* **2000**, *105*, 221–229. [CrossRef]
5. Higgins, A. Adhesive bonding of aircraft structures. *Int. J. Adhes. Adhes.* **2000**, *20*, 367–376. [CrossRef]
6. Mizukami, K.; Mizutani, Y.; Todoroki, A.; Suzuki, Y. Detection of delamination in thermoplastic CFRP welded zones using induction heating assisted eddy current testing. *NDT E Int.* **2015**, *74*, 106–111. [CrossRef]
7. Kupski, J.; de Freitas, S.T. Design of adhesively bonded lap joints with laminated CFRP adherends: Review, challenges and new opportunities for aerospace structures. *Compos. Struct.* **2021**, *268*, 113923. [CrossRef]
8. Saeed, K.; McIlhagger, A.; Harkin-Jones, E.; McGarrigle, C.; Dixon, D.; Shar, M.A.; McMillan, A.; Archer, E. Elastic Modulus and Flatwise (through-Thickness) Tensile Strength of Continuous Carbon Fibre Reinforced 3D Printed Polymer Composites. *Materials* **2022**, *15*, 1002. [CrossRef]
9. Da Silva, L.F.; Rodrigues, T.N.S.S.; Figueiredo, M.A.V.; de Moura, M.F.S.F.; Chousal, J.A.G. Effect of adhesive type and thickness on the lap shear strength. *J. Adhes.* **2006**, *82*, 1091–1115. [CrossRef]
10. Da Silva, L.F.; Carbas, R.; Critchlow, G.; Figueiredo, M.; Brown, K. Effect of material, geometry, surface treatment and environment on the shear strength of single lap joints. *Int. J. Adhes. Adhes.* **2009**, *29*, 621–632. [CrossRef]
11. Cognard, J.-Y.; Créac'Hcadec, R.; Maurice, J. Numerical analysis of the stress distribution in single-lap shear tests under elastic assumption—Application to the optimisation of the mechanical behaviour. *Int. J. Adhes. Adhes.* **2011**, *31*, 715–724. [CrossRef]
12. Naito, K.; Onta, M.; Kogo, Y. The effect of adhesive thickness on tensile and shear strength of polyimide adhesive. *Int. J. Adhes. Adhes.* **2012**, *36*, 77–85. [CrossRef]
13. Li, G.; Lee-Sullivan, P.; Thring, R.W. Nonlinear finite element analysis of stress and strain distributions across the adhesive thickness in composite single-lap joints. *Compos. Struct.* **1999**, *46*, 395–403. [CrossRef]
14. Lee, H.K.; Pyo, S.H.; Kim, B.R. On joint strengths, peel stresses and failure modes in adhesively bonded double-strap and supported single-lap GFRP joints. *Compos. Struct.* **2009**, *87*, 44–54. [CrossRef]
15. Zhao, X.; Adams, R.; da Silva, L.F. A new method for the determination of bending moments in single lap joints. *Int. J. Adhes. Adhes.* **2010**, *30*, 63–71. [CrossRef]
16. Yuuki, R.; Liu, J.-Q.; Xu, J.-O.; Ohira, T.; Ono, T. Evaluation of the fatigue strength of adhesive joints based on interfacial fracture mechanics. *Jpn. Soc. Mater. Sci. J.* **1992**, *41*, 1299–1304. [CrossRef]
17. Ryoji, Y.; Liu, J.-Q.; Xu, J.-Q.; Toshiaki, O.; Tomoyoshi, O. Mixed mode fracture criteria for an interface crack. *Eng. Fract. Mech.* **1994**, *47*, 367–377. [CrossRef]

18. Reis, P.N.; Ferreira, J.; Antunes, F. Effect of adherend's rigidity on the shear strength of single lap adhesive joints. *Int. J. Adhes. Adhes.* **2011**, *31*, 193–201. [CrossRef]
19. Quaresimin, M.; Ricotta, M. Stress intensity factors and strain energy release rates in single lap bonded joints in composite materials. *Compos. Sci. Technol.* **2006**, *66*, 647–656. [CrossRef]
20. Lee, M.; Yeo, E.; Blacklock, M.; Janardhana, M.; Feih, S.; Wang, C.H. Predicting the strength of adhesively bonded joints of variable thickness using a cohesive element approach. *Int. J. Adhes. Adhes.* **2015**, *58*, 44–52. [CrossRef]
21. Campilho, R.D.; Banea; Pinto, A.; da Silva, L.; de Jesus, A. Strength prediction of single-and double-lap joints by standard and extended finite element modelling. *Int. J. Adhes. Adhes.* **2011**, *31*, 363–372. [CrossRef]
22. Da Silva, L.F.; Öchsner, A.; Adams, R.D. *Handbook of Adhesion Technology*; Springer Science & Business Media: Berlin/Heidelberg, Germany, 2011.
23. Aydin, M.; Özel, A.; Temiz, Ş. The effect of adherend thickness on the failure of adhesively-bonded single-lap joints. *J. Adhes. Sci. Technol.* **2005**, *19*, 705–718. [CrossRef]
24. Li, Y.; Deng, H.; Takamura, M.; Koyanagi, J. Durability Analysis of CFRP Adhesive Joints: A Study Based on Entropy Damage Modeling Using FEM. *Materials* **2023**, *16*, 6821. [CrossRef]
25. Fujimoto, M.; Aoyama, S. Stress Distribution of Single Spot Welded Lap Joint under Tension-Shear Load. *J. Soc. Mater. Sci. Jpn.* **1973**, *22*, 572–578. [CrossRef]
26. Akhavan-Safar, A.; Ayatollahi, M.; da Silva, L.F.M. Strength prediction of adhesively bonded single lap joints with different bondline thicknesses: A critical longitudinal strain approach. *Int. J. Solids Struct.* **2017**, *109*, 189–198. [CrossRef]
27. Magalhães, A.G.; de Moura, M.F.S.F.; Gonçalves, J.M. Evaluation of stress concentration effects in single-lap bonded joints of laminate composite materials. *Int. J. Adhes. Adhes.* **2005**, *25*, 313–319. [CrossRef]
28. Hoang-Ngoc, C.-T.; Paroissien, E. Simulation of single-lap bonded and hybrid (bolted/bonded) joints with flexible adhesive. *Int. J. Adhes. Adhes.* **2010**, *30*, 117–129. [CrossRef]
29. Sato, M.; Koyanagi, J.; Lu, X.; Kubota, Y.; Ishida, Y.; Tay, T. Temperature dependence of interfacial strength of carbon-fiber-reinforced temperature-resistant polymer composites. *Compos. Struct.* **2018**, *202*, 283–289. [CrossRef]
30. DeVries, K.L.; Williams, M.L.; Chang, M.D. Adhesive fracture of a lap shear joint. *Exp. Mech.* **1974**, *14*, 89–97. [CrossRef]
31. Villegas, I.F.; Rans, C. The dangers of single-lap shear testing in understanding polymer composite welded joints. *Philos. Trans. R. Soc. A* **2021**, *379*, 20200296. [CrossRef]
32. Redmann, A.; Damodaran, V.; Tischer, F.; Prabhakar, P.; Osswald, T.A. Evaluation of single-lap and block shear test methods in adhesively bonded composite joints. *J. Compos. Sci.* **2021**, *5*, 27. [CrossRef]
33. Takamura, M.; Uehara, K.; Koyanagi, J.; Takeda, S. Multi-Timescale simulations of temperature elevation for ultrasonic welding of CFRP with energy director. *J. Multiscale Model.* **2021**, *12*, 2143003. [CrossRef]
34. Raza, S.F.; Khan, S.A.; Mughal, M. Optimizing the weld factors affecting ultrasonic welding of thermoplastics. *Int. J. Adv. Manuf. Technol.* **2019**, *103*, 2053–2067. [CrossRef]
35. Rani, R.; Suresh, K.; Prakasan, K.; Rudramoorthy, R. A statistical study of parameters in ultrasonic welding of plastics. *Exp. Tech.* **2007**, *31*, 53–58. [CrossRef]
36. Khan, U.; Khan, N.Z.; Gulati, J. Ultrasonic welding of bi-metals: Optimizing process parameters for maximum tensile-shear strength and plasticity of welds. *Procedia Eng.* **2017**, *173*, 1447–1454. [CrossRef]
37. Harras, B.; Cole, K.C.; Vu-Khanh, T. Optimization of the ultrasonic welding of PEEK-carbon composites. *J. Reinf. Plast. Compos.* **1996**, *15*, 174–182. [CrossRef]
38. Rashli, R.; Bakar, E.A.; Kamaruddin, S. Determination of ultrasonic welding optimal parameters for thermoplastic material of manufacturing products. *Sci. Eng.* **2013**, *64*, 19–24. [CrossRef]
39. Ogihara, S.; Koyanagi, J. Investigation of combined stress state failure criterion for glass fiber/epoxy interface by the cruciform specimen method. *Compos. Sci. Technol.* **2010**, *70*, 143–150. [CrossRef]
40. Koyanagi, J.; Ogihara, S.; Nakatani, H.; Okabe, T.; Yoneyama, S. Mechanical properties of fiber/matrix interface in polymer matrix composites. *Adv. Compos. Mater.* **2014**, *23*, 551–570. [CrossRef]
41. Morimoto, T.; Sugimoto, S.; Katoh, H.; Hara, E.; Yasuoka, T.; Iwahori, Y.; Ogasawara, T.; Ito, S. JAXA Advanced Composites Database. In *JAXA Research and Development Memorandum*; JAXA-RM-14-004; JAXA: Tokyo, Japan, 2015. (In Japanese)
42. ISO 4587:2003; Adhesives—Determination of Tensile Lap-Shear Strength of Rigid-to-Rigid Bonded Assemblies. International Organization for Standardization: Geneva, Switzerland, 2003.
43. Tong, L. Bond strength for adhesive-bonded single-lap joints. *Acta Mech.* **1996**, *117*, 101–113. [CrossRef]

Disclaimer/Publisher's Note: The statements, opinions and data contained in all publications are solely those of the individual author(s) and contributor(s) and not of MDPI and/or the editor(s). MDPI and/or the editor(s) disclaim responsibility for any injury to people or property resulting from any ideas, methods, instructions or products referred to in the content.

Article

Nondestructive Inspection and Quantification of Select Interface Defects in Honeycomb Sandwich Panels

Mahsa Khademi, Daniel P. Pulipati and David A. Jack *

Department of Mechanical Engineering, Baylor University, Waco, TX 76798, USA; mahsa_khademi1@baylor.edu (M.K.); daniel_pulipati@alumni.baylor.edu (D.P.P.)
* Correspondence: david_jack@baylor.edu

Abstract: Honeycomb sandwich panels are utilized in many industrial applications due to their high bending resistance relative to their weight. Defects between the core and the facesheet compromise their integrity and efficiency due to the inability to transfer loads. The material system studied in the present paper is a unidirectional carbon fiber composite facesheet with a honeycomb core with a variety of defects at the interface between the two material systems. Current nondestructive techniques focus on defect detectability, whereas the presented method uses high-frequency ultrasound testing (UT) to detect and quantify the defect geometry and defect type. Testing is performed using two approaches, a laboratory scale immersion tank and a novel portable UT system, both of which utilize only single-side access to the part. Coupons are presented with defects spanning from 5 to 40 mm in diameter, whereas defects in the range of 15–25 mm and smaller are considered below the detectability limits of existing inspection methods. Defect types studied include missing adhesive, unintentional foreign objects that occur during the manufacturing process, damaged core, and removed core sections. An algorithm is presented to quantify the defect perimeter. The provided results demonstrate successful defect detection, with an average defect diameter error of 0.6 mm across all coupons studied in the immersion system and 1.1 mm for the portable system. The best accuracy comes from the missing adhesive coupons, with an average error of 0.3 mm. Conversely, the worst results come from the missing or damaged honeycomb coupons, with an error average of 0.7 mm, well below the standard detectability levels of 15–25 mm.

Keywords: nondestructive testing; honeycomb core composite; defect identification; defect quantification; carbon fiber composite

Citation: Khademi, M.; Pulipati, D.P.; Jack, D.A. Nondestructive Inspection and Quantification of Select Interface Defects in Honeycomb Sandwich Panels. *Materials* **2024**, *17*, 2772. https://doi.org/10.3390/ma17112772

Academic Editors: Dimitrios Tzetzis and Raul D. S. G. Campilho

Received: 21 April 2024
Revised: 22 May 2024
Accepted: 28 May 2024
Published: 6 June 2024

Copyright: © 2024 by the authors. Licensee MDPI, Basel, Switzerland. This article is an open access article distributed under the terms and conditions of the Creative Commons Attribution (CC BY) license (https://creativecommons.org/licenses/by/4.0/).

1. Introduction

Honeycomb sandwich panels represent an integral component of structures in various industries, including aerospace, automotive, naval, sporting, and construction (see, e.g., [1,2]), owing to a variety of factors, such as their exceptional strength-to-weight ratio, insulation capabilities, and structural versatility. Comprising two thin, high-strength face sheets bonded to a lightweight core material, such as aluminum or Nomex honeycomb, these panels offer significant advantages. However, their susceptibility to defects during manufacturing or in-service poses challenges to their structural integrity and operational efficiency. One example of note was the failure of the X-33 sandwich composite liquid hydrogen tank [3]. Defects in honeycomb sandwich panels arise from various sources, including material imperfections, manufacturing process variability, environmental factors, and mechanical damage (see, e.g., [4,5]). These defects may manifest as voids, delaminations, disbonds, cracks, water ingress, or core crushes, often degrading the mechanical properties and durability of the panels (see, e.g., [6–9]). Detecting and characterizing these defects are essential for ensuring the safety, reliability, and cost-effectiveness of structures incorporating honeycomb sandwich panels.

Multiple nondestructive testing techniques exist for inspecting defects in honeycomb structures, such as eddy current testing (ECT) [10–14], ultrasonic testing (UT), phased array inspection (PA) [15], thermography testing [9,16–20], acoustic emission testing Shearography [21], and X-ray computed tomography (XCT) [22–25]. However, these methods often have limitations in terms of sensitivity, accuracy, and efficiency, particularly when dealing with hidden or subsurface defects, complex geometries, and different defect materials. As a result, there is a growing demand for advanced nondestructive evaluation (NDE) techniques capable of detecting and characterizing defects in honeycomb sandwich panels with higher sensitivity and resolution.

Many interesting studies have been conducted on debonding and defect detection in honeycomb sandwich composites using the eddy current testing method. Hagemaier [10] used eddy current conductivity testing to detect core defects in an aluminum-brazed titanium honeycomb structure. He et al. [11] employed eddy current testing to identify defects intentionally embedded between the facesheet and core, simulating delamination. However, their detection was only qualitative rather than quantitative. The study by Underhill et al. [12] investigated the use of eddy current arrays for detecting damage, including disbonding and core defects in sandwich panels with carbon fiber-reinforced polymer (CFRP) facesheets. While they successfully detected core crush or indentation in the honeycomb core, they were unable to quantify these defects. Notably, the effectiveness of defect detection relies on the probe size; where larger probes often result in blurry C-scans of defects. T. Rellinger et al. [13] presented a method combining eddy current, thermography, and 3D laser scanning for inspecting sandwich panel low-velocity impacts to simulate different types of defects. They were only able to detect and measure crushed core defects using eddy current testing without success in identifying other types of defects, such as disbonds and delaminations. Recently, Ren et al. [14] utilized eddy current testing on an aluminum honeycomb sandwich structure with CFRP panels to detect and identify core defects, including wall fractures and core wrinkles. Impact damage detection was also conducted, revealing that as impact energy increased, the affected area in the C-scan image became more blurred and the minimal signal decreased, although the area of damage did not always correspondingly increase. The eddy current method's limited penetration depth restricts its effectiveness for detecting defects deep within materials. Additionally, it is primarily suitable for conductive materials, posing challenges for inspecting non-conductive or low-conductivity materials [26].

There are also studies focused on finding defects in honeycomb sandwich panels by thermography. Qin and Bao [16] used thermography to detect defects in a honeycomb sandwich panel with an aluminum core. Their detection focused on identifying defects in the honeycomb cells near the panel's edges. Ibarra-Castanedo et al. [17] utilized three different types of thermography to investigate honeycomb sandwich structure. They concluded that pulse thermography yielded the fastest results but yielded inconsistent results due to nonuniform heating and environmental reflections. Usamentiaga et al. [18] studied holes in a reinforced honeycomb sandwich panel using active thermography. They detected the holes' presence and were able to estimate the distance from the holes to the reinforced region. Zhao et al. [19] utilized lock-in thermography for disbond detection in titanium alloy honeycomb sandwich panels. They simulated a disbond using a cylindrical air gap with an 18 mm diameter. Hu et al. [20] detected and classified different types of defects in honeycomb sandwich panels, specifically 15 mm and 30 mm diameter flaws. Their model sensitivity was around 90% for water and oil ingress and around 70% for debonding and adhesive pooling defects. The effectiveness of defect detection in thermography is compromised by issues like uneven heating and coil shielding. Moreover, the rapid decay of the temperature signals with increased detection depth limits the range and quantitative accuracy for internal defects in structures, presenting challenges in detecting micro-defects critical for health monitoring [27].

Shearography nondestructive testing (NDT) is finding considerable interest in the detection of flaws within honeycomb sandwich panels. For example, Guo et al. [21] showed that using Shearography, they could detect circular and rectangular defects, with the size of detectability being strongly correlated to the depth at which the defect occurs. In addition,

Lobanov et al. [28] showed that Shearography with a combined vacuum load is a reliable method to detect different sizes of defects in titanium honeycomb panels. Revel et al. [29] employed a Wavelet Transform algorithm to quantitatively estimate the delamination size through Shearography inspection. They tested a damaged sandwich panel consisting of a 24 mm honeycomb core and a 1.5 mm thick fiberglass skin. Their coupon featured a set of circular defects with a nominal diameter of 24 mm. The algorithm accurately assessed the defect size, yielding a result of 24.3 mm. Shearography has several drawbacks, including challenges in inspecting deep or small defects and accurately characterizing the type of defect and its impact on material integrity [30].

Ultrasound testing (UT) is often used for various defect identifications. Specifically, air-coupled ultrasound has found an increasing interest in the inspection of honeycomb sandwich structures due to the ability to inspect a part without the need for a water couplant. Zhang et al. [31] utilized this method to detect defects in a honeycomb panel interface. They could detect 30 mm and 35 mm disbonds in the interface and measure them via the K-mean clustering method. Hsu [32] utilized air-coupled ultrasound to detect various defects at the interface within a sandwich structure but did not seek to quantify them geometrically. Zhou et al. [33] investigated CFRP honeycomb sandwich panels with an aluminum core via air-coupled ultrasound. Due to the large acoustic impedance between the air–solid interface, they were presented with a poor signal-to-noise ratio (SNR). Therefore, they introduced a hybrid method using a combination of wavelength filtering and phase-coded compression to improve the SNR.

Another interesting NDT method is the immersion-type UT scanning system. Several studies have utilized this method; however, none have specifically examined honeycomb structure panels with the objective of quantifying the defect dimensions. Blanford and Jack [34] used high-frequency ultrasonic to quantify the three-dimensional damage zone in a carbon fiber laminated composite from low-velocity impacts using immersion tank inspection. Blackman et al. introduced a novel pulse–echo ultrasound technique for sizing foreign objects in carbon fiber laminates [35], and recently, a study by Pisharody et al. [36] used immersion tank inspections to perform ultrasonic scanning to assess the damage zone around an adhesive joint.

In light of the critical role that honeycomb sandwich panels play in diverse industries and the challenges posed by defects in their structural integrity, this study aims to investigate the effectiveness of single-side access, point-focused, ultrasonic NDT methods for defect detection and characterization. Specifically, the study is performed using spherically focused acoustic transducers in an immersion tank, as well as in a novel portable system [37] that provides the accuracy of an immersion tank without the need for submerging a component in water. This research seeks to enhance the understanding of defect identification in honeycomb panels and contribute to the development of reliable inspection techniques. This paper begins by outlining the manufacturing method employed for this study, followed by a concise overview of the inspection methods used for defect detection in honeycomb sandwich panels. Subsequently, the scanning setup is elucidated to provide insight into the experimental process. The next phase involves the analysis of the data captured in the preceding steps. Finally, the paper concludes by summarizing the key findings and implications drawn from the research. In this study, sizes of defects range from 5 to 40 mm and include missing adhesive, crushed core, removed core, embedded Kapton film, and embedded Polytetrafluoroethylene (PTFE, commonly termed Teflon) film. These defects are particularly challenging to detect, yet they were successfully identified and, more importantly, quantified, with an average error of only 0.6 mm over all coupons investigated, thus opening new possibilities in design and manufacturing, allowing for tighter tolerance manufacturing and better control over the required factor of safety margins.

2. Materials and Methods

2.1. Manufacturing Method

Honeycomb sandwich panels (HSP) consist of three components: two facesheets and one honeycomb core, as illustrated in Figure 1. Facesheets can be constructed from various materials such as CFRP, fiberglass reinforced polymers (FGRP), or aluminum, while the core may be composed of Nomex or aluminum. In the present study, the fabrication of HSP is made of a CFRP facesheet composed of unidirectional 250 F pre-preg composed of Toray T700 fibers supplied by Rockwest Composites (West Jordan, UT, USA) and a Nomex core that is 25 mm in thickness and a nominal cell wall thickness between 0.2 mm and 0.4 mm. The process involves two main steps: preparing two facesheets within a Carver Laboratory programmable hot press (Wabash, IN, USA) following the manufacturer's recommended cure cycle. Each facesheet is adhered to the honeycomb core using a 150 g pre-preg adhesive film from Easy Composites (Stoke-on-Trent, UK), again using the hot press to provide a controlled temperature and holding pressure. Various materials and methods are employed intentionally to simulate delamination between the facesheet and core. The unidirectional facesheets are manufactured with a layup of $[0, 45, -45, 90]_s$, where the subscript s stands for symmetric, yielding a total of 8 laminae for each facesheet with a nominal manufactured thickness of 1.2 mm. Prior to adhering the facesheets to the core, intentional defects are added, specifically those shown in Figure 2, including missing adhesive, a foreign object between the adhesive and the core, a partially damaged core, and a section of removed core. The partially damaged core is crushed with a flat bottom cylinder to one-half of the overall part thickness. These defects are selected due to their prevalence in the manufacturing process, as well as to highlight the ability to differentiate between defect types with the presented inspection method. The summary of the various defects for the 15 coupons, along with their dimensions, is provided in Table 1. The missing adhesive is created by physically removing a circular section from the adhesive film, simulating either an improper transfer of adhesive to the layup from the adhesive backing material or an air bubble preventing an adhesive from adhering to the facesheet. The foreign objects are made from either PTFE or Kapton; both materials are found within a standard manufacturing process [35]. The damaged core material is made by crushing the top surface of the core with a circular weight, simulating damage caused by improper handling during manufacturing. The final defect, the removed section of the core, simulates a purposeful removal of material as might be called out during manufacturing. The final planar dimensions of the coupon are nominally 75 mm × 75 mm, but the actual inspection area is typically much smaller, as will be noted in the following sections. Each of the defects was measured during manufacturing and prior to final assembly using a Keyence VR-3200 (Itasca, IL, USA) optical microscope to confirm proper manufacturing, record the defect for quality assurance, and capture any subtle dimension variations from the designed defect size. The dimensions for the various defects are provided in Table 1.

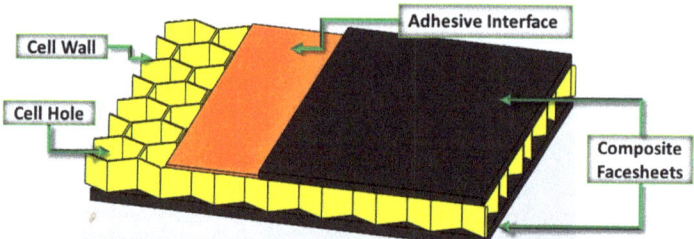

Figure 1. Representative schematic of honeycomb core material with carbon fiber facesheets.

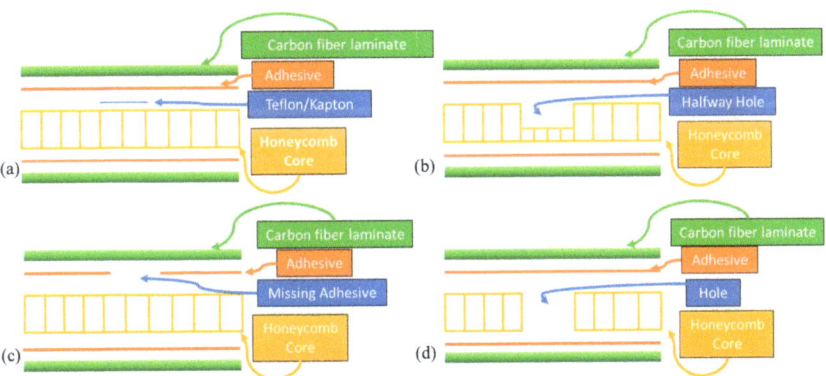

Figure 2. Representative schematic of defects placed within the manufactured coupon, (**a**) foreign object debris, (**b**) crushed honeycomb, (**c**) missing adhesive, and (**d**) removed honeycomb core.

Table 1. Test matrix of defect types and their associated planar dimensions.

Coupon Identifier	Defect Type	Diameter (mm) from Microscopy
1	Missing Adhesive	7.9
2	Missing Adhesive	19.0
3	Missing Adhesive	40.8
4	Kapton Film	4.7
5	Kapton Film	20.0
6	Kapton Film	40.3
7	Hole	5.7
8	Hole	18.8
9	Hole	34.8
10	PTFE Film	5.0
11	PTFE Film	20.0
12	PTFE Film	40.4
13	Crushed Hole	6.4
14	Crushed Hole	19.3
15	Crushed Hole	35.5

A unique feature that will be observed within this study inadvertently comes from the choice of the adhesive film. The adhesive film utilized incorporates small holes within the film, allowing air to pass between layers during manufacturing and aiding in the void removal process. These small holes appear in a regular pattern with an approximate spacing of 1 mm, as shown in Figure 3a. The microscope image of Figure 3a is after curing; thus, the resin will tend to wick along the various strands and between strands during the elevated temperature portion of fabrication. These small gaps do not impact the performance of the adhesive in the current application but are noteworthy, as they can be seen in the inspection results, as will be shown in Section 3. Of note is that these features that are 1 mm in span are smaller than the cells of the honeycomb, as shown in Figure 3b, which are typically 4.5 mm along the minor axis and 6 mm along the major axis. By observing these smaller features, one can have a sense of the effectiveness of the present system in identifying defects between the honeycomb cells and the facesheet. This high-resolution topic is briefly expanded upon in Section 3.3, where a comparison is made to the results from a roll-on adhesive, but it is not a prime focus of the current study.

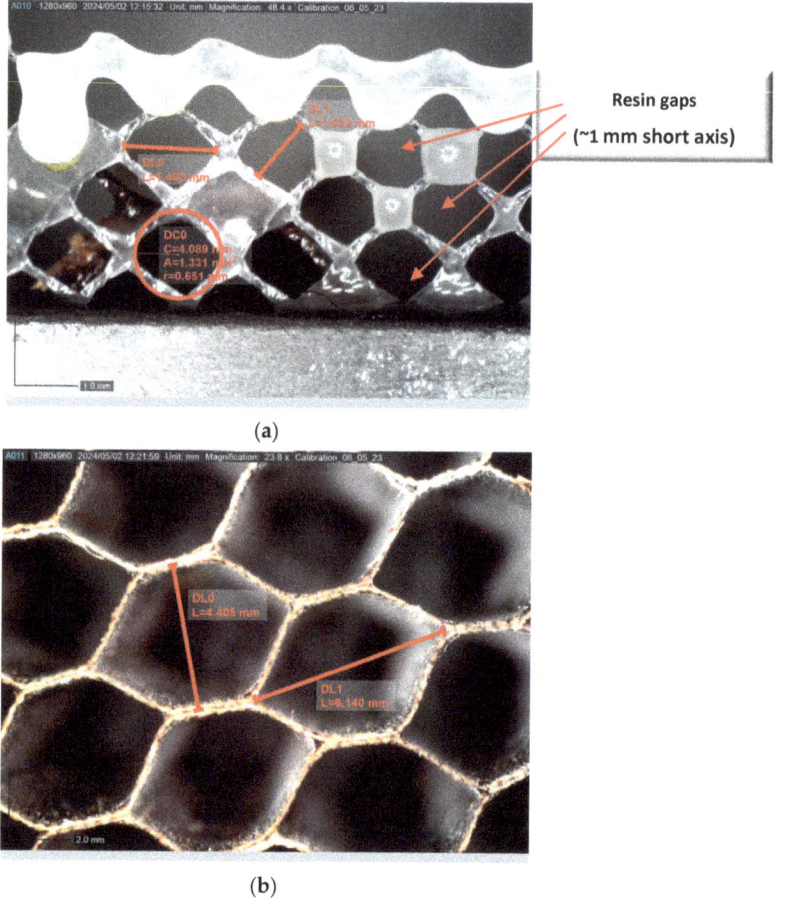

Figure 3. Microscope image of the adhesive film showing the associated punched holes allowing air pathways between layers (**a**) focusing on the resin adhesive sheet and (**b**) focusing on the honeycomb cell walls.

2.2. Inspection Methodology

2.2.1. Ultrasound Methodology

Ultrasonic testing (UT) is a nondestructive testing method commonly used for defect detection, part qualification, and evaluation of materials and structures. It relies on the propagation of high-frequency sound waves through the material being tested. These sound waves are introduced into the material using a piezoelectric transducer. All testing is performed using the same broadband spherically focused 15 MHz peak frequency transducer made by Evident (formally Olympus, headquartered in Waltham, MA, USA) with a focal length in water of 38.1 mm and a manufacturer-provided—6 dB loss beam diameter of focus, also termed the spot size, in water of 0.3 mm, which will nominally be 0.3 mm on the interface between the facesheet and the adhesive. Digitization is performed using a Focus PX, also from Evident, and spatial control is performed using Velmex (Bloomfield, NY, USA) linear translation stages. The digitizer fires a square wave pulse at 33 ns with a peak voltage of 190 V (the maximum voltage allowed by the hardware), and sampling occurs at 100 MHz. All inspections are performed with access to only a single side in what is termed pulse–echo mode. Unlike contact transducers, the spherically focused transducer is offset from the part surface. In the present study, instead of focusing on the front surface of the part, the focus is on the back surface of the facesheet using a surface offset t_{focus} of

$$t_{focus} = 2\,w_L/c_w \tag{1}$$

where c_w is the speed of sound of water and w_L is the length the signal travels in the water, given as

$$w_L = f_L - m_T \frac{c_m}{c_w} \tag{2}$$

In the above, f_L is the focal length of the transducer, 38.1 mm in the present study, c_m is the speed of sound of the CFRP of 2800 m/s, and m_T is the material thickness of 1.2 mm.

As this study uses a spherically focused transducer, an acoustic medium, specifically water, is required between the transducer and the coupon being inspected. It is commonly assumed that the highest resolution ultrasound can be obtained using an immersion tank setup, such as that shown in Figure 4a. Unfortunately, these systems are constraining due to the need to submerge a coupon, unlike the air-coupled systems. We present a compromise in the present study with the portable inspection system presented in [37] and shown in Figure 4b. This portable system allows inspections to be performed in the field or the manufacturing environment without the need for submerging a coupon.

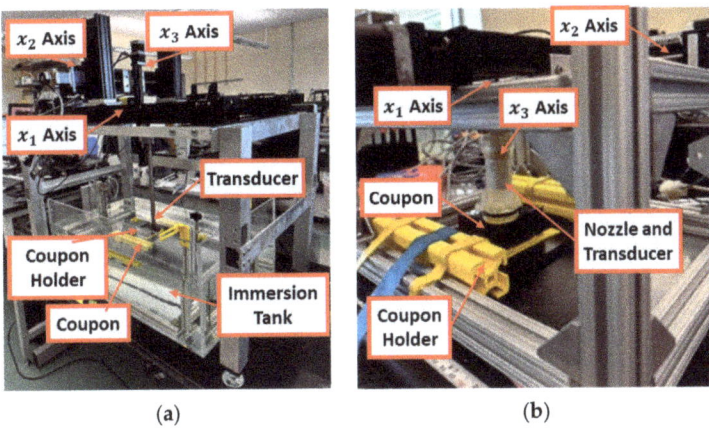

Figure 4. Custom immersion testing systems utilized in the present research. (**a**) Immersion scanning inspection system. (**b**) Portable (out-of-tank) inspection system.

2.2.2. Scanning System Set Up—Immersion Tank Inspection

Subsequently, the coupon is positioned within the immersion tank system shown in Figure 4a, where the entire surface area of each coupon undergoes scanning in a raster pattern, as depicted in Figure 5. The immersion system requires the complete submersion of the coupon, and a bagging material is placed around the edges of the honeycomb and sealed with gum tape to prevent water incursion of the coupon. The focal point of the 38.1 mm focal length transducer is at the back of the top facesheet and accounts for the impedance mismatch between the water and the composite when focusing. The raster pattern is executed in increments of 0.1 mm in both the x_1 and x_2 directions. The captured data are stored for subsequent analysis, facilitated by a custom in-house MATLAB script, discussed in Section 2.3, allowing for the identification, characterization, and interpretation of the various defects listed in Table 1 within the HSP interface.

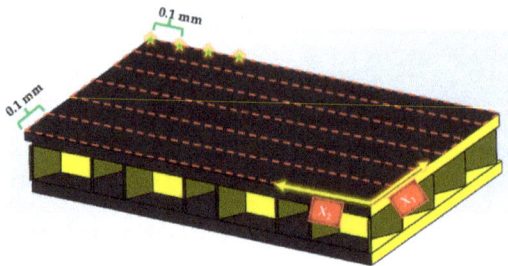

Figure 5. Raster pattern and coordinate axis utilized in scanning, with the red dashed lines indicating each of the individual scans separated by an indexing of 0.1 mm and the stars representing 4 of the over 600 points separated by 0.1 mm in each line scan.

2.2.3. Scanning System Set Up—Portable System (Out-of-Tank) Inspection

One limitation of UT immersion tank systems is their inability to be easily deployed. Therefore, this study aimed to demonstrate the effectiveness of the method employed here using a novel portable UT system, which is discussed in detail in [37]. Figure 4b illustrates the setup, wherein a fixture has been specifically designed to accommodate the transducer within a captured water column [37]. In the present study, this is the same transducer used for the immersion tank inspections. This captured water column is then placed in contact with the surface of the coupon with an acoustic membrane between the water column and the coupon surface. A thin film of gel is then used to allow for acoustic coupling between the portable housing system and the coupon being inspected. Unlike traditional contact inspection methods where the resolution is like the transducer surface area, the present instance uses a spherically focused transducer, allowing for point inspections, thus maintaining the resolution advantage of the immersion systems with the portability and infrastructure advantages of hand-held contact elements.

2.3. Data Reduction Methodology

In this investigation, data are captured by the Focus PX system purchased from Evident. The Focus PX system allows for both single-element inspections and phased array inspections with a 100 MHz digitizer with full integration capabilities for standard immersion testing systems. Data are collected and stored in the form of individual A-scans depicting the amplitude (strength) of the ultrasonic signal against the time taken for the ultrasonic pulse to traverse through the material, such as that shown in Figure 6 for Coupon 1, a sample with missing adhesive. However, in intricate structures such as the current honeycomb sandwich panels, A-scans alone are difficult to interpret and analyze. For example, Figure 6a displays A-scans taken at four different locations. The first is the blue dashed line, a typical A-scan over the defect, in this case, missing adhesive. The second is an A-scan taken directly over the cell wall but with adhesive, shown by the red dotted line. The third is a typical A-scan between the honeycomb walls but with proper bonding to the adhesive, indicated by the green line. The final A-scan, shown by a black line, is between the honeycomb walls but also in the gap between the perforations of the adhesive presented in Figure 2a. While the front wall of the top facesheet is visible in all A-scans at 0 μs, extracting useful information from a single A-scan proves challenging. Notably, the signal over the missing adhesive region, as well as in the adhesive region but between the perforations of the adhesive, the signal is similarly strong at the back wall of the facesheet near $0.8 \sim 1$ μs. Conversely, the signal over the honeycomb walls with proper adhesive is weak at the interface. This signal is only slightly different than the signal over the region with adhesive but without the honeycomb cell wall. Consequently, discerning the exact flaw location depth or identifying the back wall of the top facesheet and flaw detection becomes difficult. To aid in the analysis, an average A-scan, termed $S(t)$, over the entire scanned region of the coupon was conducted for each of the individual coupons. This is defined as

$$S(t) = \frac{1}{N_1 N_2} \sum_{k=1}^{N_1} \sum_{l=1}^{N_2} \mathcal{S}(x_{1,k}, x_{2,l}, t) \quad (3)$$

Here, $\mathcal{S}(x_{1,k}, x_{2,l}, t)$ represents the ultrasonic signal intensity measured by the transducer at the kth location along x_1 where $k \in \{1, 2, \ldots, N_1\}$, and the lth location along x_2 where $l \in \{1, 2, \ldots, N_2\}$ at a specific time t. A typical result of the average A-scan is shown in Figure 6b. The black and re-dashed lines indicate the range from which to select for later processing to gate the back wall signal.

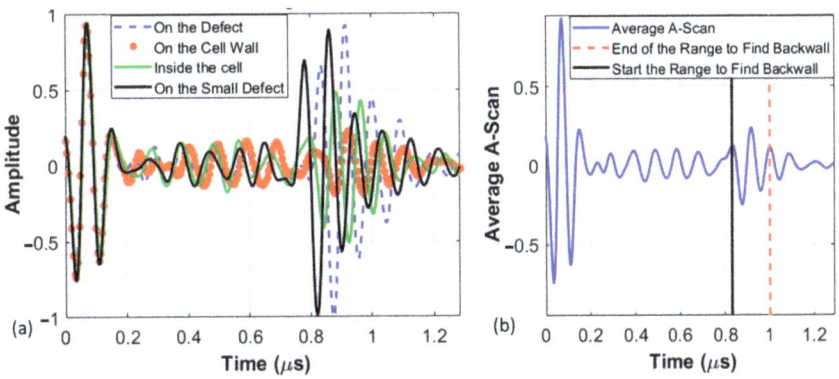

Figure 6. A-scans from Coupon 2, (**a**) at select locations and (**b**) the averaged A-scan.

After capturing the waveforms over the scan region, the time axis on each scan is shifted such that the front wall echo occurs at the same time over all spatial positions in (x_1, x_2) such that the first echo occurs at the effective time of 0 µs, as shown in Figure 6. For each spatial location $(x_{1,k}, x_{2,l})$, the scan intensity is examined to identify the initial occurrence of a value exceeding a predefined detection threshold. Subsequently, the first peak after this time value is identified as the front wall echo for that location and a third-order polynomial in (x_1, x_2) is constructed to shift the time t by $t_0(x_1, x_2)$ as $\hat{t} = t - t_0(x_{1,k}, x_{2,l})$. Thus, the ultrasonic signal intensity is investigated as $\mathcal{S}(x_{1,k}, x_{2,l}, \hat{t}(x_{1,k}, x_{2,l}))$, where $\hat{t}(x_{1,k}, x_{2,l}) = 0$ corresponds to the front coupon surface.

To enhance the quality and accuracy of the data, a frequency upsampling technique is employed using the inverse Fourier transform. The upsampled frequency spectrum can be expressed as

$$X_u(f) = \begin{cases} L \cdot X\left(\frac{f}{L}\right) & \text{for } 0 \leq f < \frac{1}{L} \\ 0 & \text{otherwise} \end{cases} \quad (4)$$

where the upsampled frequency spectrum $X_u(f)$ is obtained by zero-padding the original spectrum and scaling it accordingly. Here, the capital letter $X(f) = \mathcal{F}\{x(t(f))\}$ denotes the Fourier transform of the original signal $x(t)$, f represents frequency, and L is the upsampling factor. Subsequently, interpolation is performed in the frequency domain, followed by the application of inverse Fourier transform on the interpolated spectrum to yield the upsampled signal.

$$x_u(t) = \mathcal{F}^{-1}\{X_{interp}(f(t))\} \quad (5)$$

where $X_{interp}(f)$ is the interpolated spectrum. The results of this are shown through the B-scans shown in Figure 7, where the color bar indicates the normalized signal intensity. The B-scan is a collection of A-scans along a single axis, in this case along the index or x_1 direction. In Figure 7a, the raw scan is shown, whereas the upsampled scan is shown in Figure 7b, where the upsampling factor L is 2.

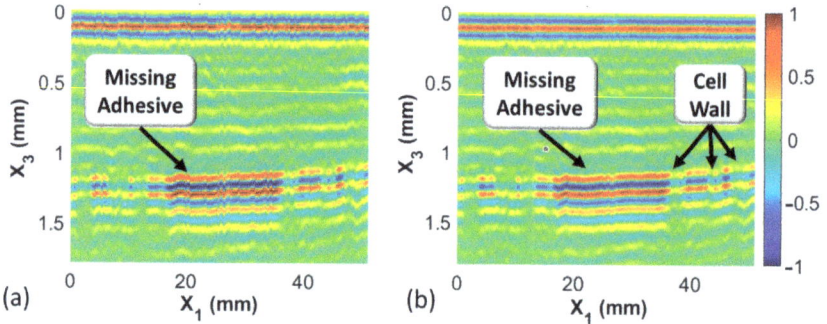

Figure 7. B-scans from Coupon 2, (**a**) without upsampling and (**b**) an upsampling factor of $L = 2$ (color bar represents the normalized signal intensity).

The next step in this study involves the application of a Gaussian smoothing filter to smooth spatial variations. This method is employed to mitigate random noise present in the A-scan data [35] as

$$\tilde{S}(x_{1,k}, x_{2,l}, \hat{t}) = \frac{1}{2\pi\sigma_{x_1}\sigma_{x_2}} \int_{-\infty}^{\infty}\int_{-\infty}^{\infty} e^{-\frac{(\tilde{x}_1 - x_{1,k})^2}{2\sigma_{x_1}^2}} e^{-\frac{(\tilde{x}_2 - x_{2,l})^2}{2\sigma_{x_2}^2}} S\left(\tilde{x}_1, \tilde{x}_2, \hat{t}(\tilde{x}_1, \tilde{x}_2)\right) d\tilde{x}_1 d\tilde{x}_2 \quad (6)$$

where the variables σ_{x_1} and σ_{x_2} tend to smooth out spatial information locally, where they are selected to be 3 times the step size in the current study. A larger value of σ leads to a wider Gaussian distribution, resulting in more extensive smoothing. Conversely, a smaller value yields a narrower Gaussian distribution, preserving more detail in the signal. Figure 8 illustrates the C-scan both before and after the Gaussian filtering. Notice that the filtering provides a crisper image and the edges of boundaries between features are more pronounced. To obtain a C-scan, the filtered acoustic signal, $\tilde{S}(x_{1,k}, x_{2,l}, \hat{t})$ is converted to a 2-dimensional representation by taking the maximum value of the signal between a prescribed range, termed a gate Δt, as

$$C(x_{1,k}, x_{2,l}) = \max_{\hat{t} \in \{t_1(x_{1,k}, x_{2,l}), t_2(x_{1,k}, x_{2,l})\}} \left\{ \left| \tilde{S}(x_{1,k}, x_{2,l}, \hat{t}) \right| \right\} \quad (7)$$

where the value for t_1 is the onset of the back wall, shown in Figure 6b, and $t_2 = t_1 + \Delta t$, where $\Delta t = 0.05$ µs. Different gate widths were investigated, but the narrow gate of 0.05 µs was found to yield an image with the highest signal contrast in the current study.

The C-scan provides a means to visualize information at a consistent depth, whereas the B-scan, as shown in Figure 7, provides a slice of information into the coupon. For example, the B-scan shown in Figure 7b can be used to detect the depth and estimate the size of a feature within a material. As can be seen in the B-scan, the interface between the facesheet and the adhesive for the honeycomb occurs around 1.2 mm into the coupon, and the width of the feature on the back wall, in this case missing adhesive, ranges from a value of $x_1 \sim 18$ mm to $x_1 \sim 38$ mm.

All analysis is implemented in the MATLAB (version 2022b, Natick, MA, USA) programming environment. This code facilitates precise measurement and analysis of the detected defects, allowing for the characterization and evaluation of their size within the scanned coupons.

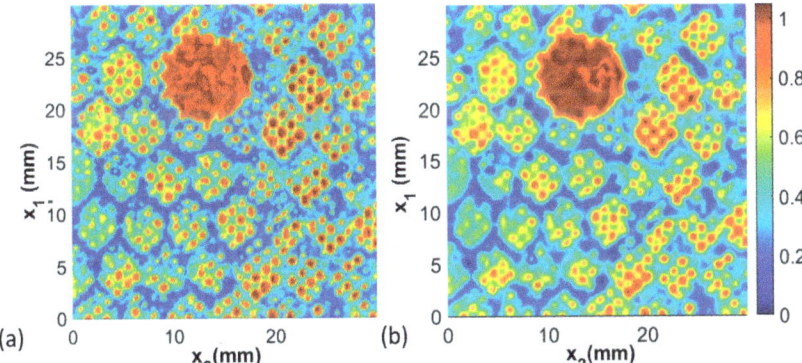

Figure 8. C-scans of Coupon 2, (**a**) prior to Gaussian filtering and (**b**) after Gaussian filtering.

3. Results

The following section provides the results from each of the studies. The results from Table 1 are presented for the immersion system, both as a function of the defect size and as a function of defect type. Next, a brief study on the subtleties between adhesive types is presented, as well as a brief study on the use of the portable inspection system. Finally, a summary of the results is presented in Section 4 to highlight the anticipated error in identifying the defect diameter, a value found to be typically less than 0.6 mm, with the inspection of Coupon 5 yielding the highest error of only 1.3 mm. The results of Section 4 also highlight that the portable system and the immersion system yield a similar error over the medium-sized, nominally 20 mm diameter, defect samples studied.

3.1. Defect Size Study

This first study presents the results for Coupons 1–3. These three coupons all have the same defect type, that of a missing adhesive, but have diameters of missing adhesive ranging from 7.9 mm to 40.8 mm. As can be seen in Figure 9, when properly gated, the defect is clearly identifiable. The diameter is manually measured three times by forming a best fit, in the visual sense, circle over the defect, and the average is then reported. For the three coupons in Figure 9 with the small (7.9 mm), medium (19.0 mm), and large (40.8 mm) diameter defects of the missing adhesive, the measured diameters using the ultrasound signal are, respectively, 8.3 mm, 19.3 mm, and 41.4 mm. This is only an error of, respectively, 0.3 mm, 0.3 mm, and 0.4 mm. Recall that the step size is 0.1 mm in both the x_1 and x_2 directions; thus, the error is on the order of 3 to 4 pixels, similar in value to the estimated beam spread of the focused signal. A study was not performed to identify if there is any difference in accuracy in the index direction x_1 or the scan direction x_2, as the accuracy of the present results is measurably better than in previous studies. It is worth noting in the figures that the missing adhesive appears as a high-intensity signal. This is due to the acoustic mismatch between the composite facesheet and the air located within the honeycomb core region. The cell walls can be clearly seen within each of the three scan data sets, as indicated by the hexagonal structure pattern throughout the image, except in the region of the missing adhesive.

3.2. Defect Type Study

The second parameter of interest is the ability to differentiate between defect types. Five different defect types were studied, and the results from the medium-sized defects, which range from 19 to 20 mm in diameter, are presented in Figures 10 and 11. The five types include two different types of foreign objects, Kapton and PTFE films, shown, respectively, in Figure 10b,e, and in Figure 11a,c. Observe that the circular defects/features are both visible in the gated C-scans on the back surface shown in Figures 10b and 11a, but it is the B-scan at the back wall of the facesheet that can be used to differentiate the material type. Specifically, the Kapton film, shown in Figure 10e, has a soft acoustic reflection at the back wall located around 1.2 mm into the coupon as sound can pass through the interface between the Kapton and the surrounding material due to the light bonding of the resin with the film. Conversely, the acoustic reflection for the PTFE insert is much higher due to the lack of bonding between the PTFE and the surrounding polymer systems, as observed in Figure 11c. The missing adhesive coupons, shown in Figure 10a,d, are the clearest to see in the study, as the acoustic echo occurs earlier in time from the interface between the back of the facesheet and the air trapped in the honeycomb. The crushed honeycomb, Figure 10c,f, and the removed honeycomb, Figure 11b,d, are indistinguishable from each other. Although there are subtle differences in the C-scans in Figures 10c and 11b, these are artifacts of the gating selected. Looking at the back wall of the B-scans in Figures 10f and 11d, it is unclear if the differences are due to the different defect types or the resulting variability in manufacturing.

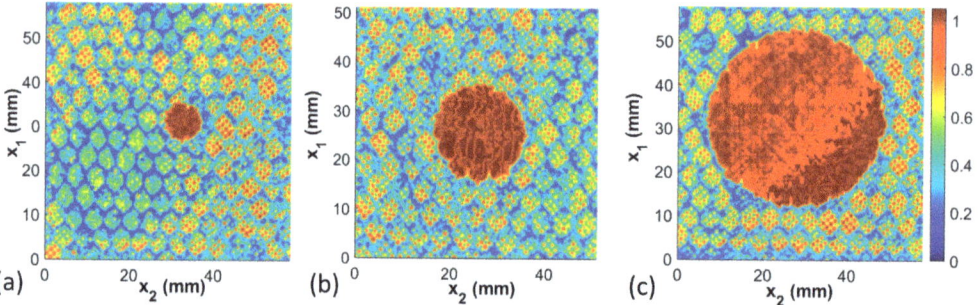

Figure 9. C-scans of Coupons 1–3, (**a**) 7.9 mm diameter defect, (**b**) 19.0 mm diameter defect, and (**c**) 40.8 mm diameter defect.

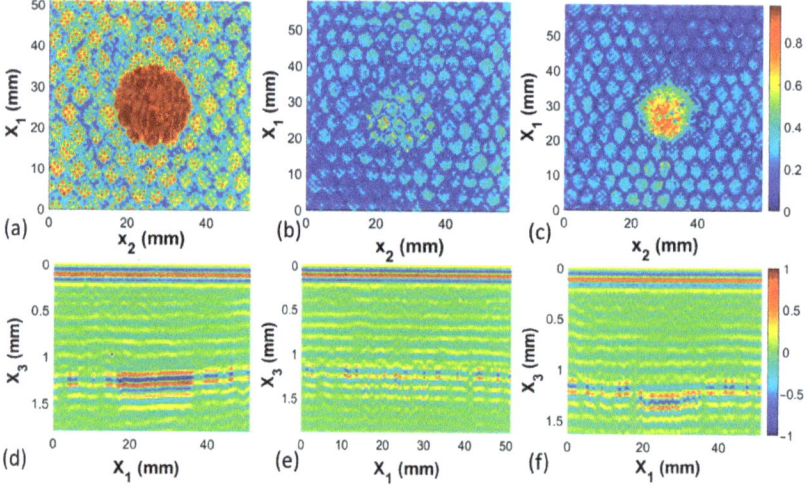

Figure 10. C-scans and B-scans for coupons with the nominal 20 mm diameter defect, (**a**) C-scan for Coupon 2, missing adhesive, (**b**) C-scan for Coupon 5, Kapton FOD, (**c**) C-scan for Coupon 8, missing honeycomb, (**d**) B-scan for Coupon 2, missing adhesive, (**e**) B-scan for Coupon 5, Kapton FOD, (**f**) B-scan for Coupon 8, missing honeycomb.

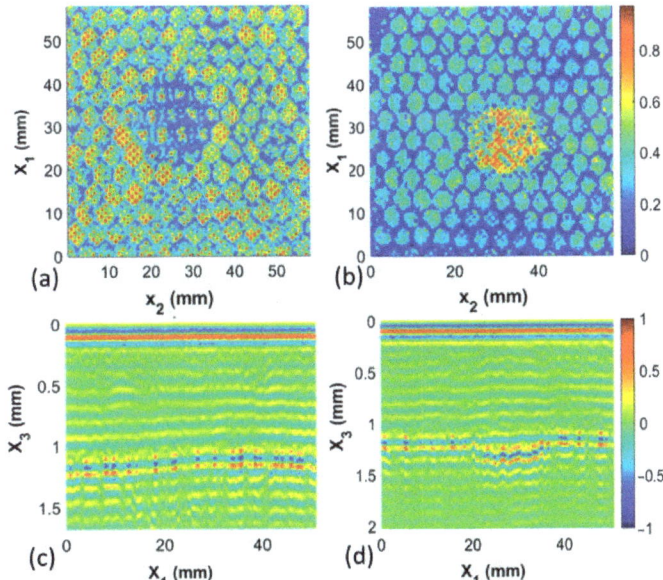

Figure 11. C-scans and B-scans for coupons with the nominal 20 mm diameter defect, (**a**) C-scan for Coupon 11, PTFE FOD, (**b**) C-scan for Coupon 14, crushed honeycomb, (**c**) B-scan for Coupon 11, PTFE FOD, (**d**) B-scan for Coupon 14, crushed honeycomb.

3.3. High-Resolution Capability and Variations in Adhesive Layer

This next study focuses on the small features present in the C-scan waveform between the honeycomb cell walls seen in each of the previously shown C-scans. A sample with a 20 mm diameter missing honeycomb defect is shown in Figure 12a. Observe the small features on the length scale of 0.5 mm to 2 mm in the figure. These features have the same spacing as the perforations shown in Figure 3 of the film adhesive and are clearly visible in the final manufactured coupon. A second coupon with the same 20 mm diameter defect was fabricated, but a roll-on adhesive was used instead of the perforated film. The roll-on adhesive is significantly thicker than the film adhesive. During the elevated temperature step of curing, the honeycomb cells are pressed into the adhesive, causing a wicking of the resin on the honeycomb cell walls. This results in an apparent wall thickness significantly larger for the roll-on adhesive than for the film adhesive, as can be observed in Figure 12b. Of note is that the honeycomb pattern continues to be quite visible, but the crispness of the cell walls no longer exists as the reflections are dominated by the resin surrounding the honeycomb walls against the facesheet surface that wicked on the cell walls. Regardless, the 20 mm diameter defect is clearly visible in both adhesive systems, as noted in Figure 12.

3.4. Comparison between Portable Inspection System and Immersion Inspection System

The final study is the comparison between the results taken from the high-resolution immersion system and the portable immersion system. The former requires the submersion of the coupon within water, whereas the portable system allows the coupon to be inspected in the field or the manufacturing environment without submersion in water. The results for Coupon 2, the missing film adhesive coupon with a 20 mm diameter defect, are shown in Figure 13a for the immersion inspection, and the portable inspection system results are shown in Figure 13b. The signal analysis methods of the captured data from both forms of data collection are identical. Observe that in both cases, the defect identified is graphically indistinguishable. Of note is that the smaller features, such as the cell walls and the perforations in the film adhesive, remain observable and quantifiable, but the resulting data set for analysis is not as clear for the portable system. Thus, the portable system can be readily used for the quantification of the primary defects sought after in the present study, but the smaller (less than 1 mm feature sizes) may be more difficult to quantify.

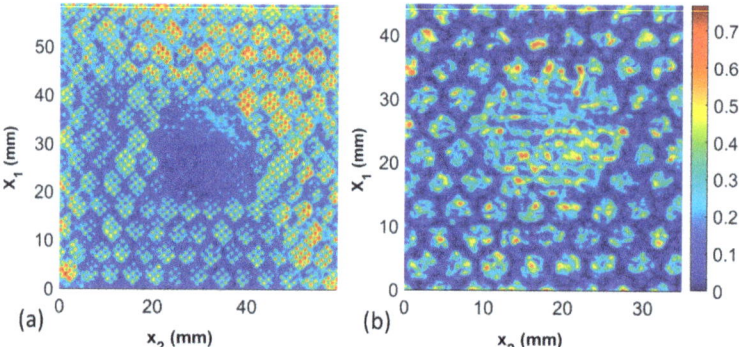

Figure 12. C-scan comparison between coupon with a nominally 20 mm diameter missing honeycomb defect made (a) with film adhesive with perforations and (b) with roll-on paste adhesive.

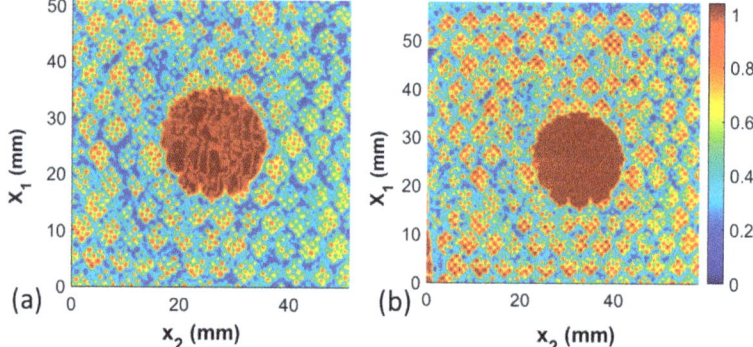

Figure 13. C-scan comparison between Coupon 2 with the inspection performed (a) in the immersion tank and (b) in the portable scanning system.

4. Discussion

The results shown in the previous section indicate the ability to capture defects that previously were considered undetectable using conventional ultrasound. Previous studies using Shearography were able to identify the existence of certain defects, specifically the largest defects investigated in the present study. Conversely, the present method allows for the detection of smaller defects and extends earlier works to allow for the quantification of the defect shape and size, as well as an empirical method to differentiate between defect types. In addition, many of the current studies focus on the probability of detection, or POD. In the present study, the medium-sized defects are often on the fringe of what is considered detectable; thus, one would expect a POD around the range of 50%. Using our presented method, we have a 100% POD over all samples studied, even the smallest defects which are considered undetectable. More importantly, we can dimensionalize the defect, something that is beyond the concept of POD but can be provided as a design tool for a structural engineer. The results for the 15 coupons from Table 1 are provided in Table 2, where the error is defined as the difference between the diameter measured using microscopy of the defect and the diameter extracted from the waveform data using the methods presented in Section 2. The error is defined as

$$Error \equiv |d_{microscope} - d_{UT\ Inspection}| \tag{8}$$

Table 2. Measurement error from ultrasonic characterization for each coupon.

Coupon	Defect Type	Diameter (mm) Microscopy	Diameter (mm) UT Results	Error (mm)
1	Missing Adhesive	7.9	8.3	0.4
2	Missing Adhesive	19.0	19.2	0.2
3	Missing Adhesive	40.8	41.1	0.3
4	Kapton Film	4.7	5.1	0.4
5	Kapton Film	20.0	21.3	1.3
6	Kapton Film	40.3	40.8	0.5
7	Hole	5.7	6.3	0.6
8	Hole	18.8	18.4	0.4
9	Hole	34.8	35.0	0.2
10	PTFE Film	5.0	4.2	0.8
11	PTFE Film	20.0	20.5	0.5
12	PTFE Film	40.4	41.3	0.9
13	Crushed Hole	6.4	7.2	0.8
14	Crushed Hole	19.3	19.1	0.2
15	Crushed Hole	35.5	34.3	1.2

The diameter reported in Table 2 is taken from the average of three tests of the same coupon to avoid any uncertainty due to human observations. A similar table is presented in Table 3, comparing the results for defect quantification from the portable system and from the immersion system. It was found that the average standard deviation by recharacterizing the captured data set for the immersion and portable system was 0.4 mm and 0.5 mm, respectively. The average error between the measured diameter between microscopy and the ultrasound inspection is 0.6 mm across all coupons for the immersion system, and it is 1.1 mm for the portable system. This is not significantly greater than the repeatability of the measurement itself of 0.4 mm. Thus, an improvement in resolution would be best obtained by refining the analysis method step that bridges the filtered and processed C-scan from Figure 8 to the quantification of the defect perimeter.

Table 3. Inspection results for coupons with a medium (nominally 20 mm) defect inspected with the portable system and the conventional immersion system.

Coupon	Defect Type	Diameter (mm) Microscopy	Diameter (mm) Immersion UT Results	Diameter (mm) Portable UT Results
2	Missing Adhesive	19.0	19.2	20.3
5	Kapton Film	20.0	21.3	20.34
8	Hole	18.8	18.4	17.07
11	PTFE Film	20.0	20.5	18.7
14	Crushed Hole	19.3	19.1	18.58

When looking at the defect quantification from the portable system and the immersion system, shown in Table 3, there is little difference between the results of the two systems. The portable UT system tended to yield slightly larger or smaller diameters across various defect types. For instance, it yielded the largest diameter measurements for missing adhesive defects. However, in PTFE, film defects showed considerable variation in diameter measurements between microscopy and portable UT. Additionally, the immersion system is more accurate than the portable system in various types of defects except Kapton film. In Kapton film, the portable system showed better results.

Of note in Table 2, there is no significant difference in the accuracy as a function of the size, where the standard deviation of error by size ranges from 0.5 mm to 0.6 mm. Conversely, there is an improvement in the accuracy for the missing adhesive coupons relative to the remaining coupons, with the missing adhesive having a standard deviation of error of 0.3 mm, whereas the standard deviation of the error for the coupons with missing honeycomb yields a value of 0.4 mm, and the remaining defect types yield a standard deviation of the error of 0.7 mm.

5. Conclusions

The presented study demonstrates the effectiveness of ultrasonic testing (UT) in differentiating between different defect types studied, including missing adhesive, embedded films, and core holes for honeycomb core structures. What we were unable to do was to differentiate between missing honeycomb and crushed honeycomb, but those two defect types are of a similar nature in that the honeycomb is not present under the surface of the CFRP facesheet. There is a high agreement between the characterized defect size measurements between the presented UT results and the microscopy images, with a typical difference between the methods of 0.6 mm. Of note, the ultrasonic immersion tank and portable systems exhibited the same inspection conclusions and both systems could identify all purposeful defects. Of interest was the unintentional identification of the film perforations that were on the order of 1 mm, a feature considerably smaller than the features of primary interest in the presented study. This is notable as it highlights the resolution capability of the presented methodology.

Overall, the findings underscore the significance of ultrasonic NDT methods in defect detection and characterization within honeycomb sandwich panels, with implications for ensuring the safety, reliability, and performance of composite structures across various industries. Moving forward, continued research and development in nondestructive testing techniques will further advance the understanding and capabilities in assessing the structural integrity of composite materials.

Author Contributions: Conceptualization, D.A.J. and D.P.P.; methodology, M.K., D.P.P. and D.A.J.; software, M.K., D.P.P. and D.A.J.; validation, M.K.; formal analysis, M.K.; investigation, M.K. and D.A.J.; resources, D.A.J.; data curation, M.K.; writing—original draft preparation, M.K. and D.A.J.; writing—review and editing, M.K., D.P.P. and D.A.J.; visualization, M.K. and D.A.J.; supervision, D.A.J.; project administration, D.A.J.; funding acquisition, D.A.J. All authors have read and agreed to the published version of the manuscript.

Funding: The research was funded in part by Verifi Technologies and Baylor University.

Institutional Review Board Statement: Not applicable.

Informed Consent Statement: Not applicable.

Data Availability Statement: The original contributions presented in the study are included in the article, further inquiries can be directed to the corresponding author.

Acknowledgments: The authors would like to thank the personnel of Baylor Materials Testing and Characterization laboratory for contributing to the experimental setup.

Conflicts of Interest: The authors declare no conflicts of interest.

References

1. Castanie, B.; Bouvet, C.; Ginot, M. Review of composite sandwich structure in aeronautic applications. *Compos. Part C Open Access* **2020**, *1*, 100004. [CrossRef]
2. Davies, J.M. *Lightweight Sandwich Construction*; John Wiley & Sons: Hoboken, NJ, USA, 2008.
3. Niedermeyer, M. X-33 LH2 Tank Failure Investigation Findings. Available online: https://ntrs.nasa.gov/citations/20030067586 (accessed on 21 March 2024).
4. Sikdar, S.; Banerjee, S. Identification of disbond and high density core region in a honeycomb composite sandwich structure using ultrasonic guided waves. *Compos. Struct.* **2016**, *152*, 568–578. [CrossRef]
5. Girolamo, D.; Chang, H.-Y.; Yuan, F.-G. Impact damage visualization in a honeycomb composite panel through laser inspection using zero-lag cross-correlation imaging condition. *Ultrasonics* **2018**, *87*, 152–165. [CrossRef] [PubMed]
6. Dong, Z.; Chen, W.; Saito, O.; Okabe, Y. Disbond detection of honeycomb sandwich structure through laser ultrasonics using signal energy map and local cross-correlation. *J. Sandw. Struct. Mater.* **2023**, *25*, 501–517. [CrossRef]
7. Uddin, M.N.; Gandy, H.T.N.; Rahman, M.M.; Asmatulu, R. Adhesiveless honeycomb sandwich structures of prepreg carbon fiber composites for primary structural applications. *Adv. Compos. Hybrid Mater.* **2019**, *2*, 339–350. [CrossRef]

8. Hay, T.R.; Wei, L.; Rose, J.L.; Hayashi, T. Rapid Inspection of Composite Skin-Honeycomb Core Structures with Ultrasonic Guided Waves. *J. Compos. Mater.* **2003**, *37*, 929–939. [CrossRef]
9. Shrestha, R.; Choi, M.; Kim, W. Thermographic inspection of water ingress in composite honeycomb sandwich structure: A quantitative comparison among Lock-in thermography algorithms. *Quant. InfraRed Thermogr. J.* **2021**, *18*, 92–107. [CrossRef]
10. Hagemaier, D.J. NDT of aluminium-brazed titanium honeycomb structure. *Non-Destr. Test.* **1976**, *9*, 107–116. [CrossRef]
11. He, Y.; Tian, G.; Pan, M.; Chen, D. Non-destructive testing of low-energy impact in CFRP laminates and interior defects in honeycomb sandwich using scanning pulsed eddy current. *Compos. Part B Eng.* **2014**, *59*, 196–203. [CrossRef]
12. Underhill, P.R.; Rellinger, T.; Krause, T.W.; Wowk, D. Eddy Current Array Inspection of Damaged CFRP Sandwich Panels. *J. Nondestruct. Eval. Diagn. Progn. Eng. Syst.* **2020**, *3*, 031104. [CrossRef]
13. Rellinger, T.; Underhill, P.R.; Krause, T.W.; Wowk, D. Combining eddy current, thermography and laser scanning to characterize low-velocity impact damage in aerospace composite sandwich panels. *NDT E Int.* **2021**, *120*, 102421. [CrossRef]
14. Ren, Y.; Zeng, Z.; Jiao, S. Eddy current testing of CFRP/Aluminium honeycomb sandwich structure. *Nondestruct. Test. Eval.* **2024**, 1–13. [CrossRef]
15. Papanaboina, M.R. Non-Destructive Evaluation of Honeycomb Sandwich Using Ultrasonic Phased Arrays. Kauno Technologijos Universitetas. 2019. Available online: https://epubl.ktu.edu/object/elaba:38191574/ (accessed on 15 April 2024).
16. Qin, Y.-W.; Bao, N.-K. Infrared thermography and its application in the NDT of sandwich structures. *Opt. Lasers Eng.* **1996**, *25*, 205–211. [CrossRef]
17. Ibarra-Castanedo, C.; Piau, J.-M.; Guilbert, S.; Avdelidis, N.P.; Genest, M.; Bendada, A.; Maldague, X.P.V. Comparative Study of Active Thermography Techniques for the Nondestructive Evaluation of Honeycomb Structures. *Res. Nondestruct. Eval.* **2009**, *20*, 1–31. [CrossRef]
18. Usamentiaga, R.; Venegas, P.; Guerediaga, J.; Vega, L.; López, I. Non-destructive inspection of drilled holes in reinforced honeycomb sandwich panels using active thermography. *Infrared Phys. Technol.* **2012**, *55*, 491–498. [CrossRef]
19. Zhao, H.; Zhou, Z.; Fan, J.; Li, G.; Sun, G. Application of lock-in thermography for the inspection of disbonds in titanium alloy honeycomb sandwich structure. *Infrared Phys. Technol.* **2017**, *81*, 69–78. [CrossRef]
20. Hu, C.; Duan, Y.; Liu, S.; Yan, Y.; Tao, N.; Osman, A.; Ibarra-Castanedo, C.; Sfarra, S.; Chen, D.; Zhang, C. LSTM-RNN-based defect classification in honeycomb structures using infrared thermography. *Infrared Phys. Technol.* **2019**, *102*, 103032. [CrossRef]
21. Guo, B.; Zheng, X.; Gerini-Romagnoli, M.; Yang, L. Digital shearography for NDT: Determination and demonstration of the size and the depth of the smallest detectable defect. *NDT E Int.* **2023**, *139*, 102927. [CrossRef]
22. Akatay, A.; Bora, M.Ö.; Fidan, S.; Çoban, O. Damage characterization of three point bended honeycomb sandwich structures under different temperatures with cone beam computed tomography technique. *Polym. Compos.* **2018**, *39*, 46–54. [CrossRef]
23. Crupi, V.; Epasto, G.; Guglielmino, E.; Mozafari, H.; Najafian, S. Computed tomography-based reconstruction and finite element modelling of honeycomb sandwiches under low-velocity impacts. *J. Sandw. Struct. Mater.* **2014**, *16*, 377–397. [CrossRef]
24. Rivera, A.X.; Venkataraman, S.; Hyonny, K.; Pineda, E.J.; Bergan, A. Characterization and Modeling of Cell Wall Imperfections in Aluminum Honeycomb Cores using X-ray CT Imaging. In Proceedings of the AIAA Scitech 2021 Forum, Virtual, 11–15 & 19–21 January 2021; American Institute of Aeronautics and Astronautics: Reston, VA, USA, 2021. [CrossRef]
25. Gebrehiwet, L.; Chimido, A.; Melaku, W.; Tesfaye, E. A Review of Common Aerospace Composite Defects Detection Methodologies. *Int. J. Res. Publ. Rev.* **2023**, *4*, 1829–1846.
26. Krautkrämer, J.; Krautkrämer, H. *Ultrasonic Testing of Materials*; Springer Science & Business Media: Berlin/Heidelberg, Germany, 2013.
27. Du, B.; He, Y.; He, Y.; Zhang, C. Progress and trends in fault diagnosis for renewable and sustainable energy system based on infrared thermography: A review. *Infrared Phys. Technol.* **2020**, *109*, 103383. [CrossRef]
28. Lobanov, L.M.; Savytskyi, V.V.; Kyianets, I.V.; Shutkevich, O.P.; Shyian, K.V. Non-Destructive Testing of Elements of Titanium Honeycomb Panels by Shearography Method Using Vacuum Load. *Tech. Diagn. Non-Destr. Test.* **2021**, *C440*, 49. [CrossRef]
29. GRevel, M.; Pandarese, G.; Allevi, G. Quantitative defect size estimation in shearography inspection by wavelet transform and shear correction. In Proceedings of the 2017 IEEE International Workshop on Metrology for AeroSpace (MetroAeroSpace), Padua, Italy, 21–23 June 2017; pp. 535–540.
30. Frohlich, H.B.; Fantin, A.V.; de Oliveira, B.C.F.; Willemann, D.P.; Iervolino, L.A.; Benedet, M.E.; Junior, A.A.G. Defect classification in shearography images using convolutional neural networks. In Proceedings of the 2018 International Joint Conference on Neural Networks (IJCNN), Rio de Janeiro, Brazil, 8–13 July 2018; pp. 1–7. [CrossRef]
31. Zhang, H.; Liu, S.; Rui, X.; Zhu, X.; Sun, J. A skin-core debonding quantitative algorithm based on hexagonal units reconstruction for air-coupled ultrasonic C-scan images of honeycomb sandwich structure. *Appl. Acoust.* **2022**, *198*, 108964. [CrossRef]
32. Hsu, D.K. Nondestructive testing using air-borne ultrasound. *Ultrasonics* **2006**, *44*, e1019–e1024. [CrossRef]
33. Zhou, Z.; Ma, B.; Jiang, J.; Yu, G.; Liu, K.; Zhang, D.; Liu, W. Application of wavelet filtering and Barker-coded pulse compression hybrid method to air-coupled ultrasonic testing. *Nondestruct. Test. Eval.* **2014**, *29*, 297–314. [CrossRef]
34. Blandford, B.M.; Jack, D.A. High resolution depth and area measurements of low velocity impact damage in carbon fiber laminates via an ultrasonic technique. *Compos. Part B Eng.* **2020**, *188*, 107843. [CrossRef]
35. Blackman, N.J.; Jack, D.A.; Blandford, B.M. Improvement in the Quantification of Foreign Object Defects in Carbon Fiber Laminates Using Immersion Pulse-Echo Ultrasound. *Materials* **2021**, *14*, 2919. [CrossRef]

36. Pisharody, A.P.; Blandford, B.; Smith, D.E.; Jack, D.A. An experimental investigation on the effect of adhesive distribution on strength of bonded joints. *Appl. Adhes. Sci.* **2019**, *7*, 6. [CrossRef]
37. Jack, D.A.; Blackman, N.J.; Blandford, B.M. System and Method for Real-Time Visualization of Foreign Objects within a Material. U.S. Patent US20210364472A1, 25 November 2021. Available online: https://patents.google.com/patent/US20210364472A1/en (accessed on 16 April 2024).

Disclaimer/Publisher's Note: The statements, opinions and data contained in all publications are solely those of the individual author(s) and contributor(s) and not of MDPI and/or the editor(s). MDPI and/or the editor(s) disclaim responsibility for any injury to people or property resulting from any ideas, methods, instructions or products referred to in the content.

Article

Effect of the Laying Order of Core Layer Materials on the Sound-Insulation Performance of High-Speed Train Carbody

Ruiqian Wang [1,2], Dan Yao [3], Jie Zhang [4,*], Xinbiao Xiao [1] and Xuesong Jin [1]

1. State Key Laboratory of Traction Power, Southwest Jiaotong University, Chengdu 610031, China
2. School of Mechanical Engineering and Rail Transit, Changzhou University, Changzhou 213164, China
3. Aviation Engineering Institute, Civil Aviation Flight University of China, Guanghan 618307, China
4. State Key Laboratory of Polymer Materials Engineering, Polymer Research Institute, Sichuan University, Chengdu 610065, China
* Correspondence: zh.receive@gmail.com

Abstract: The design of sound-insulation schemes requires the development of new materials and structures while also paying attention to their laying order. If the sound-insulation performance of the whole structure can be improved by simply changing the laying order of materials or structures, it will bring great advantages to the implementation of the scheme and cost control. This paper studies this problem. First, taking a simple sandwich composite plate as an example, a sound-insulation prediction model for composite structures was established. The influence of different material laying schemes on the overall sound-insulation characteristics was calculated and analyzed. Then, sound-insulation tests were conducted on different samples in the acoustic laboratory. The accuracy of the simulation model was verified through a comparative analysis of experimental results. Finally, based on the sound-insulation influence law of the sandwich panel core layer materials obtained from simulation analysis, the sound-insulation optimization design of the composite floor of a high-speed train was carried out. The results show that when the sound absorption material is concentrated in the middle, and the sound-insulation material is sandwiched from both sides of the laying scheme, it represents a better effect on medium-frequency sound-insulation performance. When this method is applied to the sound-insulation optimization of a high-speed train carbody, the sound-insulation performance of the middle and low-frequency band of 125–315 Hz can be improved by 1–3 dB, and the overall weighted sound reduction index can be improved by 0.9 dB without changing the type, thickness or weight of the core layer materials.

Keywords: sound insulation; sound-insulation material; sound absorption material; laying scheme; optimal design; high-speed train

Citation: Wang, R.; Yao, D.; Zhang, J.; Xiao, X.; Jin, X. Effect of the Laying Order of Core Layer Materials on the Sound-Insulation Performance of High-Speed Train Carbody. *Materials* 2023, *16*, 3862. https://doi.org/10.3390/ma16103862

Academic Editor: Raul D. S. G. Campilho

Received: 11 April 2023
Revised: 17 May 2023
Accepted: 18 May 2023
Published: 20 May 2023

Copyright: © 2023 by the authors. Licensee MDPI, Basel, Switzerland. This article is an open access article distributed under the terms and conditions of the Creative Commons Attribution (CC BY) license (https://creativecommons.org/licenses/by/4.0/).

1. Introduction

In the noise control of high-speed trains, sound-insulation design is always very important. A high-speed train's body is a complex multilayer composite structure, which includes the body profile, inner plate and core layer materials (all kinds of sound absorption/insulation materials and damping materials). Its acoustic properties are influenced by its mass, structure and material [1–3].

In order to achieve higher acoustic insulation performance, many studies were proposed. For example, from the perspective of optimizing acoustic structure, Yao et al. [4,5], taking typical cross-section mass and average sound transmission loss in the 400 to 3150 Hz frequency band as the optimization objectives, carried out lightweight sound-insulation design for the aluminum profile of the floor of a high-speed train, which improved the average sound insulation and weighted sound insulation by 0.8 dB and 1.0 dB, respectively, in the aforementioned frequency band. They also proposed a modal adaptive damping treatment optimization design method that effectively improved the acoustic and vibrational performance of the floor's structure. Lin et al. [6] optimized the acoustic and vibrational performance of aluminum profiles from the aspects of the acoustic bridge, plate thickness, structural materials, etc., and obtained a structure with better acoustic and vibrational performance. Zhang et al. [7] used the wavenumber finite element method and the wavenumber boundary element method to study the influence of core sections of rectangular, triangular and trapezoidal trusses on sound transmission loss of aluminum profiles in detail. Jacek et al. [8] proposed the composite floor with a dry floating screed and a suspended ceiling, which achieved satisfactory results in airborne and impact sound insulation. Li et al. [9] proposed an orthogonally rib-stiffened honeycomb double sandwich structure with periodic arrays of shunted piezoelectric patches, which can improve the low-frequency sound insulation of the aircraft cabin and broaden the sound-insulation bandwidth.

There are also some studies from the perspective of new materials, such as by Hu et al. [10], who analyzed the influence of honeycomb core specifications, panel materials and glass bead modification on the sound-insulation performance of honeycomb sandwich panels by using the hot-pressing method and the four-sensor impedance tube method, providing data support for the selection of train body materials. Kim et al. [11,12] evaluated the sound-insulation performance of honeycomb composite panels and discussed the feasibility of replacing traditional corrugated steel plates with honeycomb composite panels in a car body. They also improved the sound-insulation effect by placing polyurethane foam in the aluminum cavity of a high-speed train's body. Kaidouchi et al. [13] studied the sound-insulation performance of an in-plane honeycomb sandwich structure in aerospace applications and found that glass-fiber-reinforced polymer cores with fiber-reinforced plastic facing materials have better vibro-acoustic and sound transmission characteristics. Kang et al. [14] proposed an elastic layer with a low specific modulus in the middle of the core layer, which can significantly improve the broadband sound-insulation performance of the panel. Liu et al. [15] put forward the selection principle of aluminum profile surface damping material, which can predict the sound-insulation performance of an aluminum extrusion plate at the initial stage of design or put forward suggestions for improvement. Zhang et al. [16] designed a lightweight, low-frequency acoustic metamaterial (beam-like resonator) for low-frequency noise and vibration control based on the multiparameter optimization method, which has an obvious effect on the low-frequency noise and vibration control of a high-speed train's floor structure.

To sum up, the existing research mainly improves the sound-insulation performance through material modification and structural design. However, the laying order of different materials and different structures also has an important effect on the sound insulation of the whole structure, which was often ignored in the previous research. In addition, the application of a new structure and new materials also involves a lot of new problems, such as the need to redesign space and weight accounting, recheck the strength and recheck the fire rating and insulation performance. This brings higher research, development and verification costs. If the sound-insulation performance of the whole structure can be improved simply by changing the placement sequence of materials or structures, it will bring great advantages to the implementation of the scheme and cost control.

Compared with previous studies, the research in this paper is different in that it can improve the sound-insulation performance by changing only the laying sequence of core layer materials without changing the type, thickness and density of core layer materials in the structure. Relatively speaking, this approach has almost no change to the space, thickness, weight and material type of the structure. It will bring great advantages to the implementation of the program and cost control and has a high engineering practical significance. In Section 2, taking a simple sandwich composite board as an example, the influence of changing the material laying order on the overall sound-insulation characteristics is simulated and analyzed. In Section 3, experiments are carried out to verify the simulation results of the previous section. In Section 4, the research results are applied to the sound-insulation optimization design of a high-speed train's floor structure, and the optimization effect is evaluated.

2. Simulation Study on the Effect of Core Layer Material Laying Order on Sandwich Composite Panel Sound Insulation

In Figure 1, a section diagram of a high-speed train's floor structure is provided. It can be seen that the outside of the floor structure has an aluminum profile; the inside is the inner floor; and the middle is the core layer material, which is similar to a sandwich structure on the whole. Therefore, this paper first takes a simple sandwich composite board as the object to explore the effect of the laying sequence of core layer materials on the overall sound-insulation characteristics, which can provide guidance for the subsequent sound-insulation design of high-speed train floor structures.

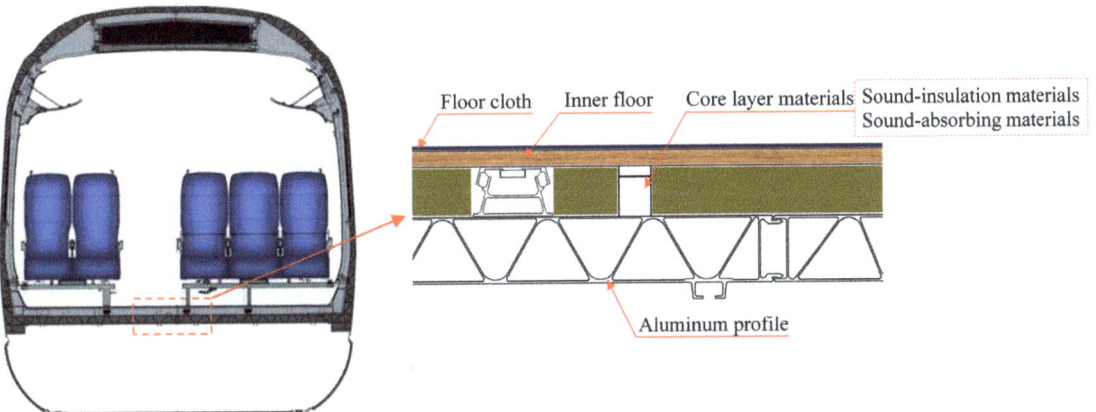

Figure 1. Section of a high-speed train's floor structure.

As shown in Figure 2, under the premise that the total thickness, weight and type of core layer materials remain unchanged, there are four different laying sequences: (1) sound insulation and absorption materials are uniformly stacked alternately; (2) sound-insulation materials and sound-absorbing materials are stacked separately; (3) the sound-insulation materials are concentrated in the middle, and the sound-absorbing material is distributed on both sides; (4) sound-absorbing materials are concentrated in the middle, and sound-insulation materials are distributed on both sides. In this section, corresponding comparative calculation conditions are set for these four situations to explore their influence on sound insulation.

2.1. Sound Insulation Prediction Model

Statistical energy analysis (SEA) is an effective method for acoustic modeling and prediction [17] that is often used in the field of sound insulation [18–20]. The whole system is composed of a number of statistical subsystems, and the external incentive is applied to obtain a quick response from the system. This is realized by using the commercial software VA One [21].

The sound cavity–sandwich plate–sound cavity insulation prediction model was established in the software VA One, as shown in Figure 3. The dimensions of both cavities are 1.0 m × 1.0 m × 1.0 m. A reverberation field is applied to one of the cavities to simulate the source excitation and to the other cavity to simulate the receiving side. The length and width of the sandwich plate are 1.0 m × 1.0 m, and it is composed of a skin substrate and core layer material (sound insulation material and sound-absorbing material), which can be realized by using the acoustic package [22–24] in the software, as shown in Figure 4. Model verification was carried out before starting the model.

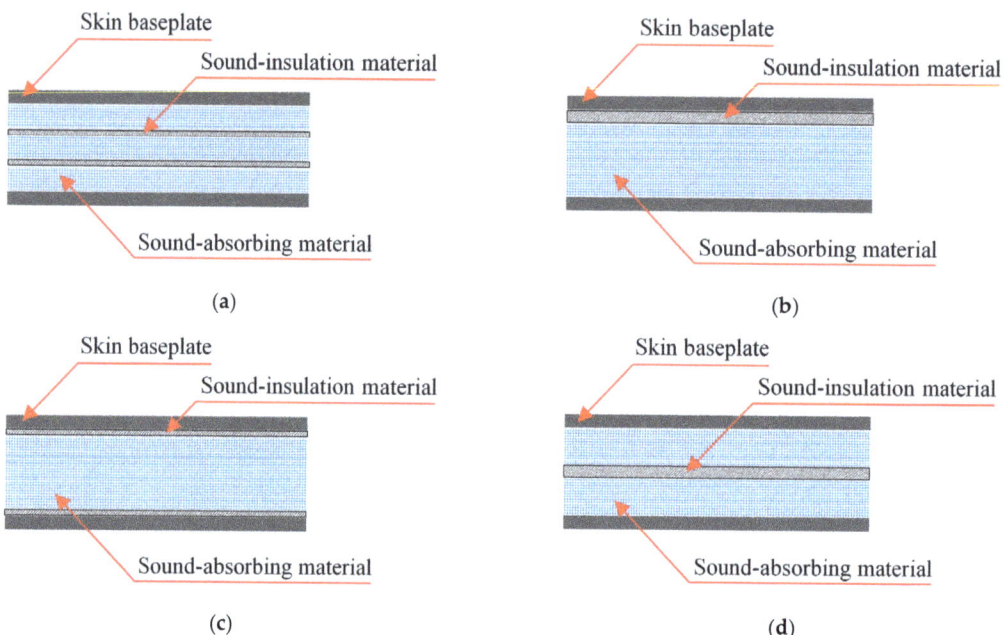

Figure 2. Four typical material laying schemes. (**a**) Scheme #1; (**b**) Scheme #2; (**c**) Scheme #3; (**d**) Scheme #4.

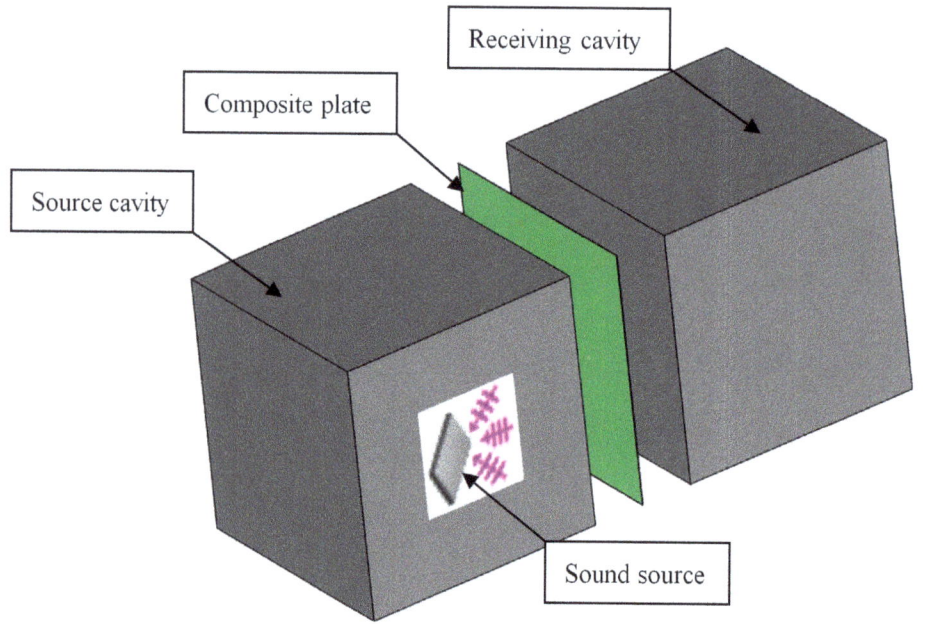

Figure 3. Sound insulation prediction model of sandwich composite plate.

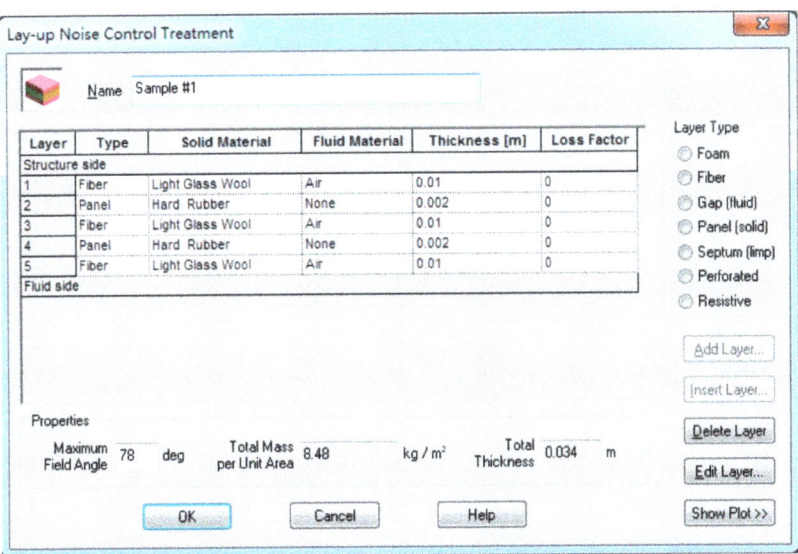

Figure 4. Sound package setup.

The sound insulation of the plate structure is defined as follows:

$$TL = L_1 - L_2 + 10\log_{10}\frac{S}{A} \tag{1}$$

where L_1 and L_2 are the average sound pressure levels of the source chamber and the receiving chamber; S is the area of the sample; and A is the sound absorption coefficient of the receiving room, which can be obtained by testing the reverberation time of the receiving room according to Equation (2).

$$A = \frac{0.16V_2}{T} \tag{2}$$

Here, V_2 is the volume of the receiving chamber, and T is the reverberation time of the receiving room. The relationship between the reverberation time and the damping loss factor of the receiving chamber is as follows:

$$\eta = \frac{2.2}{Tf} \tag{3}$$

In SEA, the difference in sound pressure level between the external and internal cavities is related to the energy density ratio of the two cavities:

$$L_1 - L_2 = 10\log_{10}\frac{E_1/V_1}{E_2/V_2} \tag{4}$$

By substituting Equations (2)–(4) into Equation (1), the sound insulation calculation results can be obtained, as shown in Equation (5).

$$TL = 10\log_{10}\left(\frac{E_1}{E_2}\frac{27.5\pi S}{\omega V_1 \eta}\right) \tag{5}$$

After obtaining the calculation result of the sound-insulation frequency curve, the weighted sound reduction index R_w is further calculated according to the standard [25], which is used as the single value evaluation quantity to evaluate the overall sound-insulation level of the sample.

2.2. Research Schemes and Results Analysis

The hard rubber and light glass wool provided in the software are used as the sound-insulation and absorption materials in the core layer, and a total of two comparison groups are set. In comparison to Group #1 and #2, a 2 mm aluminum plate and 4 mm thick plywood were used as the skin baseplate, respectively. Then, four samples with the same type of core layer materials, equal weight and thickness were set as the comparison schemes for each comparison group, and the difference was only in their laying order. Table 1 shows the basic parameters of the skin and core materials involved in the model. Tables 2 and 3, respectively, give the concrete laying schemes of the core layer materials of the two comparison groups.

Table 1. Basic parameters of materials in the model.

Parameters	Aluminum Plate	Plywood	Hard Rubber	Light Glass Wool
Density (kg/m^3)	2700	700	1100	16
Elasticity modulus (GPa)	71.0	6.0	2.3	\
Poisson's ratio	0.33	0.25	0.49	\
Porosity	\	\	\	0.99
Flow resistivity (N.s/m^4)	\	\	\	9000

Table 2. Material composition of each research scheme in Group #1.

No.	Sample #1-1	Sample #1-2	Sample #1-3	Sample #1-4
Core layer material composition	10 mm light glass wool	4 mm hard rubber	15 mm light glass wool	2 mm hard rubber
	2 mm hard rubber			
	10 mm light glass wool		4 mm hard rubber	30 mm light glass wool
	2 mm hard rubber	30 mm light glass wool		
	10 mm light glass wool		15 mm light glass wool	2 mm hard rubber

Table 3. Material composition of each research scheme in Group #2.

No.	Sample #2-1	Sample #2-2	Sample #2-3	Sample #2-4
Core layer material composition	20 mm light glass wool	2 mm hard rubber	30 mm light glass wool	1 mm hard rubber
	1 mm hard rubber			
	20 mm light glass wool		2 mm hard rubber	60 mm light glass wool
	1 mm hard rubber	60 mm light glass wool		
	20 mm light glass wool		30 mm light glass wool	1 mm hard rubber

As can be seen from Tables 2 and 3, Samples #1-1 and #2-1 are laid alternately in the core layer with sound-insulation materials and sound-absorbing materials. Sample #1-2 and #2-2 are separated soundproof materials and sound-absorbing materials; Sample #1-3 and #2-3 are soundproof materials placed in the middle of the core layer and sound-absorbing materials placed on both sides; in both Samples #1-4 and #2-4, soundproof materials are placed on both sides of the core layer, and sound-absorbing materials are concentrated in the middle of the core layer.

Figure 5a,b, respectively, show the sound-insulation calculation results of the comparison Groups #1 and #2. As can be seen from Figure 5a, the four samples have little difference at frequencies below 160 Hz and above 800 Hz, and the difference is mainly in the middle-frequency band between 200 Hz and 630 Hz. Among them, Sample #1-1 has the lowest intermediate frequency sound insulation; compared with Sample #1-1, the increase in Sample #1-3 is 9–13 dB in the frequency band 400–630 Hz. Compared with Sample #1-3, the increase in Sample #1-2 is 6–13 dB at 200–400 Hz, but a slight decrease occurs at

630–800 Hz. Compared with Sample #1-2, Sample #1-4 has an increase of nearly 2 dB at 200–800 Hz. From the perspective of the overall weighted sound reduction index R_w, the order of the four sample values is consistent with the quality of the intermediate frequency sound-insulation level. The law in Figure 5b is basically similar to that in Figure 5a, and it is not repeated.

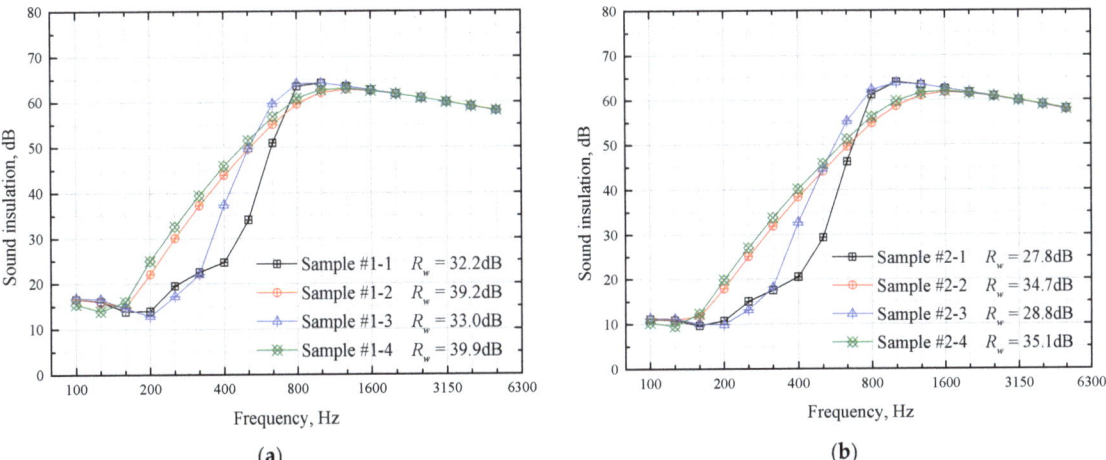

Figure 5. Calculation results of influence of material laying order on sandwich plate sound-insulation. (**a**) Group #1; (**b**) Group #2.

Furthermore, combined with the material composition of each sample, it can be found that the quality of intermediate frequency sound insulation has the greatest relationship with the concentration of sound-absorbing materials. The relationship is basically positive. Taking comparison Group #1 as an example, the total thickness of the sound-absorbing materials in 4 samples is 30 mm. Sample #1-1 is divided from 2 acoustic layers into 3 thin sound-absorbing layers with a thickness of 10 mm. Sample #1-3 is divided by a soundproof layer into 2 thin sound-absorbing layers with a thickness of 15 mm. The sound-absorbing layer of Sample #1-2 is not divided but placed separately from the sound-insulation layer without them crossing each other. The sound-absorbing layers of Sample #1-4 are also not divided but sandwiched between two soundproof layers.

The above calculation results show that for simple sandwich composite panels, the sound-insulation characteristics of the whole structure are only affected if the laying sequence is changed under the premise that the type, thickness and weight of the core layer materials are unchanged. Among them, the sound absorption material is centrally placed, significantly improving the intermediate frequency sound insulation's performance. This is caused by the variation in sound absorption properties of the material's thickness. As shown in Figure 6a,b, the measured results of the sound absorption coefficient with varying thicknesses of two porous materials are provided. It can be seen that with the increase in material thickness, the optimal frequency band of the sound absorption coefficient moves in the low-frequency direction, and the frequency band with the largest increase in the sound absorption coefficient happens to be located in the middle-frequency band of the sound absorption curve. In addition, the sound-insulation material is divided and evenly placed on both sides of the sound absorbing material, close to the skin substrate, which can further improve the level of sound insulation, but the improvement is not large.

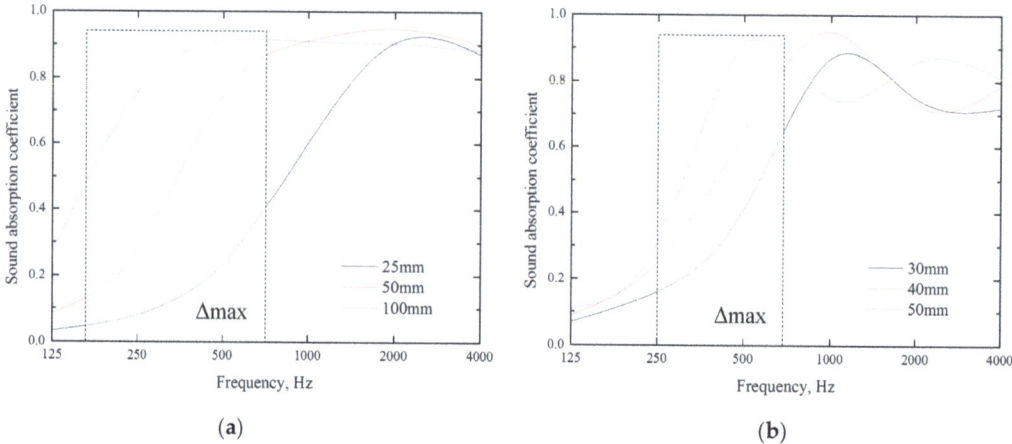

Figure 6. Variation in sound absorption coefficient with the thickness of the porous material. (**a**) Superfine glass wool (Volume–weight 60 kg/m^3); (**b**) Polyurethane foam (Volume–weight 10 kg/m^3).

3. Test Verification

Based on the double reverberation chamber method [26], the simulation calculation results in Section 2 were tested and verified. As shown in Figure 7, the test sample size was 1 m^2, which was installed in the sound-insulation hole, and the surrounding area was sealed well with oil sludge. The sound source room and the receiving room were each equipped with 6 microphones. Before the test, a technical inspection of the number and positions of loudspeakers and microphones in the source room was completed. During the test, a diffused sound field of more than 100 dB was applied in the source room by using an undirected speaker, and the average sound pressure levels L_1 and L_2 of the source room and the receiving room were measured. The test frequency ranged from 100 Hz to 5000 Hz.

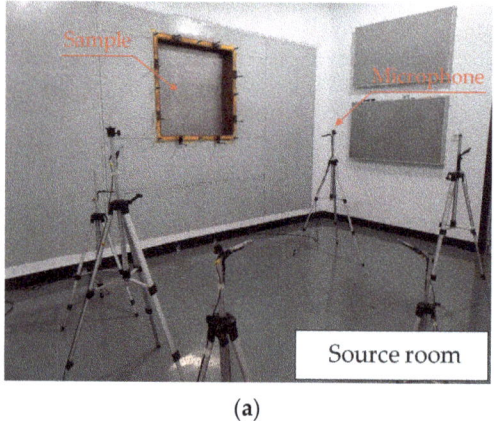
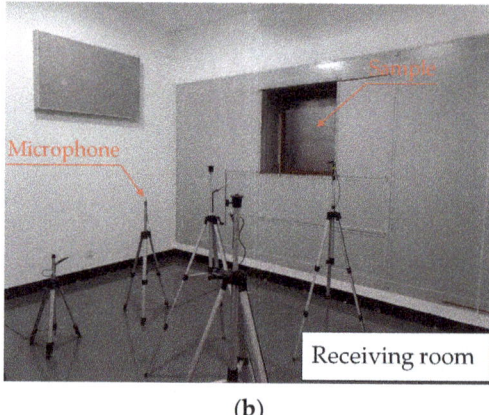

Figure 7. Sound-insulation test site. (**a**) Source room; (**b**) Receiving room.

Two test comparison groups, #3 and #4, were set up, and two samples were set up in each comparison group, as shown in Table 4. Figure 8 shows the section of the basic sound-absorbing materials and sound-insulation materials involved. Melamine and carbon fiber cotton are both soft and porous materials that are often used to fill the core layer of train body structure and can play the role of thermal insulation and sound absorption.

Soundproof pads are often used to increase the sound-insulation level for composite structures. The two types of sound-insulation pads used in this paper have different thicknesses and hardness.

Table 4. Material composition of each scheme for Groups #3 and #4.

No.	Group #3		Group #4	
	Sample #3-1	Sample #3-2	Sample #4-1	Sample #4-2
Core layer material composition	10 mm melamine	2 mm A-type soundproof pad	10 mm carbon fiber cotton	2 mm B-type soundproof pad
	2 mm A-type soundproof pad	10 mm melamine	1 mm B-type soundproof pad	10 mm carbon fiber cotton
	10 mm melamine	10 mm melamine	10 mm carbon fiber cotton	10 mm carbon fiber cotton
	2 mm A-type soundproof pad	10 mm melamine	1 mm B-type soundproof pad	10 mm carbon fiber cotton
	10 mm melamine	2 mm A-type soundproof pad	10 mm carbon fiber cotton	2 mm B-type soundproof pad

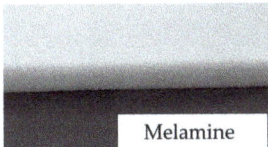
Melamine

Carbon fiber cotton
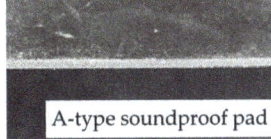
A-type soundproof pad

B-type soundproof pad

Figure 8. Cross-sections of the basic materials.

The cross-sections of the samples are shown in Figure 9. In terms of materials, the two comparison groups had different skin baseplates, sound-insulation materials and sound-absorbing materials. Among them, Group #3 was made with a 2 mm aluminum plate as the skin baseplate and an A-type soundproof pad and melamine as sound-insulation and sound-absorbing materials in the core layer. In Group #4, a 1 mm steel plate was used as the skin baseplate, and a B-type soundproof pad and carbon fiber cotton were used as the sound-insulation and sound-absorbing materials in the core layer. In terms of the laying sequence of core layer materials, Sample #3-1 and Sample #4-1 were laid alternately with sound-insulation materials and sound-absorbing materials. However, Sample #3-2 and Sample #4-2 both concentrated sound-absorbing materials in the middle of the core layer and sound-insulation materials on both sides of the core layer.

Figure 10 shows the test results of comparison Groups #3 and #4. As can be seen from Figure 10a, compared with Sample #3-1, the sound insulation of Sample #3-2 is significantly improved at 160–1000 Hz, and the improvement amount even reaches 12 dB in the frequency band 400–500 Hz. Overall, the weighted sound reduction index R_w increased by 4.8 dB. The law shown in Figure 10b is similar to that in Figure 10a, so it is not repeated.

The experimental results of the simple sandwich composite plate proved that the simulation results of the previous section are correct. This indicates that, compared with "sound insulation materials and sound absorbing materials alternately placed", the material laying scheme of "sound absorbing materials concentrated in the middle and sound insulation materials placed on both sides" is more conducive to the sound-insulation performance's improvement in the middle-frequency band and is very beneficial to the overall weighted sound reduction index's improvement.

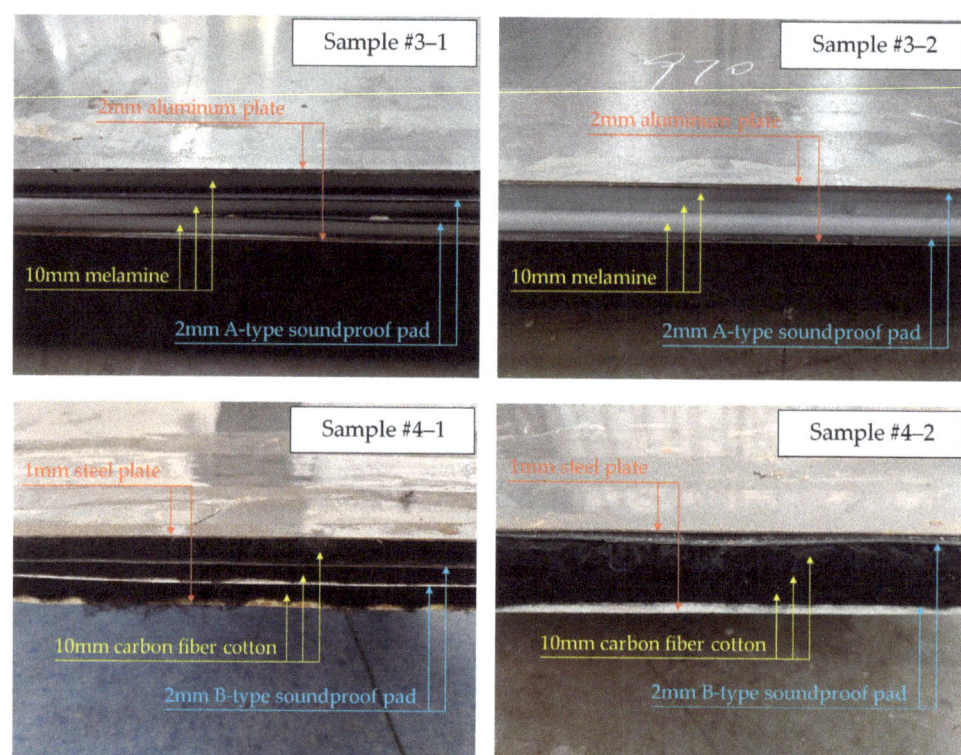

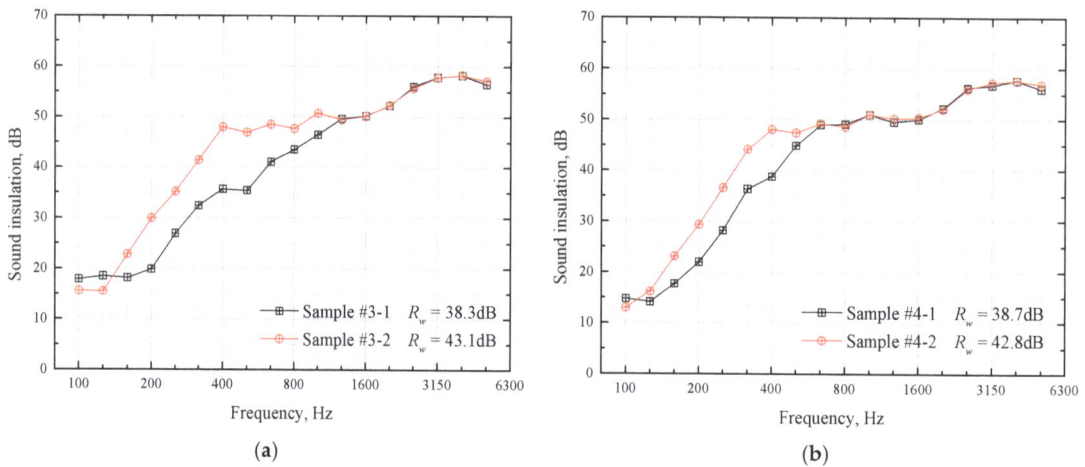

Figure 9. Cross-sections of the test samples.

Figure 10. Experimental results of Group #3 and #4. (a) Group #3; (b) Group #4.

4. Optimization Design of Floor Sound Insulation for High-Speed Train

With reference to the research conclusion of the simple sandwich composite plate mentioned above, the floor structure of a high-speed train, shown in Figure 1, was taken as an example for carrying out the sound-insulation optimization design.

As shown in Table 5, the first column provides the core layer material composition of the original Structure #0 of the high-speed train's floor. It can be seen that in the original structure, sound-insulation materials and sound absorbing materials are basically in alternate laying form; then, the second column follows the principle of "sound-absorbing materials are placed centrally in the middle and sound-isolating materials are placed on both sides", and the laying sequence of the core layer materials is adjusted to form sound-insulation optimization Structure #1. The cross-section of each structure is shown in Figure 11.

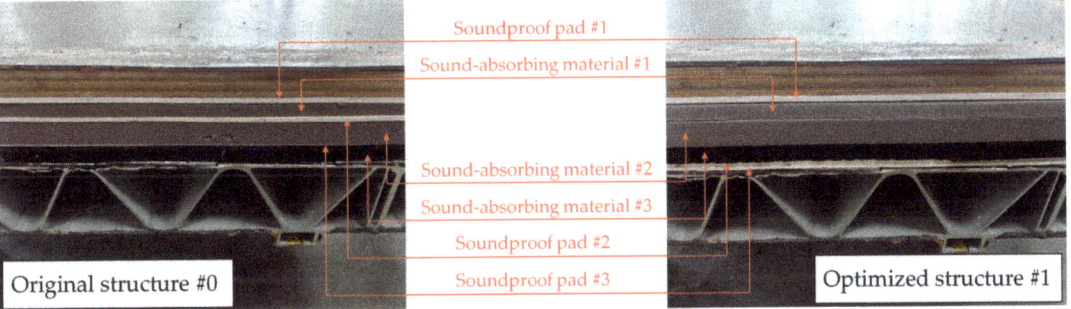

Figure 11. Cross-section of each floor structure.

Figure 12 shows the sound-insulation test results of each structure in Table 5. As can be seen from the figure, compared with the original floor Structure #0, the type, thickness and weight of the core layer materials of the floor optimization Structure #1 remain unchanged. However, the change in the materials' laying order causes the middle and low-frequency of 125–315 Hz to increase by 1–3 dB and the overall weighted sound reduction index Rw to increase by 0.9 dB.

According to the mass law, 9.9 kg of weight is needed to increase the sound insulation of this floor structure (total weight 90.6 kg) by 0.9 dB. However, by adjusting the laying sequence of the core materials, a 0.9 dB improvement in sound insulation can be achieved without adding weight, which is the main contribution of this study.

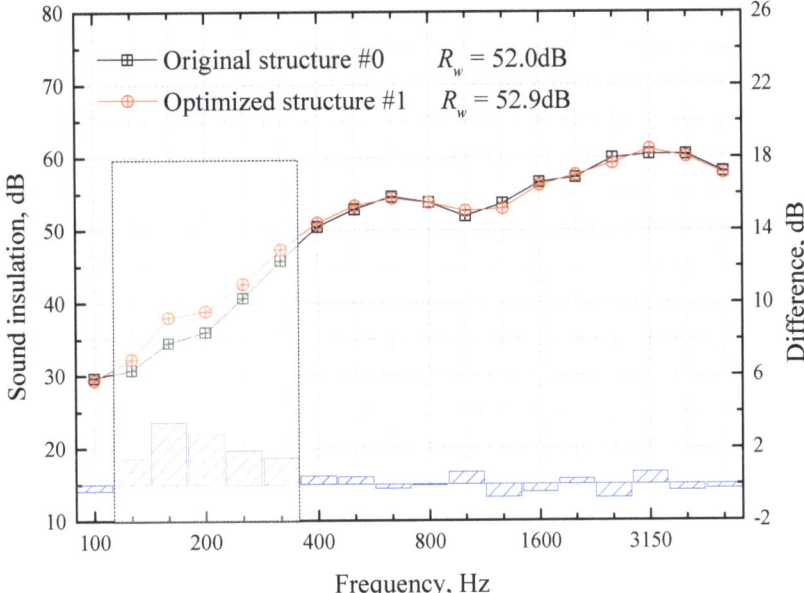

Figure 12. Sound-insulation optimization results for a high-speed train's floor structure.

Table 5. Material composition of the original and optimized floor structure of a high-speed train.

No.	Original Structure #0	Optimized Structure #1
Core layer material composition	5 mm soundproof pad #1 10 mm sound-absorbing material #1 3 mm soundproof pad #2 20 mm sound-absorbing material #2 2 mm soundproof pad #3 15 mm sound-absorbing material #3	5 mm soundproof pad #1 10 mm sound-absorbing material #1 20 mm sound-absorbing material #2 15 mm sound-absorbing material #3 3 mm soundproof pad #2 2 mm soundproof pad #3
Weight (kg)	24.6	24.6
Thickness (mm)	55.0	55.0

5. Conclusions

In this study, traditional insulation and sound-absorbing materials in a sandwich composite structure were taken as the object to study the influence of different material laying orders on the overall sound-insulation characteristics, and a high-speed train's body structure was taken as an example to implement the sound-insulation design. The results are summarized as follows:

1. For a simple sandwich composite structure, the core layer material laying strategy of "laying sound-absorbing materials in the middle and sound-insulation materials on both sides" has a better effect on sound insulation for low and medium frequencies, and the increase is even more than 10 dB around 400 Hz, compared with "alternating laying of sound-absorbing materials and sound-insulation materials". Furthermore, for the overall weighted sound reduction index R_w, it can be increased by more than 4 dB.
2. The core layer material laying strategy of "laying sound-absorbing materials in the middle and sound-insulation materials on both sides" is applied to the sound-insulation design of the high-speed train carbody. Under the premise of not changing the type, thickness and weight of materials, the low and medium frequency sound insulations of 125–315 Hz are improved by 1–3 dB, and the overall weighted sound reduction index R_w is improved by 0.9 dB.
3. By changing the laying sequence of the core layer materials, the sound-insulation performance of the composite structure can be improved without any change to the space, thickness, weight and material type of the structure. It is easy to implement, is low cost and has high engineering practical value.

Author Contributions: R.W.: conceptualization, methodology, resources, data curation, writing—original draft. D.Y.: methodology, software, writing—review and editing. J.Z.: conceptualization, methodology, writing—review and editing. X.X.: writing—review and editing. X.J.: writing—review and editing. All authors have read and agreed to the published version of the manuscript.

Funding: This research was funded by [National Natural Science Foundation of China] grant number [Nos. 52002257, U1934203], [Applied Basic Research Program of Sichuan Province] grant number [No. 2021YJ0531], [Open Project of State Key Laboratory of Traction Power] grant number [No. TPL2205] and [Changzhou Applied Basic Research Project] grant number [CJ20220020].

Institutional Review Board Statement: Not applicable.

Informed Consent Statement: Not applicable.

Data Availability Statement: The data used to support the findings of this study are available from the corresponding author upon request.

Conflicts of Interest: The authors declare that they have no known competing financial interests or personal relationships that could have influenced the work reported in this paper.

References

1. Song, S.; Lin, P.; Zhao, Y.-J.; Chen, H.-W. Prediction of sound transmission loss of composite floor structures of high speed trains. In Proceedings of the INTER-NOISE 2017-46th International Congress and Exposition on Noise Control Engineering: Taming Noise and Moving Quiet, Hong Kong, China, 27–30 August 2017.
2. Deng, T.-S.; Sheng, X.-Z.; Jeong, H.; Thompson, D.J. A two-and-half dimensional finite element/boundary element model for predicting the vibro-acoustic behaviour of panels with poro-elastic media. *J. Sound Vib.* **2021**, *505*, 116147. [CrossRef]
3. Zhang, J.; Yao, D.; Wang, R.-Q.; Xiao, X.-B. Vibro-acoustic modelling of high-speed train composite floor and contribution analysis of its constituent materials. *Compos. Struct.* **2021**, *256*, 113049. [CrossRef]
4. Yao, D.; Zhang, J.; Wang, R.-Q.; Xiao, X.-B.; Guo, J.-Q. Lightweight design and sound insulation characteristic optimisation of railway floating floor structures. *Appl. Acoust.* **2019**, *156*, 66–77. [CrossRef]
5. Yao, D.; Zhang, J.; Wang, R.-Q.; Xiao, X.-B. Vibroacoustic damping optimisation of high-speed train floor panels in low- and mid-frequency range. *Appl. Acoust.* **2020**, *174*, 107788. [CrossRef]
6. Lin, L.-Z.; Ding, Z.-Y.; Zeng, J.-K.; Zhang, C.-X. Research on the transmission loss of the floor aluminum profile for the high-speed train based on FE-SEA hybrid method. *J. Vibroeng.* **2016**, *18*, 1968–1981. [CrossRef]
7. Zhang, Y.-M.; Thompson, D.; Squicciarini, G.; Ryue, J.; Xiao, X.-B.; Wen, Z.-F. Sound transmission loss properties of truss core extruded panels. *Appl. Acoust.* **2018**, *131*, 134–153. [CrossRef]
8. Nurzyński, J.; Nowotny, Ł. Acoustic performance of floors made of composite panels. *Materials* **2023**, *16*, 2128. [CrossRef] [PubMed]
9. Li, S.; Xu, D.; Wu, X.; Jiang, R.; Shi, G.; Zhang, Z. Sound insulation performance of composite double sandwich panels with periodic arrays of shunted piezoelectric patches. *Materials* **2022**, *15*, 490. [CrossRef] [PubMed]
10. Hu, Q.-L.; Bian, G.-F.; Qiu, Y.-P.; Wei, Y.; Xu, Z.-Z. Sound insulation properties of honeycomb sandwich structure composite for high-speed train floors. *J. Text. Res.* **2001**, *42*, 75–83.
11. Seockhyun, K.; Taegun, S. Sound insulation performance of honeycomb composite panel for a tilting train. *Trans. Korean Soc. Mech. Eng. A* **2010**, *34*, 1931–1936.
12. Kim, S.H.; Seo, T.G.; Kim, J.T.; Song, D.H. Sound-insulation design of aluminum extruded panel in next-generation high-speed train. *Trans. Korean Soc. Mech. Eng. A* **2011**, *35*, 567–574. [CrossRef]
13. Kaidouchi, H.; Kebdani, S.; Slimane, S.A. Vibro-acoustic analysis of the sound transmission through aerospace composite structures. *Mech. Adv. Mater. Struct.* **2022**, *3*, 1–11. [CrossRef]
14. Kang, L.; Sun, C.; An, F.; Liu, B. A bending stiffness criterion for sandwich panels with high sound insulation and its realization through low specific modulus layers. *J. Sound Vib.* **2022**, *536*, 117149. [CrossRef]
15. Liu, X.-B.; Yang, Y.; Le, V. Airborne sound insulation of aluminum extrusion structural walls of an urban rail train. *Noise Control Eng. J.* **2014**, *62*, 47–53. [CrossRef]
16. Zhang, J.; Yao, D.; Wang, P.; Wang, R.-Q.; Li, J.; Guo, S.-Y. Optimal design of lightweight acoustic metamaterials for low-frequency noise and vibration control of high-speed train composite floor. *Appl. Acoust.* **2022**, *199*, 109041. [CrossRef]
17. Lyon, R.H.; DeJong, R.G.; Heckl, M. *Theory and Application of Statistical Energy Analysis*, 2nd ed.; Butterworth-Heinemann: Boston, MA, USA, 1995.
18. Oliazadeh, P.; Farshidianfar, A.; Crocker, M.J. Experimental study and analytical modeling of sound transmission through honeycomb sandwich panels using SEA method. *Compos. Struct.* **2022**, *280*, 114927. [CrossRef]
19. Huang, X.-F.; Lu, Y.-M.; Qu, C.; Zhu, C.-H. Study on Sound Transmission across a Floating Floor in a Residential Building by Using SEA. *Arch. Acoust. J. Pol. Acad. Sci.* **2021**, *46*, 183–194.
20. Hyoseon, J.; Carl, H. Prediction of sound transmission in long spaces using ray tracing and experimental Statistical Energy Analysis. *Appl. Acoust.* **2018**, *130*, 15–33. [CrossRef]
21. *VA One 2012 User's Guide*; The ESI Group: Rungis, France, 2012.
22. Su, J.-T.; Zheng, L.; Lou, J.-P. Simulation and Optimization of Acoustic Package of Dash Panel Based on SEA. *Shock Vib.* **2020**, *2020*, 8855280. [CrossRef]
23. Hossein, S.; Abolfazl, K.; Amin, M. A practical procedure for vehicle sound package design using statistical energy analysis. *Proc. Inst. Mech. Eng. Part D J. Automob. Eng.* **2022**, 1–16. [CrossRef]
24. Zhang, X.-X.; Wu, X.-R.; Cheng, Y.-H.; Jin, H.-Y.; Zhang, J. Application Research of Statistical Energy Analysis on Vehicle Sound Package. In *FISITA World Automotive Congress*; Springer: Berlin/Heidelberg, Germany, 2012; pp. 507–520. [CrossRef]
25. ISO 717-1:2013; Acoustics—Rating of Sound Insulation in Buildings and of Building Elements—Part 1: Airborne Sound Insulation. International Organization for Standardization: Geneva, Switzerland, 2013.
26. ISO 10140-2:2021; Acoustics—Laboratory Measurement of Sound Insulation of Building Elements—Part 2: Measurement of Airborne Sound Insulation. International Organization for Standardization: Geneva, Switzerland, 2021.

Disclaimer/Publisher's Note: The statements, opinions and data contained in all publications are solely those of the individual author(s) and contributor(s) and not of MDPI and/or the editor(s). MDPI and/or the editor(s) disclaim responsibility for any injury to people or property resulting from any ideas, methods, instructions or products referred to in the content.

Article

Experimental Analysis of the Influence of Carrier Layer Material on the Performance of the Control System of a Cantilever-Type Piezoelectric Actuator

Dariusz Grzybek

Faculty of Mechanical Engineering and Robotics, AGH University of Krakow, al. A. Mickiewicza 30, 30-059 Krakow, Poland; dariusz.grzybek@agh.edu.pl; Tel.: +48-126173080

Abstract: The subject of this article is an experimental analysis of the control system of a composite-based piezoelectric actuator and an aluminum-based piezoelectric actuator. Analysis was performed for both the unimorph and bimorph structures. To carry out laboratory research, two piezoelectric actuators with a cantilever sandwich beam structure were manufactured. In the first beam, the carrier layer was made of glass-reinforced epoxy composite (FR4), and in the second beam, it was made of 1050 aluminum. A linear mathematical model of both actuators was also developed. A modification of the method of selecting weights in the LQR control algorithm for a cantilever-type piezoelectric actuator was proposed. The weights in the R matrix for the actuator containing a carrier layer made of stiffer material should be smaller than those for the actuator containing a carrier layer made of less stiff material. Additionally, regardless of the carrier layer material, in the case of a bimorph, the weight in the R matrix that corresponds to the control voltage of the compressing MFC patch should be smaller than the weight corresponding to the control voltage of the stretching MFC patch.

Keywords: piezoelectric actuator; macro fiber composite; sandwich beam; LQR control algorithm

Citation: Grzybek, D. Experimental Analysis of the Influence of Carrier Layer Material on the Performance of the Control System of a Cantilever-Type Piezoelectric Actuator. *Materials* **2024**, *17*, 96. https://doi.org/10.3390/ma17010096

Academic Editors: Georgios C. Psarras and Raul D. S. G. Campilho

Received: 14 November 2023
Revised: 10 December 2023
Accepted: 21 December 2023
Published: 24 December 2023

Copyright: © 2023 by the author. Licensee MDPI, Basel, Switzerland. This article is an open access article distributed under the terms and conditions of the Creative Commons Attribution (CC BY) license (https://creativecommons.org/licenses/by/4.0/).

1. Introduction

A piezoelectric actuator is a device that uses the inverse piezoelectric effect to convert electrical energy into mechanical energy: because of this energy conversion, motion of the mechanical component of the actuator is generated [1]. One of the mechanical components used in piezoelectric actuators is the cantilever beam [2]. Two basic types of cantilever beam structure can be distinguished: unimorph and bimorph. The unimorph is a structure in which there is one layer of piezoelectric material and one carrier layer. The bimorph is a structure with two layers of piezoelectric material and one carrier layer [3], or with two layers of piezoelectric material alone [4]. Some researchers use the name "triple-layer" instead of the name "bimorph" [5]. In the unimorph and bimorph structures of the cantilever beam, the layers are usually glued together [6]. In the case of a structure containing a carrier layer, the motion of the cantilever beam is generated by creating tensile or compressive stresses in this carrier layer through the interaction of the piezoelectric layer (unimorph) or two layers (bimorph).

In both the unimorph and the bimorph, the piezoelectric layers can be made of different materials. The piezoelectric materials used can be divided into three main groups: (1) piezoelectric ceramics, usually lead zirconate titanate (PZT) [7]; (2) piezoelectric composites, usually type P1 macro fiber composite (MFC) made from PZT fibers and warp of nonpiezoelectric polymers [8]; and (3) piezoelectric polymers, usually polyvinylidene fluoride (PVDF) [9]. The first fundamental difference in the use of these piezoelectric materials is due to the relationship between energy conversion efficiency and brittleness. Piezoelectric ceramics are characterized by the highest energy conversion efficiency but are at the same time the most fragile compared to composites or polymers [10]. On the other hand, piezoelectric polymers are the most flexible but have the lowest energy conversion efficiency compared to ceramics and composites [9]. Composites have lower energy conversion efficiency than ceramics but are more resistant to destruction due to deformations [11]. The second fundamental difference in the use of these piezoelectric materials in the cantilever beams of the actuators results from the relationship between the direction of the stress generated in the carrier layer by the piezoelectric layer or layers and the direction of polarization of the piezoelectric layer or layers. When piezoelectric ceramics and polymers are used, the polarization direction of the piezoelectric layer or layers is perpendicular to the direction of stress generated in the carrier layer. When the composite MFC type P1 is used, the direction of stress is parallel to the direction of polarization. This difference leads to the fact that the conversion of electrical energy into mechanical energy in actuators with the use of PZT is described by the piezoelectric coefficient d_{31}, and in actuators with the use of MFC type P1 by d_{33}. Nguyen et al. [12] also noticed that MFC has better dynamic actuation than the bulk PZT type for the range of high frequency.

The carrier layer of the cantilever beam in piezoelectric actuators is made of materials that can be divided into two main groups: (1) metals, and (2) composites. The metals used in the cantilever structure are primarily aluminum alloys [13], brass [14], beryllium [15], and steel [16]. The composite used is primarily glass-reinforced epoxy composite (FR4) [17]. The use of a stiffer material in the carrier layer leads to a decrease in the value of the cantilever beam tip motion generated [18]. The tip motion of the cantilever beam made from aluminum is larger in comparison to the motion of actuators made from steel or copper; however, this difference decreases as the thickness of the carrier layer decreases [19]. In general, for the same geometrical dimension and under the same applied electric field, the lower the stiffness of the material of the carrier layer in the cantilever beam of the piezoelectric actuator, the greater displacements of this cantilever beam tip are generated. On the other hand, the application of a carrier layer with greater stiffness leads to a generation of larger blocking forces [20]. The choice of carrier layer material can also affect other areas of actuator operation [21].

Nowadays, research on control systems of cantilever-type piezoelectric actuators focuses mainly on the compensation of nonlinear phenomena: hysteresis [22] and/or creep [23]. Mathematical models of the aforementioned nonlinear phenomena, proposed by the authors, expand a linear model, which can be lumped [24] or continuous [25]. Continuous models are the direct basis for prototyping control laws, which use state space: LQR [26] and LQG [27]. In published research results, a continuous model is usually constructed for only one selected material of the carrier layer. It should be noted that the influence of the difference between the Young's modulus of the carrier layer and the piezoelectric layer is considered in energy harvesting models [28].

There are no research results presented in the available literature regarding the influence of the material of the carrier layer on the selection of weights in linear LQR control. Most often, these weights are selected by the trial-and-error method for one selected carrier material. Among the few other methods for one selected carrier material, the following can be distinguished: Ebrahimi-Tirtashi et al. [25] used Bryson's rule; Wang et al. [26] noticed that the initial values of the weights should be chosen as the desired maximum squared values under the steady states and inputs; Tian et al. [29] proposed a genetic algorithm for weights selection. In this article, an experimental analysis of the impact of the carrier layer material on actuator performance was carried out. Based on the results of laboratory experiments, a modification of Bryson's rule of weights selection in matrix R was proposed. The modification enabled effective control regardless of the material of the supporting layer, the maximum set value of the actuator displacement, and the duration of this set value at a constant level.

2. Materials and Methods

2.1. Materials

Two manufactured piezoelectric cantilever sandwich beams were the research objects. The beams differed in the material of the carrier layer. In the first beam, the carrier layer was made of glass-reinforced epoxy composite (FR4), produced by W.P.P.H.U. HATRON S.C., Kraków, Poland, and in the second beam, it was made of 1050 aluminum. The schema of the cross-section of both cantilever beams is shown in Figure 1 and a view of one of the produced cantilever beams in Figure 2.

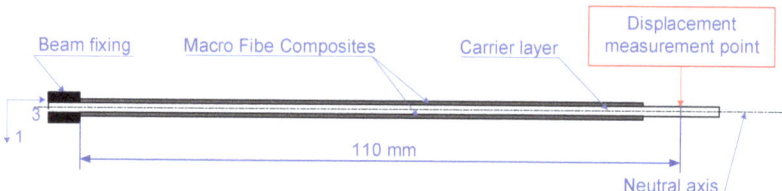

Figure 1. Schema of a cross-section of a cantilever beams: 1—longitudinal axis of the beam, 3—transverse axis of the beam.

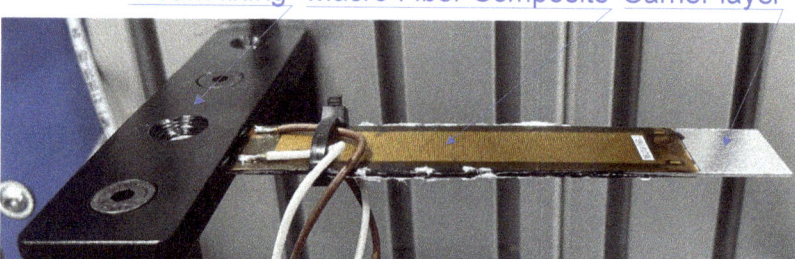

Figure 2. Produced cantilever beam containing aluminum carrier layer.

Each cantilever beam consisted of one carrier layer and two piezoelectric layers. Patches of macro fiber composite (MFC) type P1 [30], produced by Smart Material Corp., Sarasota, FL, USA, were used as piezoelectric layers. The MFC patches were symmetrically glued to both sides of the carrier layer. Epoxy adhesive DP490 [31], produced by the 3M company, Saint Paul, MN, USA, was used to create a glued connection between the MFC patches and the carrier layer. The geometric properties of the manufactured cantilever are presented in Table 1.

Table 1. Dimensions of manufactured cantilever beams (in mm).

MFC Patch			Carrier Layer		
Dimension	Symbol	Value	Dimension	Symbol	Value
Total length	l_{mfc}	100	Length	l_c	120
Total width	w_{mfc}	20	Width	w_c	20
Total thickness	t_{mfc}	0.3	Thickness	t_c	1
Active part length	l_{mfca}	85			
Active part width	w_{mfca}	14			
Active part thickness	t_{mfca}	0.18			
Passive part length	l_{mfcp}	15			
Passive part thickness	t_{mfcp}	0.12			
Distance between electrodes	t_{mfce}	0.5			

2.2. Laboratory Research Method

The motion of the beam was forced by using a system consisting of a computer with MATLAB Simulink software, an A/D board, and a voltage amplifier. The generation of control voltage waveforms, which were supplied to the MFC patch/patches, was performed in the 2019b version of the MATLAB Simulink program, in which the solver ode1 was used to perform the calculations. A fixed-step equal to 0.001 s was used in these calculations to obtain real-time calculations. The generated voltage waveforms were sent to the TD250-INV voltage amplifier, produced by PiezoDrive company, Shortland, Australia, in real time. The voltages were sent in real time using an RT_DAC/Zynq A/D board, manufactured by INTECO company Kraków, Poland, integrated with a dedicated MATLAB toolbox described in [32]. The TD250-INV voltage amplifier generated from one to two control voltages in the range from −500 V to +500 V. In all laboratory experiments, the displacement of one point in the cantilever beam structure was measured. The distance between the measured point and the beam fixing was 110 mm. The measurement system contained an LG5B65PI laser sensor of displacement, produced by BANNER company, Minneapolis, MN, USA, and the aforementioned RT_DAC/Zynq A/D board, which enabled data acquisition in real time. The LG5B65PI laser sensor had a measurement resolution equal to 40 microns for the measurement at a frequency equal to 450 Hz, and the analog linearity was ±10 microns. The measurement system schema is shown in Figure 3.

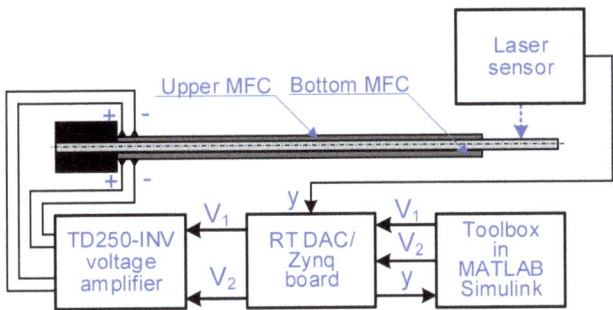

Figure 3. Schema of measurement system.

The actuator temperature was measured using the Flir E40 thermovision camera during the longest experiments, which lasted 26 s. No observable temperature changes were noted between the beginning and end of the experiment. Temperature changes affect the electrical impedance of the piezoelectric layer [33,34], but the position error of the actuator resulting from warm-up only appears where the actuator is excited for a long time, and even then, this error is very small [35].

2.3. Simulation Research Method

Simulation experiments were carried out in the 2019b version of the MATLAB Simulink program, in which the solver ode23tb was used to perform the calculations. A variable-step was used in these calculations. The ode23tb algorithm is an implementation of the TR-BDF2 method, which is a combination of trapezoidal and second-order backward differentiation [36]. The purpose of the simulations was to determine the displacement of the tip of the cantilever beam caused by an applied control voltage, of which the values were assumed in advance, to one (unimorph) or two MFC patches (bimorph). A variable-step was used in the simulation research because only such a step enabled the simulation of the operation of the actuator described by a mathematical model containing matrixes of very large sizes. The simulations did not attempt to obtain the real-time response of the modeled system.

3. Mathematical Model of Piezoelectric Actuator and Synthesis of Control System

3.1. Piezoelectric Actuator

The displacements of selected points in the cantilever beam structure were calculated using a mathematical model, which was built on the basis of two methods: Finite Element Method (FEM) and State Space Method. FEM was used because a tip mass does not occur [37]. The mathematical model was built in two stages: (1) determination of stiffness matrix $\mathbf{K_g}$ and mass matrix $\mathbf{M_g}$ for the assumed number of finite elements; and (2) determination of state matrix $\mathbf{A}$, control matrix $\mathbf{B}$, output matrix $\mathbf{C}$, and feed-through matrix $\mathbf{D}$.

The structure of the cantilever beam, which is shown in Figure 1, was divided into 48 finite elements, each of a length equal to 2.5 mm. As a result of this division, 49 nodes were created. A total of 48 nodes had two degrees of freedom, and one node, which was in the beam fixing, had zero degrees of freedom. A motion equation can be given by [25]:

$$\mathbf{M_g}\ddot{\mathbf{d}}(t) + \mathbf{C_g}\dot{\mathbf{d}}(t) + \mathbf{K_g}\mathbf{d}(t) = \mathbf{E}_1 V_1(t) + \mathbf{E}_2 V_2(t) \tag{1}$$

where $\mathbf{M_g}$ is a global mass matrix (dimensions: 96×96 for 48 nodes), $\mathbf{C_g}$ is a global damping matrix (dimensions: 96×96 for 48 nodes), $\mathbf{K_g}$ is a global stiffness matrix (dimensions: 96×96 for 48 nodes), $\mathbf{E}_1$ (dimensions: 96×1 for 48 nodes) and $\mathbf{E}_2$ (dimensions: 96×1 for 48 nodes) are localization matrixes of forces generated by the upper MFC patch and the bottom MFC patch, V_1 and V_2 are voltages applied to the upper MFC patch and the bottom MFC patch, and $\mathbf{d}$ is a vector of vertical (w) and rotational (φ) displacements: $\mathbf{d} = [w, \varphi]^T$ of node. Local mass matrixes $\mathbf{M_l}$ and local stiffness matrixes $\mathbf{K_l}$ were calculated as follows [26]:

$$\mathbf{M_l} = \mathbf{M_{lc}} + 2\mathbf{M_{lmfc}} \quad \mathbf{K_l} = \mathbf{K_{lc}} + 2\mathbf{K_{lmfc}} \tag{2}$$

where $\mathbf{M_{lc}}$ and $\mathbf{K_{lc}}$ are local mass and stiffness matrixes of the carrier layer, and $\mathbf{M_{lmfc}}$ and $\mathbf{K_{lmfc}}$ are the local mass and stiffness matrixes of MFC:

$$\mathbf{M_{lc}} = \frac{\rho_c A_c l_e}{420} \begin{bmatrix} 156 & 22l_e & 54 & -13l_e \\ 22l_e & 4l_e^2 & 13l_e & -3l_e^2 \\ 54 & 13l_e & 156 & -22l_e \\ -13l_e & -3l_e^2 & -22l_e & 4l_e^2 \end{bmatrix} \quad \mathbf{K_{lc}} = \frac{E_c \eta I_c}{l_e} \begin{bmatrix} 12 & 6l_e & -12 & 6l_e \\ 6l_e & 4l_e^2 & -6l_e & 2l_e^2 \\ -12 & -6l_e & 12 & -6l_e \\ 6l_e & 2l_e^2 & -6l_e & 4l_e^2 \end{bmatrix}$$

$$\mathbf{M_{lmfc}} = \frac{\rho_{mfc} A_{mfc} l_e}{420} \begin{bmatrix} 156 & 22l_e & 54 & -13l_e \\ 22l_e & 4l_e^2 & 13l_e & -3l_e^2 \\ 54 & 13l_e & 156 & -22l_e \\ -13l_e & -3l_e^2 & -22l_e & 4l_e^2 \end{bmatrix} \quad \mathbf{K_{lmfc}} = \frac{E_{mfc} \eta I_{mfc}}{l_e} \begin{bmatrix} 12 & 6l_e & -12 & 6l_e \\ 6l_e & 4l_e^2 & -6l_e & 2l_e^2 \\ -12 & -6l_e & 12 & -6l_e \\ 6l_e & 2l_e^2 & -6l_e & 4l_e^2 \end{bmatrix} \tag{3}$$

where ρ_c is the density of the carrier layer, A_c is the cross-section area of the carrier layer, l_e is the length of the finite element, E_c is Young's modulus of the carrier layer, I_c is the moment of inertia of the carrier layer, ρ_{mfc} is the density of the MFC patch, A_{mfc} is the cross-section area of the MFC patch, E_{mfc} is Young's modulus of the MFC patch, I_{mfc} is the moment of inertia of the MFC patch, and η is the ratio of the piezoelectric material elastic constant to that constant of the carrier layer material:

$$\eta = \frac{Y_{mfc}}{Y_c} \tag{4}$$

A global damping matrix was calculated as proportional damping in the Rayleigh form [38]:

$$\mathbf{C_g} = \alpha \mathbf{M_g} + \beta \mathbf{K_g} \tag{5}$$

where α and β are the dimensionless coefficients, which were selected experimentally.

Considering that both the upper MFC and the bottom MFC are equidistant from the neutral axis of the cantilever beam, the bending moment can be calculated in the same way for both MFCs. The bending moment per unit length generated by each MFC patch in the vertical axis (axis 1 in Figure 1) is calculated as follows:

$$M_{bi}(t) = \gamma \int_{\frac{1}{2}t_{cr}}^{\frac{1}{2}t_{cr}+t_{mfca}} d_{33a} Y_{33a} \frac{V_i(t)}{t_{mfce}} w_{mfca} y \, dy = \frac{1}{2} \gamma d_{33a} Y_{mfca} \frac{V_i(t)}{t_{mfce}} w_{mfca} \left(t_{cr} t_{mfca} + t_{mfca}^2 \right) \quad (6)$$

where γ is the ratio of the smaller Young's modulus to the larger one for a pair of two materials (the active part of the MFC patch and the carrier layer (ratio Y_c/Y_{mfca} for the composite carrier layer and Y_{mfca}/Y_c for the aluminum carrier layer)), and t_{cr} is the carrier layer thickness increased by half the thickness of the passive layer in the MFC patch. It was assumed that an equivalent concentrated force is applied, which generates the value of the bending moment calculated according to (6), at the center of gravity of each MFC patch. Therefore, the equivalent concentrated force generated by the MFC patch in the direction of axis 1, acting on the cantilever beam, can be given by

$$P_i(t) = \frac{M_{bi}(t)}{0.5(l_{mfca} + l_{mfcp})} \quad (7)$$

The point of application of the equivalent concentrated force was also assumed at the center of gravity of the MFC patch, which is located at the 20th node (distance 50 mm from beam fixing). The matrixes of forces localization for the upper MFC (E_1) and the bottom MFC (E_2) [39] are calculated as follows:

$$\begin{aligned} E_1 &= \tfrac{1}{2}\gamma d_{33a} Y_{mfca} \tfrac{1}{t_{mfce}} w_{mfca}\left(t_{cr}t_{mfca}+t_{mfca}^2\right)\left[\Theta_{1\times 38} \quad \varepsilon_1 \quad \Theta_{1\times 57}\right]^T \\ E_2 &= \tfrac{1}{2}\gamma d_{33a} Y_{mfca} \tfrac{1}{t_{mfce}} w_{mfca}\left(t_{cr}t_{mfca}+t_{mfca}^2\right)\left[\Theta_{1\times 38} \quad -\varepsilon_2 \quad \Theta_{1\times 57}\right]^T \end{aligned} \quad (8)$$

where ε_1 is a coefficient showing the contribution of the stretching MFC in generating the motion of the cantilever beam (it was assumed that the value of this parameter will be 1 in simulation studies), and ε_2 is a coefficient showing the contribution of the compressing MFC in generating the motion of the cantilever beam. It was determined in laboratory experiments that for the composite-based actuator $\varepsilon_1 = 1$ and $\varepsilon_2 = 0.36$, and for the aluminum-based actuator $\varepsilon_1 = 1$ and $\varepsilon_2 = 0.38$. The material properties used in the simulation tests are presented in Table 2.

Table 2. Material properties of manufactured cantilever beams.

Parameter		Composite-Based Actuator	Aluminum-Based Actuator
Young's modulus of carrier layer	Y_c	18.6×10^9 Pa [38]	71×10^9 Pa [26]
Density of carrier layer	ρ_c	1850 kg/m³ [38]	2710 kg/m³ [26]
Young's modulus of MFC patch	Y_{mfc}	30.336×10^9 Pa [30]	
Young's modulus of MFC of piezoceramic fibers in MFC patch	Y_{mfca}	48.3×10^9 Pa [40]	
Density of active part of MFC patch	ρ_{mfca}	5400 kg/m³ [26]	
Piezoelectric constant of MFC patch	d_{33}	400×10^{-12} C/N [30]	
Piezoelectric constant of piezoceramic fibers in MFC patch	d_{33a}	440×10^{-12} C/N [40]	

Matrixes of forces localization were used to build a state space model, which had a well-known form:

$$\begin{aligned} \dot{x}(t) &= \mathbf{A}x(t) + \mathbf{B}u(t) \\ y(t) &= \mathbf{C}x(t) + \mathbf{D}u(t) \end{aligned} \quad (9)$$

where **x** is a state vector (containing 192 state variables), **u** is an input vector, and **y** is an output vector. The matrix dimensions are as follows:

$$\mathbf{x} = \begin{bmatrix} \mathbf{d} \\ \dot{\mathbf{d}} \end{bmatrix}_{192 \times 1} \quad \mathbf{A} = \begin{bmatrix} \Theta_{96 \times 96} & \mathbf{I}_{96 \times 96} \\ -\mathbf{M}_g^{-1}\mathbf{K}_g & -\mathbf{M}_g^{-1}\mathbf{C}_g \end{bmatrix}_{192 \times 192} \quad \mathbf{B} = \begin{bmatrix} \Theta_{96 \times 1} & \Theta_{96 \times 1} \\ \mathbf{M}_g^{-1}\mathbf{E}_1 & \mathbf{M}_g^{-1}\mathbf{E}_2 \end{bmatrix}_{192 \times 2} \quad (10)$$

$$\mathbf{C} = \begin{bmatrix} \Theta_{1 \times 86} & 1 & \Theta_{1 \times 105} \end{bmatrix}_{1 \times 192} \quad \mathbf{D} = \begin{bmatrix} 0 & 0 \end{bmatrix}_{1 \times 2}$$

An output variable was the 87th state variable, which was a displacement of the 44th node in the cantilever beam structure in the direction of axis 1. The 44th node was located 110 mm from the beam fixing.

Taking into account that measurement data from the laboratory stand are available, an alternative method of modeling the actuator could be data-driven modeling [41].

3.2. Synthesis of Control System

A Linear Quadratic Gaussian (LQG) algorithm with integral feedback was used to generate two independent control voltages. The LQG consisted of a linear quadratic regulator (LQR) and a Kalman filter used to estimate the state vector. This algorithm has been extended with integral feedback. A synthesis of the control algorithm was based on the state space model (10). The basic condition for the implementation of the LQR algorithm is full controllability of the controlled object. The actuator described by (10) is fully controllable because there is at least one non-zero element in each row of a controllability matrix $\mathbf{Q}_{ctrb}$:

$$\mathbf{Q}_{ctrb} = \Phi^{-1} \mathbf{B} \quad (11)$$

where Φ is the truncated matrix consisting of n eigenvectors. The actuator model in state space (10) was extended by the additional state variable, which is the integral of the difference between the set value and the measured value of the beam tip displacement:

$$\begin{bmatrix} \dot{\mathbf{x}}(t) \\ \dot{x}_{n+1}(t) \end{bmatrix} = \begin{bmatrix} \mathbf{A} & \Theta_{192 \times 1} \\ -\mathbf{C} & 0 \end{bmatrix} \begin{bmatrix} \mathbf{x}(t) \\ x_{n+1}(t) \end{bmatrix} + \begin{bmatrix} \mathbf{B} \\ 0 \end{bmatrix} \mathbf{u}(t)$$

$$y(t) = \begin{bmatrix} \mathbf{C} & 0 \end{bmatrix} \begin{bmatrix} \mathbf{x}(t) \\ x_{n+1}(t) \end{bmatrix} + \mathbf{D}\mathbf{u}(t) \quad (12)$$

The basic condition for implementation of the Kalman filter is full observability of the controlled object. The actuator described by (10) is fully observable because there is at least one non-zero element in each column of the observability matrix $\mathbf{Q}_{obsv}$:

$$\mathbf{Q}_{obsv} = \mathbf{C}\Phi \quad (13)$$

The estimated state vector based on the Kalman filter is

$$\dot{\mathbf{x}}_{est}(t) = \mathbf{A}\mathbf{x}_{est}(t) + \mathbf{B}\mathbf{u}(t) + \mathbf{H}(y_{measured}(t) - \mathbf{C}\mathbf{x}_{est}(t)) - \mathbf{H}\mathbf{D}\mathbf{u}(t) \quad (14)$$

where $\mathbf{x}_{est}$ is the estimated state vector and **H** is the gains matrix:

$$\mathbf{H} = \mathbf{P}\mathbf{C}^T \mathbf{R}_c^{-1} \quad (15)$$

where $\mathbf{R}_c$ is the covariance matrix of measurement noise and **P** is the solution of the algebraic Ricatti equation:

$$\mathbf{AP} + \mathbf{PA}^T - \mathbf{PC}^T \mathbf{R}_c^{-1} \mathbf{CP}^T + \mathbf{Q}_c = 0 \quad (16)$$

where $\mathbf{Q}_c$ is the covariance matrix of state noise.

The final control law considering the estimated state vector is

$$u_1(t) = -\begin{bmatrix} \mathbf{K}_1 & k_{n+1,u_1} \end{bmatrix} \begin{bmatrix} \mathbf{x}_{est}(t) \\ x_{n+1}(t) \end{bmatrix} + y_{set}(t)$$
$$u_2(t) = -\begin{bmatrix} \mathbf{K}_2 & k_{n+1,u_2} \end{bmatrix} \begin{bmatrix} \mathbf{x}_{est}(t) \\ x_{n+1}(t) \end{bmatrix} + y_{set}(t) \quad (17)$$

where n is the size of the state vector, $\mathbf{K}_1$ and $\mathbf{K}_2$ are the matrixes of the state variable gains for u_1 and u_2, respectively, $k_{n+1,u1}$ and $k_{n+1,u2}$ are the gains in the integral feedback for u_1 and u_2, respectively, and y_{set} is the set value of the actuator tip displacement. The gains $\mathbf{K}_1$, $\mathbf{K}_2$, $k_{n+1,u1}$, $k_{n+1,u2}$ were calculated by the minimization of the expanded quality index:

$$J = \int_0^\infty \left(\begin{bmatrix} \mathbf{x}(t) \\ x_{n+1}(t) \end{bmatrix}^T \mathbf{Q} \begin{bmatrix} \mathbf{x}(t) \\ x_{n+1}(t) \end{bmatrix} + \mathbf{u}^T(t)\mathbf{R}\mathbf{u}(t) \right) dt \quad (18)$$

where $\mathbf{Q}$ is the positive definite or semi-definite weight matrix, and $\mathbf{R}$ is the positive definite weight matrix. The measurement system schema is shown in Figure 4.

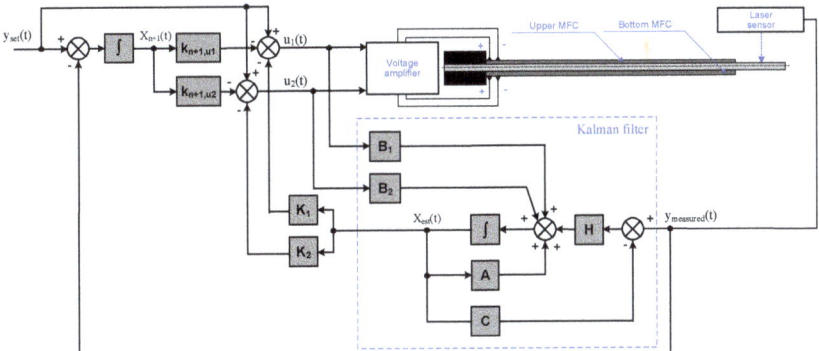

Figure 4. Schema of control system: B_1—first column of B matrix, B_2—second column of B matrix.

4. Results

Laboratory research included experiments in which the step responses of the unimorph and the bimorph with both a composite and an aluminum carrier layer were measured. The research was divided into two stages: (1) laboratory and simulation research regarding the impact of the carrier layer material on actuator performance and (2) laboratory research regarding the control system of the actuator.

4.1. Description of First Stage of Research

The first stage of research included a determination of the duration of the transition period in the creep process and a determination of the impact of the carrier layer material on actuator performance. To determine the duration of the transition period in the creep process, step responses were measured. The measurement was performed for the spike of voltage V_1 or simultaneous spikes of voltages V_1 and V_2 from 0 to the set value. The spike in voltage or voltages started in the first second and lasted for 2 s. The experiment conditions for both composite-based and aluminum-based actuators are presented in Table 3.

Table 3. Conditions of laboratory experiments to determine the duration of a transition period in creep process.

Experiment No.	1	2	3	4	5	6
Set voltage V_{1set} (V)	+500	+400	+300	+500	+400	+300
Set voltage V_{2set} (V)	0	0	0	−500	−400	−300

To determine the impact of carrier layer material on actuator performance, the supply voltage of the upper MFC patch was increased from 0 to the set value (both for unimorph and bimorph) and the simultaneous supply voltage of the bottom MFC patch was decreased from 0 to the set value (only for bimorph). The supply voltage waveforms are shown in Figure 5 (t_e is the duration time of the voltage spike). It should be noted that the upper MFC generated tensile stresses above the neutral axis (Figure 1) in the cantilever beam in both the unimorph and the bimorph. In contrast, the bottom MFC generated compressive stresses below the neutral axis in the bimorph.

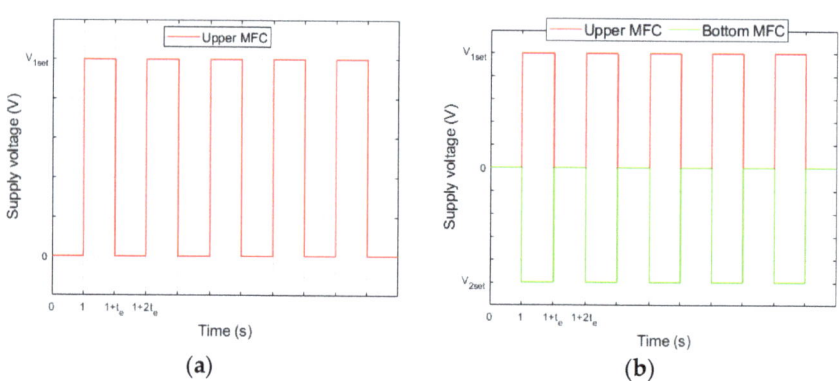

Figure 5. Supply voltage waveforms in laboratory research: (**a**) unimorph, (**b**) bimorph.

Five spikes of supply voltage V_1 for the unimorph as well as five spikes of V_1 and simultaneous V_2 for the bimorph were generated. The experiment conditions for both composite-based and aluminum-based actuators are presented in Table 4.

Table 4. Conditions of laboratory and simulation experiments to determine impact of carrier layer material on actuator performance.

	Experiment No.	V_{1set} (V)	V_{2set} (V)			t_e (s)		
Unimorph	7 to 11	+500	0	0.5	1	1.5	2	2.5
	12 to 16	+400	0	0.5	1	1.5	2	2.5
	17 to 21	+300	0	0.5	1	1.5	2	2.5
Bimorph	22 to 26	+500	−500	0.5	1	1.5	2	2.5
	27 to 31	+400	−400	0.5	1	1.5	2	2.5
	32 to 36	+300	−300	0.5	1	1.5	2	2.5

4.2. Results in First Stage of Research

Figure 6 shows the comparison of step responses obtained in laboratory experiments for both the composite and the aluminum carrier layer. In general, the duration of the transition periods is approximately the same for both the unimorph and the bimorph, as well as for the composite and aluminum carrier layers. It can be assumed that the duration of the transition periods does not exceed 0.3 s (from 1 to 1.3 s).

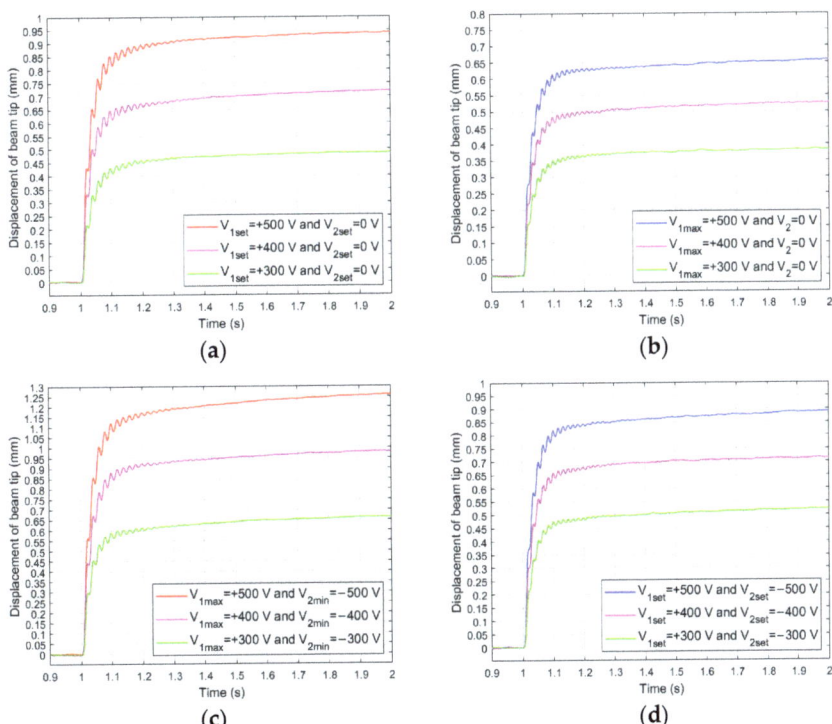

Figure 6. Step responses: (**a**) composite-based unimorph, (**b**) aluminum-based unimorph, (**c**) composite-based bimorph, (**d**) aluminum-based bimorph.

The creep process itself, however, varied depending on whether there was a composite or aluminum carrier layer. The percentage changes in the beam tip displacement in time from 1.3 s to 2 s are shown in Figure 7.

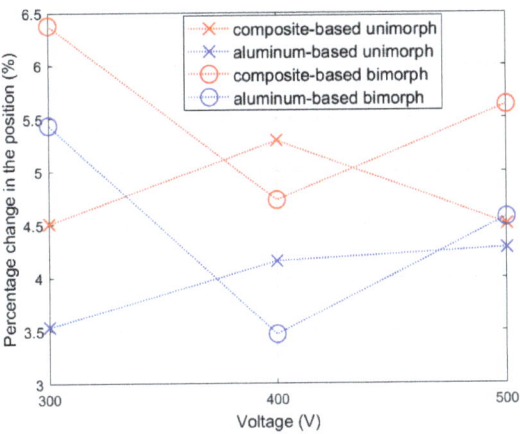

Figure 7. Percentage change in the position of the cantilever beam tip caused by the creep process.

The actuator containing a composite carrier layer exhibited significantly larger creep-induced displacements in comparison to the actuator containing an aluminum carrier layer.

Figure 8 shows the comparison of results obtained in laboratory experiments no. 7 and no. 11 for both composite-based and aluminum-based actuators.

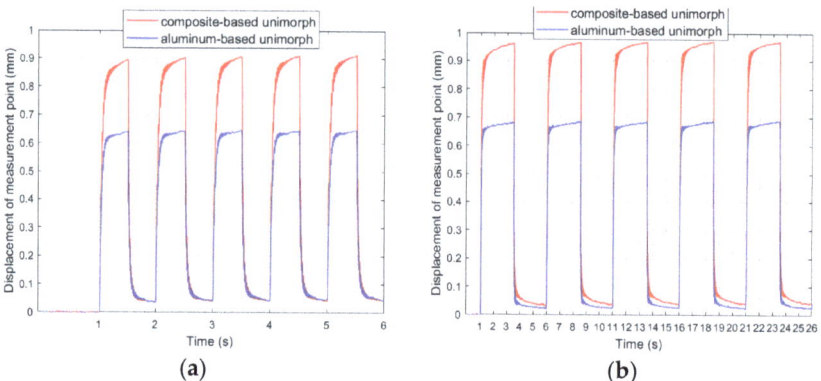

Figure 8. Step responses of unimorph for V_{1set} = +500 V: (**a**) t_e = 0.5 s, (**b**) t_e = 2.5 s.

The first observation was that there were larger displacements of the composite-based actuator compared to the aluminum-based actuator, which is consistent with the observations of other researchers regarding the influence of stiffness on the achieved displacements [18]. In experiment no. 7, the average displacement of the actuator containing an aluminum carrier layer was 70.3% of the displacement of the actuator containing a composite carrier layer, and it was 70.7% in experiment no. 11. Figure 9 shows the comparison of the results obtained in laboratory experiments no. 22 and no. 26 for bimorphs containing a composite or aluminum carrier layer.

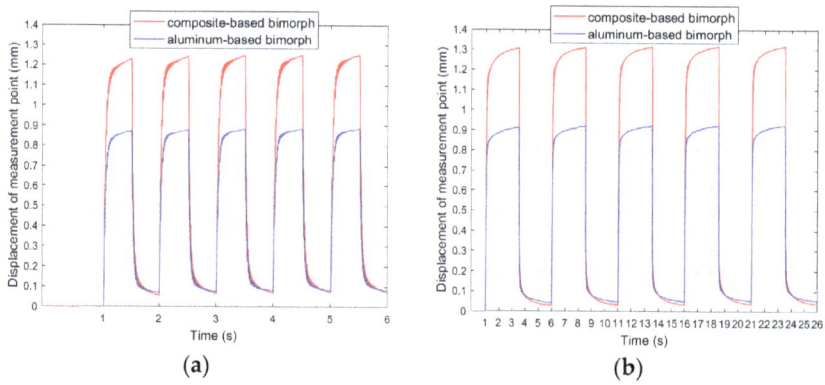

Figure 9. Step responses of bimorphs for V_{1set} = +500 V and V_{2set} = −500 V: (**a**) t_e = 0.5 s, (**b**) t_e = 2.5 s.

The composite-based actuator achieved larger displacements than the aluminum-based actuator. This difference was approximately constant for different time durations of the applied voltage spike. The average displacement of the aluminum-based actuator was 70.7% of the displacement of the composite-based actuator in experiment no. 22 and was 68.8% in experiment no. 26. On this basis, the ε_2 coefficient, which is needed in the mathematical model (Section 3.1), was determined: ε_{2com} = 0.367 for the composite carrier layer and ε_{2alu} = 0.388 for the aluminum carrier layer.

In Figures 8 and 9 it can be noticed that the actuator does not return to its initial position after the voltage spike stops. This phenomenon occurs regardless of the voltage value and the duration of the voltage spike. This is due to the phenomenon of hysteresis. Figure 10 shows the ratios of the initial positions of the composite-based actuators to the maximum displacement of these actuators.

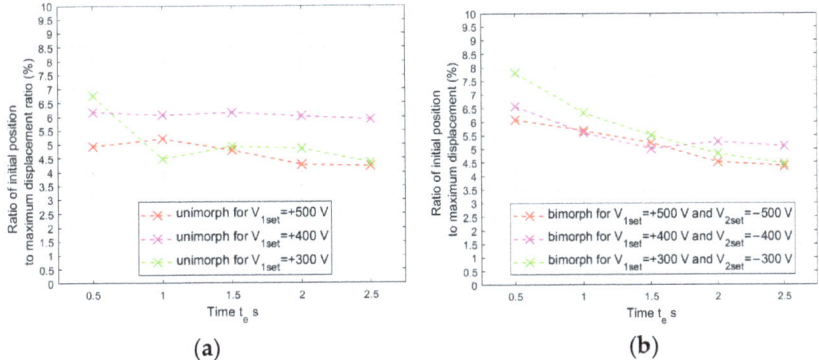

Figure 10. Ratio of initial position to maximum displacement of composite-based actuator: (**a**) unimorph, (**b**) bimorph.

In general, the position in the interval among voltage spikes (initial position) becomes a smaller and smaller part of the maximum actuator displacement as the duration of the voltage spike increases. Therefore, it can be concluded that changes in the initial position occur at a slightly slower rate than changes in the maximum position of the actuator. The initial position is, on average, from 4.69% to 6.07% of the maximum position in the case of the unimorph and from 5.18% to 5.79% in the case of the bimorph. It can be assumed that the initial position before the next voltage spike is linearly proportional to the maximum displacement of the actuator caused by the previous voltage spike. On this basis, the values of the new coefficient θ were determined for each condition, which specify linear correction of the simulated voltage values applied to the upper and bottom MFCs in the intervals between voltage spikes compared to laboratory values: for the unimorph, instead of $V_1 = 0$, it should be $V_1 = \theta V_{1set}$, and for the bimorph, instead of $V_1 = 0$, it should be $V_1 = \theta V_{1set}$ and instead of $V_2 = 0$, it should be $V_2 = \theta V_{2set}$. A similar analysis was performed for the actuator that contains an aluminum carrier layer (Figure 11). Also, for such actuators, the initial position is an approximately constant part of the maximum position. The initial position is, on average, from 5.35% to 5.75% of the maximum position in the case of the unimorph, and from 5.94% to 6.57% in the case of the bimorph. Similarly, for the composite layer, the initial position before the next voltage spike is linearly proportional to the maximum displacement of the actuator caused by the previous voltage spike. On this basis, the coefficient values of coefficient θ were determined for each condition, which specify linear correction of the simulated voltage values applied to the upper and bottom MFCs in the intervals between voltage spikes compared to laboratory values: for the unimorph, instead of $V_1 = 0$, it should be $V_1 = \theta V_{1set}$, and for the bimorph, instead of $V_1 = 0$, it should be $V_1 = \theta V_{1set}$ and instead of $V_2 = 0$, it should be $V_2 = \theta V_{2set}$.

To obtain simulation results consistent with the laboratory results, two more significant corrections were introduced to the linear mathematical model in comparison to models known from the literature. The first of these corrections was to consider the difference between the Young's modulus of the piezoelectric material and the Young's modulus of the carrier layer material. The value of the generated bending moment depends on the ratio between these Young's moduli. This relationship was introduced by using the γ coefficient in (8). This coefficient made it possible to adapt the linear model to the materials of the carrier layer, which differ in the value of Young's modulus. The second correction also concerned the generation of the bending moment: the thickness of only the piezoelectric fiber in the MFC patch was used in the model. Other researchers have used the thickness of the whole MFC patch [42] or half the thickness of the whole MFC patch [43]. A comparison of the results obtained in laboratory tests with the simulation results obtained on the basis of the modified linear model presented in Section 3.1 is shown in Figures 12 and 13.

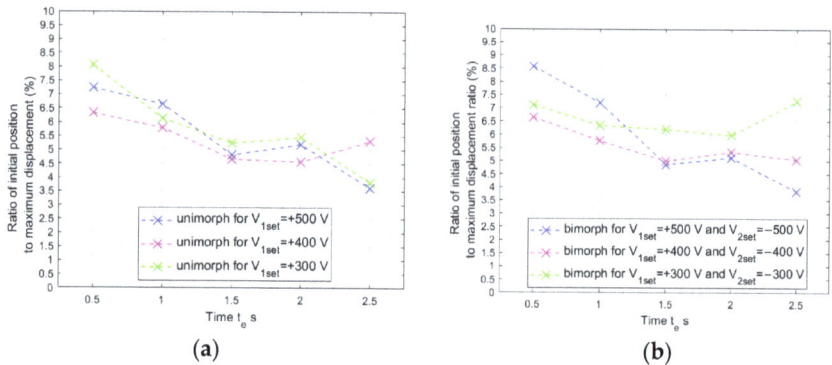

Figure 11. Ratio of initial position to maximum displacement of aluminum-based actuator: (**a**) unimorph, (**b**) bimorph.

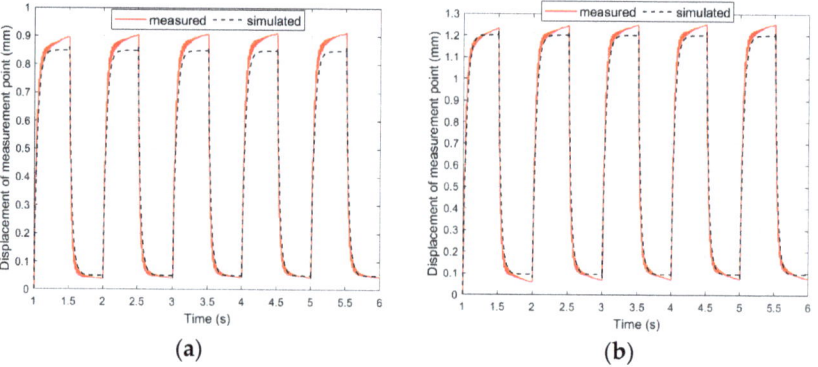

Figure 12. Comparison of simulation results and laboratory results for composite-based actuator: (**a**) unimorph, (**b**) bimorph.

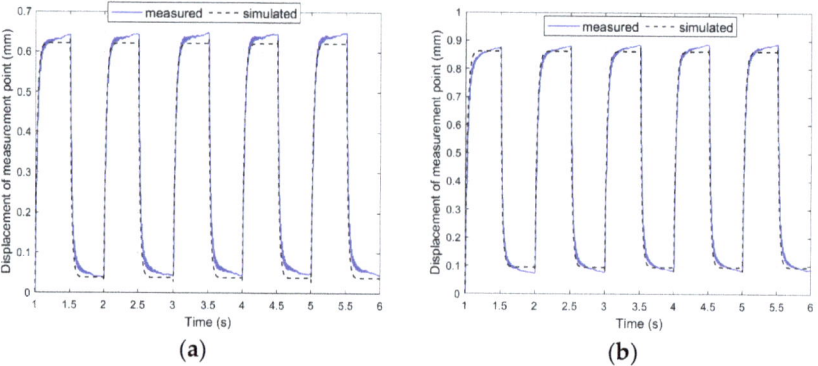

Figure 13. Comparison of simulation results and laboratory results for aluminum-based actuator: (**a**) unimorph, (**b**) bimorph.

The introduction of the first correction to the mathematical model makes it possible to adapt this model to various materials of the carrier layer. On the basis of the research, it was noticed that compliance of the simulation results with laboratory results, for the same model in the state space but for different materials of the carrier layer, can be achieved through this correction of the bending moment calculation. The introduction of the second correction allows the calculation of the bending moment, which is more consistent with the generated bending moment in the actuator beam.

4.3. Description of Second Stage of Research

The second stage of research included a determination of the impact of the material properties of the carrier layer on the weights in the quality index in the LQG control algorithm. To reduce the computational cost in the control system in the laboratory stand, the model in state space (10) was reduced to the first mode. For this purpose, the nodal displacements vector was transformed into a reduced vector:

$$\mathbf{d} = \mathbf{\Phi}_m \kappa \tag{19}$$

where $\mathbf{\Phi}_m$ is the truncated matrix and κ is the modal coordinate vector. The modal matrixes for first mode are as follows:

$$\mathbf{K}_{gm} = \mathbf{\Phi}_{m1}^T \mathbf{K}_g \mathbf{\Phi}_{m1} \quad \mathbf{M}_{gm} = \mathbf{\Phi}_{m1}^T \mathbf{M}_g \mathbf{\Phi}_{m1} \quad \mathbf{C}_{gm} = \mathbf{\Phi}_{m1}^T \mathbf{C}_g \mathbf{\Phi}_{m1} \tag{20}$$
$$\mathbf{E}_{1m} = \mathbf{\Phi}_{m1}^T \mathbf{E}_1 \quad \mathbf{E}_{2m} = \mathbf{\Phi}_{m1}^T$$

where $\mathbf{\Phi}_{m1}$ is the truncated matrix for the first mode. The model in the state space for the first mode is as follows:

$$\mathbf{x}_m = \begin{bmatrix} \kappa \\ \dot{\kappa} \end{bmatrix}_{2\times 1} \mathbf{A}_m = \begin{bmatrix} 0 & 1 \\ -\mathbf{M}_{gm}^{-1}\mathbf{K}_{gm} & -\mathbf{M}_{gm}^{-1}\mathbf{C}_{gm} \end{bmatrix}_{4\times 4} \mathbf{B}_m = \begin{bmatrix} 0 & 0 \\ \mathbf{M}_{gm}^{-1}\mathbf{E}_{1m} & \mathbf{M}_{gm}^{-1}\mathbf{E}_{2m} \end{bmatrix}_{4\times 2} \tag{21}$$
$$\mathbf{C}_m = \begin{bmatrix} \varphi_{87} & 0 \end{bmatrix} \mathbf{D}_m = \begin{bmatrix} 0 & 0 \end{bmatrix}$$

where φ_{87} is 87th element of the truncated matrix $\mathbf{\Phi}_m$. The matrixes $\mathbf{A}_m$, $\mathbf{B}_m$, $\mathbf{C}_m$, and $\mathbf{D}_m$ were introduced to Equations (12) and (14)–(18) in the laboratory research.

The waveforms of the set value of the actuator tip displacement are shown in Figure 14 (t_e is the duration time of set value spike) for both the unimorph and bimorph actuators.

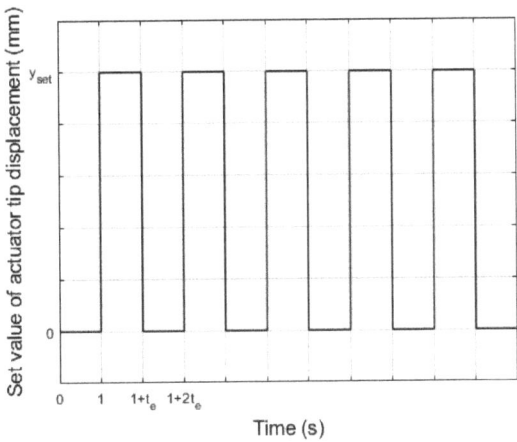

Figure 14. Waveforms of set value of actuator tip displacement for five spikes.

The first problem was to determine the set value of the actuator displacement (y_{set}) that can be achieved for the maximum (minimum) value of the control voltage without occurrence of displacement caused by the creep phenomenon. The hardware conditions, which are described in Section 2.2, showed that the maximum and minimum control voltage values were +500 V and −500 V, respectively. Values of y_{set} corresponding to ±500 V were determined experimentally based on the laboratory results, which are presented in Figure 6. On the basis of results presented in Figure 6, y_{set} values corresponding to ± 400 V and ± 300 V were also read. In this way, three values of y_{set} were established. In addition to these, one additional smaller value of y_{set} was established. The experiment conditions are presented in Table 5.

Table 5. Conditions of laboratory experiments in second stage of research.

	Experiment No.	Composite-Based Actuator		Experiment No.	Aluminum-Based Actuator	
		y_{set} (mm)	t_e (s)		y_{set} (mm)	t_e (s)
Unimorph	37 to 38	0.84	2.5	0.5	39 to 40	0.57	2.5	0.5
Unimorph	41 to 42	0.64	2.5	0.5	43 to 44	0.45	2.5	0.5
Unimorph	45 to 46	0.42	2.5	0.5	47 to 48	0.33	2.5	0.5
Unimorph	49 to 50	0.20	2.5	0.5	51 to 52	0.21	2.5	0.5
Bimorph	53 to 54	1.12	2.5	0.5	55 to 56	0.85	2.5	0.5
Bimorph	57 to 58	0.87	2.5	0.5	59 to 60	0.69	2.5	0.5
Bimorph	61 to 62	0.59	2.5	0.5	63 to 64	0.49	2.5	0.5
Bimorph	65 to 66	0.31	2.5	0.5	67 to 68	0.29	2.5	0.5

In the mathematical model (6–10) that was used to prototype control voltages u_1 and u_2, the values of the coefficients γ, ε_1, ε_{2com} and ε_{2alu} were equal to 1.

4.4. Results in Second Stage of Research

Considering Equations (18)–(21), the **R** matrix has the following form:

$$\text{For unimorph}: \mathbf{R} = [r_{11}] \quad \text{For bimorph}: \mathbf{R} = \begin{bmatrix} r_{11} & 0 \\ 0 & r_{22} \end{bmatrix} \tag{22}$$

Bryson's rule was adopted as the basis for the selection of weights. Taking into account the γ coefficient introduced in Equation (8), an analysis of the impact of reducing the maximum control voltage on the control quality was carried out. The course of the set value is shown in Figure 15.

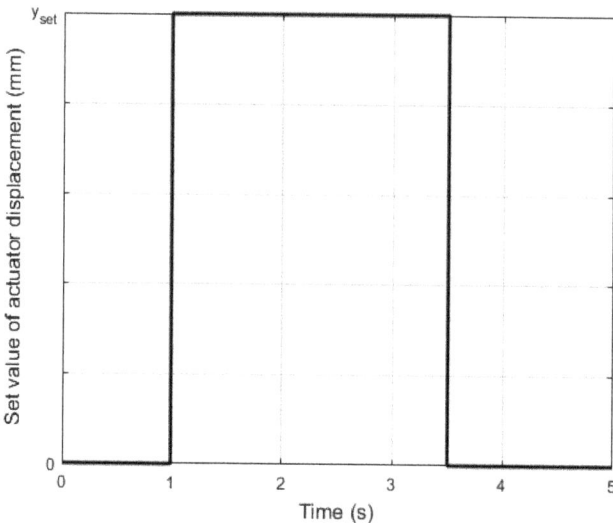

Figure 15. Waveforms of set value of actuator tip displacement for one spike.

The set value of the actuator displacement was equal to 0.84 mm for the composite-based unimorph, 1.12 mm for the composite-based bimorph, 0.57 mm for the aluminum-based unimorph, and 0.85 mm for the aluminum-based bimorph. The first two weights in the **Q** matrix were selected based on [25], and the third weight was selected using the trial-and-error method:

$$\mathbf{Q} = \begin{bmatrix} 0.01 & 0 & 0 \\ 0 & 0.01 & 0 \\ 0 & 0 & 30.33 \times 10^9 \end{bmatrix} \quad (23)$$

Figure 16 shows the impact of the value of the γ coefficient on the rising time and the overshoot of the control system output.

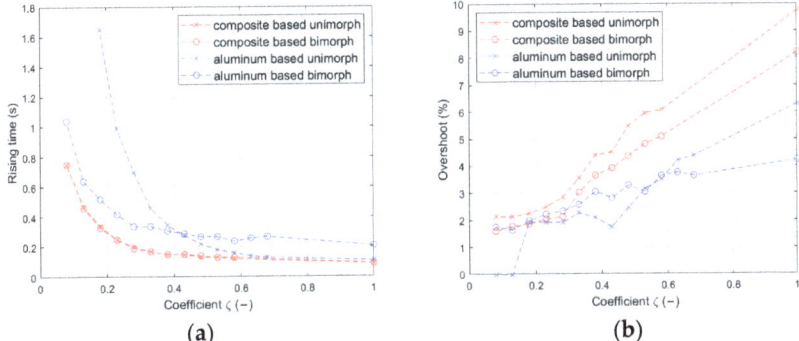

Figure 16. Impact of γ coefficient on quality indexes of control system: (**a**) on rising time, (**b**) on overshoot.

The rising time increased as the γ coefficient value decreased (Figure 16a). However, the overshoot increased as the γ coefficient value increased (Figure 16b). Hence, the choice of the γ coefficient value should be based on a compromise: on the one hand, the purpose should be to reduce the overshoot, and on the other hand, to shorten the rising time. Additionally, in the case of a bimorph, the weight in the **R** matrix that corresponds to the control voltage of the compressing MFC patch should be smaller than the weight corresponding to the control voltage of the stretching MFC patch. The following modification of Bryson's rule is proposed:

$$\begin{aligned} \text{For composite based unimorph}: r_{11} &= \frac{1}{|u_{1max}|^2 \gamma^2} \\ \text{For aluminum based unimorph}: r_{11} &= \frac{1}{|u_{1max}|^2 \gamma^2} \\ \text{For composite based bimorph}: r_{11} &= \frac{\varepsilon_1}{|u_{1max}|^2 \gamma^2} \quad r_{22} = \frac{\varepsilon_2}{|u_{2max}|^2 \gamma^2} \\ \text{For aluminum based bimorph}: r_{11} &= \frac{\varepsilon_1}{|u_{1max}|^2 \gamma^2} \quad r_{22} = \frac{\varepsilon_2}{|u_{2max}|^2 \gamma^2} \end{aligned} \quad (24)$$

The larger the value of the γ coefficient, the shorter the time it takes for the actuator to achieve the set displacement. A larger value of the γ coefficient can be used in actuator control systems with a carrier layer made of a stiffer material. This is due to the fact that the displacement caused by the creep phenomenon increases more slowly stiffer the material (compare Figure 7), which leads to a smaller increase in the overshoot. It was assumed that the γ coefficient for the composite-based actuator is equal to $Y_s/Y_{mfca} = 0.38$ (Y_s means Young's modulus of the FR4 composite) and that the γ coefficient for the aluminum-based actuator is equal to $Y_{mfca}/Y_s = 0.68$ (Y_s means Young's modulus of aluminum). On the basis of laboratory experiments in first stage of research, it was determined that for the composite-based actuator $\varepsilon_1 = 1$ and $\varepsilon_2 = 0.36$, and for the aluminum-based actuator $\varepsilon_1 = 1$ and $\varepsilon_2 = 0.38$. On the basis of the trial-and-error method, it was established that the weights in the **Q** matrix were equal to the largest value of the material constants that appear in Equation (2), which is Y_{mfc} for the composite-based actuator and Y_c for the aluminum-based actuator:

$$\text{For composite based actuator}: \mathbf{Q} = \begin{bmatrix} 30.33 \times 10^9 & 0 & 0 \\ 0 & 0 & 0 \\ 0 & 0 & 30.33 \times 10^9 \end{bmatrix}$$
$$\text{For aluminum based actuator}: \mathbf{Q} = \begin{bmatrix} 71 \times 10^9 & 0 & 0 \\ 0 & 0 & 0 \\ 0 & 0 & 71 \times 10^9 \end{bmatrix} \quad (25)$$

In all experiments, the weights in the $\mathbf{Q_c}$ matrix with dimensions 3×3 and the $\mathbf{R_c}$ matrix with dimensions 1×1, which are needed to calculate the $\mathbf{H}$ matrix in the Kalman filter, were the same (they were determined experimentally): $\mathbf{Q_c}$ = diag(1×10^{-3}, 1×10^{-3}, 1×10^{-3}) and $\mathbf{R_c}$ = 1×10^{-6}.

Figure 17 shows the measured displacement of the composite-based actuator and the generated control signals waveforms, which were obtained in the control system shown in Figure 4 for the largest set values (experiments no. 37 and no. 53).

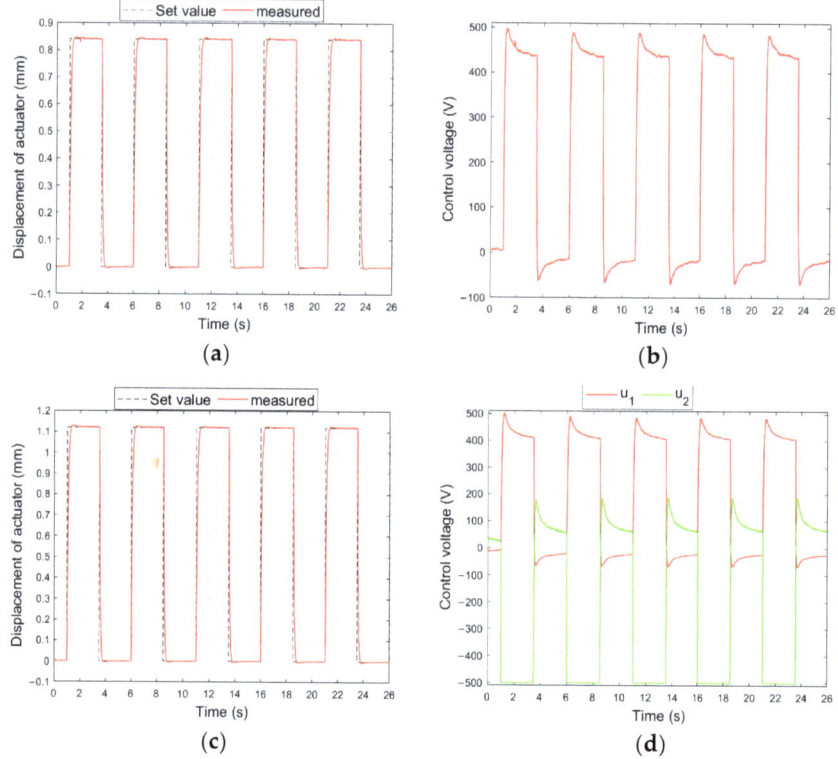

Figure 17. Control system characteristics for composite-based actuator: (**a**) unimorph displacement for y_{set} = 0.57 mm, (**b**) control voltage of unimorph, (**c**) bimorph displacement for y_{set} = 1.12 mm, (**d**) control voltages of bimorph.

Figure 18 shows the measured displacement of the aluminum-based actuator and the generated control signals waveforms, which were obtained in the control system shown in Figure 4 for the largest set values (experiments no. 39 and no. 55).

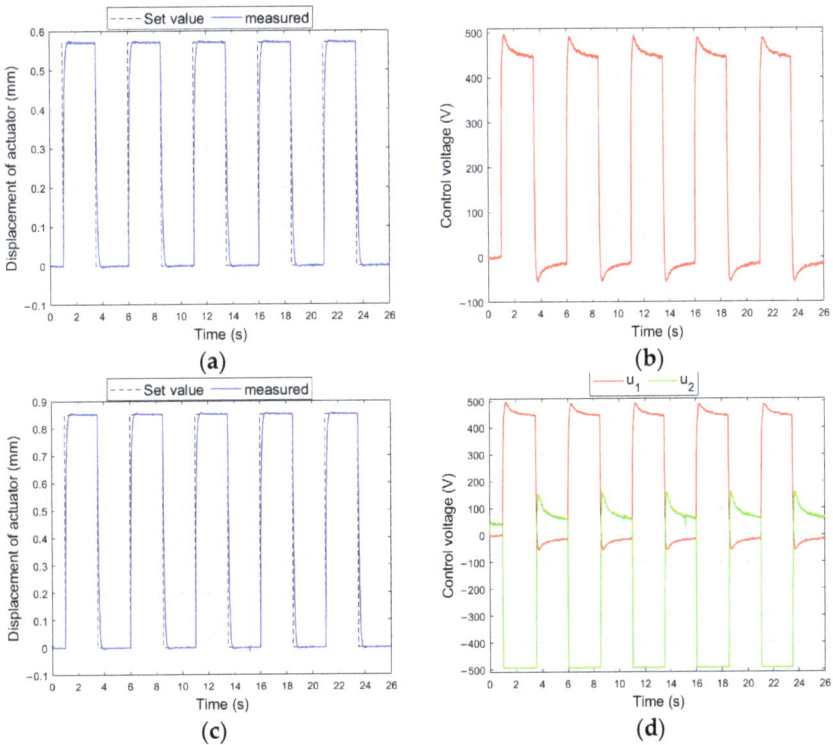

Figure 18. Control system characteristics for aluminum-based actuator: (**a**) unimorph displacement for y_{set} = 0.84 mm, (**b**) control voltage of unimorph, (**c**) bimorph displacement for y_{set} = 0.85 mm, (**d**) control voltages of bimorph.

Figure 19 shows the characteristics which were obtained in the control system shown in Figure 4 for the smallest set values of bimorph displacement (experiments no. 65 and no. 67).

To compare the control quality in all 32 laboratory experiments (Table 5), a control quality index (I_q) was determined in each of the experiments:

$$I_q = \frac{1}{y_{setmax}} \int |y_{set}(t) - y(t)| dt \qquad (26)$$

where y_{setmax} is the maximum value of the set value of the actuator tip displacement. The I_q values for each experiment are presented in Table 6.

Table 6. Value of quality index I_q.

	Experiment No.	Composite-Based		Experiment No.	Aluminum-Based	
		I_q (−)			I_q (−)	
Unimorph	37 to 38	0.837	0.822	39 to 40	0.942	0.919
	41 to 42	0.852	0.835	43 to 44	0.958	0.928
	45 to 46	0.916	0.881	47 to 48	1.011	0.963
	49 to 50	1.016	0.941	51 to 52	1.033	0.951

Table 6. *Cont.*

	Experiment No.	Composite-Based		Experiment No.	Aluminum-Based	
		I_q (−)			I_q (−)	
Bimorph	53 to 54	0.786	0.777	55 to 56	0.864	0.849
	57 to 58	0.784	0.771	59 to 60	0.878	0.857
	61 to 62	0.793	0.767	63 to 64	0.895	0.870
	65 to 66	0.859	0.816	67 to 68	0.940	0.887

The lower the value of the I_q index, the better the control quality.

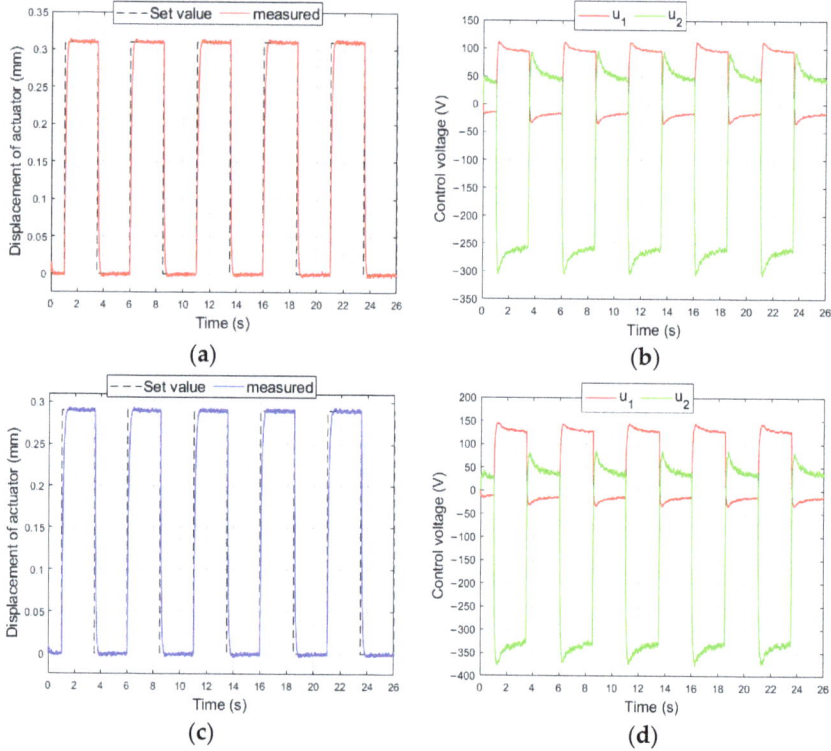

Figure 19. Control system characteristics for bimorph: (**a**) displacement of composite-based actuator for y_{set} = 0.31 mm, (**b**) control voltages of composite-based actuator, (**c**) displacement of composite-based actuator for y_{set} = 0.29 mm, (**d**) control voltages of composite-based actuator.

5. Discussion

As expected, the displacements of the composite-based actuator appeared larger compared to the aluminum-based actuator, but this difference did not increase as the time duration of the applied voltage spike increased: these differences did not exceed 3% (Figure 20).

Therefore, it can be concluded that displacements caused by the creep phenomenon of the composite-based actuator were approximately proportional to displacements of the aluminum-based actuator. These displacements were proportionally larger in the case of the composite carrier layer in comparison to the aluminum carrier layer (Figure 6).

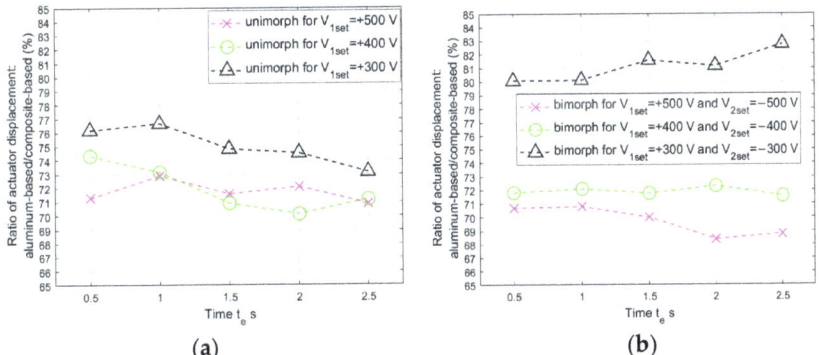

Figure 20. Aluminum-based actuator to composite-based actuator displacement ratio: (**a**) unimorph, (**b**) bimorph.

The ratios of bimorph to unimorph displacement are presented in Figure 21.

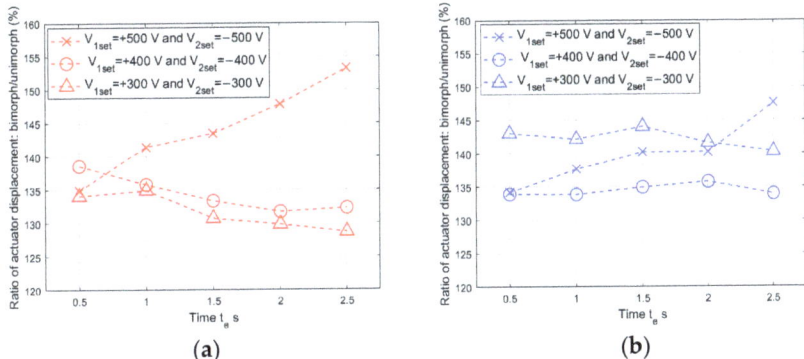

Figure 21. Bimorph actuator to unimorph actuator displacement ratio: (**a**) composite-based, (**b**) aluminum-based.

It can be noticed that the difference between the bimorph and unimorph displacement increased for the largest values of voltage spikes (V_{1set} = +500 V and V_{2set} = −500 V), as the duration of the voltage spike increased: by 18.3% for the composite-based actuator and by 13.4% for the aluminum-based actuator. In the case of voltage spikes with other tested values (V_{1set} = +400 V and V_{2set} = −400 V, V_{1set} = +300 V and V_{2set} = −300 V), this difference decreased slightly as the duration of the voltage spike increased. The average displacement ratios were determined: with a composite carrier layer it was 136.72% and with an aluminum carrier layer it was 138.83%.

Based on the results from the first stage of research, two main observations can be distinguished, which are important in the design of a linear control system of a piezoelectric actuator:

- The constant value of the control voltage causes undesirable actuator displacement, which is caused by the creep phenomenon. This is visible in Figures 6, 8 and 9;
- The control voltage of the compressing MFC should be larger than the control voltage of the stretching MFC. This observation is based on the comparison of the displacements of the unimorph and bimorph for the same carrier layer material.

These observations lead to guidelines for the determination of the weights in the **R** and **Q** matrixes:

- The use of Bryson's rule to determine the weights in the **R** matrix is not sufficient because it leads to the generation of the maximum possible control voltage, for example ±500 V in the case of the equipment presented in this article. This article proposes a modification to the method of determining the weights by introducing the ratios of the Young's modulus: see Equation (23). For the same purpose, in the **Q** matrix, the deviation from 0 of the first state variable should be limited by introducing an appropriately large weight q_{11}. Based on the results of the laboratory experiments, the article proposes a weight value q_{11} equal to the larger value of Young's modulus (either the Young's modulus value of the carrier layer material or of the piezoelectric material);
- The weight in the **R** matrix that corresponds to the control voltage of the compressing MFC patch should be smaller than the weight corresponding to the control voltage of the stretching MFC patch. This article proposes a modification to the method of determining the weights by introducing the coefficient ε_2: see Equation (24).

Based on the results of the first stage of research, it was also noted that the actuator positions in the intervals between control voltage spikes, which result from the hysteresis phenomenon, are approximately linearly dependent on the maximum displacement of the actuator. Reaching position zero in the intervals between control voltage spikes is possible by the application of a control voltage with the sign opposite to the sign of the voltage in the spikes. Obtaining position zero is possible by using a suitably large value of the weight q_{33} in the **Q** matrix. Based on the results of laboratory experiments, the article proposes a weight value q_{33} equal to the larger value of Young's modulus (either the Young's modulus value of the carrier layer material or of the piezoelectric material).

The use of modified rules for determining weights in the **R** matrix together with experimentally selected weights in the **Q** matrix enabled effective linear control of actuators for both the composite and the aluminum carrier layers, and for different values of the set value of the actuator tip displacement. First of all, it was noticed that the actuator achieved y_{set} in each of the experiments whose conditions are given in Table 5. To compare the control quality in individual experiments, the overshoot value was calculated:

$$\kappa = \left(\frac{y_{max}}{y_{steady}} 100\% \right) - 100\% \qquad (27)$$

where y_{max} is the maximum value of the actuator tip displacement and y_{steady} is the actuator tip position in a steady state after reaching y_{set} (given in Table 5). A comparison of the overshoot values in the individual experiments is shown in Figure 22.

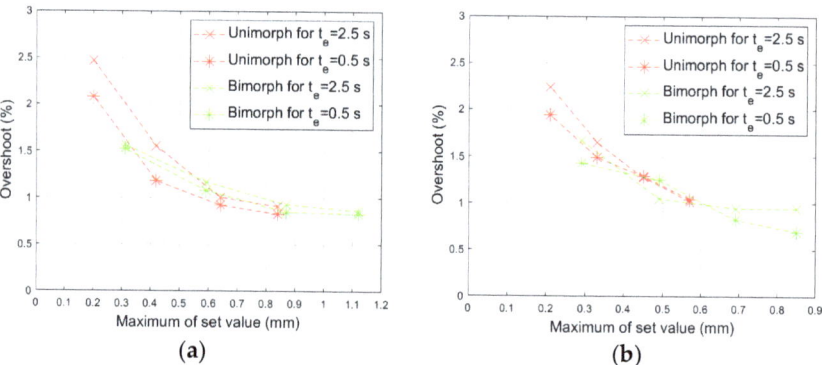

Figure 22. Overshoot: (**a**) composite-based actuators, (**b**) aluminum-based actuators.

The overshoot value increased slightly as the maximum set value decreased. However, in no experiment did it exceed 2.5%. The range of the overshoot changes in the bimorph case is smaller than in the unimorph case. Figure 23 shows the comparison of the control quality index I_q (25) in all laboratory experiments.

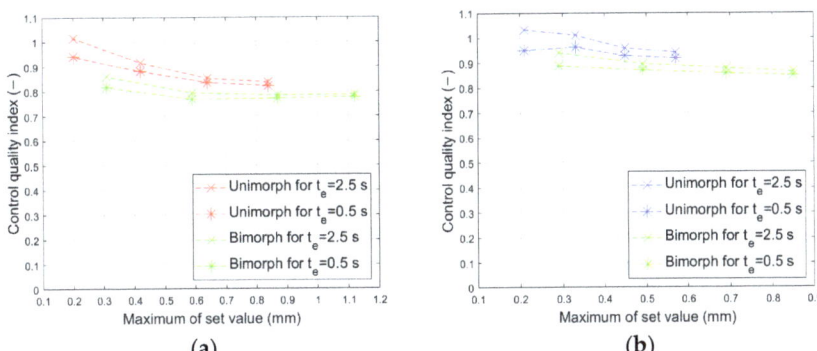

Figure 23. Quality index: (**a**) composite-based actuators, (**b**) aluminum-based actuators.

As can be seen in Figure 21, the control quality is approximately similar regardless of the material of the carrier layer, the maximum of the set value, and the duration of this maximum.

6. Conclusions

The subject of this article was an experimental analysis of the control system of a composite-based piezoelectric actuator and an aluminum-based piezoelectric actuator. Analysis was performed for both the unimorph and bimorph structures.

A modification of the method of selecting weights in the **R** matrix in the LQR control algorithm was proposed for a cantilever-type piezoelectric actuator. The weights in the R matrix for the actuator containing a carrier layer made of stiffer material should be smaller than those for the actuator containing a carrier layer made of less stiff material. Additionally, regardless of the carrier layer material, in the case of a bimorph, the weight in the **R** matrix that corresponds to the control voltage of the compressing MFC patch should be smaller than the weight corresponding to the control voltage of the stretching MFC patch.

The proposed correction of the selection of weights in the **R** matrix enables obtaining effective linear control, thanks to which displacements caused by the phenomenon of creep are eliminated. The quality of control remains approximately the same regardless of the material of the carrier layer, the maximum set value of the actuator displacement, and the duration of this set value at a constant level.

Funding: This research was funded by the AGH University of Krakow within the scope of the Research Program No. 16.16.130.942.

Institutional Review Board Statement: Not applicable.

Informed Consent Statement: Not applicable.

Data Availability Statement: Data are contained within the article.

Conflicts of Interest: The author declares no conflict of interest.

References

1. Mohith, S.; Upadhya, A.R.; Karanth, N.; Kulkarni, S.M.; Rao, M. Recent trends in piezoelectric actuators for precision motion and their applications: A review. *Smart Mater. Struct.* **2020**, *30*, 013002. [CrossRef]
2. Jin, H.; Gao, X.; Ren, K.; Liu, J.; Qiao, L.; Liu, M.; Chen, W.; He, Y.; Dong, S.; Xu, Z.; et al. Review on piezoelectric actuators based on high-performance piezoelectric materials. *IEEE Trans. Ultrason. Ferroelectr. Freq. Control* **2022**, *69*, 3057–3069. [CrossRef]
3. Yang, C.; Youcef-Toumi, K. Principle, implementation, and applications of charge control for piezo-actuated nanopositioners: A comprehensive review. *Mech. Syst. Signal Process.* **2022**, *171*, 108885. [CrossRef]
4. Chilibon, I.; Dias, C.; Inacio, P.; Marat-Mendes, J. PZT and PVDF bimorph actuators. *J. Optoelectron. Adv. Mater.* **2007**, *9*, 1939–1943.
5. Ghodsi, M.; Mohammadzaheri, M.; Soltani, P. Analysis of Cantilever Triple-Layer Piezoelectric Harvester (CTLPH): Non-Resonance Applications. *Energies* **2023**, *16*, 3129. [CrossRef]
6. Takagi, K.; Li, J.F.; Yokoyama, S.; Watanabe, R.; Almajid, A.; Taya, M. Design and fabrication of functionally graded PZT/Pt piezoelectric bimorph actuator. *Sci. Technol. Adv. Mater.* **2002**, *3*, 217–224. [CrossRef]
7. Sumit; Kane, S.R.; Sinha, A.K.; Shukla, R. Electric field-induced nonlinear behavior of lead zirconate titanate piezoceramic actuators in bending mode. *Mech. Adv. Mater. Struct.* **2022**, *30*, 2111–2120. [CrossRef]
8. Tan, D.; Yavarow, P.; Erturk, A. Nonlinear elastodynamics of piezoelectric macro-fiber composites with interdigitated electrodes for resonant actuation. *Compos. Struct.* **2018**, *187*, 137–143. [CrossRef]
9. Liu, Y.Z.; Hao, Z.W.; Yu, J.X.; Zhou, X.R.; Lee, P.S.; Sun, Y.; Mu, Z.C.; Zeng, F.L. A high-performance soft actuator based on a poly (vinylidene fluoride) piezoelectric bimorph. *Smart Mater. Struct.* **2019**, *28*, 055011. [CrossRef]
10. Akdogan, E.K.; Allahverdi, M.; Safari, A. Piezoelectric composites for sensor and actuator applications. *IEEE Trans. Ultrason. Ferroelectr. Freq. Control* **2005**, *52*, 746–775. [CrossRef]
11. Dai, Q.; Ng, K. Investigation of electromechanical properties of piezoelectric structural fiber composites with micromechanics analysis and finite element modeling. *Mech. Mater.* **2012**, *53*, 29–46. [CrossRef]
12. Nguyen, C.H.; Kornmann, X. A comparison of dynamic piezoactuation of fiber-based actuators and conventional PZT patches. *J. Intell. Mater. Syst. Struct.* **2006**, *17*, 45–55. [CrossRef]
13. Wang, H.; Xie, X.; Zhang, M.; Wang, B. Analysis of the nonlinear hysteresis of the bimorph beam piezoelectric bending actuator for the deformable mirror systems. *J. Astron. Telesc. Instrum. Syst.* **2020**, *6*, 029002. [CrossRef]
14. Mansour, S.Z.; Seethaler, R.J.; Teo, Y.R.; Yong, Y.K.; Fleming, A.J. Piezoelectric bimorph actuator with integrated strain sensing electrodes. *IEEE Sens. J.* **2018**, *18*, 5812–5817. [CrossRef]
15. Shen, D.; Wen, J.; Ma, J.; Hu, Y.; Wang, R.; Li, J. A novel linear inertial piezoelectric actuator based on asymmetric clamping materials. *Sens. Actuators A Phys.* **2020**, *303*, 111746. [CrossRef]
16. Davis, C.L.; Calkins, F.T.; Butler, G.W. High-frequency jet nozzle actuators for noise reduction. In *Smart Structures and Materials 2003: Industrial and Commercial Applications of Smart Structures Technologies*; Anderson, E.H., Ed.; Proceedings of SPIE; SPIE: Bellingham, WA, USA, 2003; Volume 5054, pp. 34–44.
17. Wood, R.J.; Steltz, E.; Fearing, R.S. Optimal energy density piezoelectric bending actuators. *Sens. Actuators A Phys.* **2005**, *119*, 476–488. [CrossRef]
18. Wang, Q.M.; Cross, L.E. Performance analysis of piezoelectric cantilever bending actuators. *Ferroelectrics* **1998**, *215*, 187–213. [CrossRef]
19. Wang, H. Analytical analysis of a beam flexural-mode piezoelectric actuator for deformable mirrors. *J. Astron. Telesc. Instrum. Syst.* **2015**, *1*, 049001. [CrossRef]
20. Mtawa, A.N.; Sun, B.; Gryzagoridis, J. An investigation of the influence of substrate geometry and material properties on the performance of the C-shape piezoelectric actuator. *Smart Mater. Struct.* **2007**, *16*, 1036. [CrossRef]
21. LaCroix, B.W.; Ifju, P.G. Investigating potential substrates to maximize out-of-plane deflection of piezoelectric macro-fiber composite actuators. *J. Intell. Mater. Syst. Struct.* **2015**, *26*, 781–795. [CrossRef]
22. Gan, J.; Zhang, X. A review of nonlinear hysteresis modeling and control of piezoelectric actuators. *AIP Adv.* **2019**, *9*, 040702. [CrossRef]
23. Jung, H.; Gweon, D.G. Creep characteristics of piezoelectric actuators. *Rev. Sci. Instrum.* **2000**, *71*, 1896–1900. [CrossRef]
24. Yang, Y.L.; Wei, Y.D.; Lou, J.Q.; Fu, L.; Tian, G.; Wu, M. Hysteresis modeling and precision trajectory control for a new MFC micromanipulator. *Sens. Actuators A Phys.* **2016**, *247*, 37–52. [CrossRef]
25. Ebrahimi-Tirtashi, A.; Mohajerin, S.; Zakerzadeh, M.R.; Nojoomian, M.A. Vibration control of a piezoelectric cantilever smart beam by $\mathcal{L}$ 1 adaptive control system. *Syst. Sci. Control Eng.* **2021**, *9*, 542–555. [CrossRef]
26. Wang, X.; Zhou, W.; Zhang, Z.; Jiang, J.; Wu, Z. Theoretical and experimental investigations on modified LQ terminal control scheme of piezo-actuated compliant structures in finite time. *J. Sound Vib.* **2021**, *491*, 115762. [CrossRef]
27. Tsushima, N.; Su, W. Concurrent active piezoelectric control and energy harvesting of highly flexible multifunctional wings. *J. Aircr.* **2017**, *54*, 724–736. [CrossRef]
28. Roundy, S.; Wright, P.K.; Rabaey, J.M. *Energy Scavenging for Wireless Sensor Networks*; Springer: New York, NY, USA, 2004.
29. Tian, J.; Guo, Q.; Shi, G. Laminated piezoelectric beam element for dynamic analysis of piezolaminated smart beams and GA-based LQR active vibration control. *Compos. Struct.* **2020**, *252*, 112480. [CrossRef]
30. Smart Material—Home of the MFC. Available online: https://www.smart-material.com/MFC-product-mainV2.html (accessed on 30 July 2022).

31. 3M™ Scotch-Weld™ Structural Adhesives. Available online: https://multimedia.3m.com/mws/media/1989908O/04-productselectionguide-epx-en.pdf (accessed on 10 September 2023).
32. RT-DAC/Zynq User's Manual. Available online: http://www.inteco.com.pl/Docs/Rtdac_Zynq.pdf (accessed on 30 July 2023).
33. Koo, K.Y.; Park, S.; Lee, J.J.; Yun, C.B. Automated impedance-based structural health monitoring incorporating effective frequency shift for compensating temperature effects. *J. Intell. Mater. Syst. Struct.* **2009**, *20*, 367–377. [CrossRef]
34. Huynh, T.C.; Kim, J.T. Quantification of temperature effect on impedance monitoring via PZT interface for prestressed tendon anchorage. *Smart Mater. Struct.* **2017**, *26*, 125004. [CrossRef]
35. Lining, S.; Changhai, R.; Weibin, R.; Liguo, C.; Minxiu, K. Tracking control of piezoelectric actuator based on a new mathematical model. *J. Micromech. Microeng.* **2004**, *14*, 1439. [CrossRef]
36. Hosea, M.E.; Shampine, L.F. Analysis and implementation of TR-BDF2. *Appl. Numer. Math.* **1996**, *20*, 21–37. [CrossRef]
37. Ismail, M.R.; Omar, F.K.; Ajaj, R.; Ghodsi, M. On the accuracy of lumped parameter model for tapered cantilever piezoelectric energy harvesters with tip mass. In Proceedings of the 2020 Advances in Science and Engineering Technology International Conferences (ASET), Dubai, United Arab Emirates, 4 February–9 April 2020; pp. 1–6.
38. Wang, B.; Luo, X.; Liu, Y.; Yang, Z. Thickness-variable composite beams for vibration energy harvesting. *Compos. Struct.* **2020**, *244*, 112232. [CrossRef]
39. Grzybek, D. Control System for Multi-Input and Simple-Output Piezoelectric Beam Actuator Based on Macro Fiber Composite. *Energies* **2022**, *15*, 2042. [CrossRef]
40. Deraemaeker, A.; Nasser, H.; Benjeddou, A.; Preumont, A. Mixing rules for the piezoelectric properties of macro fiber composites. *J. Intell. Mater. Syst. Struct.* **2009**, *20*, 1475–1482. [CrossRef]
41. Ghodsi, M.; Mohammadzaheri, M.; Soltani, P.; Ziaifar, H. A new active anti-vibration system using a magnetostrictive bimetal actuator. *J. Magn. Magn. Mater.* **2022**, *557*, 169463. [CrossRef]
42. Li, C.; Shen, L.; Shao, J.; Fang, J. Simulation and experiment of active vibration control based on flexible piezoelectric MFC composed of PZT and PI layer. *Polymers* **2023**, *15*, 1819. [CrossRef]
43. Xu, P.; Lan, X.; Zeng, C.; Zhang, X.; Liu, Y.; Leng, J. Dynamic characteristics and active vibration control effect for shape memory polymer composites. *Compos. Struct.* **2023**, *322*, 117327. [CrossRef]

Disclaimer/Publisher's Note: The statements, opinions and data contained in all publications are solely those of the individual author(s) and contributor(s) and not of MDPI and/or the editor(s). MDPI and/or the editor(s) disclaim responsibility for any injury to people or property resulting from any ideas, methods, instructions or products referred to in the content.

Article

Biocomposite Materials Derived from *Andropogon halepensis*: Eco-Design and Biophysical Evaluation

Marcela-Elisabeta Barbinta-Patrascu [1], Cornelia Nichita [2,3,*], Bogdan Bita [1,4] and Stefan Antohe [1,5,*]

1. Department of Electricity, Solid-State Physics and Biophysics, Faculty of Physics, University of Bucharest, 405 Atomistilor Street, 077125 Magurele, Romania; marcela.barbinta@unibuc.ro (M.-E.B.-P.); bogdan.bita@fizica.unibuc.ro (B.B.)
2. CTT-3Nano-SAE Research Center, Faculty of Physics, ICUB, University of Bucharest, MG-38, 405 Atomistilor Street, 077125 Magurele, Romania
3. National Institute for Chemical-Pharmaceutical Research and Development, 112 Vitan Avenue, 031299 Bucharest, Romania
4. National Institute for Lasers, Plasma and Radiation Physics, Magurele, 077125 Bucharest, Romania
5. Academy of Romanian Scientists, Ilfov Street 3, 050045 Bucharest, Romania
* Correspondence: cornelia.nichita@unibuc.ro or cornelianichita@yahoo.com (C.N.); santohe@solid.fizica.unibuc.ro or s_antohe@yahoo.com (S.A.)

Abstract: This research work presents a "green" strategy of weed valorization for developing silver nanoparticles (AgNPs) with promising interesting applications. Two types of AgNPs were phyto-synthesized using an aqueous leaf extract of the weed *Andropogon halepensis* L. Phyto-manufacturing of AgNPs was achieved by two bio-reactions, in which the volume ratio of (phyto-extract)/(silver salt solution) was varied. The size and physical stability of *Andropogon*—AgNPs were evaluated by means of DLS and zeta potential measurements, respectively. The phyto-developed nanoparticles presented good free radicals-scavenging properties (investigated via a chemiluminescence technique) and also urease inhibitory activity (evaluated using the conductometric method). *Andropogon*—AgNPs could be promising candidates for various bio-applications, such as acting as an antioxidant coating for the development of multifunctional materials. Thus, the *Andropogon*-derived samples were used to treat spider silk from the spider *Pholcus phalangioides*, and then, the obtained "green" materials were characterized by spectral (UV-Vis absorption, FTIR ATR, and EDX) and morphological (SEM) analyses. These results could be exploited to design novel bioactive materials with applications in the biomedical field.

Keywords: silver nanoparticles; "green" synthesis; *Andropogon halepensis*; biocomposites; antioxidant activity; spider silk; *Pholcus phalangioides*

Citation: Barbinta-Patrascu, M.-E.; Nichita, C.; Bita, B.; Antohe, S. Biocomposite Materials Derived from *Andropogon halepensis*: Eco-Design and Biophysical Evaluation. *Materials* 2024, 17, 1225. https://doi.org/10.3390/ma17051225

Academic Editor: Raul D. S. G. Campilho

Received: 8 February 2024
Revised: 1 March 2024
Accepted: 3 March 2024
Published: 6 March 2024

Copyright: © 2024 by the authors. Licensee MDPI, Basel, Switzerland. This article is an open access article distributed under the terms and conditions of the Creative Commons Attribution (CC BY) license (https://creativecommons.org/licenses/by/4.0/).

1. Introduction

Nanotechnology, a cutting-edge multidisciplinary field in science and technology, has been penetrating more and more into fields such as cosmetics, medicine, or agriculture. In recent years, special attention has been paid to the development of new "green" nano-strategies to design materials with unusual properties by combining bio-organic and inorganic matter. Thus, current trends in nanotechnology are the use of bio-methods based on *Green Chemistry* principles (involving nontoxic precursors and mild reaction conditions) and the usage of bioresources, including bacteria, fungi, viruses, yeasts, algae, and vegetal extracts, that act as nano-factories for the "green" preparation of nanoparticles (NPs) [1]. Harnessing the huge potential of plants and plant waste for the "green" synthesis of metal nanoparticles is a current trend in nanotechnology. Recent reports demonstrated that leaf extracts are the best option for the biosynthesis of silver nanoparticles (AgNPs) due to the presence of phyto-compounds such as polyphenols, flavonoids, terpenoids, aliphatic alcohols, and aldehydes, which act both as bioreducing and capping agents in AgNPs synthesis. This process, called AgNPs phyto-synthesis, is based on *Green Chemistry* principles; therefore, it is a nontoxic, low-cost, and harmless process that uses natural resources, avoiding the use of hazardous ingredients [2,3]. Phyto-fabricated AgNPs (phyto-AgNPs) possess interesting bioactivities (e.g., antioxidant, antimicrobial, and antiproliferative qualities, among others) that make them suitable for biomedical applications [2,4,5]. Composite materials containing phyto-AgNPs have been applied as coatings for medical devices, medical fabrics, or scaffolds, diabetic socks, wound dressings, and tissue engineering and regeneration [2]. Biocomposites based on biocompatible polymers and phyto-AgNPs have antibacterial properties and may help protect a scaffold against infection [6].

The present work aimed to design "green" biocomposites based on a natural polymeric material, spider silk, and silver nanoparticles phyto-generated from an *Andropogon halepensis* aqueous extract. The development of these composites consisted of the following steps: (i) the preparation of the *Andropogon halepensis* aqueous extract from fresh leaves, (ii) the phyto-synthesis of AgNPs through two bio-reactions, (iii) the treatment of spider silk with the phyto-nanometallic particles, and (iv) the biophysical-chemical characterization of the *Andropogon*-derived materials.

Through this study, we wanted to show the huge value of natural resources, whether they are of vegetable (weeds) or insect origin (spider webs), in developing valuable hybrid materials for various applications. In this way, the use of "green" technologies and bio-waste will contribute to our planet's safety, preserving "green" trees, "blue" waters, and blue sky.

Andropogon halepensis L., commonly known as Johnsongrass, is a warm-season-perennial grass, native to the Mediterranean region, which belongs to the Poaceae family [7]. This plant is known as one of the most common and troublesome weeds, with a worldwide distribution [8], meaning that it is an abundant free natural source which could be used both as a bioreducing and capping agent to develop metallic nanoparticles. The chemical composition of *Andropogon halepensis* L. can vary depending on its growth stage, environmental conditions, and geographic location. Its main chemical components are carbohydrates (represented by cellulose, hemicellulose, and sugars), proteins, lipids, and various phytochemicals, including phenolic compounds and flavonoids. The specific phenolic compounds found in *Andropogon halepensis* L. are ferulic acid, caffeic acid, p-coumaric acid, quercetin, and tannins [9–12]. Huang et al. reported for the first time, in 2010, the presence of three flavonoids—apigenin, tricin, and luteolin—in the aerial parts of this plant [13]. Also, the grass *Andropogon halepensis* L. contains essential minerals such as potassium, magnesium, and calcium and trace elements like iron, zinc, manganese, chromium, and copper [6]. *Andropogon halepensis* possesses various biological properties, among which we can mention its invasive character, adaptability to different types of soil, and drought resistance [14–16]. Regarding its pharmaceutical properties, the studies carried out by Shah et al. (2021) found antioxidant, antimicrobial, and antidiabetic properties in the methanolic extract obtained from the rhizomes of *A. halepensis* L. [12].

Thus, an important, *key point* in our study is to demonstrate that invasive plants can be converted into materials (AgNPs) beneficial for human health. Another *key point* in this work is the development of new materials through the functionalization of a natural biomaterial—that is, spider silk—with *Andropogon*-derived AgNPs.

Spider silk is a valuable bionic material in nature, with fascinating properties which are more performant than most artificial materials [17,18].

Spiders produce cobwebs to catch their prey, thus securing their food. Also, spiders use their webs for shelter, and they deposit their eggs in them. With the help of their cobwebs, spiders can move from one place to another and also "fly", in a process called "ballooning", during which spiders move through the air thanks to the electric fields detected by their electroreceptors [19].

Spider silk is an antimicrobial, hypoallergenic, and completely biodegradable natural proteic material [18]. Spider webs have been used since ancient times, such as by the Greeks and the Romans, who stopped battle wounds from bleeding by covering them with spider silk [20]. Spider webs are still used nowadays as a hemostatic agent and in wound healing [21]. Moreover, spider silk is also used as a bioactive material for tissue engineering and drug delivery due to its excellent biocompatibility and biodegradability [22].

As spider silk has as its main component a protein (spidroin), it is biocompatible, nontoxic, and fully biodegradable [23]. Spider silk is an ideal material for wound suture and prosthesis since human tissues can absorb its degradation products.

In addition, spider silk has been used in the textile field to make clothes that are similar to worm silk but have a better performance in terms of being lightweight, non-breakable, with good breathability, strong water absorption, and UV resistance qualities. Despite these interesting features, the production efficiency of spider silk is low, and spider silk clothing is a luxury [23,24] because spider webs must be collected from many spiders, and, in addition, spiders are cannibals and should not be grown in close proximity.

Moreover, spider silk–metallic nanoparticles composites have been developed and used for biomedical applications, including as antimicrobial and anticoagulant agents [25].

As it is known, biocomposites combine the properties of all their components. In this regard, taking into account the properties of phyto-AgNPs and spider webs, the current study proposes the functionalization of spider webs (arising from the spider *Pholcus phalangioides*) with phyto-AgNPs obtained from *Andropogon halepensis*, as mentioned above.

To our knowledge, there is no prior report on the use of the weed *Andropogon halepensis* to "green" develop biocomposites with spider silk. In this paper, the obtained biocomposites were biophysico-chemically characterized, and their potential use in the biomedical field is discussed.

2. Materials and Methods

2.1. Materials

Rutin trihydrate ($\geq$94.0%), gallic acid ($\geq$97.5%), Folin–Ciocalteu's phenol reagent, sodium carbonate anhydrous ($\geq$99.0%), hydrochloric acid (HCl, 37%), aluminum chloride (99.99%), silver nitrate ($AgNO_3$), urea (99.5%), sodium acetate ($\geq$99.0%), luminol (5-amino-2,3-dihydro-phthalazine-1,4-dione), dimethyl sulfoxide (DMSO, $\geq$99.9%), hydrogen peroxide (H_2O_2, 30%), hydroxy methyl aminomethane base (TRIS $\geq$ 99.8%), and methanol ($\geq$99.9%) were purchased from Merck Company (Darmstadt, Germany). Urease from Jack Bean was acquired from Fisher Scientific (Oxford, UK), and the conductivity standard solution utilized (1413 μS/cm) was purchased from Hanna Instruments. For the preparation of the vegetal aqueous extract, fresh leaves of *Andropogon halepensis* L. (Johnsongrass) were harvested on June 2022, from the Prahova county (Romania) (44°50′03″ N 25°53′57″ E).

2.2. Preparation of Andropogon-Derived Materials

2.2.1. Preparation of the Phyto-Extract

Fresh leaves of *Andropogon halepensis* were washed many times with tap water to remove any dust and then rinsed in distilled water. The cleaned leaves were further chopped into small pieces, immersed in boiling distilled water, and then boiled for 15 min. The mass ratio of *plant leaves/distilled water* reached a value of 1:4. The obtained phyto-extract (referred to as EAh) was filtered through Whatman filter paper no. 1 and kept in the freezer until use.

2.2.2. Biological Nano-Synthesis of Silver Nanoparticles

In this study, two types of silver nanoparticles were nano-synthesized, using the prepared phyto-extract, through the following bioreducing reactions:

<u>Bio-Reaction 1</u>: A volume of 1 mL of the prepared *A. halepensis* extract was mixed with 1 mL of 1 mM $AgNO_3$ aqueous solution, in the dark, under continuous magnetic stirring (VIBRAX stirrer, Milian, OH, USA, 200 rpm), at room temperature. After 24 h, the color stabilized as a dark-brown-green color, indicating the completion of the phyto-synthesis

of the AgNPs, a fact which was further demonstrated by UV-Vis absorption spectroscopy. The obtained AgNPs were named AgNPs_1.

<u>Bio-Reaction 2</u>: A volume of 1 mL of the prepared phyto-extract was mixed with 100 mL of 1 mM AgNO$_3$ aqueous solution, in the dark, under continuous magnetic stirring (VIBRAX stirrer, Milian, OH, USA, 200 rpm), at room temperature. After 45 min, the color of the mixture changed. After 24 h, the color stabilized as a reddish brown, indicating the completion of NPs synthesis, a fact which was further proved by UV-Vis absorption spectroscopy. These AgNPs were named AgNPs_2.

The phyto-molecules from the *Andropogon halepensis* extract gave up electrons to silver ions (arising from the AgNO$_3$ solution), reducing and then surrounding them, forming nanoparticles. Thus, the EAh acted both as the bioreducing and capping agent for the silver ions.

The biosynthesized AgNPs samples were also monitored after 18 months. The "new" synthesized nanoparticles were referred to as AgNPs_1n and AgNPs_2n, and the "old" ones were referred to as AgNPs_1o and AgNPs_2o.

2.2.3. Preparation of Biocomposites Based on Silk and Phyto-Extract/Phyto-Nanometals

The spider cobwebs used in these experiments were harvested, using a long stick, from the spider *Pholcus phalangioides* (common name: "Daddy" Long Legs [26]) (from Bucharest, Romania), which is a common Romanian spider with extremely long legs (Figure 1). The cobwebs were washed in a mild detergent, then rinsed several times with distilled water, and left to dry at room temperature.

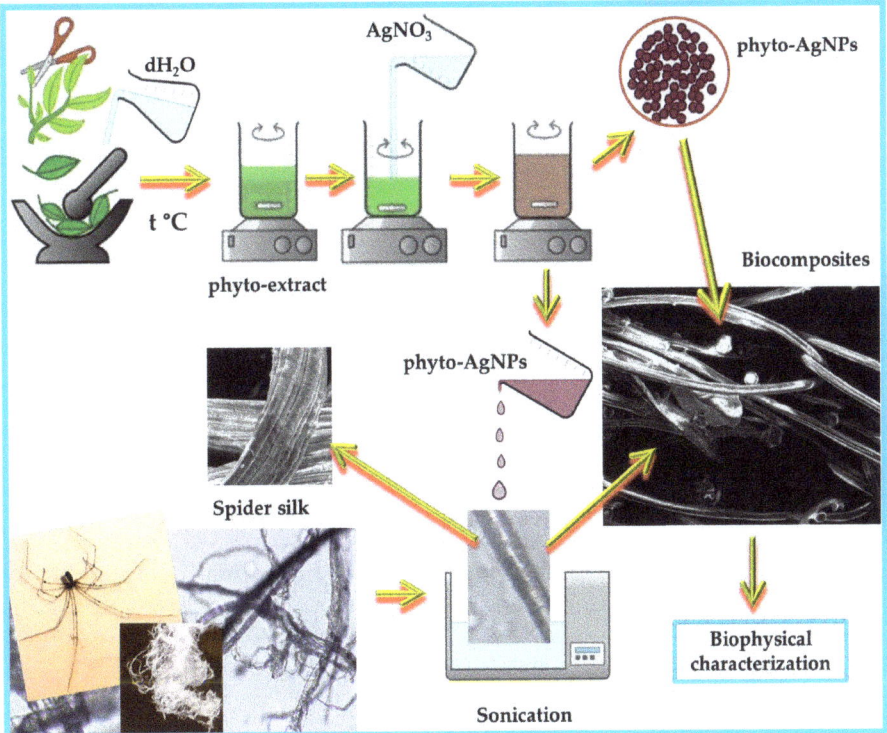

Figure 1. Schematic representation of the eco-design and preparation of the materials derived from *Andropogon halepensis* L. The figure was created with Chemix (https://chemix.org/, accessed on 25 January 2024) and with PowerPoint and Paint 3D. This figure also contains images taken by us with a camera, an optical microscope, and SEM.

The spider web samples were divided into many experimental batches: (1) untreated (sample S), (2) treated with EAh (sample S_EAh), (3) treated with AgNPs_1 (sample S_AgNPs_1), and (4) treated with AgNPs_2 (sample S_AgNPs_2).

The codes of the spider web samples studied are presented in Table 1. The treatment was performed by ultrasound irradiation in a water bath (BRANSON 1210, Marshall Scientific, Hampton, NH, USA) for 60 min (with a break after 30 min). Then, the spider webs were left for another 5 h at room temperature in the suspensions in which they had been immersed, then they were removed from the suspensions and left to dry, and then they were analyzed by SEM/EDS, UV-Vis, and FTIR-ATR spectroscopy.

Table 1. Names and description of the *Andropogon halepensis*-derived samples.

Sample Name	Description
EAh	Aqueous extract of *Andropogon halepensis* L. leaves
AgNPs_1o	Silver nanoparticles phyto-synthesized from *Andropogon halepensis* extract, using method 1 (the "old" sample)
AgNPs_1n	Silver nanoparticles phyto-synthesized from *Andropogon halepensis* extract, using method 1 (the "new" sample)
AgNPs_2o	Silver nanoparticles phyto-synthesized from *Andropogon halepensis* extract, using method 2 (the "old" sample)
AgNPs_2n	Silver nanoparticles phyto-synthesized from *Andropogon halepensis* extract, using method 2 (the "new" sample)
S	Spider silk (untreated) collected from the spider *Pholcus phalangioides*
S_EAh	Spider silk treated with *Andropogon halepensis* extract
S_AgNPs_1o	Spider silk treated with AgNPs_1o
S_AgNPs_1n	Spider silk treated with AgNPs_1n
S_AgNPs_2o	Spider silk treated with AgNPs_2o
S_AgNPs_2n	Spider silk treated with AgNPs_2n

Table 1 displays the abbreviations of all the samples obtained during this research. The schematic representation of the eco-design of all the samples is shown in Figure 1.

2.3. Physical-Chemical Characterization of Andropogon-Derived Materials

A JASCO UV-Vis V-570 spectrophotometer (Jasco International Co., Ltd., Tokyo, Japan) was used to record the **UV-Vis absorption** spectra of the phyto-extract and silver nanoparticles in the 200–800 nm wavelength and at a scanning rate of 1 nm/s, at 25 °C. The UV-Vis absorbance spectra of the spider silk-based materials were recorded on the JASCO UV–Vis 530 spectrophotometer with an integrating sphere accessory.

The chromatic parameters of the pristine and treated silk samples were measured on the same equipment using the JASCO's color diagnosis system (CIE LAB system of colors).

The following parameters were calculated: L*, a*, and b*, using the CIE LAB system of colors. The three coordinates of the CIE LAB system represent the following:

- L* represents the lightness (L* = 0 indicates black, L* = 100 indicates white);
- a* indicates the relative position between green and red (negative values of a* indicate a green color, and positive values indicate a red color);
- b* represents the relative position between blue and yellow (negative values of b* indicate the color blue, and positive values indicate the color yellow).

Based on the different values in the color parameters between the reference (untreated spider silk) and the sample (treated spider silk), the color difference (ΔE^*) of the sample from the reference was calculated using the following equation [27,28]:

$$\Delta E^* = [(\Delta L^*)^2 + (\Delta a^*)^2 + (\Delta b^*)^2]^{1/2} \quad (1)$$

Fourier-transform IR (FTIR) spectra were collected using a Perkin Elmer Spectrum 400 instrument with an **attenuated total reflectance (ATR)** diamond crystal, in a transmittance mode. Scans in the 4000–650 cm^{-1} range were accumulated for each spectrum at a spectral resolution of 4 cm^{-1}.

Further investigations into the morphostructural aspects of the sample were conducted using the Apreo S ThermoFisher scanning electron microscope (**SEM**), with a field emission gun (FEG), operating at 10 kV in a secondary electron mode, equipped with a 9 μm aperture and a sample-to-detector distance of 9 mm. To facilitate imaging, the samples were affixed to a conductive self-adhesive carbon tape and securely positioned on the aluminum SEM holder cap. In the case of liquid samples, a drop-casting method was employed for deposition. Elemental analysis was performed using energy dispersive X-ray spectroscopy (**EDX**) with the EDAX TEAM system, using a 10 kV excitation voltage and a 70 mm^2 area detector to gain valuable insights into the samples' composition and structural characteristics.

The average size of particles (Zav) was determined, in triplicate, by **dynamic light scattering (DLS)**, and the results were reported as the mean values ± S.D. **DLS measurements** were performed on a Zetasizer Nano ZS (Malvern Instruments Inc., Worcestershire, UK), at a temperature of 25 °C, using 173° backscatter angle detection, assuming a laser beam at a wavelength of 633 nm. The polydispersity index, PdI, as the indicator of the size distribution's width, was also determined.

Zeta potential (ξ, mV) measurements were performed in triplicate, at 25 °C, in an appropriate device, the Zetasizer Nano ZS (Malvern Instruments Ltd., Malvern, UK), by applying an electric field across the analyzed samples. Zeta potential measurements were carried out in ultrapure water, at a pH of 6.98. The ξ values were reported as mean ± S.D.

2.4. Bio-Evaluation of the Andropogon-Derived Samples

2.4.1. Total Polyphenol Content (TPC)

The **total polyphenol content (TPC)** of the vegetal extract was determined by means of the UV-Vis spectrophotometric method using the Folin–Ciocalteu assay [29,30]. In this assay, the Folin–Ciocalteu reagent, a mixture of phosphotungstic ($H_3PW_{12}O_{40}$) and phosphomolybdic ($H_3PMo_{12}O_{40}$) acids, is generally reduced to blue oxides of tungsten (W_8O_{23}) and molybdenum (Mo_8O_{23}) during phenol oxidation. This reaction takes place in alkaline conditions provided by a sodium carbonate solution. Briefly, in our experiment, a volume of 100 μL of vegetal extract was mixed with 0.5 mL of Folin–Ciocalteu phenol reagent (Merck, Darmstadt, Germany), followed by the addition of 2 mL of an anhydrous sodium carbonate (Na_2CO_3 99% purity, Merck Company) solution (20%). The above-mentioned mixture was incubated for 60 min, and the optical absorbance was recorded at 760 nm using a JASCO UV-Vis V-570 spectrophotometer (Jasco International Co., Ltd., Tokyo, Japan). The TPC values were calculated using a gallic acid calibration curve, and the results were expressed as mg gallic acid (≥97.5% purity, Merck Company) equivalent/g dry extract (mg GAE g^{-1}) [31,32]. All the experiments were replicated three times.

2.4.2. Total Flavonoid Content (TFC)

The **total flavonoid content (TFC)** of the vegetal extract was estimated using an aluminum chloride colorimetric assay, as previously described [33]. Briefly, an aliquot of 5 mL of the *A. halepensis* extract or of AgNPs was mixed with 5.0 mL of sodium acetate (≥99.0% purity, Merck Company) 100 g/L, 3.0 mL of AlCl$_3$ (99.99% purity, Merck Company) 25 g/L, and filled up to 25 mL with methanol (≥99.9% purity, Merck Company) in a volumetric flask. After incubation at room temperature for 30 min, the optical absorbances of the reaction mixtures were recorded at 430 nm using the JASCO UV-Vis V-570 spectrophotometer (Jasco International Co., Ltd., Tokyo, Japan). The TFC was determined using a rutin calibration curve, and the results were expressed as mg rutin trihydrate (≥94.0% purity, Merck Company) equivalent/g dry extract (mg RE g^{-1}). All the experiments were performed in triplicate.

2.4.3. In Vitro Antioxidant Activity Assay

The in vitro non cellular antioxidant activity of the *Andropogon halepensis* extract and its phyto-derived nanoparticles was evaluated by means of the chemiluminescence (CL) technique, using the procedure described elsewhere [3,34]. The CL experiments were carried out on a Sirius Luminometer Berthold–GmbH (Pforzheim, Germany), using luminol-H_2O_2 as a generator system (in TRIS-HCl pH 8.65) for reactive oxygen species. The value of the in vitro antioxidant activity (%AA) for each sample was calculated using the following equation:

$$AA(\%) = \frac{I_0 - I_S}{I_0} \times 100 \qquad (2)$$

where I_0 = CL intensity for control (the reaction mixture without sample) at t = 5 s; and I_s = CL intensity in the presence of the tested sample, at t = 5 s. For each sample, the %AA results were reported as the mean values of three CL assays.

2.4.4. Investigation of Urease Inhibitory Activity of Andropogon-Derived AgNPs

The urease-inhibiting activity of *Andropogon* metallic nanoparticles was evaluated by measuring electrical conductivity (Cobra3 Chem-Unit, Phywe System GmbH, Göttingen, Germany), as described in [34,35]. Urease (urea amidohydrolase, EC 3.5.1.5) is the enzyme that catalyzes the hydrolysis of urea into ammonia and carbon dioxide; the ions arising from this reaction's products increase a medium's conductivity. The resulting conductivity value is closely correlated with the rate of urea hydrolysis reaction. Therefore, the reaction rate can be evaluated as the variation in time of conductivity. A volume of 50 µL of urease solution [in 50% glycerine (1000 U/mL)] was added to 40 mL of 1.6% urea solution, under continuous stirring. Then, after 300 s, a volume of 1 mL of the sample was added into this reaction mixture. The electrical conductivity evolution was monitored on the PC using the measureAPP software (https://www.phywe.com/sensors-software/measurement-software-apps/measureapp-the-free-measurement-software-for-all-devices-and-operating-systems_2274_3205/, accessed on 1 April 2023).

3. Results and Discussion

3.1. Evaluation of Total Phenolic and Total Flavonoid Contents of Andropogon halepensis Extract

Phenolic compounds are bioactive phyto-molecules that have grafted one or multiple hydroxyl groups on an aromatic ring and are responsible for scavenging free radicals, which explains their antioxidant properties. Flavonoids are one type of phenolic compounds with a wide range of bioactivities.

The TPC of the Andropogon halepensis aqueous leaf extract, determined using the Folin–Ciocalteu method, was 20.83 ± 0.96 mg GAE/g dry extract. Shah et al. had previously reported a TPC value of 20.3 ± 1.5 mg GAE/g dry extract for the aerial parts of *S. halepense* aqueous extract [36], and Mohammed et al. [37] reported a polyphenol content of 20.78 ± 1.83 mg GAE/g dry extract for an aqueous extract of aerial parts of Beta vulgaris.

The TFC value of the *Andropogon halepensis* aqueous leaf extract was TFC = 9.12 ± 0.62 mg RE/g dry extract, a value which is comparable with the results for other extracts produced [38].

3.2. Optical Characterization of Andropogon-Derived Materials

The samples were optically characterized by UV-Vis absorption spectra and also by FTIR ATR spectroscopy. Moreover, chromatic characterization was carried out.

The UV-Vis absorption spectra (Figure 2) of *A. halepensis*-derived silver nanoparticles showed a single strong peak for each type of AgNPs, located at 402 nm, 419 nm, 423 nm, and 447 nm for AgNPs_2n, AgNPs_1n, AgNPs_2o, and AgNPs_1o, respectively. In this wavelength range of 400–500 nm, the vegetal extract did not show any peaks. These findings are consistent with those found in previous works [39–41]. The above-mentioned maxima were only due to metallic NPs formation and are generally called SPR (surface plasmon Resonance) bands, which are produced by the electromagnetic field which induces the collective oscillation of the electrons in the conduction band on the nanoparticles' surface [42]. According to the Mie theory [43], the presence of only one SPR band indicates the formation of spherical AgNPs. The SPR peaks of our old samples broadened and shifted to higher wavelengths, due to their aggregation tendency, indicating an increase in AgNPs size. On the contrary, narrow peaks at shorter wavelengths indicated a decrease in AgNPs size [44]. Thus, the dimensions of AgNPs_1n and AgNPs_2o appeared to be very close. These assumptions were further confirmed by DLS, zeta potential measurements, and SEM analysis. In addition, the phyto-extract showed UV-Vis absorption bands characteristic for polyphenolic compounds (335 nm), and for carbohydrates, and the aromatic amino acid residues of proteins (247–270 nm) [34].

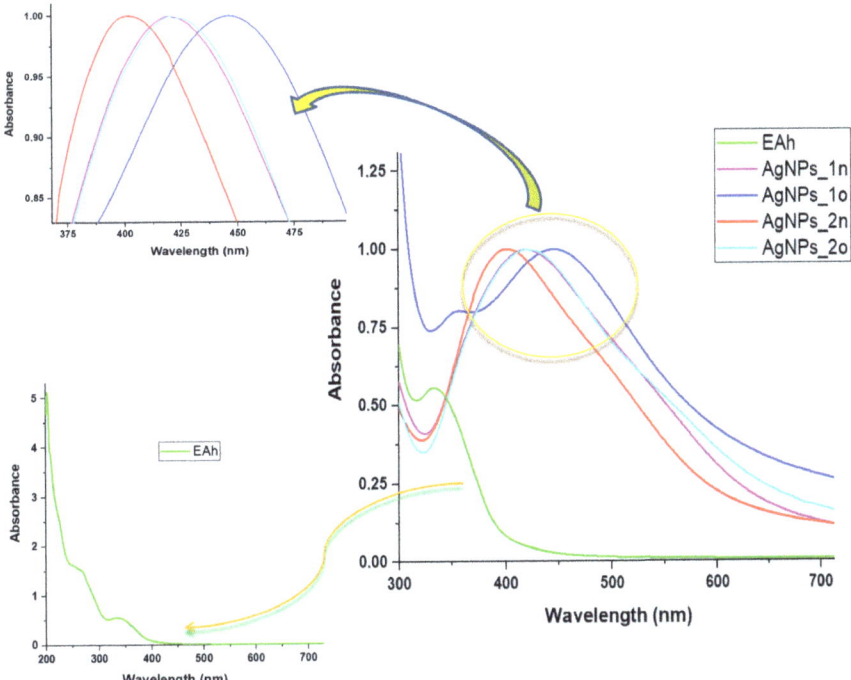

Figure 2. Comparative presentation of UV-Vis absorption spectra of *Andropogon*-derived samples. All AgNPs spectra were normalized at their maximum. The top inset shows the SPR bands of the *Andropogon*-derived AgNPs. The bottom inset shows the spectrum of *Andropogon halepensis* aqueous extract (EAh).

The chromatic parameters and the color change values (ΔE^*) of the silk materials are displayed in Table 2. For the silk samples treated with *A. halepensis* extract and its derived AgNPs, the value of L^* decreased compared to the pristine spider silk samples (S), indicating that the whiteness of S decreased after the treatments. The values of the chromatic parameters a^* and b^* of the spider silk samples have increased after treatment, indicating a slight shift to red and yellow, respectively. So, the color of the samples was in the yellow–greenish–green range. The color difference (ΔE^*) had small values after the treatment of fresh *Andropogon*-derived AgNPs. Greater values of ΔE^* were obtained for spider silk treated with 18 months-aged AgNPs (S_AgNPs_1o and S_AgNPs_2o).

Andropogon-derived AgNPs can be used in the treatment of spider silk–based fabrics/materials when the aim is not to change the color of the fibers but to preserve the color and only slightly intensify its tones. This can be an important aspect in their use in the conservation/treatment of the colored objects, including, for example, the treatment of heritage objects, as mentioned in [45].

Table 2. Chromatic parameters of untreated and treated spider silk samples.

Color Parameters	Spider Silk Sample					
	S	S_EAh	S_AgNPs_1n	S_AgNPs_1o	S_AgNPs_2n	S_AgNPs_2o
L*	98.05	97.74	97.00	88.77	96.72	91.15
a*	−6.97	−5.25	−5.21	−5.25	−5.19	−6.63
b*	5.53	7.63	5.75	5.76	5.75	6.95
ΔL*	-	−0.31	−1.05	−9.28	−1.33	−6.9
Δa*	-	1.72	1.76	1.72	1.78	0.34
Δb*	-	2.1	0.22	0.23	0.22	1.42
ΔE*	-	2.73	2.06	9.44	2.23	7.05

The UV-Vis absorption spectra of the silk materials in our study presented the specific signatures of phyto-nanosilver in the case of the spider silk treated with silver nanoparticles (Figure 3). After treatment with the vegetal extract EAh, the spectrum of spider silk shifted to longer wavelengths. A bathochromic shift was also observed in the case of the spectra of the spider silk treated with aged silver nanoparticles compared to fresh ones.

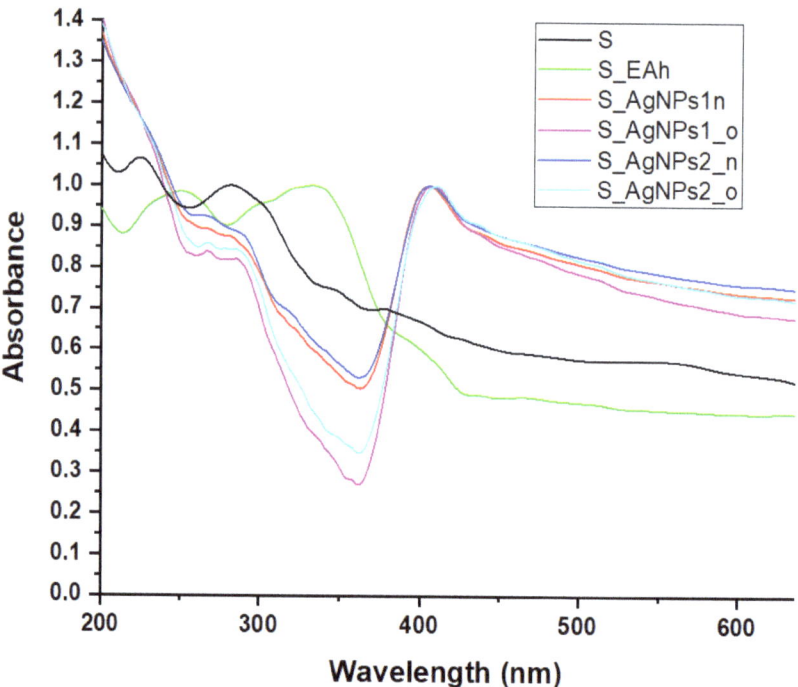

Figure 3. Comparative presentation of UV-Vis absorption spectra of the spider silk samples untreated (S) and treated with plant extract (S_EAh) or with silver nanoparticles obtained by *Bio-Reaction 1* (S_AgNPs_1n and S_AgNPs_1o) and *Bio-Reaction 2* (S_AgNPs_2n and S_AgNPs_2o).

The *Andropogon*-derived samples were further investigated with FTIR ATR spectroscopy (Figures 4 and 5). The attributions of the main FTIR bands are displayed in Tables S1 and S2 (Supplementary Material).

The FTIR ATR spectra provided useful information about the phyto-compounds (created from the leaf extract) wrapping the surface of nanoparticles (see Figure 4).

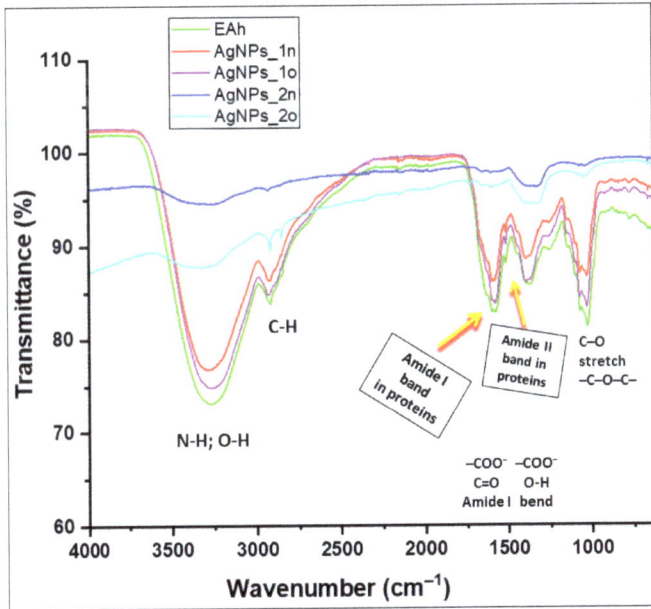

Figure 4. Comparative presentation of FTIR ATR spectra of *Andropogon halepensis* extract (EAh) and the derived AgNPs phyto-synthesized through *Bio-Reaction 1* (AgNPs_1n and AgNPs_1o) and *Bio-Reaction 2* (AgNPs_2n and AgNPs_2o). The index "n" refers to the "new" synthesized nanoparticles, while the index "o" refers to the "old" (18 months aged) ones.

The intense band centered at 3277 cm^{-1} in the FTIR ATR spectrum of the phyto-extract (assigned to bending and stretching vibrations of hydroxyl groups intermolecularly hydrogen-bonded in phenolic compounds and polysaccharides and also to the stretching vibrations of the primary and secondary amines) shifted to 3286 cm^{-1} and 3319 cm^{-1} in the FTIR ATR spectra of AgNPs_1n and AgNPs_2n, respectively. These findings suggest that phyto-synthesized silver nanoparticles carry, on their metallic surface, amino and hydroxyl groups associated through hydrogen bonding.

The samples EAh, AgNPs_1n, and AgNPs_2n presented FTIR ATR bands attributed to a C–H anti-symmetric stretching vibration, located at wavenumbers 2920, 2925, and 2930 cm^{-1}, respectively, and also the C–H symmetrical stretch vibration of alkyl chains located at wavenumbers 2845, 2851, and 2861 cm^{-1}, respectively (see Table S1 in the Supplementary Material).

In the fingerprint region, the narrow and medium band at 1369 cm^{-1} (assigned to carboxylates; phenol or tertiary alcohol, O–H bend; primary or secondary, O–H in-plan bend; C–H bend) in the EAh spectrum was weakened after the addition of silver nitrate and shifted to 1398 cm^{-1} in the AgNPs_1n spectrum and to 1354 cm^{-1} in the AgNPs_2n spectrum. Moreover, this band broadened in the case of AgNPs_2n.

The peak at 1261 cm^{-1}, ascribed to a primary or secondary O–H in-plan bend, aromatic ethers, and aryl–O stretching, was weakened after the bio-reaction in the case of AgNPs_1n and disappeared in the case of AgNPs_2n.

After the bioreducing reaction, the band at 1030 cm^{-1} observed in the spectrum of EAh (characteristic for the stretching mode of the –C–O group of polysaccharides and chlorophyll, the –C–O–C– group in ethers and secondary alcohols, and C–O bending in esters) shifted to 1074–1040 cm^{-1} in the AgNPs_1n spectrum. In addition, these bands were weakened in the nanoparticle spectrum.

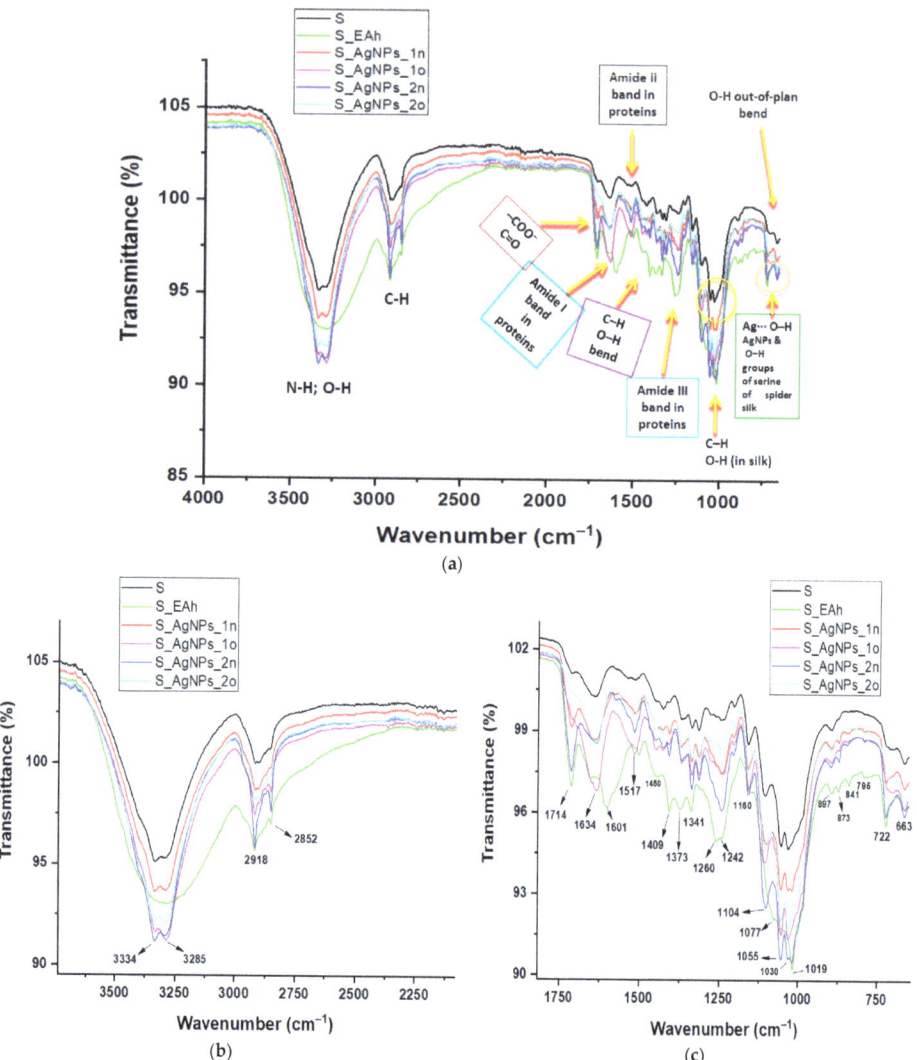

Figure 5. Comparative presentation of FTIR ATR spectra of the spider silk samples untreated (S) and treated with plant extract (S_EAh) or with silver nanoparticles obtained by *Bio-Reaction 1* (S_AgNPs_1n and S_AgNPs_1o) and *Bio-Reaction 2* (S_AgNPs_2n and S_AgNPs_2o) (**a**). Insets show the magnified regions of the FTIR ATR spectra of the biocomposites (**b**,**c**).

The characteristic vibration band for amide I (due to carbonyl stretch in proteins) in the spectrum of the vegetal extract shifted towards longer wavenumbers after the addition of the silver nitrate solution. Moreover, this band weakened and broadened in the AgNPs_2n spectrum. In addition, the small peak assigned to amide II due to N–H bending and C–N stretching in proteins disappeared in the AgNPs_2n spectrum.

As observed, the shape of the FTIR ATR spectra of the silver nanoparticles obtained through bio-reaction (1)—AgNPs_1n and AgNPs_1o—closely resembled the spectrum of the phyto-extract EAh, while the FTIR ATR spectra of the silver nanoparticles obtained through bio-reaction (2)—AgNPs_2n and AgNPs_2o—were much different from those of the samples previously mentioned.

After 18 months, some spectral changes occurred in the spectrum of the silver nanoparticles.

FTIR ATR spectra confirmed the development of phyto-AgNPs. This analysis suggested the involvement of phyto-molecules originated from the *Andropogon halepensis* extract, especially polyphenols, flavonoids, proteins, carboxylic acids, polysaccharides, chlorophylls, esters, and ethers, in the bioreduction of silver ions and the development of AgNPs. The obtained silver nanoparticles carried on their surfaces various functional groups (e.g., hydroxyl, carbonyl, and amino) associated by hydrogen bonding.

A comparative presentation of the FTIR ATR spectra of the untreated and treated spider silk samples is given in Figure 5.

After treatment with the phyto-extract, the FTIR ATR spectrum of spider silk underwent important changes in the wavenumber region of 3500–3000 cm^{-1} (Figure 5b), indicating the role of hydroxyl groups (belonging to the phenolic compounds and flavonoids from the phyto-extract) in the surface coverage of the spider silk fibers. Moreover, in the fingerprint region (Figure 5c), there were many changes in the Amide I, II, and III bands in the spider silk proteins. In addition, more bands appeared in the region 897–722 cm^{-1} in the spectrum of S_EAh, indicating the presence of hydroxyl groups at the surface of the spider silk fibers.

Similar changes occurred in the same regions in the spectra of the spider silk treated with metallic nanoparticles.

It can be observed that there were changes in the spectra of spider silk treated with the "old" AgNPs compared to those of silk treated with the fresh ones (Figure 5a–c).

3.3. Evaluation of the Zeta Potential of the Phyto-Derived Metallic Nanoparticles

Zeta potential measurements revealed the negative charges of the phyto-nanometallic particles due to the presence of carboxylate and hydroxyl groups on their surfaces, arising from proteins and polyphenols, respectively (as demonstrated by UV-Vis and FTIR ATR spectroscopy). Figure 6 displays a comparative presentation of the zeta potential values for the *Andropogon*-derived AgNPs. The particles AgNPs_2 were more stable than AgNPs_1, since they had a more negative value for the zeta potential; therefore, electrostatic repulsion was more pronounced in this case. AgNPs_1n presented a moderate physical stability ($\xi = -17.7 \pm 7.41$ mV), while AgNPs_2n showed a good stability ($\xi = -29.3 \pm 7.70$ mV). These values obtained for our developed AgNPs are in line with other reports in the scientific literature. For instance, Vishwasrao et al. prepared AgNPs from sapota (*Manilkara zapota*) pomace extract, with a moderate stability ($\xi = -13.41 \pm 0.02$ mV) [46]. Malaka et al. [47] synthesized AgNPs from *Zea mays* husk extract, with a zeta potential value of -28.7 mV.

After 18 months, the phyto-nanometallic particles had aggregated, a fact shown by the decrease in ξ magnitude.

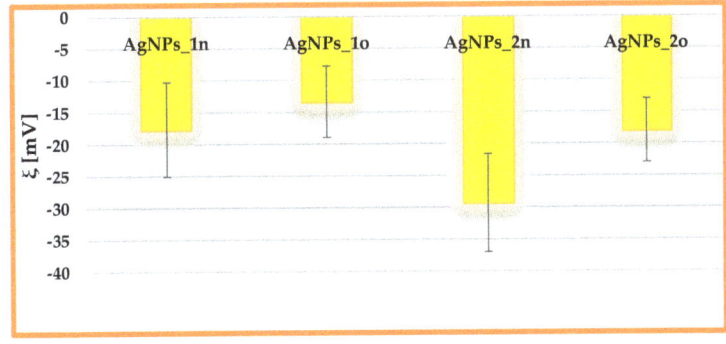

Figure 6. Comparative presentation of the electrokinetic potential of the *Andropogon*-derived silver nanoparticles. The AgNPs samples obtained through Bio-Reaction 1 (AgNPs_1n and AgNPs_1o) are placed next to those obtained through Bio-Reaction 2 (AgNPs_2n and AgNPs_2o) by alternating the aged samples with the fresh ones.

3.4. Size, Morphological, and Compositional Characterization of Spider Silk Biocomposites and Their Building Blocks

The average particle size (Zav, nm) values of the *Andropogon*–derived nanoparticles and the polydispersity indices (PdI), estimated by dynamic light scattering (DLS) measurements, are displayed in Figure 7. The values of Zav were in the range 24–68 nm, with a PdI index between 0.267 and 0.329 (Figure 7a). The smallest size was recorded for AgNPs_2n (24.36 nm), which was also the most stable sample (see Section 3.3).

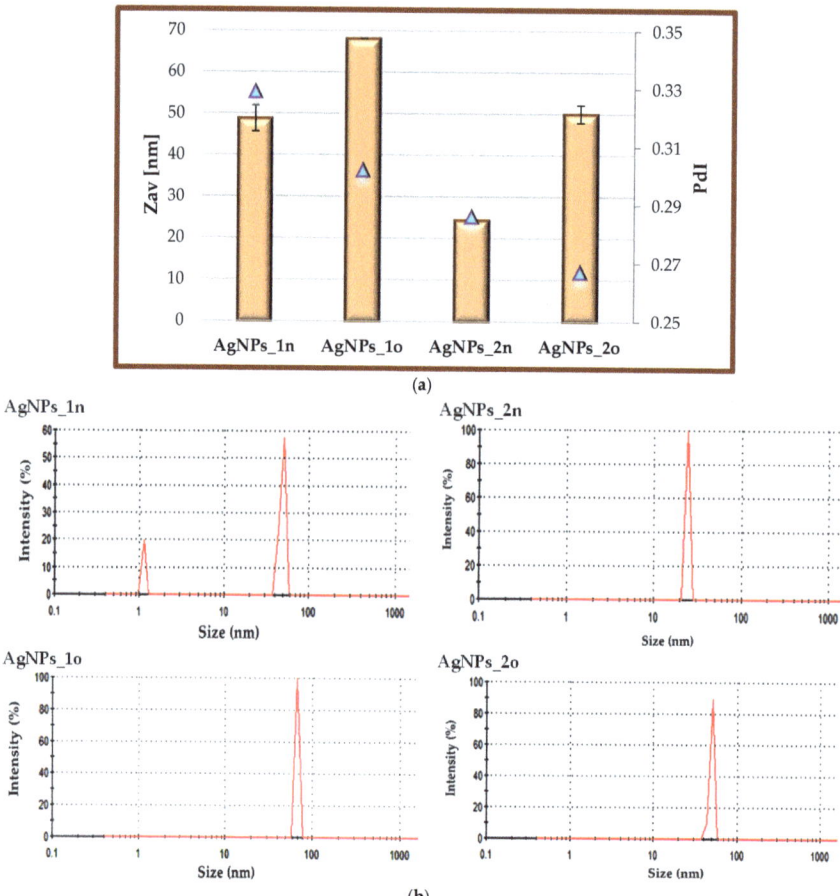

Figure 7. (a) Comparative presentation of the average particle size (Zav, nm) and PdI index of "green"-developed silver nanoparticles, estimated by dynamic light scattering (DLS) measurements; (b) Size distribution profiles of particle population for all types of phyto-developed AgNPs. For comparison, the aged AgNPs samples (AgNPs_1o and AgNPs_2o) are arranged next to the fresh ones (AgNPs_1n and AgNPs_2n).

Figure 7b displays a narrow size distribution profile of the particle population for all the types of phyto-developed AgNPs. It can be seen that the size of both types of nanoparticles increased after 18 months.

The obtained results are within the range of AgNPs size reported for AgNPs synthesized from *Azadirachta Indica* L. leaves extract (20–50 nm) [48] and from aqueous extracts of aerial parts of *Astragalus spinosus* (10–60 nm) [49].

The morphological aspects of the samples were analyzed by SEM analysis (Figure 8). The SEM micrographs revealed spherical metallic nanoparticles (AgNPs_1 and AgNPs_2), confirming that the *A. halepensis* extract acted both as a bioreducing and capping agent in the development of the silver nanoparticles. The spherical shape of the phyto-developed AgNPs was predicted by the UV-Vis absorption spectra (see Section 3.2). Similar results were also reported by Sathishkumar et al. for the "green" synthesis of silver nanoparticles using *Morinda citrifolia* L. [50]. The SEM image of the sample S_EAh showed silk bundle fibers with a phyto-matrix. The silk treatment with the silver nanoparticles under ultrasound irradiation led to silk nanofibers functionalized with AgNPs. The SEM micrographs showed the presence of AgNPs on the surface of the treated silk nanofibers: S_AgNPs_1n, S_AgNPs_1o, S_AgNPs_2n, and S_AgNPs_2o. The functionalization of spider silk was more effective when freshly prepared AgNPs were used compared to aged ones (see samples S_AgNPs_1n and S_AgNPs_2n).

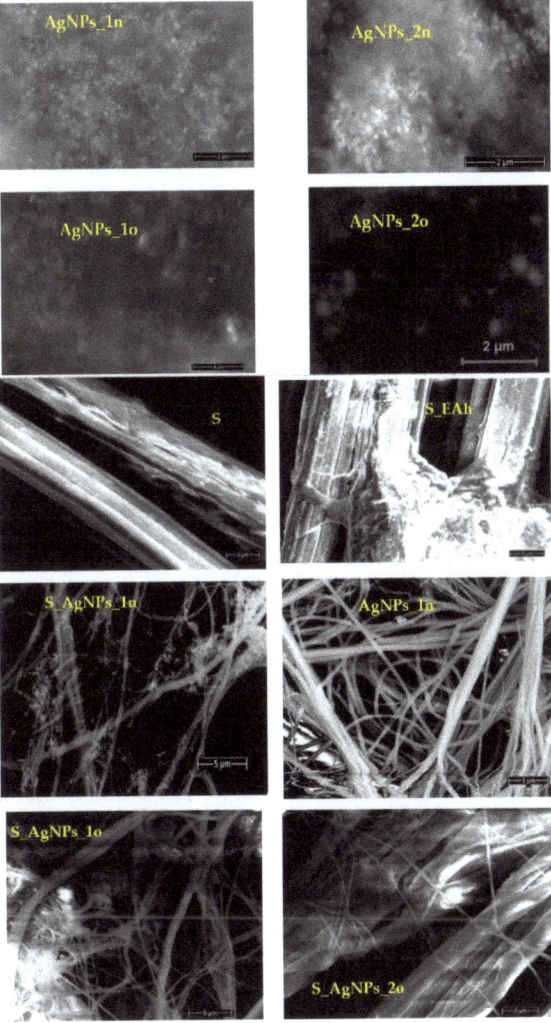

Figure 8. The SEM images of the phyto-developed AgNPs (AgNPs_1n, AgNPs_1o, AgNPs_2n, and AgNPs_2o) and of the spider silk fibers (S) and spider silk biocomposites with plant extract (S_EAh) or with silver nanoparticles (S_AgNPs_1n, S_AgNPs_1o, S_AgNPs_2n, and S_AgNPs_2o).

The EDX spectra (Figure S1, Supplementary Material) evidenced the presence of silver in the phyto-nanometallic samples (AgNPs_1n, AgNPs_1o, AgNPs_2n, and AgNPs_2o) and in the silk-biocomposites S_AgNPs_1n, S_AgNPs_1o, S_AgNPs_2n, and S_AgNPs_2o. Other chemical elements (O, Na, P, K, and Cl) came from the plant extract.

3.5. Investigation of Antioxidant Activity of Andropogon Extract and Its Derived AgNPs

The in vitro antioxidant activity of the *A. halepensis* extract and the derived phyto-metallic particles was evaluated using the chemiluminescence technique. The AA% values varied in the range 74–81%, and the results are displayed in Figure 9. The ability of the *Andropogon* extract to scavenge free radicals was due to the presence of bioactive molecules such as polyphenols and flavonoids (see Sections 3.1 and 3.2). The good antioxidant properties of the developed AgNPs were retained even after 18 months. Only slight changes in the antioxidant activity values occurred.

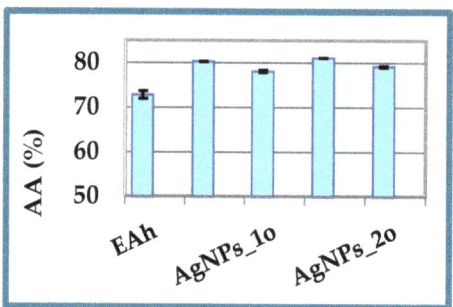

Figure 9. The antioxidant activity of the *Andropogon halepensis* extract (EAh) and the phyto-metallic particles obtained by *Bio-Reaction 1* (AgNPs_1n and AgNPs_1o) and *Bio-Reaction 2* (AgNPs_2n and AgNPs_2o), estimated using the chemiluminescence technique.

The antioxidant activity of the phyto-developed silver nanoparticles reached values greater than the activity of the precursor extract due to the presence, on their surface, of phyto-compounds (capping agents) from the vegetal extract and due to a nanosized effect that generates many reaction centers for the capturing of free radicals. This behavior of phyto-nanometallic particles compared to their phyto-extract precursors was highlighted in our previous studies [3,34]. In the current experiment, the best antioxidant properties were obtained for AgNPs_2n due to two key factors: (1) a good physical stability (see Figure 6) and (2) the smallest dimension (see Figure 7). The CL results were in good agreement with the DLS and zeta potential measurements.

3.6. Investigation of Effects of Andropogon Extract and Derived AgNPs on Urease Activity

The potential urease-inhibiting effect of the *Andropogon halepensis* extract and its developed phyto-metallic particles was estimated by means of the conductometric technique (Figure S2, Supplementary Material), which is a very simple approach.

Urea is an important organic molecule used in dermatology as a moisturizer and a keratolytic agent as well as in wound healing [51]. Urea is broken down by the enzyme urease into carbon dioxide and ammonia, resulting in the alkalization of the medium. Urease has been identified as a virulence factor for several microbial pathogens [52], being involved in many diseases' pathogenesis in both humans and animals. Thus, finding novel urease inhibitors is an important research topic.

In the absence of a sample, during the first 300 s of the reaction evolution in our experiment (Figure S1, in Supplementary Material), the conductivity continuously increased due to the formation of ions during the hydrolysis reaction of urea as a result of the action of urease. After the addition of the samples, at t = 300 s, the conductivity drastically increased at this point due to the introduction of ions, originating from the sample, in the reaction medium. After this moment, if the conductivity increased, the sample either did not have any inhibitory actions (in the case of EAh) or had a slight inhibitory effect (see AgNPs_1n). In the case of sample AgNPs_2n, no further increase in conductivity was observed because the enzyme was inhibited and no transformation of the substrate (urea) was possible. The results shown in Figure S2 (Supplementary Material) reveal that the vegetal extract exhibited no inhibitory action on urease, while the phyto-derived AgNPs presented an inhibitory effect.

As seen in Figure S2 (Supplementary Material), AgNPs_1n showed a slight inhibitory action, while AgNPs_2n exhibited a strong inhibitory action against urease. Practically, urease lost its enzymatic activity in the presence of AgNPs_2n, and no further urea transformation was possible. This behavior could be explained by the smallest size of AgNPs_2n, allowing for a better interaction with urease's active site, which could have changed urease's conformation and then blocked this active site with nanoparticles [53].

The functionalization of spider silk with AgNPs_2n, which have an inhibitory effect on urease, could be exploited in the development of medical textiles that can be applied on skin treated with urea, thus preventing the latter's decomposition.

4. Conclusions

This study presented a simple eco-design of biocomposites containing silk webs from spider *Pholcus phalangioides* and silver-based nanoparticles phyto-synthesized (using two phyto-reactions) from an aqueous extract of *Andropogon halepensis* leaves, a common weed. To the best of our knowledge, *Andropogon halepensis* has not been previously used for the development of such biocomposites.

Through this study, we highlighted the importance of using natural resources such as spider silk and plants, even invasive ones, to build biocomposites with potential applicability in the biomedical field.

The phyto-synthesis of the two types of AgNPs was demonstrated using spectral methods (UV-Vis and FTIR ATR) and by means of SEM analysis. The FTIR ATR investigation showed the involvement of organic molecules such as polyphenols, flavonoids, proteins, carboxylic acids, polysaccharides, chlorophylls, esters, and ethers, originating from the *Andropogon halepensis* extract, in the "green" synthesis of silver nanoparticles. The nanoscale dimension of the phyto-metallic particles was demonstrated by the DLS measurements and confirmed by the SEM images. Their zeta potential values indicated moderate-to-good physical stability. Certain changes in the properties of the obtained AgNPs were observed after 18 months, especially regarding their size and zeta potential values. On the contrary, their antioxidant properties did not undergo significant changes. The biophysical characterization of the phyto-metallic nanoparticles showed that sample AgNPs_2 (with the smallest ratio of phyto-extract/AgNO$_3$) presented the best results, including in terms of its antioxidant properties and urease inhibitory activity.

The *Andropogon*-derived metallic nanoparticles were used to develop biocomposites with *Pholcus phalangioides* spider silk by means of a simple "green" bottom–up approach. The optical characterization (UV-Vis and FTIR ATR) and SEM analysis demonstrated the functionalization of spider silk with *Andropogon*-derived AgNPs. The FTIR ATR analysis suggested the key role of proteins, phenolic compounds, and flavonoids in the development of biocomposites.

The biophysical characterization of the phyto-metallic nanoparticles showed that the fresh AgNPs obtained via the second type of bio-reaction (sample AgNPs_2n) presented the best results in terms of physical stability, antioxidant activity, and urease-inhibiting activity. Moreover, this sample was the most efficient in the functionalization of spider silk. Thus, the spider silk-based composites developed in this study obtained by means of silk's functionalization with AgNPs_2 could be further exploited in development of medical textiles that can be applied on skin treated with urea, thus preventing urea decomposition, and assuring the scavenging of free radicals. Moreover, our "green"-developed biocomposites can be applied also in various fields, such as biomedicine, environmental remediation, and nanotechnology, where these materials could potentially make significant contributions.

Supplementary Materials: The following supporting information can be downloaded at https://www.mdpi.com/article/10.3390/ma17051225/s1: Table S1: FTIR ATR band assignment for vegetal extract and phyto-derived metallic nanoparticles; Table S2: FTIR ATR band assignment for untreated and treated spider silk; Figure S1: The EDX spectra of all the samples: *Andropogon halepensis* extract (EAh) and *Andropogon*-derived AgNPs (AgNPs_1n, AgNPs_1o, AgNPs_2n, and AgNPs_2o), spider silk (S), and spider silk composites with plant extract (S_EAh) or with silver nanoparticles (S_AgNPs_1n, S_AgNPs_1o, S_AgNPs_2n, and S_AgNPs_2o), at 10 kV; and Figure S2: Investigation of urease inhibitory action of the *Andropogon halepensis* extract and the developed phyto-metallic nanoparticles, estimated by the conductometric method. Refs. [54–61] are cited in the Supplementary Materials.

Author Contributions: Conceptualization, M.-E.B.-P., C.N. and S.A.; investigation, M.-E.B.-P., C.N. and B.B.; methodology, M.-E.B.-P., C.N., B.B. and S.A.; supervision, M.-E.B.-P., C.N. and S.A.; writing—original draft, M.-E.B.-P., C.N. and S.A.; writing—review and editing, M.-E.B.-P., C.N., B.B. and S.A. All authors have read and agreed to the published version of the manuscript.

Funding: C.N. acknowledges the Ministry of Research, Innovation and Digitization, CNCS—UEFISCDI, for the financial support, with project PN 23280501/2023.

Institutional Review Board Statement: Not applicable.

Informed Consent Statement: Not applicable.

Data Availability Statement: The data were included in the text.

Acknowledgments: This paper was supported by the Romanian National Authority for Scientific Research, the Ministry of Research, Innovation and Digitization, CNCS—UEFISCDI, Project PN 23280501/2023.

Conflicts of Interest: The authors declare no conflicts of interest.

References

1. Selmani, A.; Kovačević, D.; Bohinc, K. Nanoparticles: From synthesis to applications and beyond. *Adv. Colloid Interface Sci.* **2022**, *303*, 102640. [CrossRef] [PubMed]
2. Mohanta, Y.K.; Mishra, A.K.; Panda, J.; Chakrabartty, I.; Sarma, B.; Panda, S.K.; Chopra, H.; Zengin, G.; Moloney, M.G.; Sharifi-Rad, M. Promising applications of phyto-fabricated silver nanoparticles: Recent trends in biomedicine. *Biochem. Biophys. Res. Commun.* **2023**, *688*, 149126. [CrossRef] [PubMed]
3. Khan, F.; Shariq, M.; Asif, M.; Siddiqui, M.A.; Malan, P.; Ahmad, F. Green Nanotechnology: Plant-Mediated Nanoparticle Synthesis and Application. *Nanomaterials* **2022**, *12*, 673. [CrossRef]
4. Zgura, I.; Badea, N.; Enculescu, M.; Maraloiu, V.A.; Ungureanu, C.; Barbinta-Patrascu, M.-E. Burdock-Derived Composites Based on Biogenic Gold, Silver Chloride and Zinc Oxide Particles as Green Multifunctional Platforms for Biomedical Applications and Environmental Protection. *Materials* **2023**, *16*, 1153. [CrossRef] [PubMed]
5. Barbinta-Patrascu, M.-E.; Bacalum, M.; Antohe, V.A.; Iftimie, S.; Antohe, S. Bio-nanoplatinum phyto-developed from grape berries and nettle leaves: Potential adjuvants in osteosarcoma treatment. *Rom. Rep. Phys.* **2022**, *74*, 601.
6. Liu, X.; Rapakousiou, A.; Deraedt, C.; Ciganda, R.; Wang, Y.; Ruiz, J.; Gu, H.; Astruc, D. Multiple applications of polymers containing electron-reservoir metal-sandwich complexes. *Chem. Commun.* **2020**, *56*, 11374–11385. [CrossRef] [PubMed]
7. Khayal, A.Y.; Ullah, I.; Nughman, M.; Shah, S.M.; Ali, N. Elemental, antimicrobial and antioxidant activities of a medicinal plant *Sorghum halepense*. *Pure Appl. Biol. PAB* **2019**, *8*, 795–803. [CrossRef]
8. Travlos, I.S.; Montull, J.M.; Kukorelli, G.; Malidza, G.; Dogan, M.N.; Cheimona, N.; Antonopoulos, N.; Kanatas, P.J.; Zannopoulos, S.; Peteinatos, G. Key Aspects on the Biology, Ecology and Impacts of Johnsongrass [*Sorghum halepense* (L.) Pers] and the Role of Glyphosate and Non-Chemical Alternative Practices for the Management of This Weed in Europe. *Agronomy* **2019**, *9*, 717. [CrossRef]
9. Baerson, S.R.; Dayan, F.E.; Rimando, A.M.; Nanayakkara, N.D.; Liu, C.J.; Schroder, J.; Fishbein, M.; Pan, Z.; Kagan, I.A.; Pratt, L.H.; et al. A functional genomics investigation of allelochemical biosynthesis in *Sorghum bicolor* root hairs. *J. Biol. Chem.* **2008**, *283*, 3231–3247. [CrossRef]
10. Liu, Y.A.N.; Zhang, C.; Wei, S.; Cui, H.; Huang, H. Compounds from the subterranean part of Johnsongrass and their allelopathic potential. *Weed Biol. Manag.* **2011**, *11*, 160–166. [CrossRef]
11. Czarnota, M.A.; Rimando, A.M.; Weston, L.A. Evaluation of root exudates of seven sorghum accessions. *J. Chem. Ecol.* **2003**, *29*, 2073–2083. [CrossRef] [PubMed]

12. Shah, M.A.R.; Khan, R.A.; Ahmed, M. Sorghum halepense (L.) Pers rhizomes inhibitory potential against diabetes and free radicals. *Clin. Phytoscience* **2021**, *7*, 19. [CrossRef]
13. Huang, H.; Liu, Y.; Meng, Q.; Wei, S.; Cui, H.; Zhang, C. Flavonolignans and other phenolic compounds from Sorghum halepense (L.) Pers. *Biochem. Syst. Ecol.* **2010**, *38*, 656–658. [CrossRef]
14. Peerzada, A.M.; Ali, H.H.; Hanif, Z.; Bajwa, A.A.; Kebaso, L.; Frimpong, D.; Iqbal, N.; Namubiru, H.; Hashim, S.; Rasool, G.; et al. Eco-biology, impact, and management of Sorghum halepense (L.) Pers. *Biol. Invasions* **2017**, *16*, 955–973. [CrossRef]
15. Loddo, D.; Masin, R.; Otto, S.; Zanin, G. Estimation of base temperature for Sorghum halepense rhizome sprouting. *Weed Res.* **2012**, *52*, 42–49. [CrossRef]
16. Rambabu, B.; Patnaik, K.R.; Srinivas, M.; Abhinayani, G.; Sunil, J.; Ganesh, M.N. Evaluation of central activity of ethanolic flower extract of Sorghum halpense on albino rats. *J. Med. Plants.* **2016**, *4*, 104–107.
17. Xie, B.; Wu, X.; Ji, X. Investigation on the Energy-Absorbing Properties of Bionic Spider Web Structure. *Biomimetics* **2023**, *8*, 537. [CrossRef] [PubMed]
18. Römer, L.; Scheibel, T. The elaborate structure of spider silk: Structure and function of a natural high performance fiber. *Prion* **2008**, *2*, 154–161. [CrossRef]
19. Morley, E.L.; Robert, D. Electric Fields Elicit Ballooning in Spiders. *Curr. Biol.* **2018**, *28*, 2324–2330. [CrossRef]
20. Salehi, S.; Koeck, K.; Scheibel, T. Spider Silk for Tissue Engineering Applications. *Molecules* **2020**, *25*, 737. [CrossRef]
21. Johansson, J.; Rising, A. Doing What Spiders Cannot a Road Map to Supreme Artificial Silk Fibers. *ACS Nano* **2021**, *15*, 1952–1959. [CrossRef] [PubMed]
22. Petrovic, S.M.; Barbinta-Patrascu, M.-E. Organic and Biogenic Nanocarriers as Bio-Friendly Systems for Bioactive Compounds' Delivery: State-of-the Art and Challenges. *Materials* **2023**, *16*, 7550. [CrossRef] [PubMed]
23. Gu, Y.; Yu, L.; Mou, J.; Wu, D.; Zhou, P.; Xu, M. Mechanical properties and application analysis of spider silk bionic material. *e-Polymers* **2020**, *20*, 443–457. [CrossRef]
24. V&A (Victoria and Albert Museum). Golden Spider Silk. Available online: https://www.vam.ac.uk/articles/golden-spider-silk (accessed on 29 January 2024).
25. Kiseleva, A.P.; Kiselev, G.O.; Nikolaeva, V.O.; Seisenbaeva, G.; Kessler, V.; Krivoshapkin, P.V.; Krivoshapkina, E.F. Hybrid Spider Silk with Inorganic Nanomaterials. *Nanomaterials* **2020**, *10*, 1853. [CrossRef] [PubMed]
26. Bielecki, K. Spider Silk Structures as Seen with a Scanning Electron Microscope. *Bios* **2009**, *80*, 170–175. [CrossRef]
27. Nam, S.; Hillyer, M.B.; Condon, B.D.; Lum, J.S.; Richards, M.N.; Zhang, Q. Silver nanoparticle-infused cotton fiber: Durability and aqueous release of silver in laundry water. *J. Agric. Food Chem.* **2020**, *68*, 13231–13240. [CrossRef]
28. Wilfred, I. Society of Dyers and Colourists. In *Colour for Textiles: A User's Handbook*; Society of Dyers and Colourists: Bradford, UK, 1993; ISBN 0901956562.
29. Singleton, V.L.; Orthofor, R.; Raventos, R.M.L. Analysis of total phenols and other oxidation substrates and antioxidants by means of Folin–Ciocâlteu reagent. *Methods Enzymol.* **1999**, *299*, 152–178.
30. Geremu, M.; Tola, Y.B.; Sualeh, A. Extraction and determination of total polyphenols and antioxidant capacity of red coffee (*Coffea arabica* L.) pulp of wet processing plants. *Chem. Biol. Technol. Agric.* **2016**, *3*, 25. [CrossRef]
31. Nichita, C.; Stamatin, I. The Antioxidant Activity of The Biohybrides Based on Carboxylated/Hydroxylated Carbon Nanotubes-Flavonoid Compounds. *Dig. J. Nanomater. Biostructures* **2013**, *8*, 445–455.
32. Nichita, C.; Al-Behadili Mikhailef, F.; Vasile, E.; Stamatin, I. Silver nanoparticles synthesis. Bioreduction with gallic acid and extracts from *Cyperus rotundus* L. *Dig. J. Nanomater. Biostructures* **2020**, *15*, 419–433. [CrossRef]
33. Romanian Pharmacopoeia Commission National Medicines Agency. *Romanian Pharmacopoeia*, 10th ed.; Medical Publishing House: Bucharest, Romania, 1993; p. 335.
34. Barbinta-Patrascu, M.-E.; Chilom, C.; Nichita, C.; Zgura, I.; Iftimie, S.; Antohe, S. Biophysical insights on Jack bean urease in the presence of silver chloride phytonanoparticles generated from *Mentha piperita* L. leaves. *Rom. Rep. Phys.* **2022**, *74*, 605.
35. Barbinta-Patrascu, M.-E. Biogenic nanosilver from *Cornus mas* fruits as multifunctional eco-friendly platform: "green" development and biophysical characterization. *J. Optoelectron. Adv. Mater.* **2020**, *22*, 523–528.
36. Shah, M.A.R.; Khan, R.A.; Ahmed, M. Phytochemical analysis, cytotoxic, antioxidant and anti-diabetic activities of aerial parts of Sorghum halepense. *Bangladesh J. Pharmacol.* **2019**, *14*, 144–151. [CrossRef]
37. Mohammed, E.A.; Abdalla, I.G.; Alfawaz, M.A.; Mohammed, M.A.; Al Maiman, S.A.; Osman, M.A.; Yagoub, A.E.A.; Hassan, A.B. Effects of Extraction Solvents on the Total Phenolic Content, Total Flavonoid Content, and Antioxidant Activity in the Aerial Part of Root Vegetables. *Agriculture* **2022**, *12*, 1820. [CrossRef]
38. Khorasani Esmaeili, A.; Mat Taha, R.; Mohajer, S.; Banisalam, B. Antioxidant Activity and Total Phenolic and Flavonoid Content of Various Solvent Extracts from In Vivo and In Vitro Grown *Trifolium pratense* L. (Red Clover). *BioMed Res. Int.* **2015**, *2015*, 643285. [CrossRef] [PubMed]
39. Bindhani, B.K.; Panigrahi, A.K. Biosynthesis and Characterization of Silver Nanoparticles (SNPs) by using Leaf Extracts of *Ocimum Sanctum* L (Tulsi) and Study of its Antibacterial Activities. *J. Nanomed. Nanotechnol. S.* **2015**, *6*, 2157–7439. [CrossRef]
40. Ashraf, J.M.; Ansari, M.A.; Khan, H.M.; Alzohairy, M.A.; Choi, I. Green synthesis of silver nanoparticles and characterization of their inhibitory effects on ages formation using biophysical techniques. *Sci. Rep.* **2016**, *6*, 20414. [CrossRef]

41. Syafiuddin, A.; Salmiati-Hadibarata, T.; Salim, M.R.; Kueh, A.B.H.; Sari, A.A. A purely green synthesis of silver nanoparticles using *Carica papaya*, *Manihot esculenta*, and *Morinda citrifolia*: Synthesis and antibacterial evaluations. *Bioprocess Biosyst. Eng.* **2017**, *40*, 1349–1361. [CrossRef]
42. Eustis, S.; El-Sayed, M.A. Why gold nanoparticles are more precious than pretty gold: Noble metal surface plasmon resonance and its enhancement of the radiative and nonradiative properties of nanocrystals of different shapes. *Chem. Soc. Rev.* **2006**, *35*, 209–217. [CrossRef]
43. Kora, A.J.; Beedu, S.R.; Jayaraman, A. Size-controlled green synthesis of silver nanoparticles mediated by gum ghatti (*Anogeissus latifolia*) and its biological activity. *Org. Med. Chem. Lett.* **2012**, *2*, 17. [CrossRef]
44. Alharbi, N.S.; Alsubhi, N.S.; Felimban, A.I. Green synthesis of silver nanoparticles using medicinal plants: Characterization and application. *J. Radiat. Res. Appl. Sci.* **2022**, *15*, 109–124. [CrossRef]
45. Lite, M.C.; Constantinescu, R.R.; Tănăsescu, E.C.; Kuncser, A.; Romaniţan, C.; Lăcătuşu, I.; Badea, N. Design of Green Silver Nanoparticles Based on *Primula Officinalis* Extract for Textile Preservation. *Materials* **2022**, *15*, 7695. [CrossRef]
46. Vishwasrao, C.; Momin, B.; Ananthanarayan, L. Green Synthesis of Silver Nano-particles Using Sapota Fruit Waste and Evaluation of Their Antimicrobial Activity. *Waste Biomass Valorization* **2018**, *10*, 2353–2363. [CrossRef]
47. Malaka, R.; Subramanian, S.; Muthukumarasamy, N.P.; Hema, J.A.; Sevanan, M.; Sambandam, A. Green synthesis of silver nanoparticles using *Zea mays* and exploration of its biological applications. *IET Nanobiotechnol.* **2016**, *10*, 288–294. [CrossRef]
48. Dakal, T.C.; Kumar, A.; Majumdar, R.S.; Yadav, V. Mechanistic basis of antimicrobial actions of silver nanoparticles. *Front. Microbiol.* **2016**, *7*, 1831. [CrossRef] [PubMed]
49. Ghabban, H.; Alnomasy, S.F.; Almohammed, H.; Al Idriss, O.M.; Rabea, S.; Eltahir, Y. Antibacterial, Cytotoxic, and Cellular Mechanisms of Green Synthesized Silver Nanoparticles against Some Cariogenic Bacteria (*Streptococcus mutans* and *Actinomyces viscosus*). *J. Nanomater.* **2022**, *2022*, 9721736. [CrossRef]
50. Sathishkumar, G.; Gobinath, C.; Karpagam, K.; Hemamalini, V.; Premkumar, K.; Sivarama-krishnan, S. Phyto-synthesis of silver nano scale particles using *Morinda citrifolia* L. and its inhibitory activity against human pathogens. *Coll. Surf. B.* **2012**, *95*, 235–240. [CrossRef]
51. Verzì, A.E.; Musumeci, M.L.; Lacarrubba, F.; Micali, G. History of urea as a dermatological agent in clinical practice. *Int. J. Clin. Pract.* **2020**, *74*, 187. [CrossRef] [PubMed]
52. Mora, D.; Arioli, S. Microbial Urease in Health and Disease. *PLOS Pathog.* **2014**, *10*, e100447. [CrossRef] [PubMed]
53. Ponnuvel, S.; Subramanian, B.; Ponnuraj, K. Conformational change results in loss of enzymatic activity of jack bean urease on its interaction with silver nanoparticle. *Protein J.* **2015**, *34*, 329–337. [CrossRef]
54. Coates, J. Interpretation of Infrared Spectra, a Practical Approach. In *Encyclopedia of Analytical Chemistry*; Meyers, R.A., Ed.; John Wiley & Sons Ltd.: Chichester, UK, 2000; pp. 1–23.
55. Kiseleva, A.; Nestor, G.; Östman, J.R.; Kriuchkova, A.; Savin, A.; Krivoshapkin, P.; Krivoshapkina, E.; Seisenbaeva, G.A.; Kessler, V.G. Modulating Surface Properties of the *Linothele fallax* Spider Web by Solvent Treatment. *Biomacromolecules* **2021**, *22*, 4945–4955. [CrossRef] [PubMed]
56. Goto, H.; Kikuchi, R.; Wang, A. Spider Silk/Polyaniline Composite Wire. *Fibers* **2016**, *4*, 12. [CrossRef]
57. Gorinstein, S.; Park, Y.S.; Heo, B.G.; Namiesnik, J.; Leontowicz, H.; Leontowicz, M.; Ham, K.S.; Cho, J.Y.; Kang, S.G. A comparative study of phenolic compounds and antioxidant and antiproliferative activities in frequently consumed raw vegetables. *Eur. Food Res. Technol.* **2009**, *228*, 903–911. [CrossRef]
58. Barbinta-Patrascu, M.-E.; Badea, N.; Bacalum, M.; Ungureanu, C.; Suica-Bunghez, I.R.; Iordache, S.M.; Pirvu, C.; Zgura, I.; Maraloiu, V.A. 3D hybrid structures based on biomimetic membranes and *Caryophyllus aromaticus*—"Green" synthesized nano-silver with improved bioperformances. *Mater. Sci. Eng. C* **2019**, *101*, 120–137. [CrossRef] [PubMed]
59. Seo, S.J.; Das, G.; Shin, H.S.; Patra, J.K. Silk Sericin Protein Materials: Characteristics and Applications in Food-Sector Industries. *Int. J. Mol. Sci.* **2023**, *24*, 4951. [CrossRef] [PubMed]
60. Barbinta-Patrascu, M.-E.; Nichita, C.; Badea, N.; Ungureanu, C.; Bacalum, M.; Zgura, I.; Iosif, L.; Antohe, S. Biophysical aspects of bio-nanosilver generated from *Urtica dioica* Leaves and *Vitis vinifera* fruits' extracts. *Rom. Rep. Phys.* **2021**, *73*, 601.
61. Shameli, K.; Ahmad, M.B.; Jazayeri, S.D.; Shabanzadeh, P.; Sangpour, P.; Jahangirian, H.; Gharayebi, Y. Investigation of antibacterial properties silver nanoparticles prepared via green method. *Chem. Cent. J.* **2012**, *6*, 73. [CrossRef]

Disclaimer/Publisher's Note: The statements, opinions and data contained in all publications are solely those of the individual author(s) and contributor(s) and not of MDPI and/or the editor(s). MDPI and/or the editor(s) disclaim responsibility for any injury to people or property resulting from any ideas, methods, instructions or products referred to in the content.

Article

Enhancing Sustainability and Antifungal Properties of Biodegradable Composites: Caffeine-Treated Wood as a Filler for Polylactide

Aleksandra Grząbka-Zasadzińska [1], Magdalena Woźniak [2], Agata Kaszubowska-Rzepka [1], Marlena Baranowska [3], Anna Sip [4], Izabela Ratajczak [2] and Sławomir Borysiak [1,*]

1. Institute of Chemical Technology and Engineering, Poznan University of Technology, Berdychowo 4, 60-965 Poznan, Poland; aleksandra.grzabka-zasadzinska@put.poznan.pl (A.G.-Z.); kaszubowskaagata@gmail.com (A.K.-R.)
2. Department of Chemistry, Poznan University of Life Sciences, Wojska Polskiego 75, 60-625 Poznan, Poland; magdalena.wozniak@up.poznan.pl (M.W.); izabela.ratajczak@up.poznan.pl (I.R.)
3. Department of Silviculture, Poznan University of Life Sciences, Wojska Polskiego 42, 60-625 Poznan, Poland; marlenab@up.poznan.pl
4. Department of Biotechnology and Food Microbiology, Poznan University of Life Sciences, Wojska Polskiego 48, 60-625 Poznan, Poland; anna.sip@up.poznan.pl
* Correspondence: slawomir.borysiak@put.poznan.pl

Citation: Grząbka-Zasadzińska, A.; Woźniak, M.; Kaszubowska-Rzepka, A.; Baranowska, M.; Sip, A.; Ratajczak, I.; Borysiak, S. Enhancing Sustainability and Antifungal Properties of Biodegradable Composites: Caffeine-Treated Wood as a Filler for Polylactide. *Materials* 2024, *17*, 698. https://doi.org/10.3390/ma17030698

Academic Editor: Raul D. S. G. Campilho

Received: 3 January 2024
Revised: 26 January 2024
Accepted: 30 January 2024
Published: 1 February 2024

Copyright: © 2024 by the authors. Licensee MDPI, Basel, Switzerland. This article is an open access article distributed under the terms and conditions of the Creative Commons Attribution (CC BY) license (https:// creativecommons.org/licenses/by/ 4.0/).

Abstract: This study investigates the suitability of using caffeine-treated and untreated black cherry (*Prunus serotina* Ehrh.) wood as a polylactide filler. Composites containing 10%, 20%, and 30% filler were investigated in terms of increasing the nucleating ability of polylactide, as well as enhancing its resistance to microorganisms. Differential scanning calorimetry studies showed that the addition of caffeine-treated wood significantly altered the crystallization behavior of the polymer matrix, increasing its crystallization temperature and degree of crystallinity. Polarized light microscopic observations revealed that only the caffeine-treated wood induced the formation of transcrystalline structures in the polylactide. Incorporation of the modified filler into the matrix was also responsible for changes in the thermal stability and decreased hydrophilicity of the material. Most importantly, the use of black cherry wood treated with caffeine imparted antifungal properties to the polylactide-based composite, effectively reducing growth of *Fusarium oxysporum*, *Fusarium culmorum*, *Alternaria alternata*, and *Trichoderma viride*. For the first time, it was reported that treatment of wood with a caffeine compound of natural origin alters the supermolecular structure, nucleating abilities, and imparts antifungal properties of polylactide/wood composites, providing promising insights into the structure-properties relationship of such composites.

Keywords: polylactide; biocomposites; wood; filler modification; caffeine; structure; biological resistance

1. Introduction

An increasing interest in ecological products requires new, effective, and sustainable materials that reduce global pollution but also have additional user-friendly features. Polylactide (PLA) is a well-known biodegradable polymer that not only gets a lot of attention in the scientific world, but is also becoming more popular among consumers. Although biocompatible and biodegradable, PLA often does not provide the application properties required by users. One way out of this problem is to prepare composite materials. Therefore, many attempts have been made to produce fully biodegradable, sustainable, and satisfactory PLA-based composites in terms of functional properties.

PLA/wood composites are an emerging group of these kinds of materials. For this purpose, numerous species and sources including pine wood [1], hardwood and softwood pulps [2,3], wood flour [4], blends of aspen, oak, pine, basswood [5], and sawdust [6] have been used. Of course, the environmental impact of PLA/wood composites depends on various factors, such as the source of the wood. The use of non-native invasive species may also be an interesting solution to this problem [7]. Given the high prices of PLA, the use of relatively inexpensive by-products of wood processing and wood sourced from responsible forestry (removing invasive plant species) is believed to be more profitable from an economic point of view.

Composites of PLA and wood offer some attractive properties. They were found to be a waste source for the production of high quality 3D polymeric materials [6], reduce production costs, and compensate for the deficiencies of PLA [3,5,8,9]. Despite these obvious advantages, there are also some important issues regarding the use of wood in polymer composites. Achieving a uniform dispersion of wood particles within the PLA matrix may be challenging, and non-homogenous dispersion can lead to weak points, reducing the overall strength and performance of the composite. Moreover, wood tends to absorb moisture, which can lead to swelling, dimensional instability, and susceptibility to microbial attack or decay. Another aspect is ensuring good compatibility and adhesion between the hydrophilic nature of wood fibers and the hydrophobic nature of PLA, crucial to maintaining the mechanical properties of the composites [10,11]. This is also associated with the formation of a proper supermolecular structure of the material. The nucleating abilities and structural aspects influence the mechanical properties of the composite and the enhanced crystalline structure often results in improved mechanical performance, increased heat resistance, and barrier properties [12,13].

The main advantage of PLA composites with wood filler is their biocompatibility and biodegradability under composting conditions. However, this is also their disadvantage: PLA-based materials lack biological protection and thus, are prone to destruction by microorganisms, also during shelf life [14].

There are several strategies that are employed to enhance the biological resistance of PLA/wood composites against microorganisms. Busolo et al. [15] demonstrated the effectiveness of PLA compounds with silver-based engineered clays against *Staphylococcus aureus*. Harnet et al. [16] further improved this protection by using multilayer films made of polyethylene-functionalized chloroethylene on adhesive materials to inhibit the spread of *Escherichia coli*. The scope of antimicrobial protection was expanded by modifying the PLA surface with water-resistant polymer brushes, which significantly inhibited bacterial adhesion [17]. Finally, Ahmed et al. [18] explored the use of plastic PLA films based on essential oils to suppress *S. aureus* and *Campylobacter jejuni*, with cinnamon and clove oil films showing the highest antimicrobial activity. These approaches can be grouped into the following types: (a) chemical treatments that include applying fungicides, biocides, or other chemical treatments that can help inhibit microbial growth and decay in the composite material; (b) natural antimicrobials like essential oils, extracts from plants, or other biobased substances added to the composite materials that can deter microbial attack; (c) surface coatings with antimicrobial properties applied to the surface of composites that can provide an extra layer of protection against microorganisms; (d) nanoparticles with antimicrobial properties; (e) filler modifications through processes like acetylation, heat treatment, or chemical modification [19,20]. The last method mentioned, chemical modification of wood seems to be particularly interesting. Depending on the reagents used, it can be relatively inexpensive and enhance the resistance of the composite to decay and microbial attack.

Caffeine has been found to be an effective and environmentally friendly wood protection agent against decay, fungi, and termites [21,22]. Caffeine has been found to interact with wood components with varying degrees of binding strength depending on the specific component. These interactions are influenced by the composition of wood hemicellulose, with certain sugar monomers and polymers showing higher binding potential [23]. Caffeine treatment at a concentration of 2% has been shown to be effective against wood-destroying fungi and at 1% in some cases [21]. It has also been found to improve the fungistatic properties of thermally modified pinewood [22].

Understanding and optimizing all of the above-mentioned aspects of PLA composites is essential for tailoring their properties to meet specific application requirements, ensuring better performance and functionality of the material in various industries such as packaging, biomedical devices, automotive components, and more. In an ideal situation, the filler used for PLA composites should not only enhance the crystallization process, but also provide antimicrobial protection. The use of caffeine-treated wood as a filler for a PLA matrix has not yet been reported and shows great economic and ecological potential. Consequently, the objective of this investigation was to determine whether the addition of caffeine-treated or untreated black cherry wood, a nonnative invasive wood species in Polish ecosystems, has a positive influence on the nucleating behavior of PLA composites and can provide antifungal protection against bacteria and fungi.

2. Materials and Methods

2.1. Materials

Polylactide (PLA) 2500HP (Nature Works, Blair, NE, USA) was used as a polymer matrix. The content of D-lactide in this PLA is 0.5% [24]. The wood material was obtained from black cherry (*Prunus serotina* Ehrh.) shrubs collected in 2022, in the Arboretum in Zielonka (52°33′18.2″ N 17°05′49.7″ E). The wood particles had sizes between 0.5 and 1.0 mm. Caffeine in form of powder with purity $\geq$ 99% (Sigma Aldrich, Darmstadt, Germany) was used for wood treatment.

2.2. Wood Treatment

Black cherry wood was treated with caffeine as previously described in the paper by Tomczak et al. [25]: a 2% aqueous solution of caffeine was prepared and mixed with wood at 20 °C for 2 h. After treatment, the wood was dried until a constant mass was achieved.

2.3. Microorganism Preparation

The antimicrobial activity of composite samples was assessed against the microorganisms listed in Table 1. Before the test, the bacterial strains were cultured in nutrient broth (Oxoid, UK) at a temperature of 35 $\pm$ 2 °C for 24 h. Next, petri dishes with Mueller-Hilton Agar (Graso Biotech, Owidz, Poland) were inoculated with a suspension of these cultures at a concentration of 10^6 CFU/mL. In the antifungal test, discs with a diameter of 10 mm were cut from the 7-day-old plate fungal cultures grown in YGC (Yeast Extract Glucose Chloramphenicol) medium and placed in the center of petri dishes with Sabouraud Dextrose Agar (Graso Biotech, Poland).

Table 1. Microorganisms used as indicators to assess the antimicrobial activity of the tested materials.

Microorganisms	Category	Origin
Listeria innocua	Gram-positive bacteria non-pathogenic bacteria with strong adhesive properties found in many production plants;	ATCC 33090
Listeria monocytogenes	pathogenic bacteria; parasite of animals and humans;	ATCC 19111
	Fungi	
Fusarium oxysporum	toxin-producing fungi	environmental isolate
Fusarium culmorum	toxin-producing fungi	environmental isolate
Alternaria alternata	fungi that cause spoilage of various raw materials and food products.	environmental isolate
Trichoderma viride	fungi used in biological protection of plants against pathogenic organisms	environmental isolate

2.4. Preparation of Composites

Before any processing, the materials were first dried in a convection dryer at 65 °C for at least 12 h.

Polylactide composites with wood filler were obtained using a two-step process. First, the corotating twin screw extruder was used. The diameter of the screws was 16 mm and the L/D ratio was 40. The compression ratio, that is, the height of the screw channel in the feeding zone to its height in the dosing zone, was 2:1. The temperatures of the extruder zones were in the range of 190–205 °C. In this process, the filler was introduced into the dosing zone using an external hopper at a speed of 8 rpm. The speed of the extruder screws was 80 rpm. The composites were extruded through an extrusion head with a circular nozzle with a diameter of 3 mm, cooled on a silicone conveyor, and then cut to obtain granules.

The second processing step included injection molding. An ENGEL 80/20 HLS injection molding machine (Schwertberg, Austria) was used. The clamping force of the machine was 200 kN and the diameter of the screw was 22 mm. The temperatures of the injection molding machine zones were set in the range of 190–210 °C. The injection speed was 75 mm/s, the pressing time was 6 s, and the cooling time was 35 s. A single-cavity mold was used as a cold runner with a square socket with dimensions of 100 mm × 100 mm × 2.6 mm (length × width × thickness). An external cooling unit was used to maintain the mold temperature at 35 °C. At least 10 specimens of each composite were molded.

A diagram of the workflow carried out in this research is presented in Figure 1 and the composites prepared in this study are presented in Table 2.

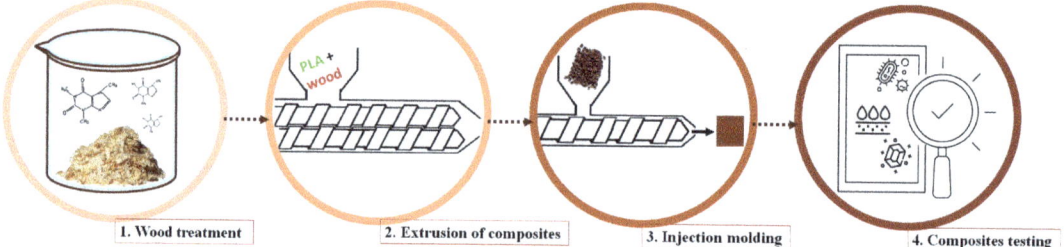

Figure 1. Workflow of the research.

Table 2. Prepared PLA-based samples.

Polymer Matrix	Filler Type	Filler Content	Sample Name
PLA	-	0%	PLA
	untreated wood	10%	PLA + 10 W
		20%	PLA + 20 W
		30%	PLA + 30 W
	caffeine-treated wood	10%	PLA + 10 CW
		20%	PLA + 20 CW
		30%	PLA + 30 CW

2.5. X-ray Diffraction

The supermolecular structure of the materials was analyzed by means of wide angle X-ray scattering—XRD (Rigaku, Tokyo, Japan). CuKα radiation at 40 kV and 30 mA anode excitation was used. The XRD patterns were recorded for 2θ angles from 5 to 40° in steps of 4°/min. The diffraction patterns were used to define the supermolecular structure of the materials, including the polymorphic form of the fillers used. The characteristic peaks were assigned to respective crystal planes.

2.6. Differential Scanning Calorimetry

The thermal characteristics of the produced composites were assessed through differential scanning calorimetry (DSC) using a Netzsch DSC 200 instrument (Netzsch, Bavaria, Germany), operating in a nitrogen atmosphere. To conduct non-isothermal crystallization studies, the samples underwent an initial heating phase from 40 to 200 °C at a rate of

20 °C/min, followed by a 3-min dwell at this temperature to eliminate any prior thermal or mechanical influences. Subsequently, the samples were cooled to 40 °C at a rate of 5 °C/min and held at this temperature for 1 min. This entire procedure was repeated. The enthalpy of the crystallization (H) values determined were employed to calculate the degree of crystallization (crystal conversion), denoted as α, using Equation (1):

$$\alpha = \frac{\int_0^t \left(\frac{dH}{dt}\right) \times dt}{\int_0^\alpha \left(\frac{dH}{dt}\right) \times dt} \tag{1}$$

Based on curves of α = f(t), the half-time of crystallization (t0.5) was determined when the crystal conversion reached 50%. The degree of crystallinity (X_c) of the materials was calculated according to Equation (2).

$$X_c = \left(\frac{\Delta Hm - \Delta Hcc}{\Delta Hm° \times \left(1 - \frac{\%wt\ filler}{100}\right)}\right) \times 100 \tag{2}$$

where: ΔHm is the melting enthalpy, ΔHcc is the enthalpy of the phase produced during cold crystallization, ΔHm° is the melting enthalpy of a 100% crystalline polylactide (93.0 J/g [26]) and %wt. filler is the percentage of filler weight.

In addition, the melting (Tm), crystallization (Tc), and cold crystallization (Tcc) temperatures were defined.

2.7. Fourier Transformation Infrared Spectroscopy

Untreated and caffeine-treated wood were mixed with KBr (Merck KGaA, Darmstadt, Germany) at a 1:200 weight ratio and analyzed in the form of pellets using a Nicolet iS5 Fourier transform spectrophotometer (Thermo Fisher Scientific, Waltham, MA, USA).

2.8. Thermogravimetric Analysis

Thermogravimetric curves for the composites (samples of approximately 10 mg) were obtained using a NETZSCH TG 209F3 apparatus (Netzsch, Germany). Measurements were carried out at a heating rate of 10 °C/min over the temperature range of 50 to 500 °C under nitrogen flow (20 mL/min).

2.9. Polarized Light Microscopy

The isothermal crystallization of the PLA in the presence of untreated and treated wood was conducted using a Linkam TP93 hot stage optical microscope (Linkam, Redhill, UK) and a Nikon Eclipse polarizing optical microscope, model LV100POL (Nikon, Melville, NY, USA) equipped with a Panasonic CCD GP-KR222 camera (Panasonic, Newark, NJ, USA).

The PLA granules were cut into small pieces (ca. 30 µg). One piece was placed on a microscopic glass and then a few particles of filler were strategically placed on its surface. The samples underwent initial heating to 200 °C at a rate of 20 °C/min and were maintained at this temperature for 3 min to erase any prior thermal and/or mechanical effects. After melting, the sample was pressed with another microscopic glass. The samples were then cooled at 10 °C/min to 136 °C, at which point isothermal crystallization of the PLA was possible. Photographs were taken every 5 s and analyses were performed using the ToupView program (ToupTek, Hangzhou, China). On that basis, the size and growth rate of the PLA spherulites was determined.

2.10. Contact Angle Measurements

The measurements of the contact angle were performed using a Dataphysics OCA 200 instrument (DataPhysics Instruments, Filderstadt, Germany). All measurements were carried out at 21 ± 0.5 °C. A drop of 0.2 mL of water was dispensed from the capillary in a controlled manner and placed onto the solid surface of the sample.

2.11. Antimicrobial Properties

PLA-based samples (sterilized by autoclaving at 121 °C for 15 min) of approximately 1 × 1 cm were put on petri dishes with Mueller-Hilton Agar (Graso Biotech, Poland) and Sabouraud Dextrose Agar (Graso Bioetch, Poland) inoculated with bacterial or fungal strains. The samples tested with bacteria were incubated at 35 ± 2 °C for 24 h, while the samples tested with fungi were kept at a temperature of 30 ± 2 °C for 28 days. After incubation, the antimicrobial activity of the tested materials was assessed. For this purpose, measurements of the inhibition zones formed around the polymeric samples were made. Measurements were performed using a Computer Scanning System (MultiScaneBase v14.02). The results were expressed in millimeters.

For the antifungal activity testing, mold growth was assessed on the 7th, 14th, 21st, and 28th day of incubation. It was also observed whether the mold grew on the tested samples. The degree of colonization of the materials tested by fungi was determined according to the four-level scale presented in the paper by Tomczak et al. [25].

In the second set of antimicrobial experiments, the test samples were placed on the surface of nutrient agar (Oxoid, UK) and then covered with Mueller-Hilton Agar medium (Graso Biotech, Poland) inoculated with 10^6 CFU/mL of the tested bacterial strains. The plates were incubated under the same conditions as in the first series of experiments. The antibacterial activity of the tested materials was also evaluated in a similar way.

3. Results and Discussion

3.1. Structure of Lignocellulosic Materials

In the first step, black cherry wood fillers (both untreated and caffeine-treated) used for the formation of composites were tested in terms of supermolecular structure, using the XRD method.

As shown in Figure 2, both fillers exhibited a typical structure for cellulosic materials: the wide peaks were present at 2θ 15.7° and 22.5°. However, in untreated wood, some smaller peaks, around 15° and 30°, were observed that may be a result of the presence of substances of low molecular weight naturally occurring in wood [27]. These peaks were not registered on the diffractogram of the caffeine-treated wood, meaning that they had been removed during treatment. The crystallinity degree of both types of filler was comparable, around 25%.

Then, PLA and its composites were tested. The obtained diffraction patterns are also presented in Figure 2.

The diffractogram of PLA shows a wide amorphous halo with a maximum intensity around 2θ = 16°, but no crystalline peaks were present. In all composite samples, the amorphous halo was present, but only in the sample containing 30% of untreated wood was a small crystalline peak at 2θ = 16.3° present. It corresponds to the planes (110) and (200) of the ortho-rhombic α-crystalline phase of PLA [28]. Wood is known to be a nucleating agent [29,30], but in this case, it is rather a result of composite processing—at a higher filler loading, shear forces occurring during injection molding are higher than in case of lower filler composites, enhancing the crystallization process [31,32]. A wide maximum around 2θ = 22.5°, more pronounced for composites with a high content of wood (un- and treated), resulted from the presence of cellulose, a component of lignocellulosic mass [30].

The treatment of cherry wood with caffeine caused changes in its chemical structure, which was confirmed by FTIR analysis, presented in form of spectrum in Figure 3.

The most important changes in the spectrum of treated wood compared to the spectrum of untreated wood confirm the presence of bands characteristic for caffeine molecules. The peaks at 1703 and 1650 cm^{-1} can be attributed to the vibrations of C=O and/or the N-H bonds of the acetamide group of amide I [25,33]. In turn, the peaks at 1555 and 765 cm^{-1} can be ascribed to the vibration of the N–H and/or C–N groups of amide II [34,35].

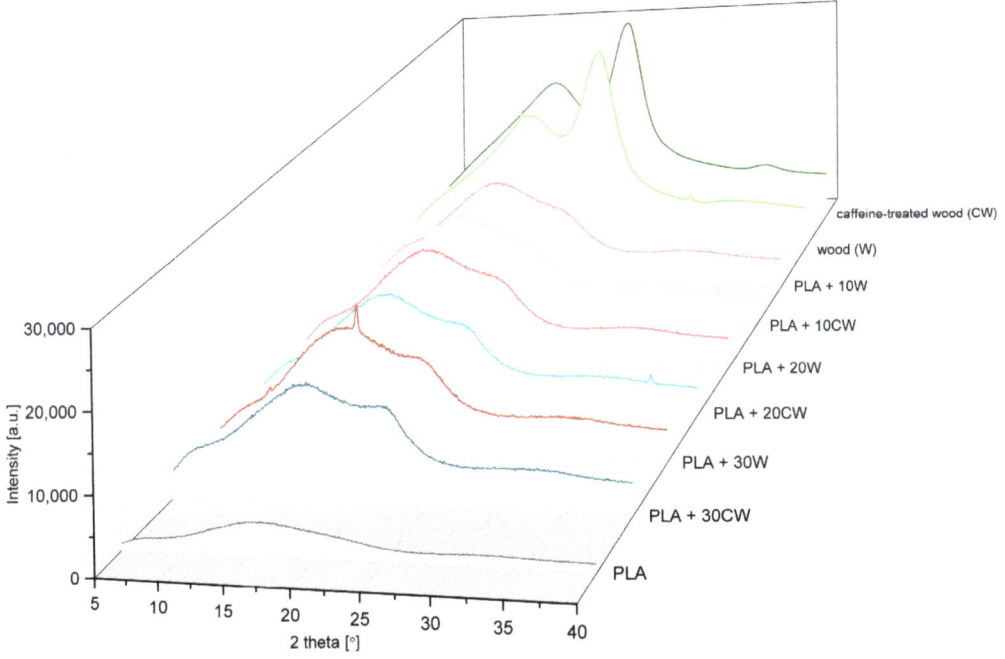

Figure 2. Diffractograms of fillers, PLA matrix and composites.

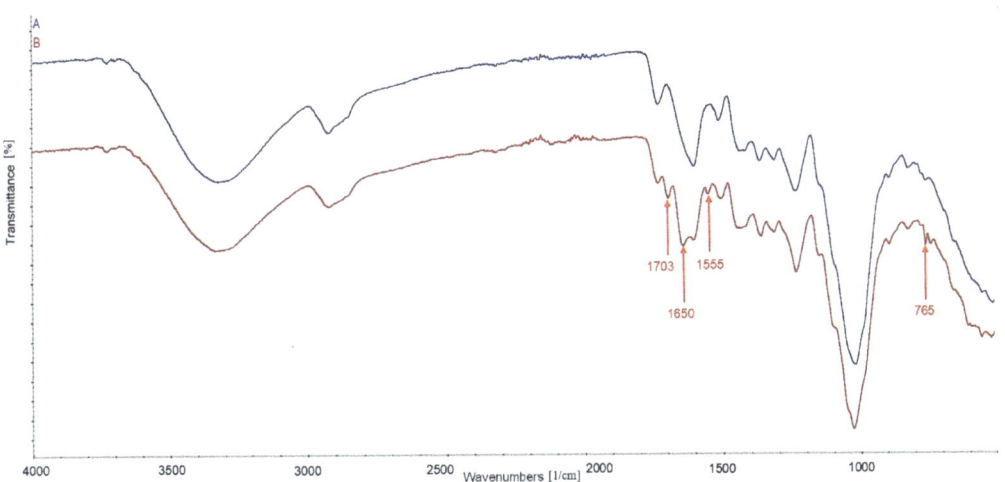

Figure 3. Spectra of untreated cherry wood (A) and cherry wood treated with caffeine (B).

3.2. Nucleation Ability of Wood Fillers and Crystallization Process in Composites

A differential scanning calorimetry study was carried out to determine the phase transitions that take place in polylactide in the presence of a filler, un- and treated with caffeine. As a result of the experiment, thermograms showing heating and cooling curves were obtained. Exemplary thermograms of PLA and composites are shown in Figure 4, while all data collected from the second runs are summarized in Table 3.

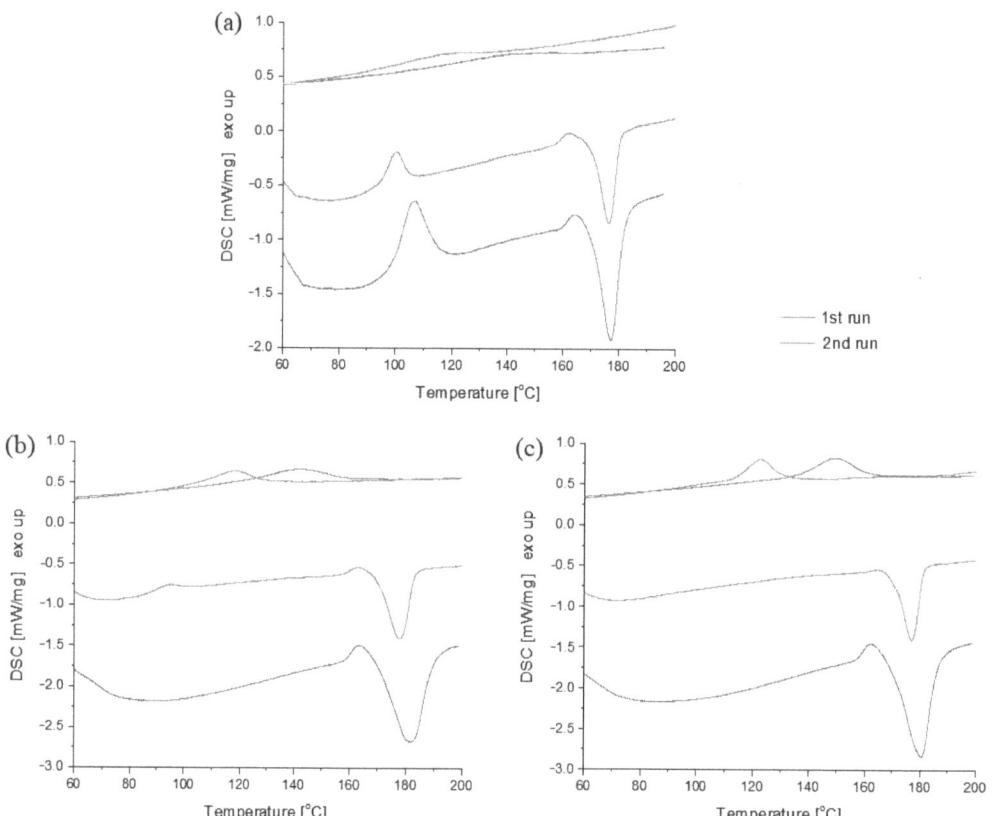

Figure 4. Exemplary thermograms of: (**a**) PLA, (**b**) PLA + 20 W, and (**c**) PLA + 20 CW.

Table 3. Summarized data from second runs of DSC.

Sample	Tcc [°C]	Tm [°C]	Tc [°C]	ΔHm [J/g]	Xc [%]
PLA	100.3	176.5	nd	37.39	28
PLA + 10 W	97.8	178.3	95.9	58.72	60
PLA + 20 W	95.0	177.7	102.5	46.23	56
PLA + 30 W	96.5	177.7	102.6	46.65	64
PLA + 10 CW	98.4	177.6	99.9	53.01	45
PLA + 20 CW	nd	176.9	104.9	38.29	52
PLA + 30 CW	nd	176.3	106.9	37.35	58

nd—not detected.

For polylactide, cold crystallization and melting peaks were observed, but no crystallization occurred during cooling. This is the typical behavior of this polymer, which was also described in other papers. The temperatures at which the phase transitions occurred were also in line with those of the literature [36,37].

The presence of cold crystallization is a characteristic feature of PLA. The fact that PLA does not fully crystallize during cooling makes it possible that during subsequent heating, below the melting point, disordered polymer chains can rearrange into ordered crystalline structures. The degree and rate of cold crystallization in PLA varies according to the cooling rate, processing conditions, molecular weight, and the presence of additives and nucleating agents [38].

In the case of composites, the cold crystallization peak was present in all samples containing untreated wood and in the PLA with 10% CW. It can also be seen that for PLA, the Tcc was greater than 100 °C, while for composites it was in the range of 95.0–98.4 °C. This suggests that the energy required for the sample to recrystallize is lower if it contains filler, thus it enhances the crystallization process.

The addition of untreated wood was not found to have any important influence on the melting temperature, ranging from 176.5 °C to 178.3 °C. It was proven that the stress applied to the polymer melt during processing induces crystallization, which in turn influences the melting temperature. Also, the crystal perfection influences this parameter: the better the formation of crystal structures, the higher the melting temperature [39,40]. Nonetheless, no changes suggesting differences in crystal ideality were observed in the present study.

The most noticeable changes were observed for the crystallization temperature. The unfilled PLA did not crystallize during cooling at all. This is a well-known fact that affects its application potential. It also shows that the cooling rate of PLA used in this study (5 °C/min) was still too fast to obtain a semicrystalline structure. However, the addition of fillers was favorable, inducing the crystallization process, which took place at 95.9 °C for PLA + 10 W and at around 102.5 °C for PLA with 20 W and 30 W. If caffeine-treated wood was used, the Tc parameter varied from 99.9 °C up to 106.9 °C, for 10 CW and 30 CW, respectively.

When differences in the crystallization behavior of samples are taken into account, it is no surprise that the crystallinity degree parameter was also highly affected by the amount and type of filler used. The degrees of crystallinity calculated indicate that the addition of each type of filler caused an increase of the parameter from 28% for PLA to 45–64% for composites. In composites with treated wood, the following relationship was observed: the higher the wood content, the greater the degree of crystallinity. This is also consistent with the values of Tc. In contrast, in composites with untreated wood, Xc did not depend on the filler content, but it was comparable, in the range of 56–64%.

The relationship between the crystallinity degree of a polymer composite and the filler content is often complex and may not be strictly proportional. It is a delicate balance between an increase in the number of nucleating points and a decrease in the polymer mobility. Several factors contribute to this nonlinear behavior, and the effect of filler content on crystallinity can vary depending on the specific characteristics of the polymer, the filler, and the processing conditions.

The uneven morphology of caffeine-treated wood offers more nucleating sites than the smooth surface of the wood particles. The number of nucleating sites at which the formation of spherulites begins is higher when more filler is present in the sample. Therefore, the crystallinity degree of the materials increases, which, from a technological point of view, is an important aspect. The ability to achieve a higher degree of crystallinity during a shorter time, without the need to anneal, enables easier production of elements using injection molding, extrusion, or blow molding processes [41,42].

On the basis of the thermograms, crystallization half-times were calculated, according to Equation (1). The curves used for calculations along with crystallization half-times are presented in Figure 5.

For composites with untreated fillers, half times were in the range of 4.0–4.4 min, while for modified composites, they were 3.1–3.8 min. It can be concluded that modification of the filler strongly affects the crystallization of PLA, causing a shortening of the conversion time and thus, indicating an acceleration of the crystallization process. There was also a tendency for composites with the lowest amount of filler to have the highest conversion time.

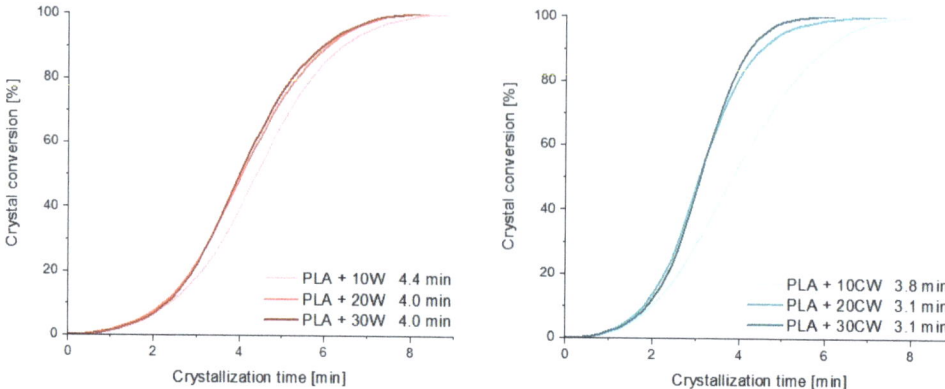

Figure 5. Crystal conversion curves.

The results of thermal analysis suggest that the modified filler offers more nucleating sites at which the crystallization process may begin. This is supported by the findings of Tomczak et al. [25], who stated that the surface of caffeine-treated wood is less smooth than that of pristine wood and has some grooves that may facilitate the formation of crystalline structures. This is confirmed by shorter crystallization half-times and higher crystallization temperature values. All of the fillers used were heterogeneous nucleants for the PLA matrix. When high loads of caffeine-treated wood were used (20% and 30%), crystallization during cooling was sufficient and no cold crystallization occurred. Furthermore, the crystallization temperature of the composites differed, reaching the highest values for PLA + 20 CW and PLA + 30 CW, once again proving the nucleating effect.

A study was carried out using a polarizing microscope to check the effect of the filler on the crystallization of polymer matrix. Photos taken during the experiment under isothermal conditions are shown in Figure 6.

For the unfilled PLA matrix and PLA with untreated wood, the crystallization process took place in bulk. However, in the composite samples with treated filler (PLA + CW), the formation of spherulites was observed not only in bulk, but also on the surface of the caffeine-treated wood. Interestingly, the use of that filler resulted in the formation of transcrystalline structures, TCL (shown with blue arrows). In addition, for the PLA + CW samples, a larger number of spherulites were formed. This suggests that this type of filler was a better nucleant for the polylactide matrix.

Observations made by means of microscopy support the findings of the DSC analysis (especially the conversion rate and crystallization temperature) that showed that caffeine-treated wood is an active nucleant for PLA. As a result, a well pronounced formation of the transcrystalline layer was observed. This may be the result of the removal of low-mass molecules present in untreated wood and, therefore, the incorporation of some roughness on the surface of the filler. The following analysis of the literature shows that surface topography may affect the nucleation of polymers in two main ways. First, the thermal stress that develops at the filler/polymer interface might induce the local orientation of the polymer chain segments, which enables easier nucleation. Second, the nucleation rate of the composite was found to be determined by changes in the free energy barrier and the diffusion activation energy under different interfacial interactions [43]. Crystallization on a smooth surface requires a higher energy, and when grooves are present, the free energy barriers necessary to initiate the nucleation process are lower [44,45].

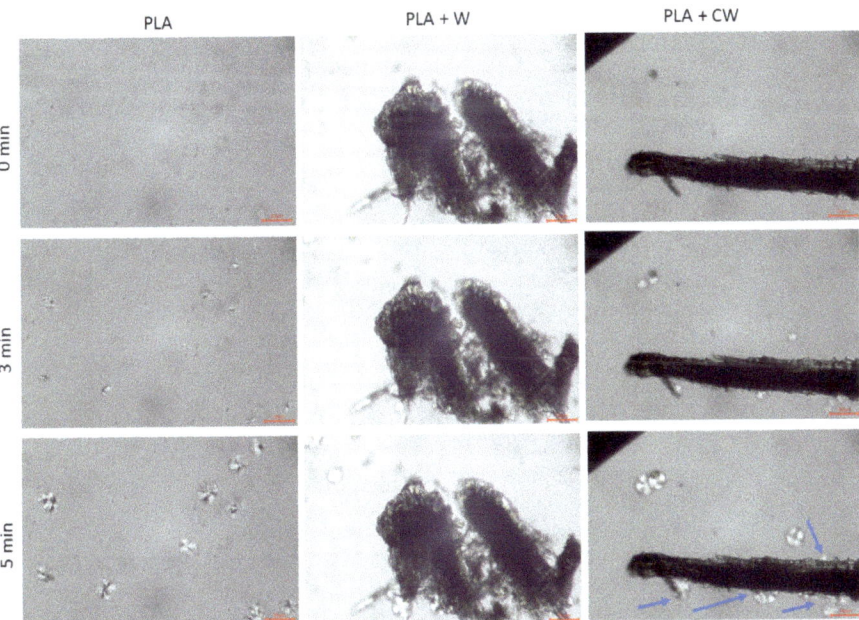

Figure 6. PLM photographs of unfilled PLA and PLA with wood and caffeine-treated wood. The scale bar is 50 μm.

3.3. Thermogravimetric Analysis

A thermogravimetric analysis of PLA and composite samples was performed to define the thermal stability of the samples. The recorded thermograms are shown in Figure 7, while Table 4 summarizes the temperatures at which 10%, 50%, and 90% of the mass of the samples were lost, along with their char content.

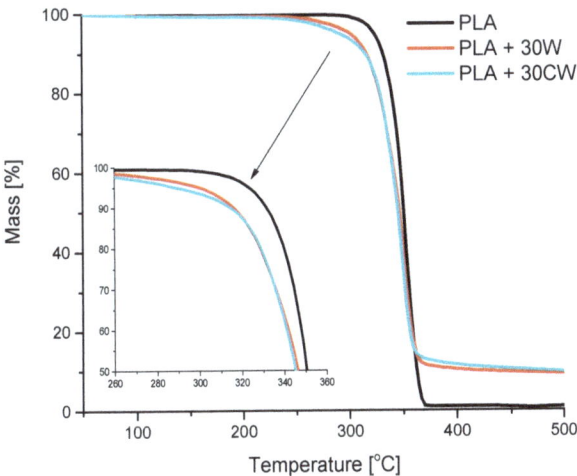

Figure 7. Thermogravimetric curves for PLA and composites with 30% filler.

Table 4. Temperature at which specific mass loss occurred and char content.

Sample	Temperature [°C]			Char Content at 500 °C [%]
	at 10% Mass Loss	at 50% Mass Loss	at 90% Mass Loss	
PLA	332	350	362	0.9
PLA + 30 W	316	345	410	9.0
PLA + 30 CW	315	347	471	9.8

All of the TG curves had similar shapes. The decomposition of the samples occurred in one step, with a characteristic inflection of the curve at around 300 °C and at approximately 360 °C. These are values typically observed for PLA. During its thermal decomposition, lactide is released first and then the higher cyclic oligomers [46].

Taking into consideration the temperatures at which 10% and 50% mass loss occurred, it can be seen that the values observed for the composites were slightly lower than those observed for the unfilled samples, especially at the first threshold. Other studies reported similar observations about PLA with wood fibers [47,48] and wood flour [49]. This behavior is ascribed to the presence of wood and its main constituent, cellulose, which begins to degrade at lower temperatures [50].

At a mass loss of 90%, there was a large difference between the PLA, PLA + 30 W, and PLA + 30 CW samples. The temperature observed for PLA was 48 °C and 109 °C lower than for the composite with wood and treated wood, respectively. This interesting observation can be discussed in terms of wood particles being a physical obstacle to effective heat transfer. The untreated wood present in the composite simply hinders heat transfer. In the case of the modified filler, the caffeine treatment removed some of the low-mass particles present in the wood and thus improved the interaction between the filler and the matrix, making it more difficult for heat to be transferred through the material. Furthermore, the amounts of char present in the residual sample were ten times higher for composites than for PLA. This is in line with other data from the literature [51]. It results from the thermal decomposition of the lignin (one of main constituents of wood) that was found responsible for the formation of char during thermal decomposition of wood, at a temperature over 400 °C [52]. Data analysis also shows that in composites with modified filler less wood was decomposed, proving that caffeine treatment can 'protect' wood from heat.

Overall, what is most important, under typical processing conditions, that is, up to 250 °C, was that the mass loss for both composites did not exceed 2%.

3.4. Contact Angle Measurements

Contact angle measurements were taken to determine the wettability of the composites obtained. Based on the photos obtained (Figure 8), contact angles were determined.

The contact angle measured for the PLA, 72.2°, is similar to that presented in other articles [13]. It can be seen that the unfilled PLA, as well as all composites, are within the range that classifies materials as hydrophilic. When 10% and 20% of untreated wood is added to the PLA, the composites produced have comparable values of contact angle polymer matrix (68.6–75.3° vs. 72.2° for PLA). A higher content of untreated wood causes the material to be more hydrophilic. This is rather expected because the wood itself is a hydrophilic material. The composite with the smallest amount of caffeine-treated filler had a contact angle similar to that of unfilled polymer wood. On the other hand, increasing the content of modified filler in the composite increased the contact angle. These materials were characterized with lower hydrophilicity than the PLA (contact angle in the range of 82.4–86.6°).

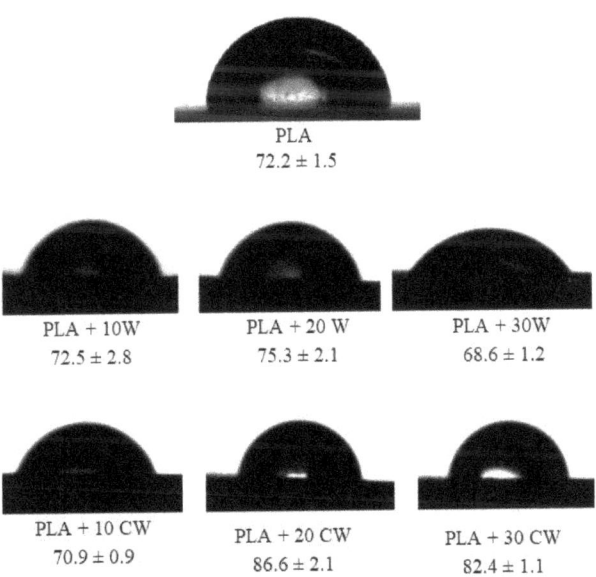

Figure 8. Water drops deposited on samples and calculated, mean values of contact angle.

Yet optimistically, these results still remain within the range typical of a hydrophilic material. However, they show that the incorporation of properly modified wood into PLA does not have to result in increasing the hydrophilicity of the material. This may arise from interactions between the constituents of wood and the modifier, caffeine. Treatment of the filler could have resulted in the formation of more hydrogen bonds between cellulose and caffeine molecules [23]. Additionally, research conducted by Tavagnacco et al. [53] confirmed that hemicellulose has a significant affinity for caffeine, which also creates additional bonds.

3.5. Antimicrobial Properties

Antimicrobial tests were performed to assess whether the addition of caffeine-treated wood to the PLA matrix altered the susceptibility of the material to microorganisms. Data regarding antifungal properties are summarized in Table 5 and exemplary photos of samples are shown in Figure 9.

Table 5. Resistance of PLA and its composites to fungi, tested after 21 days.

Fungi	PLA	PLA + 10 W	PLA + 20 W	PLA + 30 W	PLA + 10 CW	PLA + 20 CW	PLA + 30 CW
F. oxysporum	2	2	2	0	0	0	0
F. culmorum	2	2	2	0	0	0	0
A. alternata	2	2	2	1	0	0	0
T. viride	2	2	2	2	2	0	0

2: over 66% of the sample covered with fungal mycelium; 1: less than 33% sample covered with mycelium; 0: no signs of fungal mycelium on the sample.

Regardless of the bacterial strain used (*L. innocua* or *L. monocytogenes*), none of the composite samples showed antibacterial activity.

The results presented in Table 5 indicate that the PLA and composites with untreated wood showed no activity against the tested fungal strains. More than 66% of the surface of the PLA and composites with 10% and 20% wood was colonized by *F. culmorum*. Each sample containing caffeine-treated wood and PLA with 30 CW filler showed no mold growth on their surface. A similar behavior was observed for *A. alternata*: all composites with caffeine-treated filler remained unaffected by fungi, while the unfilled PLA and materials with 10% and 20% untreated wood were significantly infected. The composite with 30% wood was also covered with mold, but less so (33% of the sample). All other fungi were responsible for mold growth, but only in the PLA sample and in the composites with untreated wood. The presence of filler and crystallinity degree were reported to affect the biodegradation process of PLA and its susceptibility to microorganisms [54–56], but in this study, it was caffeine treatment that affected antifungal properties.

F. culmonorum

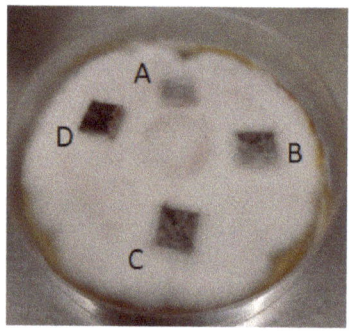 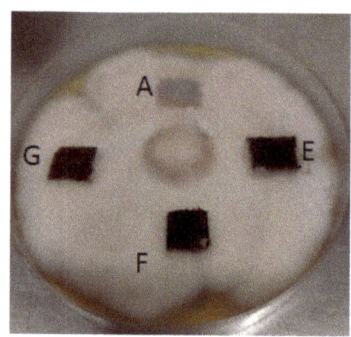

T. viride

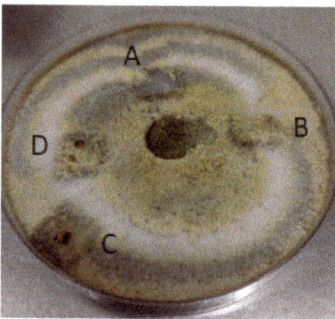 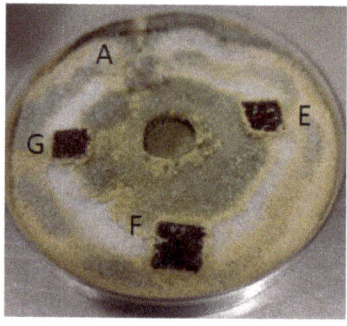

Figure 9. Samples tested against *F. culmonorum* and *T. viride* after 21 days. Samples in figures are named as follows: A: PLA, B: PLA + 10 W, C: PLA + 20 W, D: PLA + 30 W, E: PLA + 10 CW, F: PLA + 20 CW, G: PLA + 30 CW. (*F. culmorum*: over 60% of the surfaces of samples A, B, and C are covered with fungal mycelium; traces of fungal mycelium are visible on the surface of samples D, E, F, and G; *T. viride*: over 60% of the surfaces of samples A, B, C, and D are covered with fungal mycelium; traces of fungal mycelium are visible on the surface of samples E, F, and G).

Caffeine has previously been shown to have an antifungal effect not only in wood but also in polymer composites [21,22,25,57]. However, these previous studies have been performed for wood or wood-polypropylene composites. This is the first study on caffeine-treated wood used as a filler for a PLA matrix, which exhibits antifungal properties. This antimicrobial behavior is attributed to interactions between the cell walls of fungi and caffeine, since caffeine inhibits the synthesis of chitinases, enzymes responsible for the synthesis of cell walls of fungi [58,59]. The interesting thing is that the composite with untreated black cherry wood (PLA + 30 W) also showed antifungal behavior against *F. oxysporum* and *F. culmorum* (Table 5). This may be due to the fact that black cherry is a rich source of phenolic compounds [60] that were found to have antimicrobial properties [61,62].

This is a very important finding, as the fungi used in this research produce toxins (*Fusarium* spp.), cause the spoilage of various raw materials and food products (*A. alternata*), or are used in the biological protection of plants against pathogenic organisms (*T. viride*). As a consequence, these results provide new information that wood-filled composites, based on the bioderived polymer matrix, PLA, can be successfully and relatively easily modified in order to obtain materials that can withstand the negative influence of microorganisms.

4. Conclusions

This work deals with the preparation and in-depth characterization of polylactide biocomposites filled with black cherry wood treated with caffeine. This is the first study to show an important relationship between the structure of a polylactide filled with a wood product treated with a natural compound (caffeine) and the performance, including the antifungal properties.

Based on the findings presented in this study, it can be concluded that black cherry wood treated with caffeine is an effective heterogeneous nucleant for polylactide. The degree of crystallinity of composites with caffeine-treated wood was definitely higher than that of PLA, and increased with the wood filler content. The most noticeable changes in the crystallization temperature towards higher values and a reduction in the half-time of crystallization were observed for the modified filler. Unlike untreated wood, the use of a caffeine-treated wood filler resulted in the formation of transcrystalline structures. Interestingly, the introduction of this filler into the polylactide matrix also decreased its hydrophilicity. The experiment demonstrated that the incorporation of black cherry wood treated with caffeine into the polylactide matrix results in a material exhibiting antifungal properties against *F. oxysporum*, *F. culmorum*, *A. alternata*, and *T. viride*.

These findings underscore the potential of black cherry wood, especially when treated with caffeine, as a versatile and multifunctional filler to improve the crystallinity, structure, and antifungal properties of polylactide-based materials. This research contributes valuable information on the development of sustainable and functional biocomposites for diverse applications in the fields of materials science and antimicrobial engineering.

Author Contributions: A.G.-Z.: Conceptualization, performing experiments, results development, writing—original draft preparation; M.W.: preparation and characterization of fillers, manuscript preparation; A.K.-R.: performing experiments; M.B.: wood preparation; A.S.: preparation and carrying out antimicrobial tests; I.R.: preparation and characterization of fillers, supervision of filler modification; S.B.: results discussion, coordination of all tasks in the paper, writing—review and editing, supervision. All authors have read and agreed to the published version of the manuscript.

Funding: The research has received financial support of Ministry of Education and Science (Poland).

Institutional Review Board Statement: Not applicable.

Informed Consent Statement: Not applicable.

Data Availability Statement: Data are contained within the article.

Conflicts of Interest: The authors declare no conflicts of interest.

References

1. Dobrzyńska-Mizera, M.; Knitter, M.; Woźniak-Braszak, A.; Baranowski, M.; Sterzyński, T.; Di Lorenzo, M.L. Poly(l-Lactic Acid)/Pine Wood Bio-Based Composites. *Materials* **2020**, *13*, 3776. [CrossRef] [PubMed]
2. Liu, L.; Lin, M.; Xu, Z.; Lin, M. Polylactic acid-based wood-plastic 3D printing composite and its properties. *BioResources* **2019**, *14*, 8484–8498. [CrossRef]
3. Peltola, H.; Pääkkönen, E.; Jetsu, P.; Heinemann, S. Wood based PLA and PP composites: Effect of fibre type and matrix polymer on fibre morphology, dispersion and composite properties. *Compos. Part A Appl. Sci. Manuf.* **2014**, *61*, 13–22. [CrossRef]
4. Kim, D.; Andou, Y.; Shirai, Y.; Nishida, H. Biomass-based composites from poly (lactic acid) and wood flour by vapor-phase assisted surface polymerization. *ACS Appl. Mater. Interfaces* **2011**, *3*, 385–391. [CrossRef] [PubMed]
5. Pilla, S.; Gong, S.; O'Neill, E.; Yang, L.; Rowell, R.M. Polylactide-recycled wood fiber composites. *J. Appl. Polym. Sci.* **2009**, *111*, 37–47. [CrossRef]
6. Narlıoğlu, N.; Salan, T.; Alma, M.H. Properties of 3D-printed wood sawdust-reinforced PLA composites. *BioResources* **2021**, *16*, 5467. [CrossRef]
7. Medved, S.; Tomec, D.K.; Balzano, A.; Merela, M. Alien Wood Species as a Resource for Wood-Plastic Composites. *Appl. Sci.* **2021**, *11*, 44. [CrossRef]

8. Pilla, S.; Gong, S.; O'Neill, E.; Rowell, R.M.; Krzysik, A.M. Polylactide-pine wood flour composites. *Polym. Eng. Sci.* **2008**, *48*, 578–587. [CrossRef]
9. Li, X.; Yu, J.; Meng, L.; Li, C.; Liu, M.; Meng, L. Study of PLA-based Wood-Plastic Composites. In Proceedings of the 3rd International Conference on Mechanical Engineering and Materials (ICMEM 2022), Nanchang, China, 18–19 November 2022; Journal of Physics: Conference Series. p. 012029.
10. Lee, S.-Y.; Kang, I.-A.; Doh, G.-H.; Yoon, H.-G.; Park, B.-D.; Wu, Q. Thermal and Mechanical Properties of Wood Flour/Talc-filled Polylactic Acid Composites: Effect of Filler Content and Coupling Treatment. *J. Thermoplast. Compos. Mater.* **2008**, *21*, 209–223. [CrossRef]
11. Dalu, M.; Temiz, A.; Altuntaş, E.; Demirel, G.K.; Aslan, M. Characterization of tanalith E treated wood flour filled polylactic acid composites. *Polym. Test.* **2019**, *76*, 376–384. [CrossRef]
12. Grząbka-Zasadzińska, A.; Klapiszewski, Ł.; Borysiak, S.; Jesionowski, T. Thermal and mechanical properties of silica–lignin/polylactide composites subjected to biodegradation. *Materials* **2018**, *11*, 2257. [CrossRef]
13. Grząbka-Zasadzińska, A.; Odalanowska, M.; Borysiak, S. Thermal and mechanical properties of biodegradable composites with nanometric cellulose. *J. Therm. Anal. Calorim.* **2019**, *138*, 4407–4416. [CrossRef]
14. Turalija, M.; Bischof, S.; Budimir, A.; Gaan, S. Antimicrobial PLA films from environment friendly additives. *Compos. Part B Eng.* **2016**, *102*, 94–99. [CrossRef]
15. Busolo, M.A.; Lagaron, J.M. Antimicrobial biocomposites of melt-compounded polylactide films containing silver-based engineered clays. *J. Plast. Film. Sheeting* **2013**, *29*, 290–305. [CrossRef]
16. Harnet, J.-C.; Guen, E.; Ball, V.; Tenenbaum, H.; Ogier, J.; Haikel, Y.; Vodouhê, C. Antibacterial protection of suture material by chlorhexidine-functionalized polyelectrolyte multilayer films. *J. Mater. Sci. Mater. Med.* **2009**, *20*, 185–193. [CrossRef]
17. Verma, M.; Biswal, A.K.; Dhingra, S.; Gupta, A.; Saha, S. Antibacterial response of polylactide surfaces modified with hydrophilic polymer brushes. *Iran. Polym. J.* **2019**, *28*, 493–504. [CrossRef]
18. Ahmed, J.; Hiremath, N.; Jacob, H. Antimicrobial, Rheological, and Thermal Properties of Plasticized Polylactide Films Incorporated with Essential Oils to Inhibit *Staphylococcus aureus* and *Campylobacter jejuni*. *J. Food Sci.* **2016**, *81*, E419–E429. [CrossRef] [PubMed]
19. Hill, C.A. *Wood Modification: Chemical, Thermal and other Processes*; John Wiley & Sons: Hoboken, NJ, USA, 2007.
20. Shao, L.; Xi, Y.; Weng, Y. Recent Advances in PLA-Based Antibacterial Food Packaging and Its Applications. *Molecules* **2022**, *27*, 5953. [CrossRef] [PubMed]
21. Pánek, M.; Borůvka, V.; Nábělková, J.; Šimůnková, K.; Zeidler, A.; Novák, D.; Černý, R.; Kobetičová, K. Efficacy of Caffeine Treatment for Wood Protection—Influence of Wood and Fungi Species. *Polymers* **2021**, *13*, 3758. [CrossRef] [PubMed]
22. Kwaśniewska-Sip, P.; Cofta, G.; Nowak, P.B. Resistance of fungal growth on Scots pine treated with caffeine. *Int. Biodeterior. Biodegrad.* **2018**, *132*, 178–184. [CrossRef]
23. Kobetičová, K.; Ďurišová, K.; Nábělková, J. Caffeine Interactions with Wood Polymers. *Forests* **2021**, *12*, 533. [CrossRef]
24. Oguz, H.; Dogan, C.; Kara, D.; Ozen, Z.T.; Ovali, D.; Nofar, M. Development of PLA-PBAT and PLA-PBSA bio-blends: Effects of processing type and PLA crystallinity on morphology and mechanical properties. *AIP Conf. Proc.* **2019**, *2055*, 030003. [CrossRef]
25. Tomczak, D.; Woźniak, M.; Ratajczak, I.; Sip, A.; Baranowska, M.; Bula, K.; Čabalová, I.; Bubeníková, T.; Borysiak, S. Caffeine-treated wood as an innovative filler for advanced polymer composites. *J. Wood Chem. Technol.* **2023**, *43*, 271–288. [CrossRef]
26. Garlotta, D. A Literature Review of Poly(Lactic Acid). *J. Polym. Environ.* **2001**, *9*, 63–84. [CrossRef]
27. Lersten, N.R.; Horner, H.T. Calcium oxalate crystal types and trends in their distribution patterns in leaves of Prunus (Rosaceae: Prunoideae). *Plant Syst. Evol.* **2000**, *224*, 83–96. [CrossRef]
28. Krikorian, V.; Pochan, D.J. Poly (l-Lactic Acid)/Layered Silicate Nanocomposite: Fabrication, Characterization, and Properties. *Chem. Mater.* **2003**, *15*, 4317–4324. [CrossRef]
29. Cosnita, M.; Cazan, C.; Duta, A. Effect of waste polyethylene terephthalate content on the durability and mechanical properties of composites with tire rubber matrix. *J. Compos. Mater.* **2017**, *51*, 357–372. [CrossRef]
30. Borysiak, S.; Grząbka-Zasadzińska, A.; Odalanowska, M.; Skrzypczak, A.; Ratajczak, I. The effect of chemical modification of wood in ionic liquids on the supermolecular structure and mechanical properties of wood/polypropylene composites. *Cellulose* **2018**, *25*, 4639–4652. [CrossRef]
31. Wang, J.; Bai, J.; Zhang, Y.; Fang, H.; Wang, Z. Shear-induced enhancements of crystallization kinetics and morphological transformation for long chain branched polylactides with different branching degrees. *Sci. Rep.* **2016**, *6*, 26560. [CrossRef] [PubMed]
32. Garbarczyk, J.; Borysiak, S. Kompozyty polipropylenu z włóknami celulozowymi. Cz. 1. Wpływ warunków wytłaczania i wtryskiwania na strukturę matrycy polipropylenowej. *Polimery* **2004**, *49*, 541–546. [CrossRef]
33. Woźniak, M.; Gromadzka, K.; Kwaśniewska-Sip, P.; Cofta, G.; Ratajczak, I. Chitosan–caffeine formulation as an ecological preservative in wood protection. *Wood Sci. Technol.* **2022**, *56*, 1851–1867. [CrossRef]
34. Belščak-Cvitanović, A.; Komes, D.; Karlović, S.; Djaković, S.; Špoljarić, I.; Mršić, G.; Ježek, D. Improving the controlled delivery formulations of caffeine in alginate hydrogel beads combined with pectin, carrageenan, chitosan and psyllium. *Food Chem.* **2015**, *167*, 378–386. [CrossRef] [PubMed]
35. Kwaśniewska-Sip, P.; Woźniak, M.; Jankowski, W.; Ratajczak, I.; Cofta, G. Chemical changes of wood treated with caffeine. *Materials* **2021**, *14*, 497. [CrossRef] [PubMed]

36. Mysiukiewicz, O.; Barczewski, M. Crystallization of polylactide-based green composites filled with oil-rich waste fillers. *J. Polym. Res.* **2020**, *27*, 374. [CrossRef]
37. Wang, Y.; Liu, C.; Shen, C. Crystallization behavior of poly(lactic acid) and its blends. *Polym. Cryst.* **2021**, *4*, e10171. [CrossRef]
38. Xin, J.; Meng, X.; Xu, X.; Zhu, Q.; Naveed, H.B.; Ma, W. Cold Crystallization Temperature Correlated Phase Separation, Performance, and Stability of Polymer Solar Cells. *Matter* **2019**, *1*, 1316–1330. [CrossRef]
39. Lunt, J. Large-scale production, properties and commercial applications of polylactic acid polymers. *Polym. Degrad. Stab.* **1998**, *59*, 145–152. [CrossRef]
40. Beauson, J.; Schillani, G.; Van der Schueren, L.; Goutianos, S. The effect of processing conditions and polymer crystallinity on the mechanical properties of unidirectional self-reinforced PLA composites. *Compos. Part A Appl. Sci. Manuf.* **2022**, *152*, 106668. [CrossRef]
41. Huang, T.; Yamaguchi, M. Effect of cooling conditions on the mechanical properties of crystalline poly(lactic acid). *J. Appl. Polym. Sci.* **2017**, *134*, 44960. [CrossRef]
42. Simmons, H.; Tiwary, P.; Colwell, J.E.; Kontopoulou, M. Improvements in the crystallinity and mechanical properties of PLA by nucleation and annealing. *Polym. Degrad. Stab.* **2019**, *166*, 248–257. [CrossRef]
43. Ming, Y.; Zhou, Z.; Hao, T.; Nie, Y. Polymer Nanocomposites: Role of modified filler content and interfacial interaction on crystallization. *Eur. Polym. J.* **2022**, *162*, 110894. [CrossRef]
44. Wang, B.; Wen, T.; Zhang, X.; Tercjak, A.; Dong, X.; Müller, A.J.; Wang, D.; Cavallo, D. Nucleation of Poly(lactide) on the Surface of Different Fibers. *Macromolecules* **2019**, *52*, 6274–6284. [CrossRef]
45. Wang, C.; Fang, C.Y.; Wang, C.Y. Electrospun poly(butylene terephthalate) fibers: Entanglement density effect on fiber diameter and fiber nucleating ability towards isotactic polypropylene. *Polymer* **2015**, *72*, 21. [CrossRef]
46. Kopinke, F.D.; Remmler, M.; Mackenzie, K.; Möder, M.; Wachsen, O. Thermal decomposition of biodegradable polyesters—II. Poly(lactic acid). *Polym. Degrad. Stab.* **1996**, *53*, 329–342. [CrossRef]
47. Zhang, L.; Lv, S.; Sun, C.; Wan, L.; Tan, H.; Zhang, Y. Effect of MAH-g-PLA on the Properties of Wood Fiber/Polylactic Acid Composites. *Polymers* **2017**, *9*, 591. [CrossRef]
48. Awal, A.; Rana, M.; Sain, M. Thermorheological and mechanical properties of cellulose reinforced PLA bio-composites. *Mech. Mater.* **2015**, *80*, 87–95. [CrossRef]
49. Tao, Y.; Wang, H.; Li, Z.; Li, P.; Shi, S.Q. Development and Application of Wood Flour-Filled Polylactic Acid Composite Filament for 3D Printing. *Materials* **2017**, *10*, 339. [CrossRef]
50. Renneckar, S.; Zink-Sharp, A.; Ward, T.C.; Glasser, W.G. Compositional analysis of thermoplastic wood composites by TGA. *J. Appl. Polym. Sci.* **2004**, *93*, 1484–1492. [CrossRef]
51. Huda, M.S.; Drzal, L.T.; Misra, M.; Mohanty, A.K. Wood-fiber-reinforced poly(lactic acid) composites: Evaluation of the physicomechanical and morphological properties. *J. Appl. Polym. Sci.* **2006**, *102*, 4856–4869. [CrossRef]
52. Poletto, M.; Zattera, A.J.; Santana, R.M.C. Thermal decomposition of wood: Kinetics and degradation mechanisms. *Bioresour. Technol.* **2012**, *126*, 7–12. [CrossRef] [PubMed]
53. Tavagnacco, L.; Engström, O.; Schnupf, U.; Saboungi, M.-L.; Himmel, M.; Widmalm, G.; Cesàro, A.; Brady, J.W. Caffeine and Sugars Interact in Aqueous Solutions: A Simulation and NMR Study. *J. Phys. Chem. B* **2012**, *116*, 11701–11711. [CrossRef]
54. Gorrasi, G.; Pantani, R. Hydrolysis and Biodegradation of Poly (lactic acid). *Synth. Struct. Prop. Poly (Lact. Acid)* **2018**, *279*, 119–151.
55. Tsuji, H. Hydrolic degradation. In *Poly(Lactic Acid)*; John Wiley & Sons: Hoboken, NJ, USA, 2022; pp. 467–516.
56. Kale, G.; Kijchavengkul, T.; Auras, R.; Rubino, M.; Selke, S.E.; Singh, S.P. Compostability of bioplastic packaging materials: An overview. *Macromol. Biosci.* **2007**, *7*, 255–277. [CrossRef]
57. Arora, D.S.; Ohlan, D. In vitro studies on antifungal activity of tea (*Camellia sinensis*) and coffee (*Coffea arabica*) against wood-rotting fungi. *J. Basic Microbiol.* **1997**, *37*, 159–165. [CrossRef]
58. Adams, D.J. Fungal cell wall chitinases and glucanases. *Microbiology* **2004**, *150*, 2029–2035. [CrossRef]
59. Pánek, M.; Šimůnková, K.; Novák, D.; Dvořák, O.; Schönfelder, O.; Šedivka, P.; Kobetičová, K. Caffeine and TiO2 Nanoparticles Treatment of Spruce and Beech Wood for Increasing Transparent Coating Resistance against UV-Radiation and Mould Attacks. *Coatings* **2020**, *10*, 1141. [CrossRef]
60. Brozdowski, J.; Waliszewska, B.; Gacnik, S.; Hudina, M.; Veberic, R.; Mikulic-Petkovsek, M. Phenolic composition of leaf and flower extracts of black cherry (*Prunus serotina* Ehrh.). *Ann. For. Sci.* **2021**, *78*, 66. [CrossRef]
61. Daglia, M. Polyphenols as antimicrobial agents. *Curr. Opin. Biotechnol.* **2012**, *23*, 174–181. [CrossRef]
62. Takó, M.; Kerekes, E.B.; Zambrano, C.; Kotogán, A.; Papp, T.; Krisch, J.; Vágvölgyi, C. Plant Phenolics and Phenolic-Enriched Extracts as Antimicrobial Agents against Food-Contaminating Microorganisms. *Antioxidants* **2020**, *9*, 165. [CrossRef]

Disclaimer/Publisher's Note: The statements, opinions and data contained in all publications are solely those of the individual author(s) and contributor(s) and not of MDPI and/or the editor(s). MDPI and/or the editor(s) disclaim responsibility for any injury to people or property resulting from any ideas, methods, instructions or products referred to in the content.

MDPI AG
Grosspeteranlage 5
4052 Basel
Switzerland
Tel.: +41 61 683 77 34

Materials Editorial Office
E-mail: materials@mdpi.com
www.mdpi.com/journal/materials

Disclaimer/Publisher's Note: The statements, opinions and data contained in all publications are solely those of the individual author(s) and contributor(s) and not of MDPI and/or the editor(s). MDPI and/or the editor(s) disclaim responsibility for any injury to people or property resulting from any ideas, methods, instructions or products referred to in the content.

www.ingramcontent.com/pod-product-compliance
Lightning Source LLC
LaVergne TN
LVHW070224100526
838202LV00015B/2084